In accordance with the latest syllabus prescribed by the Council for the Indian School Certificate Examinations, New Delhi.

MOST LIKELY

ISC QUESTION BANK

BIOLOGY

CLASS XII

By

A PANEL OF AUTHORS

OSWAL PUBLISHERS

1/12, Sahitya Kunj, M.G. Road, Agra - 282002

EDITION
YEAR 2021

DISCLAIMER

With the ambition of providing standard academic resources, we have exercised extreme care in publishing the content. In case of any discrepancies in the matter, we request readers to excuse the unintentional lapse and not hold us liable for the same. Suggestions are always welcome.

ISBN : 978-93-91184-36-0

PUBLISHED BY

1/12, Sahitya Kunj, M.G. Road, Agra - 282002

(0562) 2527771-4, +91 7534077222

info@oswalpublishers.in

www.oswalpublishers.com

The cover of this book has been designed using resources from Freepik.com

Printed At Upkar Printing Unit, Agra

PREFACE

It is a matter of immense pride for us to present our 'MOST LIKELY ISC QUESTION BANK' series, especially prepared for students appearing for Board examinations in the oncoming year.

This book is a perfect capsule for building self-confidence during exam. preparation. Based on chunking strategy, the 'Categorywise–Chapterwise' format with its exhaustive set of questions allow the students to cover every category in a chapterwise manner.

Covering easy categories first boosts student's self-esteem and the ascending score braces them to take up challenging categories without fear. This prepares the student to see the exam paper as achievable at all times.

With its simple language and style, it is a one-stop solution for smart study. We are confident that the book will enable the candidates to develop a better understanding of the curriculum and help them organize their learning process. This book shall definitely prove to be a fruitful tool for the students and encourage them towards scholastic excellence.

Constructive suggestions for further improvement of the book are always welcome.

Note: Questions marked with '*' are frequently asked in previous years board examinations.

—**Publisher**

LET'S GO ORGANIC

Growing microgreens can be a good, easy and economical way for a student to start practicing plantation at home.

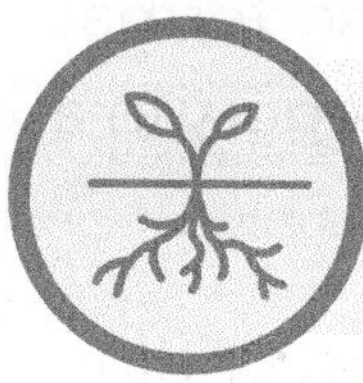

Microgreens are highly nutritious plants that are harvested after the sprout-stage and just before the maturity phase.

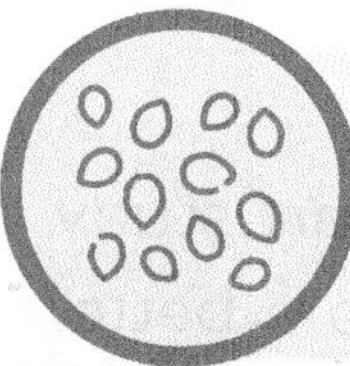

Microgreens may include mustard, coriander, basil seeds, dried peas, beans, broccoli, radish, spinach, beetroot, cauliflower and others.

It is more convenient to start growing microgreens that require low maintenance like mustard, coriander, peas, basil and others.

Start with finding a flat tray or a shoe-box or any tetra-pack carton.

Make holes in the box for easy drainage of water.

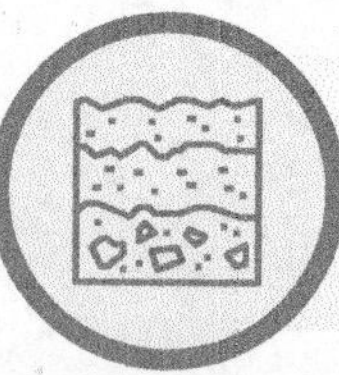

Spread two inches of soil in the box. You can get ready-made soil from the garden or nursery. Sprinkle your desired seeds over the soil.

You can spread a mixture of coco-peat and fertile topsoil or a stack of moist tissue papers over it. Remember to keep the tissue paper moist till the seeds start germinating.

Cover the box with newspaper and keep it in a place where there is no direct sunlight.

Spray water twice a day atleast.

On third and fourth day, once the leaves start to appear, start exposing the germinating microgreens to indirect sunlight.

You can try to expose them to direct sunlight for not more than 4 hours.

By ninth or tenth day, your microgreens will be ready, harvest them!

Microgreens can be eaten with sandwiches, wraps, salads, smoothies, juices or as garnishers on pizza, soup and curries.

CONTENTS

Note: Questions marked with '*' are frequently asked in previous years board examinations.

SUGGESTIONS FOR STUDENTS

1. Read the question paper carefully and follow the instructions given.
2. Attempt all parts and sub-parts of a question in proper sequence.
3. Prepare short notes of the concepts using reference books, magazines, journals, projects etc. as it is easier to study.
4. Study all topics thoroughly.
5. Avoid selective learning.
6. Solve previous years' question papers and get the answers checked by the subject teacher to assess the performance.
7. Practice self-explanatory diagrams with proper labelling, arrows and headings.
8. Pay attention to the correct structure and shape of different parts.
9. All parts and sub parts of a question should be attempted together in the same sequence.
10. Write answers point-wise rather than in the form of an essay.
11. Lay stress on understanding the terms/concepts instead of rote learning.
12. Learn main laws, principles and key-words/acronyms with proper understanding and use of keywords.
13. Learn the biological names and scientific terms with correct spellings.
14. Differences must be written in tabular form and should be compatible.
15. The process or phenomena should be learn in the correct sequence.
16. Differences should always be compatible.
17. Avoid writing examples as differences.
18. Practice cross with the help of different characteristic features in three stages :

    (a) Parents (b) F1 generation (c) F2 generation.
19. Study all topics thoroughly.
20. Draw neat labelled diagrams.

Word of Advice

Chapter 1. Reproduction in Organisms

Topics Found difficult By Students
- ➢ Meiocytes and gametes; Ovary and ovum/ovule.
- ➢ Oestrous cycle

1. Majority of the students missed out on the keywords in the term *'Clone'* like Genetically & Morphologically identical individuals and wrote about identical individuals.
2. Many students were confused between *Meiocytes* and *Gametes*. Some only mentioned haploid and diploid without mentioning the actual chromosome number in the cell.
3. Some students wrote about *heat period for seasonal breeders* but did not mention the particular period in a year.
4. Many students wrote about *fertilisation* but missed writing the meiosis or gamete formation part. Some students mentioned only one process, either gamete formation or fertilisation. Some did not mention the haploid number from each parent and diploid in the offspring.
5. Some candidates were unable to write a compatible difference between the *Oestrous and Menstrual cycle,* a few others wrote the opposite answer. Some candidates were able to explain the menstrual cycle but did not seem to be clear about the oestrous cycle as they wrote that it operates in frog, reptiles, etc. The word *'non-primates'* was not mentioned in many cases and many of those who mentioned non-primates did not write *mammals*. A few candidates wrote the duration of oestrous cycle as 10 days, 13 days, 23 days, etc. Some candidates, instead of writing the difference, wrote the names of the stages.
6. Majority of the candidates did not use the term *natural death* while defining *life span* but defined it as *the period between birth and death*. Some candidates defined it as *an average life of an individual.*
7. Majority of the candidates were unclear of the concept of *haploid* and *diploid*. They were not clear that *every somatic cell has diploid number of chromosomes.*
8. Many candidates, instead of writing *Strobilanthes kunthiana,* wrote Bamboo as the flower which blooms after every 12 years.

Chapter 2. Sexual Reproduction in Flowering Plants

Topics Found difficult By Students
- ➢ Parthenocarpy / Parthenogenesis /Apomixis.
- ➢ Types of placenta and types of ovules.

1. Many students, instead of describing the features of flowers pollinated by insects, described those of wind-pollinated flowers. Some students wrote about the advantages of cross-pollination.
2. Some students made errors in the diagram by drawing the *L.S of anther* instead of *T.S of the anther*; some drew stamens or whole flower. Few did incorrect labelling.
3. Many students did not make a clear-cut distinction between *cellular* and *nuclear endosperms* based on *cytokinesis* and *karyokinesis*. They did not know the *examples of cellular* and *helobial type.*
4. A few students were confused between *endospermic and non-endospermic seeds.*
5. Most candidates, instead of writing valid points like wastage of pollen grains and to ensure pollination, wrote general answers, for the statement *'Pollen grains are produced in large number'*. Some candidates wrote about how the wind helps in the process of pollination.
6. A few candidates wrote the features of *anemophilous* flowers, confusing them with *entomophilous* and mentioned the role of insects. In a few answer scripts, contrivances of cross-pollination were written in detail.

7. Many candidates, instead of drawing the embryo sac of angiosperm drew the entire anatropous ovule.
8. For *polyembryony*, several candidates wrote *"formation of many embryos"* without mentioning plant, animal, ovule or seed, which made the definition vague and incomplete.
9. A few candidates were confused between the terms *'Porogamy'* and *'Mesogamy'*.
10. A number of the candidates, instead of describing *parthenocarpy*, described *parthenogenesis*. Many of them defined it as the formation of embryo without fertilisation or development of animals without fertilisation. Some candidates gave the example of the honey bee.

Chapter 3. Human Reproduction

Topics Found difficult By Students

- Location of accessory glands of male reproductive system and their function.
- Spermiogenesis / Spermatogenesis

1. Many students got mixed up in the *growth and development of the follicles*. Some students drew the diagram of the ovule, T.S of an ovary of flower and mammalian ovum instead of human ovary. A few only explained the parts without making the diagram. In several cases, the diagrams were not neat, and the labelling was incorrect.
2. The *physico-chemical events during fertilisation* were written in random order, *e.g.* capacitation was discussed before insemination.
3. Many students described *embryonic development* instead of *fertilisation*. Several students described the process of double fertilisation in plants or gametogenesis. Some drew the female reproductive system only. A few students discussed the menstrual cycle.
4. Some candidates, instead of drawing the *L.S. of the human testis, drew the T.S. of the human testis.*
5. Students are confused about *menarche* and *menopause* concept.
6. Instead of the scientific term *parturition* many candidates used general term *'delivery'*.
7. Some candidates defined *spermatogenesis* instead of *spermiogenesis*. A few candidates used the term division in place of transformation of spermatids to form spermatozoa.

Chapter 4. Reproductive Health

Topics Found difficult By Students

- Reproductive fitness and reproductive health.
- Amenorrhea / Lactation
- Pedigree analysis

1. While defining *reproductive health*, many candidates did not mention the word reproduction, thereby defining general health.
2. The role of *amniocentesis* was incorrectly given by some students as *determination of sex, which is actually a misuse of this technique*. A few students defined amniocentesis simply as a prenatal diagnostic technique, without mentioning the role of amniotic fluid.
3. Majority of the candidates did not know the meaning of *cryptorchidism, oligospermia*, etc. Some candidates wrote the causes of infertility in general without referring specifically to males.
4. While defining *amenorrhea*, some candidates used the term permanent stoppage of menstruation, which implies menopause while some defined lactation. A few of them did not use the words stoppage of menstruation anywhere in the definition.
5. Some candidates omitted the word *'deficiency'* and wrote Immuno virus as a full form of *AIDS*.
6. Most of the candidates wrote *intra vitro* instead of *in vitro* for abbreviation *IVF*.
7. Many students misspelt the biological name of the bacteria for *STDs*. Some students only wrote the name of the species without writing the name of the genus. A few students wrote vague symptoms such as fever, pain, headache, etc.

Chapter 5. Principles of Inheritance and Variation

Topics Found difficult By Students

- Definitions in Genetics.
- Mendel's Laws, Features of pea plant.
- Punnett square, Back cross and test cross.
- Pleiotropic, multiple allelism and Polygenic inheritance.
- Example of cross showing complete linkage.

1. Many students were confused between *test cross* and *back cross*.
2. Many students confused *homologous chromosomes* with terms *homozygous* and *heterozygous*. Some students defined *homologous chromosomes* in terms of *homologous organs* and wrote chromosomes having similar structure and different functions, not mentioning the keyword corresponding chromosomes.
3. Many students were unable to recognize *monohybrid* or *dihybrid cross*. Several students did not understand how many or what types of alleles were there in the gametes, like only R or y, RR, Yy.
4. Some students wrote *1:2:1 or 1:3:3:1 for phenotypic ration in dihybrid cross*.
5. For the case study related to *inheritance of blood group,* majority of the candidates write the possible blood groups separately without using Punette square method. Correct alleles were not given. Many candidates mentioned the occurrence of all the four blood groups in the progeny instead of writing the type of alleles.
6. Very few candidates drew the Punnette square to explain the carrier of *colour blindness* as an example. Many candidates wrote the genotype and phenotype of the parents but not the progeny.
7. Some wrote symptoms of alkaptonuria *e.g.,* black urine for phenylketonuria.
8. Some candidates described *sex determination* in Drosophila instead of honey bees. A few described the division of labour in honeybees but did not mention the mechanism of sex determination. Many missed the point that *'unfertilised eggs develop into males by parthenogenesis'*.

Chapter 6. Molecular Basis of Inheritance

Topics Found difficult By Students

- Abiogenesis / Biogenesis
- Avery, McCleod, McCarty's experiment, Griffith's experiment.
- Structure of DNA, Nucleosome.
- Transcription in prokaryotic cell, Lac - operon.

1. Instead of writing *triplet nature of codon,* many students gave the contribution as, concept of genetic code/ discovered DNA/ developed method of DNA fingerprinting.
2. Many students wrote the contribution of *Sanger* as, discovered nucleic acid/proteins/proposed Recapitulation theory instead of DNA sequencing/protein sequencing.
3. Many students did not describe the structure of a *nucleosome,* instead discussed nucleotides of DNA. Some students drew the nucleus. A few described prokaryotic/eukaryotic cells.
4. Majority of the candidates were unable to write the correct *length of the DNA*.
5. Some candidates, instead of describing the process of *post-transcriptional processing,* described transcription.
6. Many candidates, instead of describing *Avery, McLeod and McCarty's experiment,* described *Griffith's experiment or Meselson and Stahl's experiment*. Some candidates mentioned the *use of DNAase, RNAase and protease but did not correlate it with the corresponding observations*.
7. The concept of a *Recon* being the smallest unit capable of undergoing recombination was not clear to many candidates instead of 'recombination', which was the key word, candidates wrote 'replication'.
8. Many candidates wrote only *'alleles'* instead of *'Multiple Alleles'*. Some wrote 'polygenic inheritance' when asked about an alternative form of the single gene which influences the same character and produces different expressions in different individuals of a species.

Chapter 7. Evolution

Topics Found difficult By Students

- Protobionts and Eobionts.
- Lamarckism and Darwinism, Industrial melanism.
- Gene flow and genetic drift/genetic erosion.
- Characteristics of Dryopithecus, Chromosomal and anatomical features of apes and man.

1. Many students were confused with the spellings of *Dryopithecus, Ramapithecus, Cro-magnon* and *Neanderthalman.*
2. Majority of the students described or used the terms Monkeys to Apes, Cro-magnon and Neanderthal man, instead of *Dryopithecus.*
3. Majority of students were unable to study *pedigree chart* correctly and did not gave correct answers related to them.
4. Most of the students attempted to explain the elongation of the giraffe neck according to Lamarckism. They were confused between *Darwinian and Lamarckian interpretation of giraffe neck elongation.*
5. Instead of *non-living matter* students wrote *pre-existing life* in context to *abiogenesis.*
6. Several candidates were not able to write correct characterstics of *Ramapithecus.* They used vague terminologies, like less or more cranial capacity. Some candidates wrote that *these ancestors existed during the times of God Ram.* A few candidates gave the characteristics of *Dryopithecus.*
7. Most candidates instead of mentioning the approximate height of *Cro-Magnon man,* in feet wrote just tall height. Several candidates wrote the *incorrect cranial capacity.*
8. Some candidates were confused between *homologous and analogous organs.*
9. *Drawbacks of Darwinism* were mixed up with *Lamarckism's.* Some candidates explained the long neck of a giraffe.
10. *Hardy Weinberg principle* was not explained by many candidates. Only the equation was given. In some cases, *'=1'* was not written. The term population was not mentioned by some candidates. A few candidates wrote the definition of gene pool/genetic erosion.
11. Some candidates wrote, *'frequency of one phenotype is high'* for *directional* instead of *one extreme phenotype.* Some candidates explained the differences through examples but gave similar examples for both.
12. Most candidates wrote *cranial capacity as 'large' and 'small', without giving the exact values.*
13. Many candidates wrote *'First cell'* instead of *precursor/protocell/cell like* when asked about *protobionts.* Some were confused between *eobionts* and *protobionts.*
14. Some candidates defined *fossils* as */remains of dead organisms'* without any reference to prehistoric plants and animals.
15. Majority of the students wrote more about reproduction, sperm count, healthy females, etc., while explaining reproductive fitness. Several students described reproductive fitness concerning reproductive health. Very few students could interpret reproductive fitness in terms of Darwinian fitness and give the example accordingly.

Chapter 8. Human Health and Diseases

Topics Found difficult By Students

- Types of immunity, Primary and secondary lymphoid organs.
- Action of T cells to antigens.
- Name of vaccines and the diseases they prevent.
- Source of opium, Cannabinoids and Barbiturates.

1. Several students mixed up *antibody* with *antibiotic.*
2. Many students were confused between *histamine* and *antihistamine.*
3. Some gave the names of medicines like Avil, Cetirizine, Allegra w.r.t. allergies. A few mentioned IgG and IgM also.
4. Most of the students wrote about the *effects of drug addiction* like hallucination and sleepiness instead of giving reasons for drug addiction. Some students gave different reasons for drug addiction but referred to the same concept in different words. A few students confused drug addiction with alcoholism.

5. Several students repeated the same point in different ways for *the effects of alcoholism*, e.g. alcohol affects different organs, alcohol affects the liver, it affects the brain and so on. Many students wrote causes instead of effects of alcoholism.
6. Most of the students confused about the concept of *primary and secondary lymphoid organs*. Many students wrote spleen as a primary lymphoid organ while some wrote B & T lymphocytes. Some also wrote maturation in secondary instead of proliferation and transfer. Some students mentioned that lymph is produced in primary organs and is stored in secondary organs. A few students *interpreted both Primary and Secondary organs as primary and secondary immune response.*
7. Majority of the candidates were able to write the correct name for the *diagnostic test for AIDS*. Some candidates wrote Widal test, AIDS test, PAP smear test, LISA/ALIZA etc. A few candidates wrote the wrong order for ELISA.
8. Some candidates wrote *Cannabis/ Erythroxylum/* Datura as the source of opium instead of *Papaver somniferum.*
9. Some students misunderstood the *concept of humoral and cell-mediated immunity for active and passive immunity* and defined them accordingly. Some candidates simply wrote produced antibodies and did not produce antibodies without mentioning T cells and B cells.
10. Many candidates wrote *cancerous for benign tumour and non-cancerous for malignant tumour*, which is incorrect.
11. Many candidates misspelt the biological name of *Salmonella, e.g., Salmonela, Semlonala,* etc. and some wrote only the generic name. The scientific name, according to the rules of Binomial nomenclature, was not written by many candidates.

Chapter 9. Strategies for Enhancement in Food Production

Topics Found difficult By Students
- Pisciculture and Ichthyology.
- Multiple Ovulation Embryo transfer technology and artificial insemination.
- Significance of SCP.
- Tissue culture technique / Plant hybridization.
- Golden rice.

1. Some students ignored the word *poultry* and discussed *dairy farm management*. Some discussed the economic advantages derived from poultry farms. Several students repeatedly used the same concept in different ways e.g. give food, nutritious food, clean water, filtered water and so on. A few students mentioned breeding and hatching techniques.
2. Instead of writing about *tissue culture technique* in plants, students wrote advantages and uses of tissue culture and the technique of hybridisation in plants.
3. Several students defined *fortification* instead of *Biofortification*.
4. Many students wrote that *golden rice* contains high level of protein. Some students wrote about *Vitamin A* but very few mentioned the introduction of genes of *β-carotene*.
5. Many candidates wrote that *bagging* is essential to protect the plant from damage. Some candidates got confused with artificial insemination, emasculation and hybridization.
6. Instead of *MOET*, which is used in cattle, many candidates described GIFT and ZIFT in humans. Many confused *MOET with Artificial Insemination*. Some candidates failed to explain the formation of many embryos.
7. For *applications of tissue culture*, many students repeated the same points in different terms *e.g.*, improving insect resistance, pest resistance, and fungal attack resistance and so on.

Chapter 10. Microbes in Human Welfare

Topics Found difficult By Students
- Role of microbes in human welfare.
- Integrated Pest Management.
- Infertility in humans and infertility in soil.

1. Many students interpreted *single-cell protein* as *single-celled protein*. Several students did not mention the significance of a single cell protein. Some students wrote that it is a source of food or used as a monoclonal antibody.

2. Answers such as, *"used to kill insects, spore is toxic, spore kills, chokes the gut"*, were given for the spores of *Bacillus thuringiensis* used as bio insecticide.
3. Majority of the candidates mis-spelt *Streptococcus*, while some were either unable to write the use of *Streptokinase* or mentioned that it is used in dairy products instead of in curd, in bakeries, in production of antibiotics, for clotting instead of clot dissolving.
4. Some mis-spelt about *Statins* and wrote the commercial product of *Monascus purpureus;* Some did not write the use of *Monascus purpureus* or its use as immunosuppressant and some unable to write the *scientific name of Lactic acid Bacteria.*

Chapter 11. Biotechnology : Principles and Processes

Topics Found difficult By Students

- Contribution of Scientists Sanger.
- DNA finger printing technique, Nomenclature of RE.
- RNA interference; Artificial chromosomes.
- Blue-white method of selection of recombinants.

1. Several students expanded *EFB as Extended Forest Board, Elimination of the fittest, Energy Food Bureau.* Some students expanded B as *Biology* instead of *Biotechnology.*
2. Many students wrote about *electroporation* instead of *electrophoresis.* Many of those students, who discussed electrophoresis, ignored the importance of the electric field in the process of separation of biomolecules.
3. Many candidates described the process of *DNA replication instead of DNA fingerprinting.*
4. Some wrote the full form of *PCR* as *Police Control Room / Private Community Rational / Pulmonary Cardio Resuscitation.*
5. While describing PCR, very few candidates mentioned the exact temperatures which cause Denaturation, Annealing and Extension. Only a few candidates mentioned the role of Taq polymerase in this technique.
6. Majority of the candidates described the occurrence of blue and white moths and the role of natural selection. Candidates were confused between recombinants and non-recombinants. Several candidates wrote that *the recombinant colonies turn blue and non-recombinant colonies remain white,* which was just the opposite of the correct answer. Many candidates did not mention *the role of the gene lacZ and the enzyme beta-galactosidase* in this process.
7. Many students wrote that *restriction enzymes* cut the DNA into fragments that could be separated by gel electrophoresis, without mentioning the production of sticky ends which are *used to ligate foreign DNA.* Some students drew PBR 322 plasmid to explain restriction enzymes.
8. Most of the students mixed up the mechanism of restriction endonucleases with its definition. Many students did not write the rules of their nomenclature correctly. Some students wrote that restriction endonucleases were used to cut DNA but did not mention at specific sites. Some described the rules of binomial nomenclature, instead of restriction endonucleases.
9. Some students confused in *PCR* and *RE or Molecular scissors.* A few students gave very vague features like, easy availability or it being harmless, ignoring the major features like the origin of replication, selectable markers, multiple cloning sites, etc.
10. When asked about vector-less methods ,some students mentioned *Agrobacterium*, pBR 322, BAC and YAC as tools of gene transfer without realizing that these were vectors,. Some students did not write both the methods completely. A few students wrote $CaCl_2$ and heat shock as different points.

Chapter 12. Biotechnology and its Applications

Topics Found difficult By Students

- Electrophoresis and electroporation.
- Acronym PCR, SCID, GEAC, EFB.
- Gene therapy / SCID.
- Transgenic animals.
- Bio patent /Bio piracy.

1. Many students mentioned the role of GEAC as monitoring research on environment and ecology, without discussing the role of genetically modified organisms. Some of them wrote self-devised expansions of the term like *"Gas Emission Association Community"*.
2. Majority of the candidates did not mention the key words *legal rights* in the definition of *Bio patent*. Most of the candidates wrote general answers without mentioning the biological discovery or product. Some candidates were confused between bio patent and bio piracy. A few candidates wrote the definition of patent, without relating it to biological inventions.
3. Students made mistakes in correlating *SCID* with *ADA*. Some misspelt *Adenosine Deaminase*. Instead of writing the deficiency of T-cell and B-cells 'both', vaguely wrote that the immune system does not work properly, or deficiency of lymphocytes, some wrote about genetic counselling.

Chapter 13. Organisms and Populations

Topics Found difficult By Students

- Contribution of Scientists R. Mishra.
- Cannabinoids and Cannibalism.
- Biotic potential and carrying capacity.
- Distinguishing reasons for population explosion in India and in the world.

1. Majority of the students took the word literally and defined *niche* as the lower part of the body. Most students could not distinguish between habitat and niche. The key word functional role was missing in several answer scripts.
2. Many students wrote incorrect definition of *carrying capacity* like, the capacity of a population to carry the load by organisms or the ability of an area to carry individuals. Several students did not write the keyword maximum, only mentioning the number of individuals an environment can support. Some students mistook carrying capacity as the ability of the uterus to carry the foetus during gestation.
3. A few candidates were confused between *proto-co-operation* and *commensalism*.
4. Some candidates gave contribution of scientist *R. Mishra* such as, Father of Biology /Father of Green Revolution/ Father of Plant Embryology.

Chapter 14. Ecosystem

Topics Found difficult By Students

- Detritus, Standing crops, Food chain.
- Stenothermal and cold-blooded organisms.
- Primary and secondary succession.

1. Some wrote *climax stage* was achieved quickly because it was fast or because it came after secondary.
2. Many students wrote about *standing crop* as crops which remain standing in fields/stand erect/never bend with wind, etc. Several students did not mention organic/biomass.
3. Most students confused stenothermal with *poikilothermic* and *homothermic*. Many students, instead of writing the operative word narrow range wrote low/high/hot/cold, etc.
4. Most of the students defined the *dispersal of seeds* and *fruits* instead of their significance. Some students interpreted dispersal as the transfer of pollen grains and adaptations in seeds for dispersal.
5. Many students drew the food chain or food web while some wrote the Calvin Cycle, instead of *carbon cycle*. In many cases, correct arrows were not shown. Some students wrote in a paragraph. A few students used graph paper to attempt this question as the question mentioned graphic representation. In many cases, the cycle was not in proper form.
6. Many candidates, instead of writing the names of the organisms like decomposers, top carnivores, wrote producers, predators, carnivores.
7. Majority of the candidates were not clear about the *role of predators*. Most candidates vaguely wrote that the predators disturb the ecosystem. Some candidates gave the reason as uncontrolled hunting by the predators, not of the predators. In some answer scripts there was no mention of increase of herbivores to exhaust vegetation leading to death of herbivores.

8. Several students, instead of writing *dead organic matter*, wrote only *organic matter* w.r.t. *detritus*. A few students wrote *Detrivores*.

Chapter 15. Biodiversity and its Conversation

Topics Found difficult By Students
- In situ and Ex situ strategies for conservation of biodiversity.
- Hotspots /Biosphere reserves, Ramsar sites.

1. Most students wrote terms such as cryptopreservation, cold storage and cryogenesis, instead of cryopreservation. Some students repeated the term germplasm preservation which is scientific term used for a very low temperature.
2. Most of the students described the *Rivet Popper hypothesis* by taking the example of an aeroplane without explaining the corresponding ecology terms. Several students did not write the role of keystone species. Some students wrote that it was proposed by Haeckel.
3. Some candidates, instead of defining *hotspots* defined *biosphere reserves*. Some candidates did not write the keyword endemic species.
4. Very few candidates, while defining *Ramsar sites*, wrote the key term *wetlands*. A few defined it as a sacred site related to God Ram.
5. Some candidates did not write the catchword endangered while defining *Red Data Book*.
6. *In situ* and *Ex situ* modes of conservation were mixed up. Some candidates wrote general points. Several candidates discussed the importance of biodiversity and the consequences of loss of biodiversity.

Chapter 16. Environmental Issues

Topics Found difficult By Students
- Acronyms BOD, PAR
- Global warming / Ozone hole

1. Majority of the students were unable to express the main agenda of the *Montreal Protocol*.
2. Majority of the candidates were mystified between *global warming* and *ozone hole*. Some candidates wrote about ozone depletion and acid rain. A few candidates, instead of writing the effects of global warming, wrote its causes.
3. Most candidates wrote about firecrackers and no horn zones in noise *pollution*. Several candidates wrote the same points in different words and some wrote general points.
4. Majority of the candidates, instead of defining *BOD*, expanded the acronym. Most of the candidates could not write the *significance of BOD in an aquatic ecosystem. Several candidates were not as clear as to how BOD indicates the level of pollution in a water body.*
5. Some candidates mentioned incorrect names of the famous personalities associated with *Chipko Movement*.
6. Most of the candidates considered the word *Joint Forest Management* in literal terms and wrote that all the forests of the country should be joined together. Many candidates could not explain the collaborative efforts between the Forest Department (Government) and Local Tribes/villagers for sustainable development of forest without damaging the ecosystem.

Miscellaneous

1. Many candidates wrote the causative agents without Binomial Nomenclature. Spellings were incorrect in several cases. Preventive measures were either missed or written incorrectly.
2. Names of scientists were spelt incorrectly by a number of candidates. Some candidates did not write the full name correctly.
3. Many students wrote repeated points in answers, by change in terms or language only to increase the length of an answer.

Word of Advice

Chapter at a Glance

| Set 1 |

Chapter 1. Reproduction in Organisms

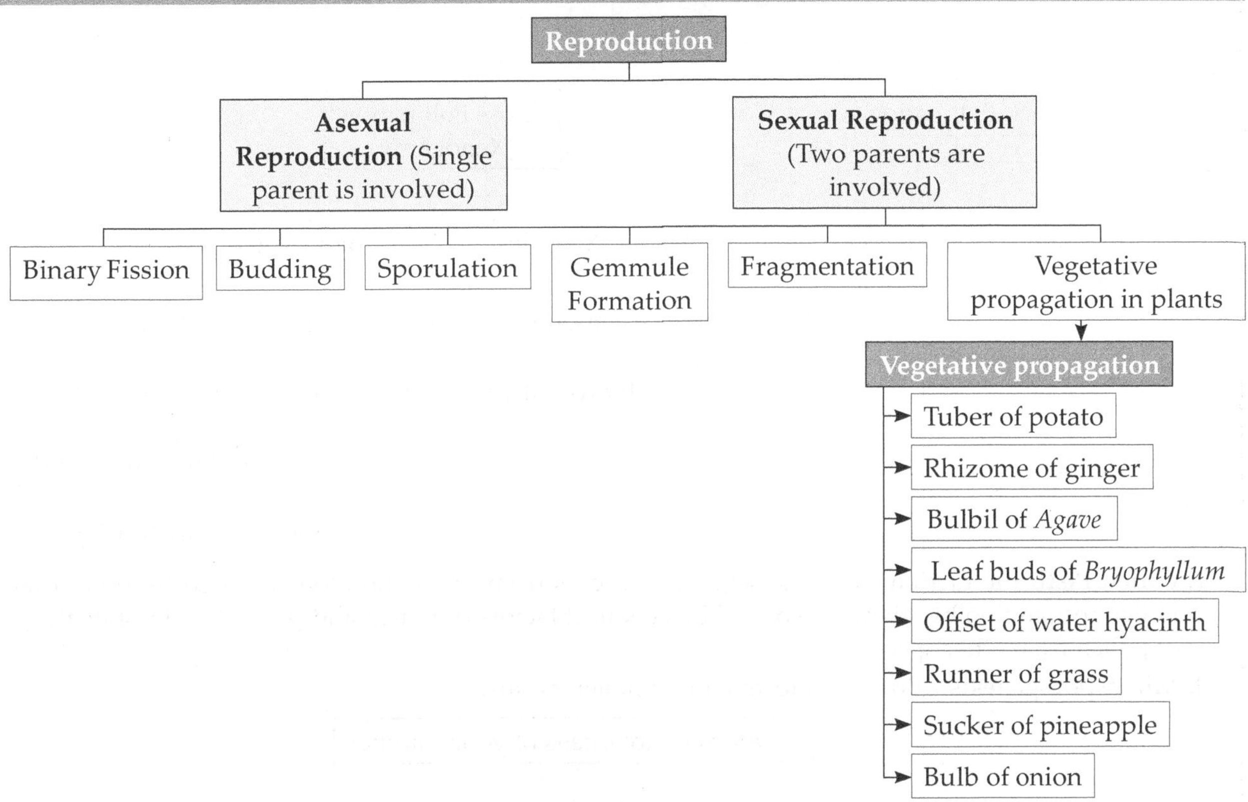

1. **Life span**—The maximum number of years that humans or animals can live.
2. **Life expectancy**—Life expectancy is defined as the age to which a person is expected to live, or the remaining number of years a person is expected to live.
3. **Clone**—Clonning is the process of generating a genetically identical copy of a cell in an organism.

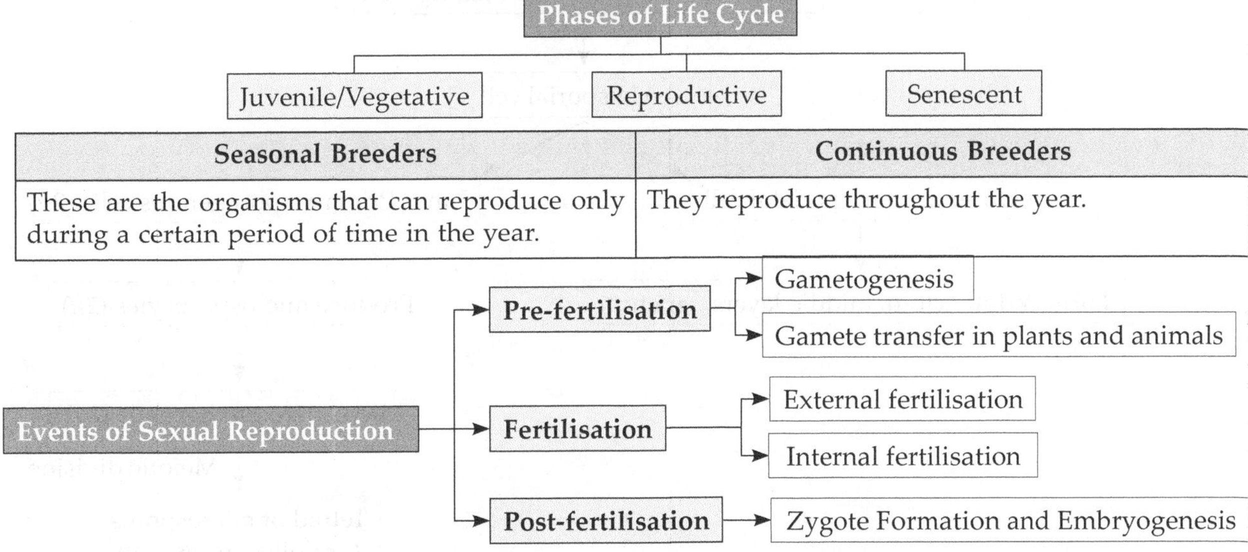

Seasonal Breeders	Continuous Breeders
These are the organisms that can reproduce only during a certain period of time in the year.	They reproduce throughout the year.

5. **Parthenogenesis :** It is a form of reproduction in which an egg can develop into embryo without being fertilised by the sperm.

Chapter 2. Sexual Reproduction in Flowering Plants

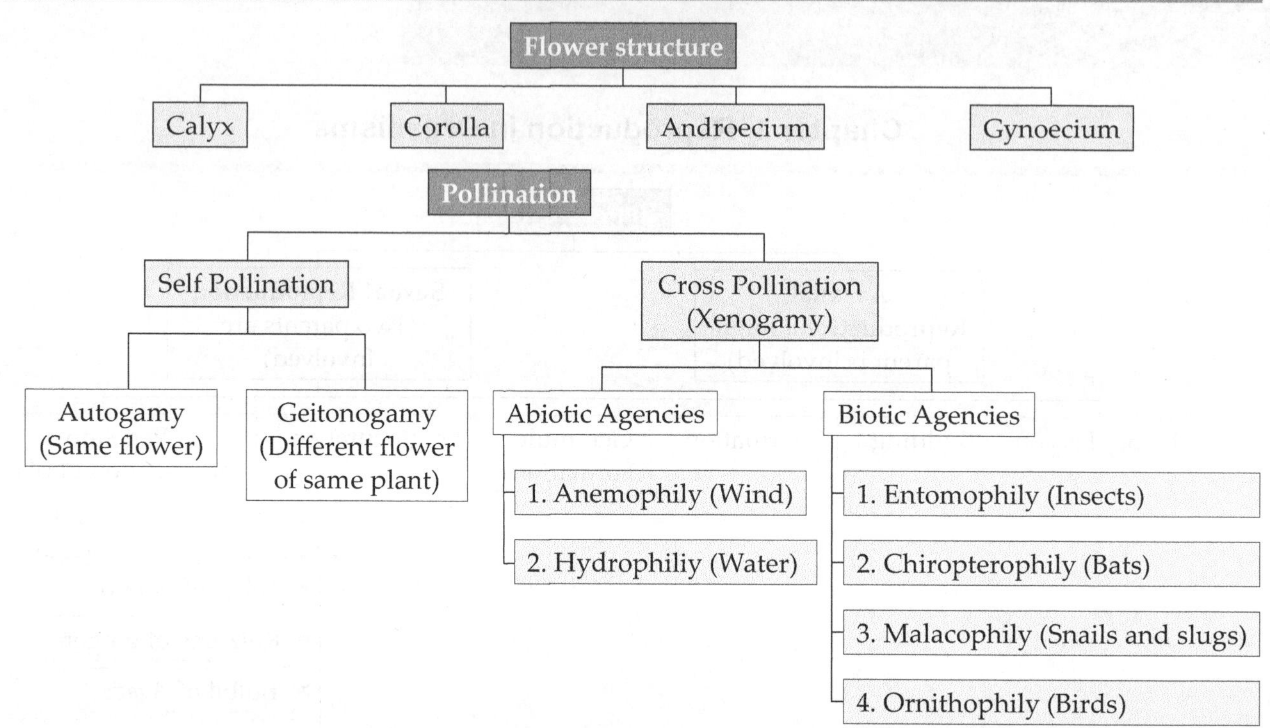

1. **Pollen-Pistil interaction**—It is the sequential events from pollination to before-fertilisation. It includes pollen recognition by stigma, mediated by chemical factors of pollen and pistil. This inhibits the growth of pollen from other species.
2. **Microsporogenesis**—Process of formation of pollen grains.

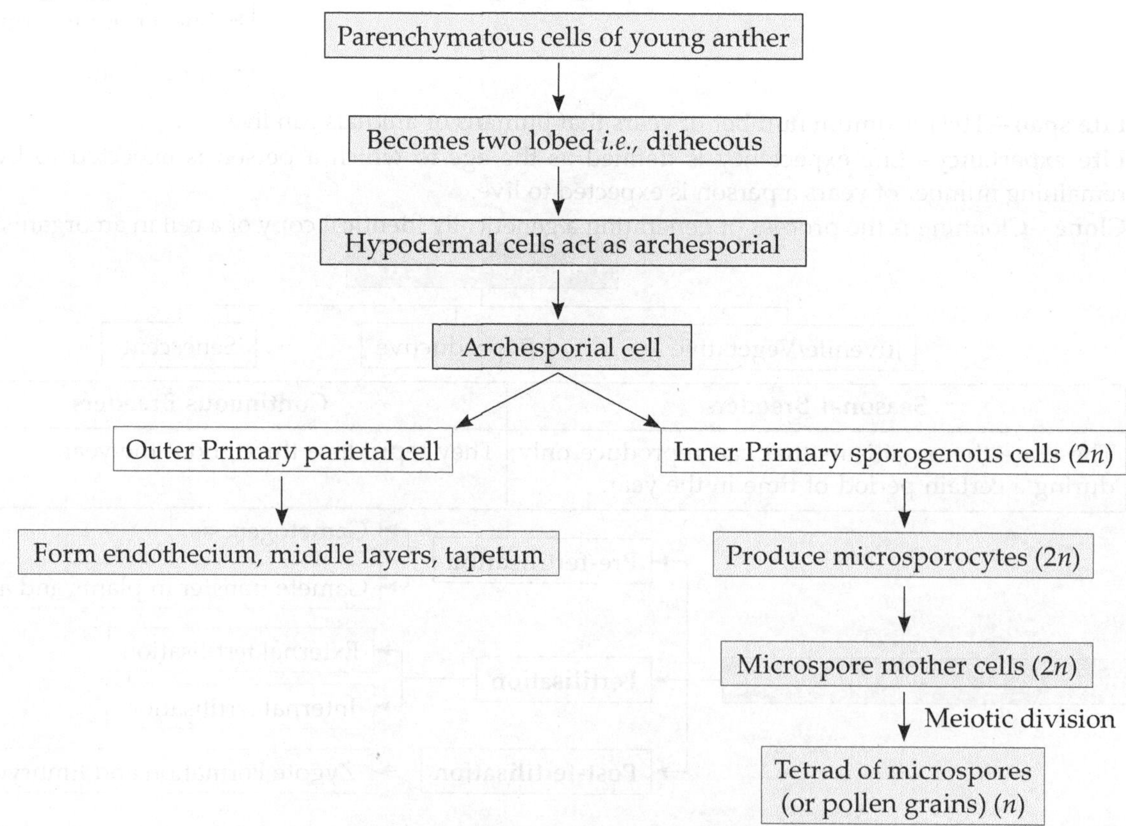

3. Megasporogenesis—Process of development of egg or ova.

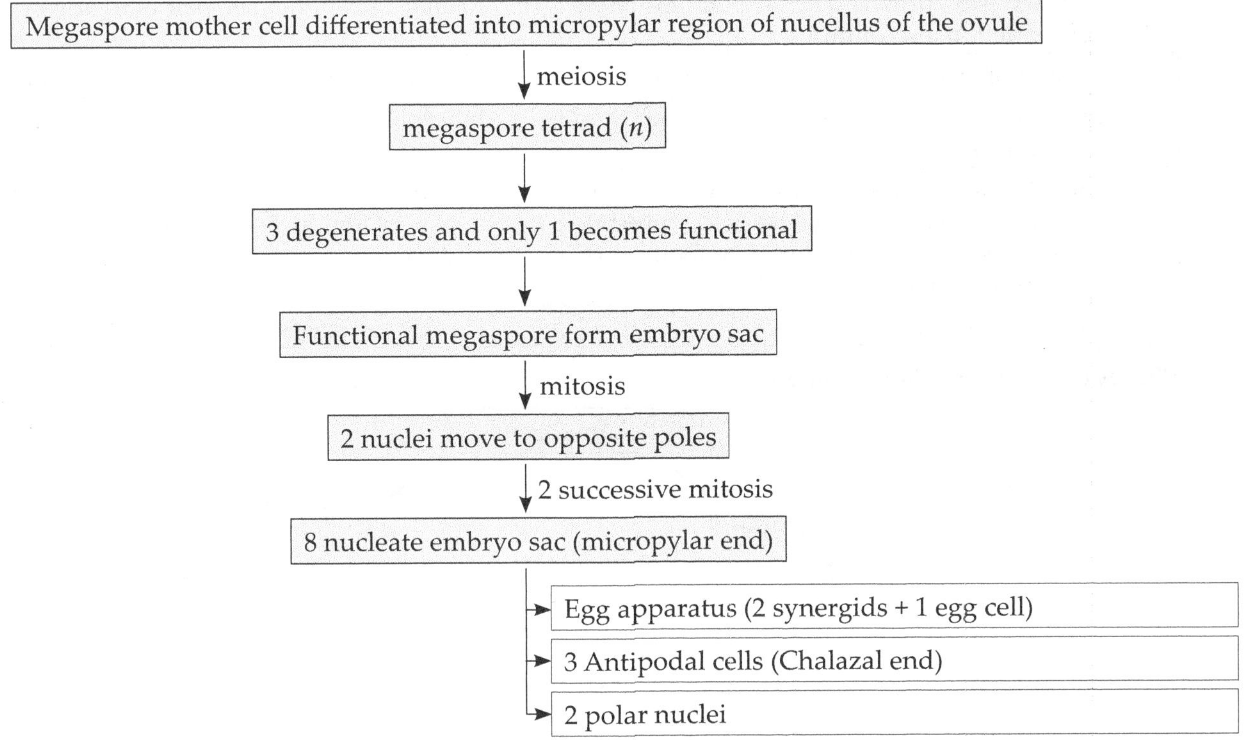

6. **Process of fertilisation**—

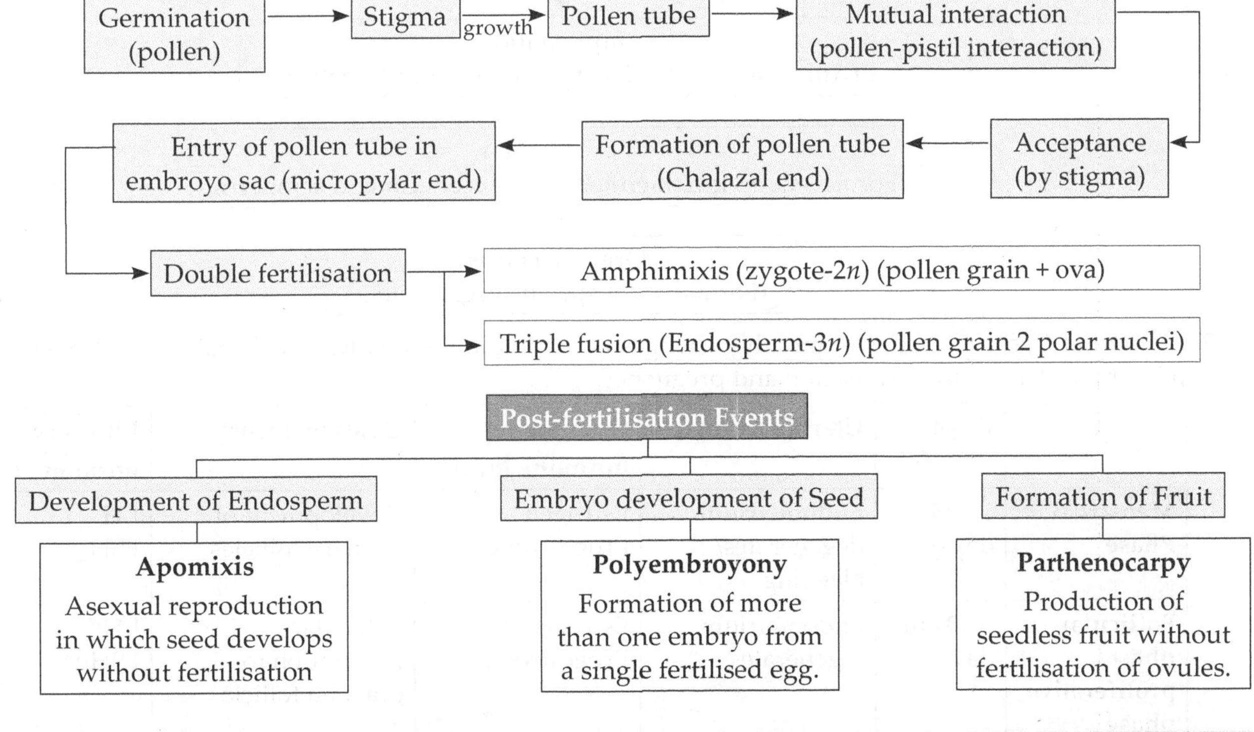

Post-fertilisation Events

Development of Endosperm	Embryo development of Seed	Formation of Fruit
Apomixis Asexual reproduction in which seed develops without fertilisation	**Polyembroyony** Formation of more than one embryo from a single fertilised egg.	**Parthenocarpy** Production of seedless fruit without fertilisation of ovules.

Chapter 3. Human Reproduction

1. Humans are sexually reproducing and viviparous organisms. They show remarkable differences in the reproductive system of male and female. Human reproduction is any form of sexual reproduction resulting in human fertilisation. It typically involves sexual intercourse between a man and a woman. These are specialised reproductive cells called gametes, created in a process called meiosis.

Male reproductive system	Female reproductive system
(i) Primary sex organs *i.e.,* a pair to testes suspended in a scrotum.	(i) Primary sex organ i.e., pair of ovaries.
(ii) Consist of external and internal genitalia.	(ii) Consists of external and Internal genitalia
(a) External genitalia-Penis urethra and scrotum	External genitalia consists of vulva only.
(b) Internal genitalia—Testes, epididymis, vas deferens, seminal vesicles, prostate glands urethra, a pair of bulbourethral glands and seminal vesicles	Internal genitalia consist of a pair of ovaries, a pair of oviducts (fallopian tubes), uterus and vagina.
(iii) Male gametes called sperm.	Female gametes called ova.

2. Phases of embryonic development

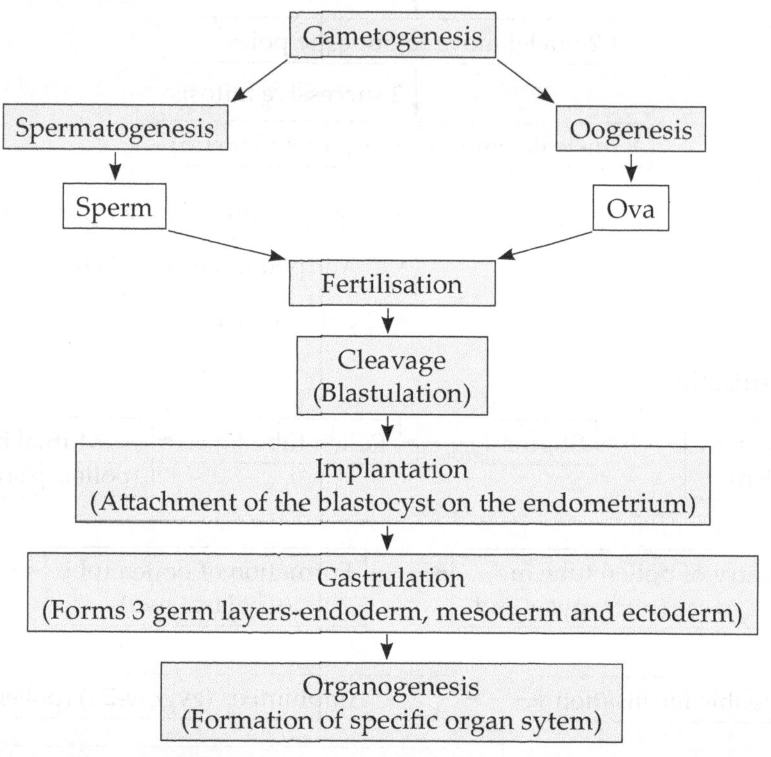

3. Menstrual Cycle—A series of monthly changes occurring in the ovaries and female genital tract which prepare the female for fertilisation and pregnancy.

	Days	**Uterine Events**	**Ovarian hormone level**	**Ovarian events**	**Pituitary hormone level**
Menstrual Phase	1-5 (5 days)	Endometrium degenerates/ bleeding	Estrogen__ Progestrone__	Development of primary follicles	LH__ FSH__
Follicular phase/ proliferative phase	6-13 (10 days)	Endometrium regenerates	Estrogen^^ Progestrone__	Follicular maturation to graafian follicle	LH^ FSH^^
Ovulatory Phase	About 14/15th day	Endometrium thickens	Estrogen^^ Progesterone^	Ovulation (graafian follicle ruptures to release ova)	LH^^^^ FSH^^
Luteal Phase/ secretory phase	15 to 28 (14 days)	Endometrium thicknes	Estrogen^ Progesterone^^	Corpus luteum formation	LH^^ FSH^^

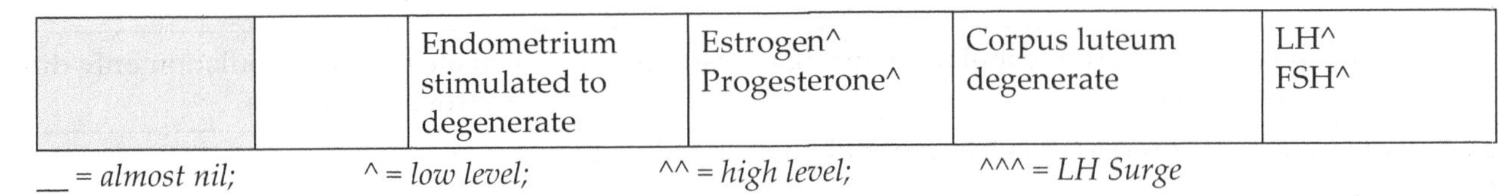

		Endometrium stimulated to degenerate	Estrogen^ Progesterone^	Corpus luteum degenerate	LH^ FSH^

__ = almost nil; *^ = low level;* *^^ = high level;* *^^^ = LH Surge*

4. Fertilisation

Approach of sperm to ovum
(Ejaculation; Capacitation; release of ovum from ovary)

↓

Penetration by sperm
(Fertilizin-antifertilizin interaction, acrosomal reaction; completion of first meiotic division)

↓

Cortical reaction
(to check the entry of more sperms)

↓

Fusion of gametic nuclei
(Second maturation division of egg; amphimixis; formation of zygote)

5. Implantation

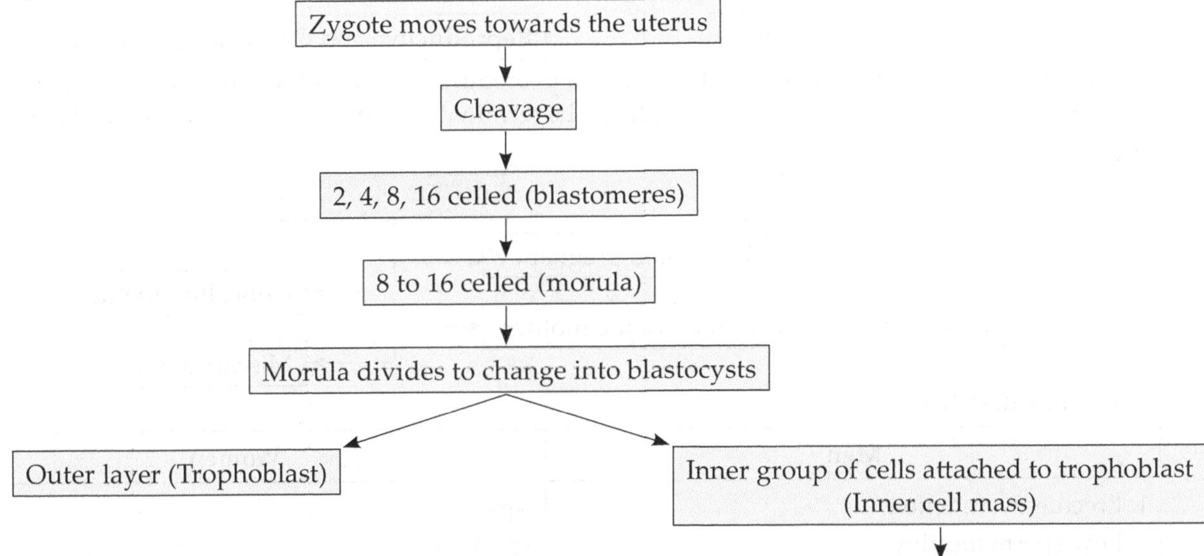

Gametogenesis	Spermatogenesis	Oogenesis
The process of the formation of haploid gametes from diploid germ cells by meiosis in the gonads.	The process of the formation of haploid sperm from the dipolid spermatogonia by meiosis in testes.	The process of formation of haploid egg from diploid Oogonia by meiosis in the ovaries.

Menstrual cycle	Oestrous cycle
1. It occurs in monkeys, apes and human beings.	It occurs in cow, dog.
2. Cycle consists of 3 phases.	Its consists of a short period of oestrous or heat cycle.
3. Blood flows in the last few days of the cycle. The broken endometrium is passed out during menstruation.	It is 12-24 hours followed by anoestrous or passive period.

4. Female does not permit copulation during menstrual phase of the cycle.	Female permits copulation only during oestrous period.

6. **Parturiton**—The process of delivery of fully developed foetus.

Signals for parturition

- Fully developed foetus
- Release of oxytocin from posterior pituitary
- Mild uterine contractions

7. The mammary glands of female start producing milk, at the end of pregnancy by the process of **lactation.** The milk produced during the initial few days of lactation is called **colostrum,** which contain several antibodies.
8. Placenta is attached with foetus with an umbilical cord that transport food and oxygen to embryo.
9. **Placental hormones**—hCG (human chorionic gonadotropin), hPL (human Placental Lactogen) and relaxin.

Chapter 4. Reproductive Health

1. **Reproductive Health** means that total well being in all aspects of reproduction *i.e.,* physical, emotional, behavioural and social.

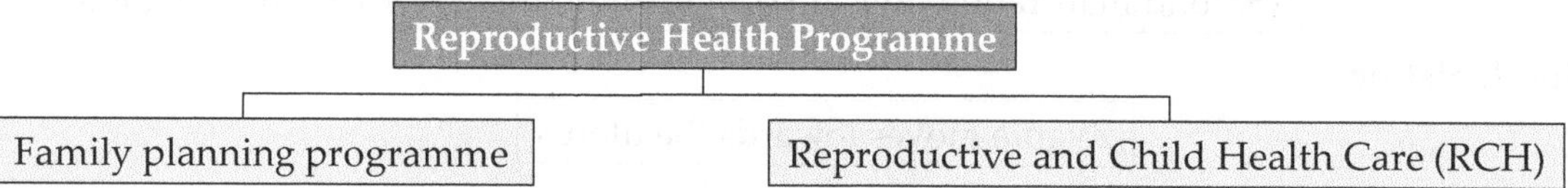

2. **Medical Termination of Pregnancy (MTP)**—It is a procedure of terminating pregnancy using medicines. In the early stages of pregnancy (7-9 week). It can be terminated with the help of medicine. After that, the surgical process is needed.

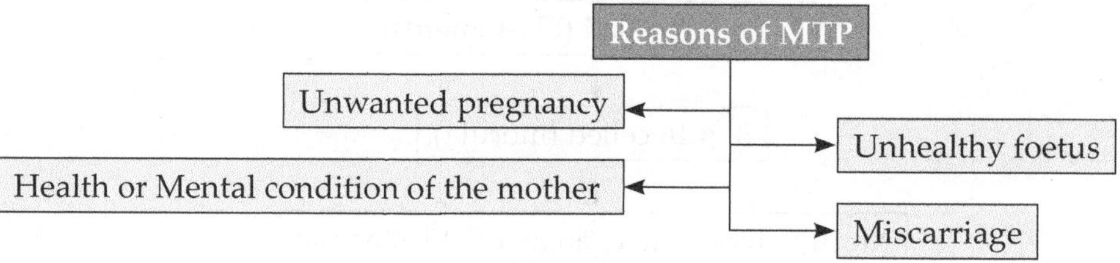

3. **Causes of Infertility**

Men	Women
1. Erectile dysfunction	Age
2. Low sperm motility	Smoking
3. Genetic problem	Excess alcohol use
4. Acidity in urethra	Stress
5. Shape of the sperm	Poor diet
6. Sperm count	Athletic training
7. STDs (Sexually Transmitted Diseases)	Being overweight or under weight
8. Impotency	Sexually Transmitted Infections (STIs)

4. **Amniocentesis**—A process in which amniotic fluid is sampled using a hollow needle inserted into the uterus, to screen for abnormalities in the developing fetus.
5. **Assisted Reproductive Technologies (ARTs)**
   (a) **IVF** – In Vitro Fertilisation
   (b) **IUT** – Intra Uterine Transfer
   (c) **ZIFT** – Zygote Intrafallopian Transfer
   (d) **ICSI** – Intracytoplasmic Sperm Injection
   (e) **GIFT** – Gamete Intrafallopian Transfer
   (f) **AI** – Artifical Insemination
   (g) **IUI** – Intra Uterine Insemination

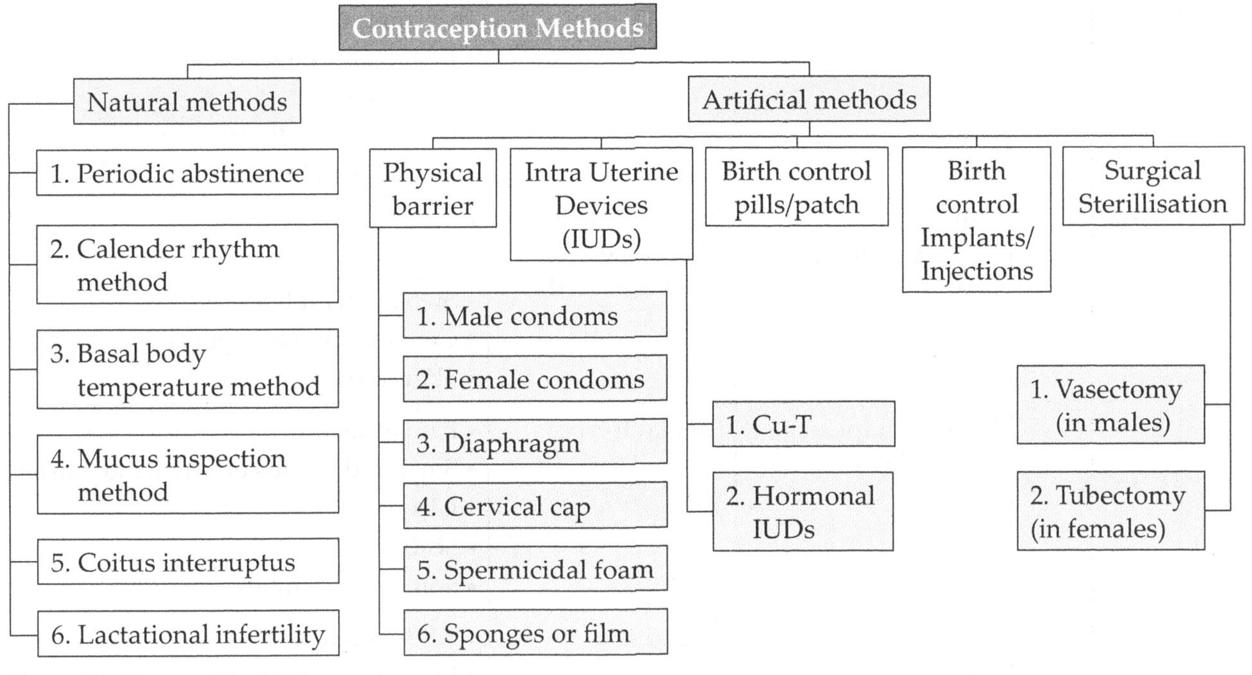

6. **Sexually Transmitted Diseases**

Disease	Causative organism	Diseases	Symptoms
Bacteria	*Neisseria gonorrhoeae*	Gonorrhoea	**In Males :** • Painful urination • Pus discharge from penis • Pain or swelling in one testicle **In Females :** • Increased vaginal discharge • Painful urination • Vaginal bleeding between periods • Abdominal or pelvic pain
	Tereponema pallidum	Syphills	**Primary syphills :** Appearance of chancre (sore) **Secondary Syphilis** • A skin rash that causes small, reddish-brown sores • Sores in mouth, vegaina, or anus • Swollen glands
	Chlamydia trachomatis	Chlamydiasis	• Painful urination • Vaginal discharge in women • Discharge from the penis in men • Painful sexual intercourse in women • Bleeding between periods • Testicular pain in men
Parasite	*Trichomonas vaginalis*	Trichomoniasis	**In Males :** • Itching or irritation inside the penis • Burning after urination or ejaculation • Discharge from the penis

			In Females : • Itching, burning, redness or soreness of the genitals • Discomfort with urination • A change in their vaginal discharge
Virus	Human Immunodeficiency Virus (HIV)	Acquired Immune Deficiency Syndrome (AIDS)	• Fever • Chills • Swollen lymph nodes • General aches and pains • Skin rash
	Herpes Simplex Virus (HSV)	Genital Herpes	• Pain or itching • Small red bumps or tiny white blisters • Ulcers • Scabs • Sores near genitals
	Hepatitis B Virus (HBV)	Hepatitis B	• Abdominal pain • Dark urine • Joint pain • Jaundice like symptoms
	Human Papilloma virus (HPV)	Genital warts	• Appearance of flat lesions, small cauliflower-like bumps or tiny stem-like protrusions. • In women, on vulva, near anus, on cervix or in vagina. • In men, on penis and scrotum or around anus

Chapter 5. Principles of Inheritance and Variations

1. Heredity refers to the genetic traits passed down by our biological parents.

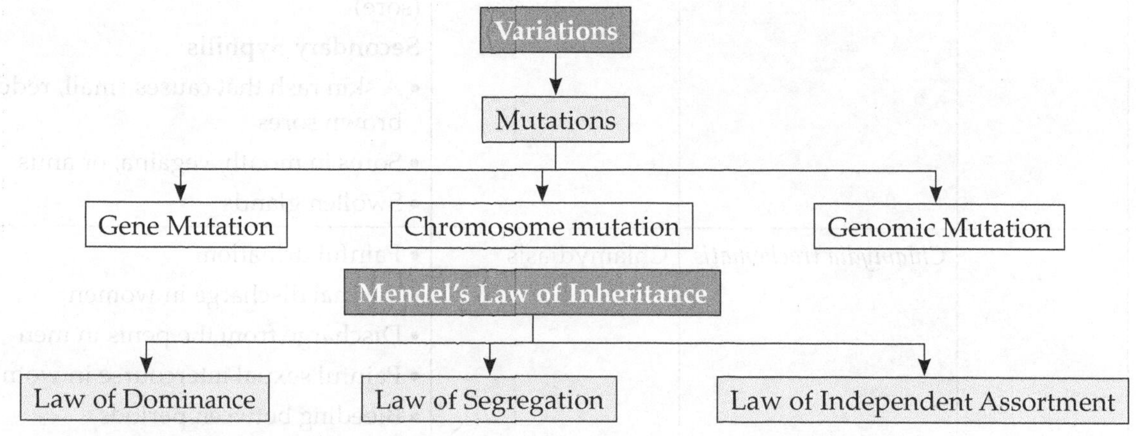

2. Seven pair of contrasting characters selected by Mendel in *Pisum sativum* (pea plant)

S.No.	Characters	Contrasting pairs (Allelic pairs)	
		(Dominant)	**(Recessive)**
1.	Form of seed	Round (R)	Wrinkled (r)

2.	Colour of cotyledons	Yellow (Y)	Green (y)
3.	Colour of flower	Purple (P)	White (w)
4.	Form of pod	Inflated (I)	Constricted (i)
5.	Colour of pod	Green (G)	Yellow (g)
6.	Position of flower	Axial (A)	Terminal (a)
7.	Height of plant (Length of stem)	Tall (T)	Dwarf (t)

3. Types of chromosomes

Autosomes Chromosome	Sex Chromosome
Chromosomes other than sex chromosomes, are autosomes *i.e.* 22 pair autosomes.	Humans and most other mammals have two sex chromosomes, X and Y.

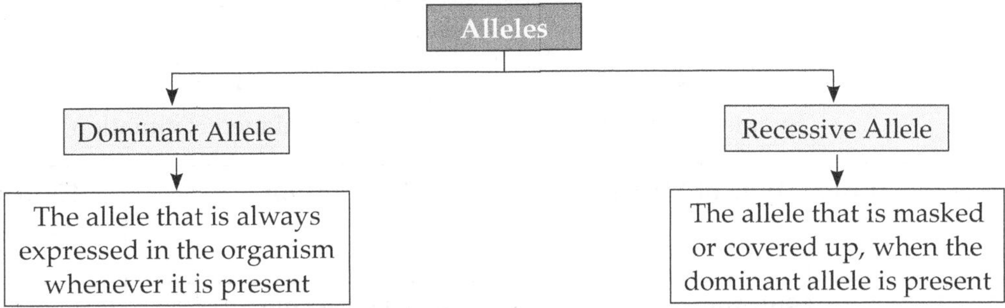

Homozygus Individual	Heterozygous Individual
Both the alleles of a character are similar. E.g., TT, tt	It carries dissimilar alleles. E.g., Tt
Either the pure dominant or pure recessive individual.	Impure dominant individual.

Genotype	Phenotype
Genotype refers to the genetic code of the individual.	Phenotype is the expression of the genotype *i.e.*, visible to other people and can be observed.

Monohybrid Cross	Dihybrid Cross
It is a cross between two parents different in a single pair of contrasting character.	It is cross between two parents different in two alternatives of two different traits characters.

Back Cross	Test Cross
It is the cross between F1 hybrid with any of the parent.	It is the cross between the dominant phenotype with the recessive phenotype.

4. **Punnett square** is a graphical representation to predict the genotypes of a particular cross on breeding experiment.

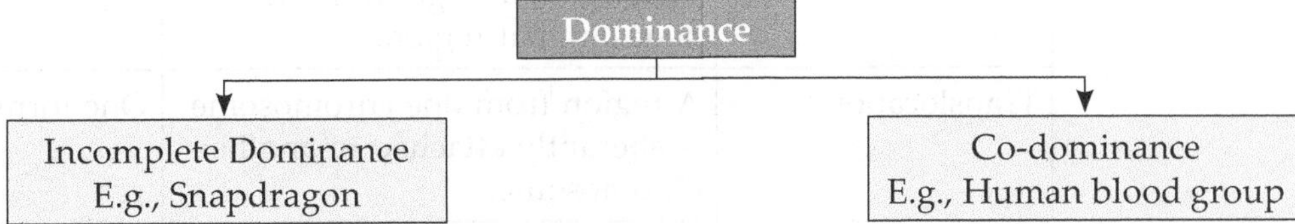

5. **Polygenic inheritance** is determined by additive effects of two or more genes on a single characters.

   E.g., Skin colour in human.

6. **Pedignee analysis**—A record of inheritance of certain genetic traits for two or more generations presented in the form of a diagram or family tree.

7. **Pleiotropy** refers as when one gene which effects multiple characters.
   E.g., 1. Phenyl Ketonuria (PKU) in human beings.
   2. Starch synthesis in pea seeds.
8. Sex linked inheritance was discovered by Morgan that is males and females differ in their sex chromosome combination.
9. **Sex Determination**

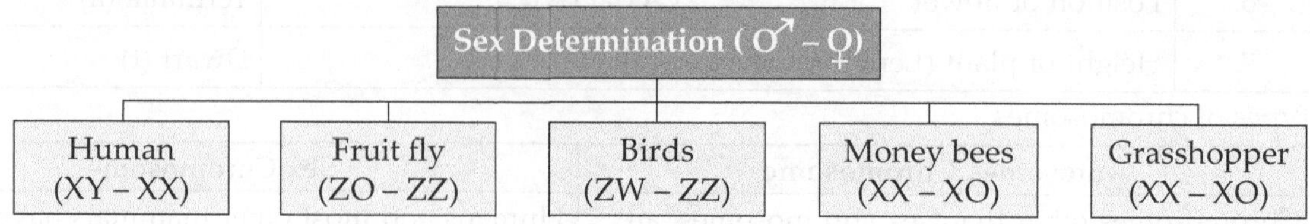

10. **Types of Mutation**

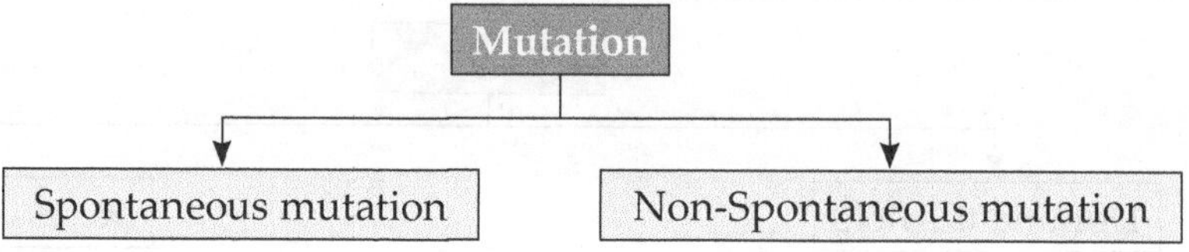

Class of Mutation	Type of Mutation	Description	Human Disease(s) Linked to This Mutation
1. Point mutation	Substitution	One base is incorrectly added during replication and replaces the pair in the corresponding position on the complementary strand.	Sickle-cell anemia
	Insertion	One nucleotides are inserted into replicating DNA, often resulting in a frameshift.	One form of betathaiassemia
	Deletion	One nuclcetides "skipped" during replication or otherwise excised. Often resulting in a frameshift.	Systic fibrosis
2. Chromosomal mutation	Inversion	One region of a chromosome is fipped and reinserted.	Opitz-Kaveggia syndrome
	Deletion	A region of a chromosome is lost resulting in the absence of all the genes in that area.	Cri du chat syndrome
	Duplication	A region of a chromosome is repeated, resulting in an increase in dosage from the genes in that region.	Some cancers
	Translocation	A region from one chromosome is aberantly attached to another chromosome.	One form of leukemia
3. Copy number variation	Gene amplification	The number of tandem copies of a locus is increased.	Some breast cancers
	Expanding trinucleotide repeat	The normal number of repeated trinucleotide sequences is expanded.	Fragile X syndrome Huntington's disease

Chapter 6. Molecular Basis of Inheritance

1. Types of Genetic Material

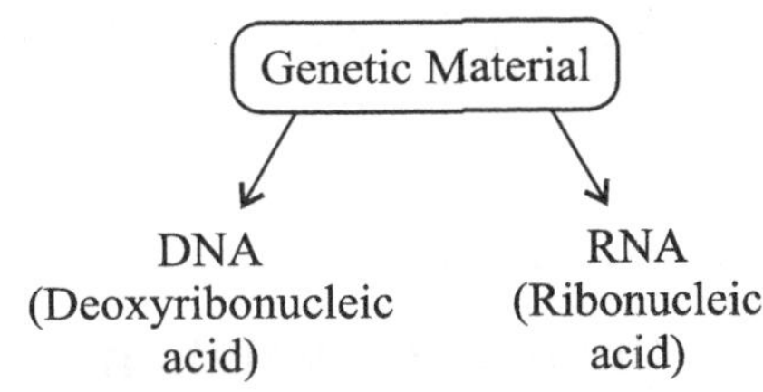

Name of Experiment	Scientist	Findings
1. Griffith's experiment (1028)	Griffith	The type II-R had been "transformed" into the lethal III-S strain by a "transforming principle" that was somehow part of the dead III-S strain bacteria
2. Hershey and Chase's experiment (1952)	Hershey and Chase.	The genetic material of phage is DNA, not protein DNA went when it existed the protein coat and entered the bacteria, they used radioactive phosphorus-Labeled phages bacteria.
3. Avery, McLeod and McCarty experiment. (1944)	Avery, McLeod and McCarty.	They identified DNA as the "transforming principle" while studying *Streptococcus pneumonia,* bacteria that can cause pneumonia.

2. Types of RNA

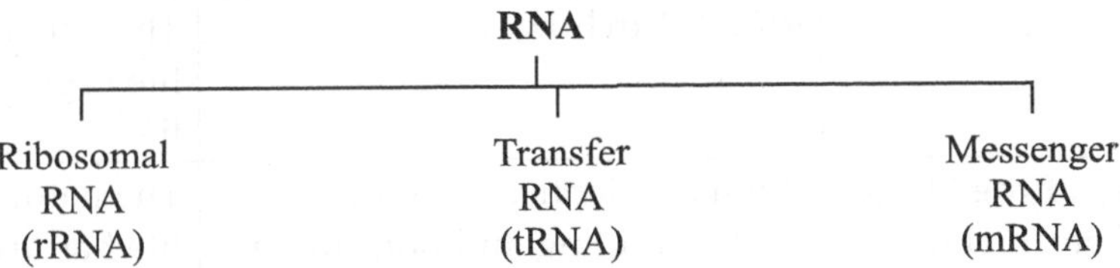

3. **Central Dogma :** DNA to RNA to protein

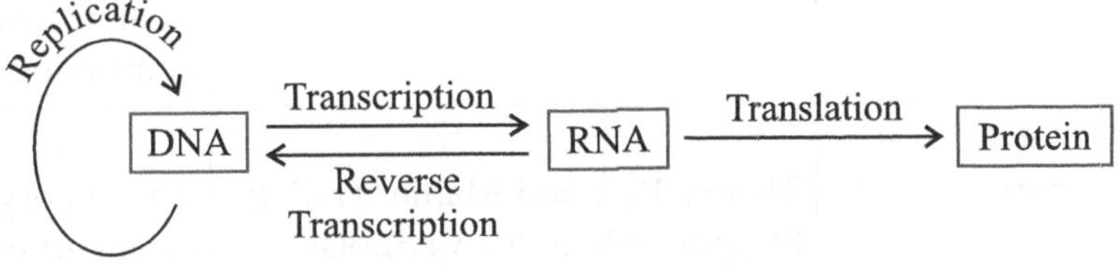

4. Human Genome Project

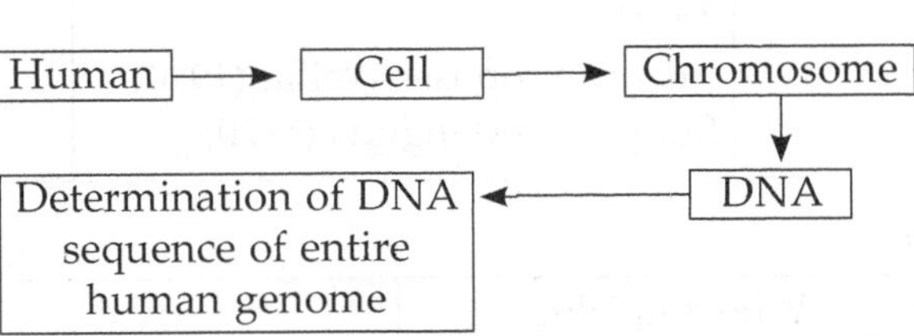

5. **Rice genome project** has been utilised in evolutionary analysis, rice gemomics of functional genonic studies. This was used and found beneficial for bringing the knowledge gap between genotype phenotype as well as facilitating molecular breeding via gene design in rice.

Intron	Exon	Cistron
Introns are the non-coding sequences of a gene. They do not appear in mature mRNA.	Exons are coding sequences gene. They collectively make the final RNA molecule.	It is the gene a real sense capable of synthesising a polypeptide chain of the enzyme.

Lac Operon	
ON	OFF
1. Absence of lactose 2. Regulator binds with represssor protein 3. No enzyme release	Presence of lactose. Repressor mRNA binds with inducer to inactivate it. Activate gene z — β-galactosidase y — permease a — Transacetylase

Chapter 7. Evolution

1. **Evolution** is the process of change in all forms of life through numerous generations. The researchers have put forward various theories on the origin of life.

2. **Origin of Life**

Name of Theory	Who and When	Objective
1. Theory of Special Creation	Father Suarez	'God' the supreme power has created the earth, water, sun, moon and above all; Adam and Eve were the first man and woman on the earth.
2. Theory of Spontaneous Generation	Aristotle	Living creatures could arise from the non-living matter due to the influence of heat and moisture.
3. Theory of Biogenesis	Rudolf Virchow, 1858	This theory demonstrated that life orginates from pre-existing life.
4. Modern Theory of the Origin of Life or Oparin-Haldane theory of the origin of life	Russian scientist A.I. Oparin (1923) and England born Indian scientist J.B.S. Haldane (1928).	The natural process by which life has arisen from non-living matter, such as simple matter, such as simple organic compounds.
5. Origin of Eukaryotes **(i) Evolutionary theory**	Theory Raff and Mahler (1972); Supporter-Vzzell and Spolstary 1974).	This theory explains the origin of eukaryotic cells.
(ii) Symbiotic theory	Theory Proposer-Wallin (1920); Supporter-Mangulis (1970)	

3. **Some Important Theories**

Theory	Who and When	Objective
1. Drawinism theory	English naturalist Charles Drawin (1859)	Species of an organism arise and develop through naturally selected inherited variations that increase the individual ability to compete, survive, and reproduce in its riche.
2. Modern synthetic theory	T. Dobzhansky, J.B.S. Haldane, R.A. Fisher, Sewall wright, Stebbins, Ernst Mayr (1930-1940).	Genetic changes occurring in the population leads to the formation of new species. It also describes the genetic population or Mendelian population, gene pool and gene frequency.

3. Hardy- Weinberg's Theory	G.H. Hardy and wilhelm Weinberg (1908)	Genetic variation in a population will remain constant from one generation to the next in the absence of disturbing factors.
4. Theory of recapitulation or Biogenetic law or Embyological parallelism	Ernst Haeckel (1866)	The theory describes that before they are born, organisms pass through developmental stages that look like adult animals of other species, in roughly the same order that these other species split off during evolution.
5. Lamarck's theory	Lamarck (1809)	All the physical changes occurring in a individual during its lifetime are inherited by its offspring. E.g. evaluation of the long neck of a griaffe, the resistance of mosquitoes to DDT and resistance of bacteria to antibiotics.

4. Evidences of Evolution

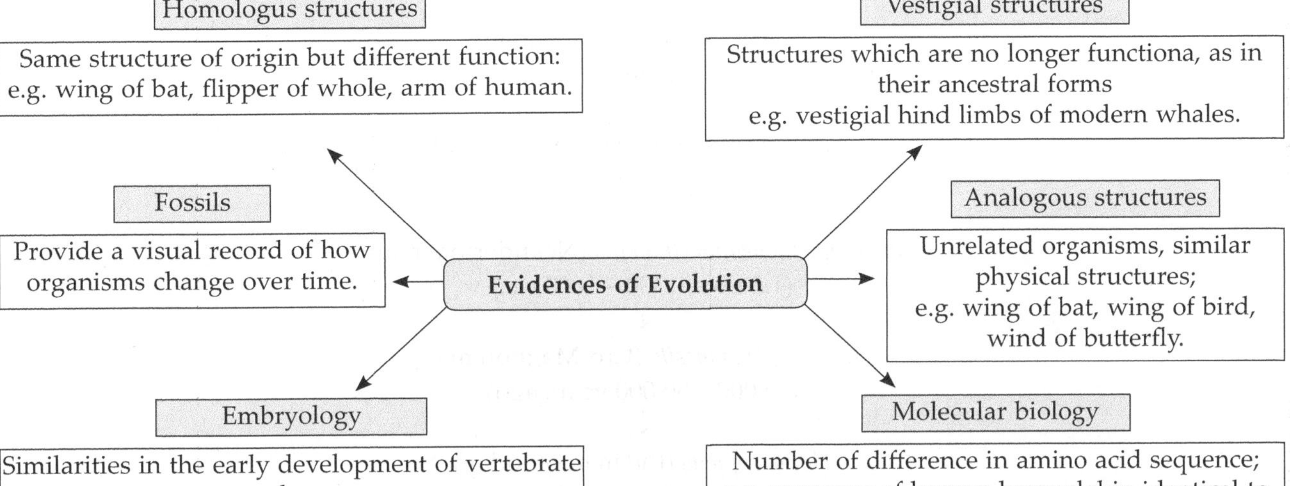

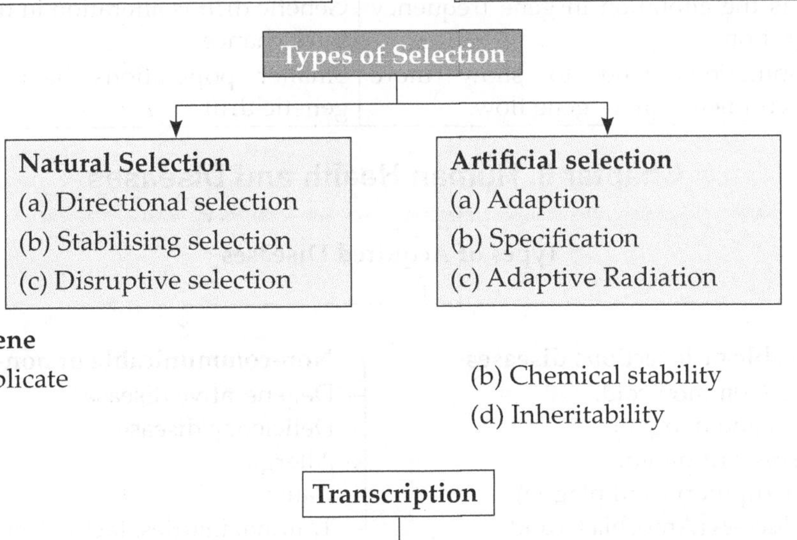

5. Properties of Gene

(a) Ability to replicate
(b) Chemical stability
(c) Mutability
(d) Inheritability

6. Transcription

Transcription
- Initiation
- Elongation
- Termination

7. Human Evolution

Dryopithecus- Apes
(20 mya)
↓
Shivapithecus
(5.1 mya)
↓
Ramapithecus
(14-15 mya)
↓
Australopithecus africanus
(5 mya)
A. robustus
A. boisei
↓
Homo habilis (the tool maker)
(2 mya)
↓
Homo erectus (Java ape man)
(1.7 mya)
↓
Homo erectus pekinesis (peking man)
(0.6 mya)
↓
Homo sapiens neanderthalenis (Neanderthal man)
(1.5 mya, 1300 – 1600 c.c.)
↓
Homo sapiens fossils (Cro-Magnon man)
(35,000 – 50,000 years ago)
↓
Homo sapiens (the man of today)
(25000 years)

Gene flow	Genetic drift
1. Gene flow is the alteration in gene frequency due to migration.	Genetic drift is alteration in the gene frequency by pure chance.
2. Larger populations tend to show more migrations and hence more gene flow.	Smaller populations have greater chances of genetic drift

Chapter 8. Human Health and Diseases

Types of Acquired Diseases

- **Communicable or infectious diseases**
  - *Viral diseases* (common cold, chikengunya and dengue)
  - *Bacterial diseases* (typhoid, pneumonia, diphteria and plague)
  - *Protozonal diseases* (Amoebiasis and malaria)
  - *Helminitic diseases* (Ascariasis and filariasis)
  - *Fungal diseases* (ring worm)
- **Non-communicable or non-infectious diseases**
  - Degenerative disease
  - Deficiency disease
  - Allergies
  - Cancer
  - Trauma, injuries, factures etc.

Chapter at a Glance

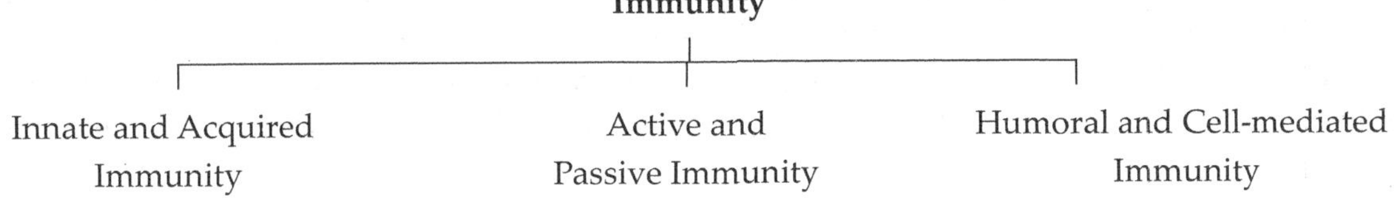

1. **Interferons**—Interferons are a group of signalling proteins, made and released by host cells in response to the presence of several viruses.
2. **Types of Antibody**

Antibody

IgG IgA IgM IgD IgE

Vaccination	Immunisation
Vaccination is the term used for giving a vaccine *i.e.,* the process of introducing a substance to a host.	Immunisation is the term used for the process of both getting the vaccine and becoming immune to the disease as a result of the vaccine

3. **Allergy**—An allergy occurs when the body's immune system sees a substance as harmful and act against it. The substances that cause allergic reactions are allergens.

Primary Lymphoid	Secondary Lymphoid
Bone Marrow and thymus	Lymph nodes, spleen, tonsils

Modes of Transmission of AIDS

Blood — Semen of pre-seminal fluid — Rectal fluid — Vaginal fluids — Breast milk

4. **ELISA**—(Enzyme-Linked Immunosorbent Assay) - It is a plate-based assay technique designed for detecting and quantifying soluble substances such as peptides, proteins, antibodies, and hormones.
5. **Types and effect of Drugs :**

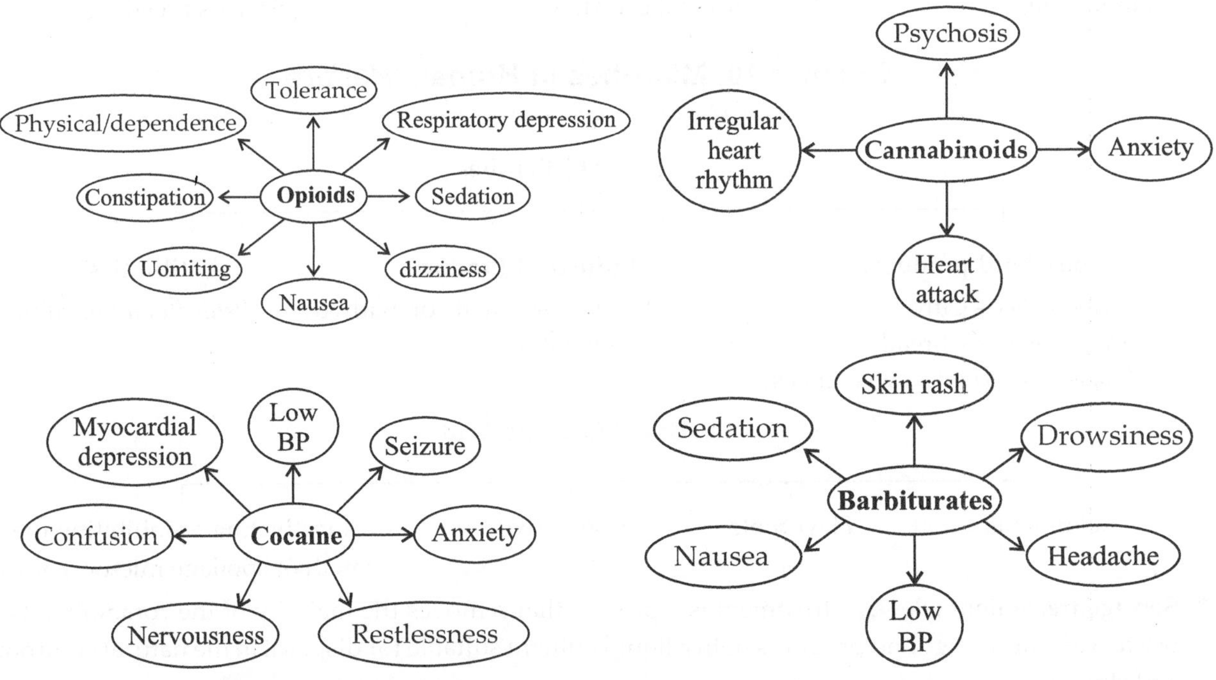

Chapter 9. Strategies for the Enhancement in Food Production

1. Food is necessary for the development and survival of all organisms. To cater to the need of the increasing population, it is necessary to increase food production.

2. **Plant breeding :** It is the science of changing the traits of plants to produce desired characteristics. It has been used to improve the quality of nutrition in products for humans and animals.
3. **The Green Revolution** is a great increase in the production of food grains (such as rice and wheat) due to the introduction of high-yielding varieties, with the help of use of pesticides and better management techniques.
4. **Tissue culture** is the growth of tissues or cells in an artificial medium separate from the parent organism. This technique is also called **micropropagation**. This is typically facilitated via the use of a liquid, semi-solid, or solid growth media, such as broth or agar.
5. **Single-cell protein** is defined as the biomass or protein extract from pure or mixed cultures of algae, yeasts, fungi or bacteria that may be used as an ingredient or a substitute for protein-rich foods and is suitable for human consumption or as animal feeds.
6. **Biofortification** is the idea of breeding crops to increase their nutritional value. This can be done either through conventional selective breeding or through genetic engineering.
7. **Animal Husbandry :** The branch of agriculture concerned with the care and breeding of domestic animals such as cattle, elogs, sheep and horses.
8. **Dairy Farm Management :** It is the management of animals for milk and other dairy products to increase the yield and improve the quality of milk.
9. **Poultry farming** is raising of birds domestically or commercially, primarily for meat and eggs like chickens, turkeys, ducks, and geese (primary importance), while guinea fowl and squabs (local interest).
10. **Apiculture :** Bee-keeping or apiculture is the maintenance of hives of honeybees for the production of honey. Bee-keeping can be practiced in any area where there are sufficient bee pastures of some wild shrubs, fruit orchards and cultivated crops.
11. **Pisciculture :** Fish farming is known as *pisciculture.* It includes rearing, raising or breeding of aquatic life.
12. **Animal Breeding** is an important aspect of animal husbandry, which aims at increasing the yield of animals and improving the desirable qualities of the produce.
13. **Methods of Breeding**

    (a) Inbreeding (b) Out breeding (c) Cross breeding

Chapter 10. Microbes in Human Welfare

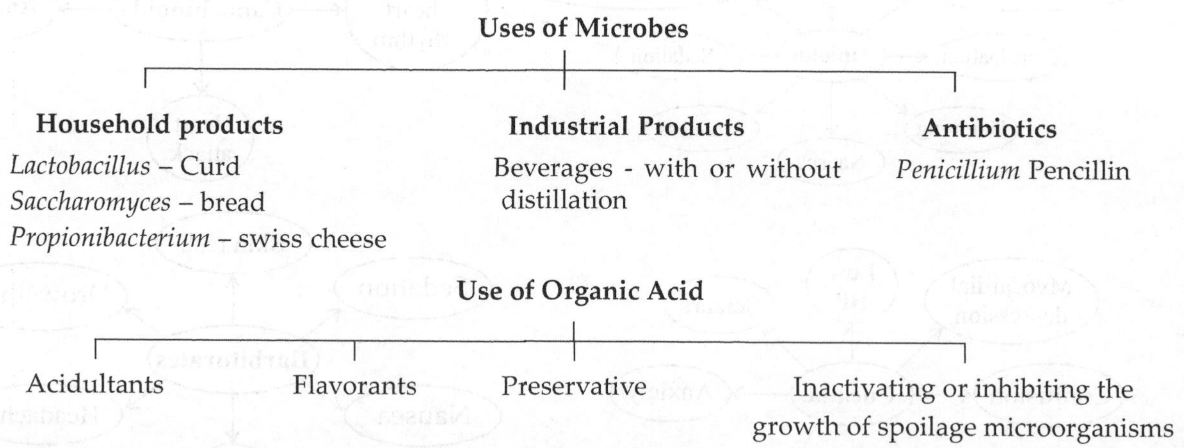

2. **Sewage treatement :** Sewage treatment is a process that removes the majority of the contaminants from waste water or sewage and produces both a liquid effluent suitable for disposal in the natural environment and sludge.

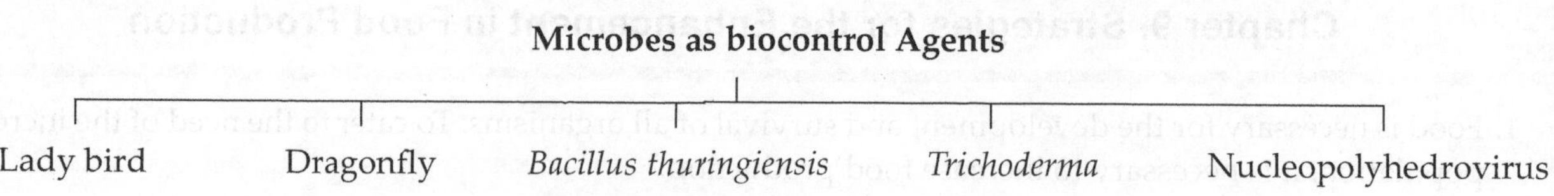

Chapter at a Glance

2. Sewage Treatment Process

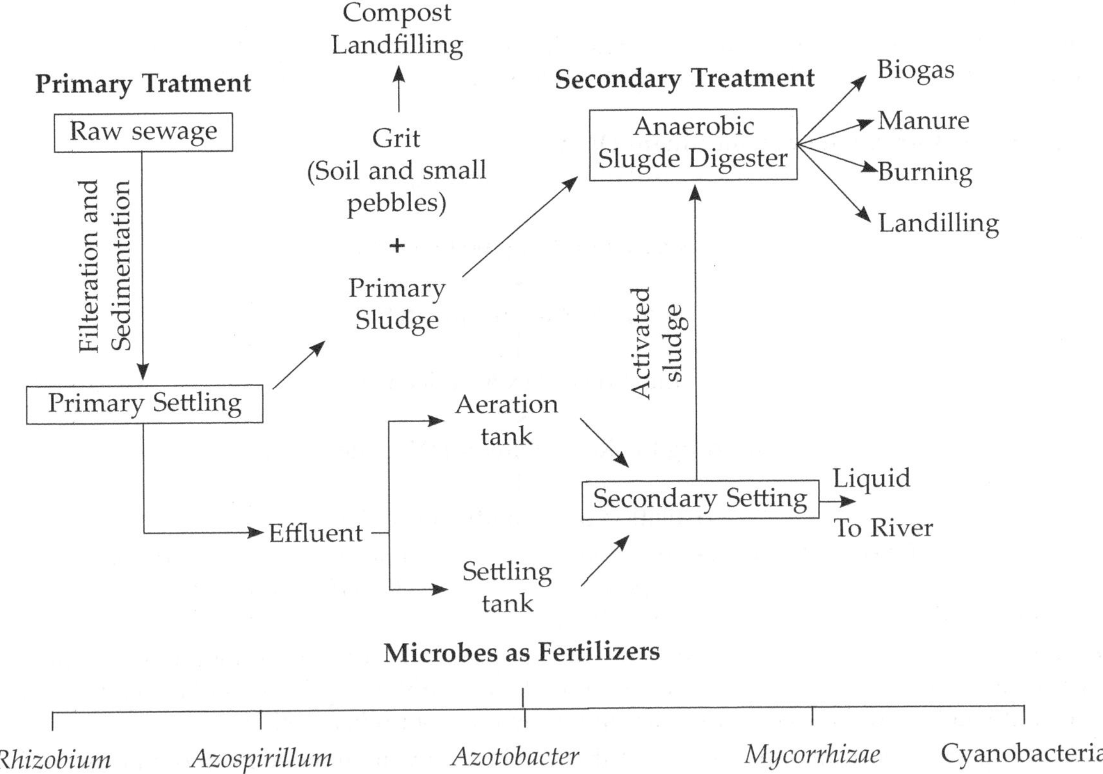

3. **Integrated Pest Management (IPM) :** Integrated Pest Management, also known as Integrated Pest Control is a broad-based approach for economic control of pests. IPM aims to suppress pest populations below the economic injury level.
4. **Harmful effects of chemical pesticides :** Pesticides can cause short-term adverse health effects, called acute effects, as well as chronic adverse effects that can occur months or years after exposure. Examples of acute health effects include stinging eyes, rashes, blisters, blindness, nausea, dizziness, diarrhoea and death.

Chapter 11. Biotechnology : Principles and Processes

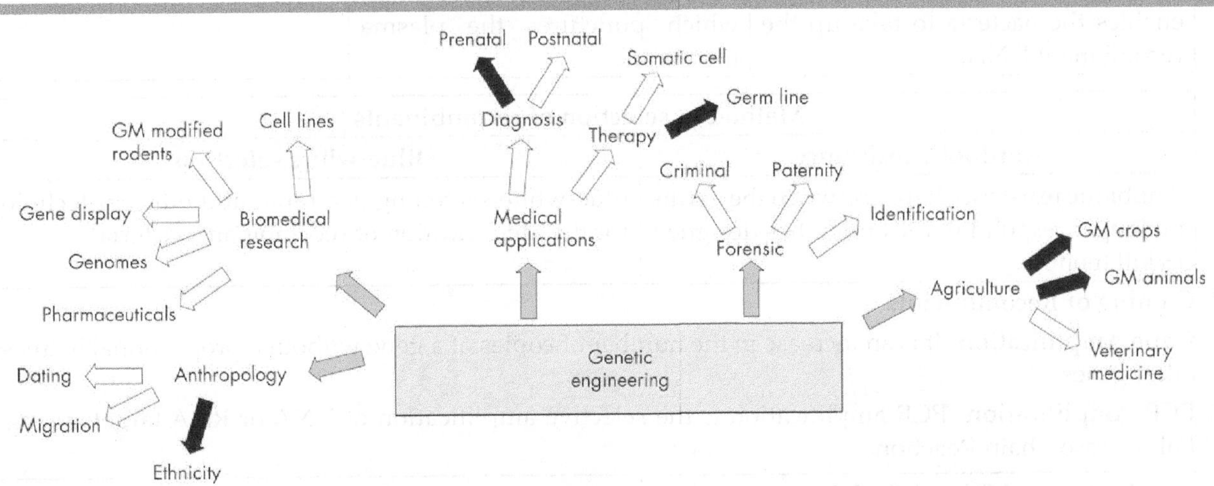

1. **Biotechnology** is defined as the technology that utilises biological systems, living organisms or their parts to develop or create different products.
2. The two Principles of Biotechnology are:
   (a) Genetic Engineering
   (b) Chemical Engineering

3. Genomic (chromosomal) DNA from either bacteria, plant or animal cell can be isolated by the lysis method.
4. **Electrophoresis** is an electrokinetic process that separates charged particles in a fluid using a field of electrical charge. It is used to separate protein molecules or DNA depending on the type and size of the molecules.
5. **Steps for the formation of recombinant DNA**

Isolating Genetic Material
↓
Restriction Enzyme Digestion
↓
Using PCR for Amplification
↓
Ligation of DNA Molecules
↓
Inserting the Recombinant DNA into a host
↓
Isolation of Recombinant Cells.

6. The **Restricted Enzyme Catalyses** the *sequence-specific hydrolysis* of double-stranded DNA. Restriction enzymes occur in microorganisms as part of *restriction–methylation systems* consisting of DNA-cleaving enzyme/DNA-methylating enzyme pairs that recognise a common sequence.
7. **Role of Ligase-** Enzymes that are capable of catalysing the reaction of joining two large molecules by establishing a new chemical bond, generally with concomitant hydrolysis of a small chemical group on one of the bulky molecules or simply linking of two compounds together.
8. **A Cloning Vector** is a small piece of DNA that can be stably maintained in an organism, and into which a foreign DNA fragment can be inserted for cloning purposes. **For examples**-pBR322, Agroviruses, Bacterial Artificial Chromosome (BAC), Yeast Artificial Chromosome (YAC).

Method of Selection of Recombinants		
Temperature Shock	**Microinjection**	**Gene gun**
Recombinant DNA can then be forced into such cells by incubating the cells with recombinant DNA on ice, followed by placing them briefly at 42ºC (heat shock) and then putting them back on ice which enables the bacteria to take up the recombinant DNA.	A cell is held on a glass pipette with a diameter much smaller than the cell, under a microscope and foreign DNA is injected directly into the nucleus using an incredibly fine micropipette, which punctures the plasma membrane.	This technique fires microscopic gold or tungsten particles coated with foreign DNA at cells using a compressed air gun (burst of helium)

Method of selection of recombinants	
Antibiotic resistance	**Blue-white selection**
Antibiotic resistance happens when the germs no longer respond to the antibiotics designed to kill them	Blue-white screening is a rapid and efficient technique for the identification of recombinant bacteria.

9. **Cloning of Recombinants**

   Gene Amplification- It is an increase in the number of copies of a gene without a proportional increase in other genes.
10. **PCR Amplification-** PCR amplification is the selective amplification of DNA or RNA targets using the Polymerase Chain Reaction

Features of Stirred Tank Bioreactors	Features of Sparged Tank Bioreactors
1. Heat transfer	Mixing
2. Mass transfer	Oxygen availability throughout the Bioreactor.
3. Aeration	
4. Mixing for Homogenisation	

Chapter 12. Biotechnology and its Applications

1. **Applications of Biotechnology** in health and agriculture :
   (a) Human Insulin
   (b) Vaccine Production
   (c) Stem Cell Technology.
   (d) Gene Therapy
   (e) Genetically Modified Organisms (GMOs)
2. **Applications :**

Bt crops	Transgenic animals	Biosafety issues	Biopiracy	Bio Patents
Bt Crops are transgenic crops that produce the same toxin as the bacterium *Bacillus thuringiensis* in the plant cell, thereby, protecting the crops from pests.	Transgenic animals are animals (most common mice) that have a foreign gene deliberately inserted into their genome.	They refer to the procedures, policies, and principles to be adopted to safeguard the environment and the human population.	Biopiracy designates the illegitimate appropriation of genetic resources and/or associated traditional knowledge.	The patents of living organisms can include plant and animal species, and related biological and biotechnology-enabled inventions.

3. **Applications of Biotechnology in Agriculture :**
   (a) For the production of ***crops tolerant*** to abiotic stresses (cold, drought, salt, heat).
   (b) **Pest resistant crops** (primarily cotton and maize), have been genetically modified so they are toxic to certain insects. Example : Bt cotton
   (c) RNA interference (RNAi) is generally accepted as a powerful tool for manipulating gene expression and perform analyses of their functions and incognita genes such as proteases, gland proteins and peroxiredoxins which showed efficient gene suppression using the same method.
   (d) **Crops with enhanced nutritional value-** Golden rice is a genetically modified, biofortified crop. It is genetically modified to produce beta carotene, which is not normally produced in rice. Beta carotene is converted into Vitamin A when metabolized by the human body.
4. **Application of biotechnology in Medicine-**
   (a) Insulin : Insulin is a peptide hormone produced by beta cells of the pancreatic islets; it is considered to be the main anabolic hormone of the body.
   (b) Gene therapy :
   (i) **Treatment of SCID**

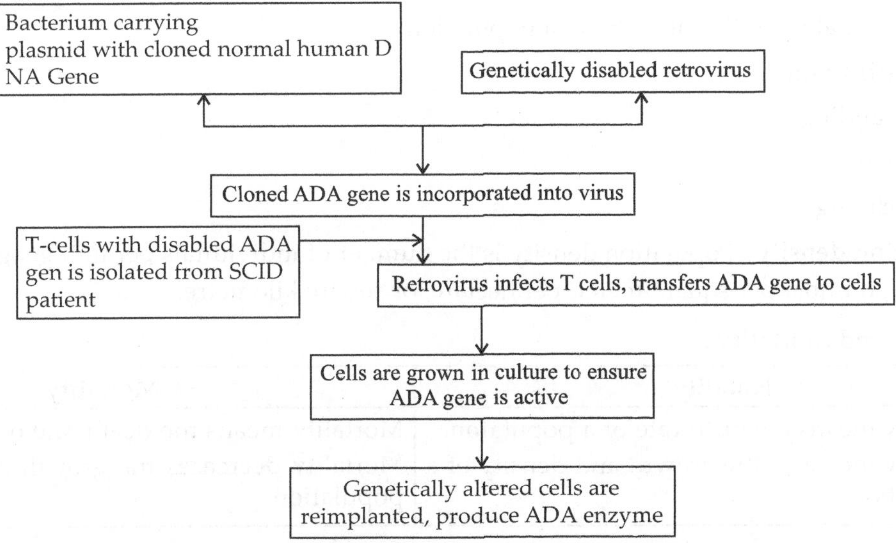

(ii) **Molecular diagnosis by PCR-** It is a molecular diagnostic testing technique of identifying targeted genetic sequence of DNA or RNA; e.g. of a cancer gene.

(iii) **ELISA-** It is a test that detects and measures antibodies in the blood. This test can be used to determine the antibodies related to certain infectious conditions.

(iv) **Use of DNA/RNA probe-** A probe is a single-stranded sequence of DNA or RNA used to search for its complementary sequence in a sample genome. The probe is placed into contact with the sample under conditions that allow the probe sequence to hybridise with its complementary sequence.

5. **Transgenic animals for bioactive products-** Many transgenic animals are designed to increase our understanding of how genes contribute to the development of diseases. The transgenic animals for bioactive products like alpha-1-antitrypsin for emphysema, alpha-lactalbumin etc.
6. **Vaccine safety testing-** Before a vaccine is ever recommended for use, it is tested in labs. This process can take several years. FDA uses the information from these tests to decide whether to test the vaccine with humans.
7. **Chemical safety testing-** Chemical safety assessment, are done to test toxicity manufactured products that may pose a threat to the health of humans or the environment.
8. **Study of disease-** Epidemiology is the study of how often diseases occur in different groups of people and why.
9. **Role of GEAC (Genetic Engineering Appraised Committee)**—It is a body under the Ministry of Environment and Forests (India) and is responsible for the approval of genetically engineered products in India.

Chapter 13. Organisms and Populations

1. **Ecology-** The branch of biology that deals with the relations of organisms to one another and their physical surroundings are known as ecology.
2. **Major biomes of India-**

(a) **Tropical rain forest :** Tropical rainforests are rainforests that occur in areas of tropical rainforest climate in which there is no dry season.

(b) **Deciduous forest :** A deciduous forest is a biome dominated by deciduous trees which lose their leaves seasonally.

(c) **Desert and sea coasts :** Coastal deserts are found on coasts, where the land meets the ocean and sea coasts are the part of the land adjoining or near the sea.

3. **Population attributes-**

(a) **Sex ratio :** Male is to Female ratio in a population.

(b) **Age distribution**

(i) Expanding

(ii) Stable

(iii) Declining

(c) **Population density :** Population density is the number of individuals per unit geographical area, for example, number per square meter, per hectare, or square kilometre.

(d) **Natality and Mortality :**

Natality	Mortality
1. Natality means the birth rate of a poputaton.	Mortality means the death rate of a population.
2. Natality increases the growth and density of a population.	Mortality decreases the growth and density of a population

(e) **Emigration and Immigration :**

Emigration	Immigration
Emigration means the movement of people outside the country.	Immigration means the movement of people into a country.

(f) **Carrying capacity**- The number of individuals for a particular species that the environment carries or sustain.

4. Population density is often measured in three different ways:

   (a) Arithmetic density

   (b) Physiological density

   (c) Agricultural density

5. **Factors affecting population :**

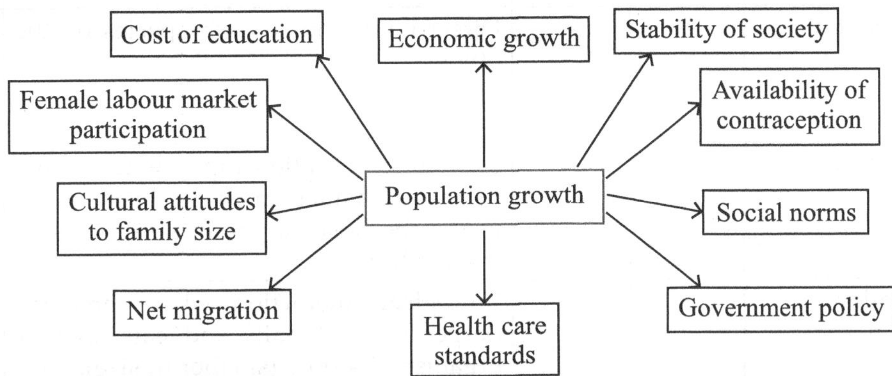

Poputation Growth	Population growth equation
1. Population growth is change in population over time or the change in the number of individuals in a population per unit time. 2. Propulation Growth Rate is the rate at which the number of individuals in a population increases	A simple equation to show population growth as : Change is population size. = (Births + Immigration) – (Deaths + Emigration) Expressing Population Changes as a Percentage.

6. **Growth model.**

Exponential Growth	Logistic Growth
$P = P_0 ert$ where as, Exponential Growth Population size Time P = total population after time t P_0 = starting population r = % rate of growth t = time e = Euler's number	$y = \dfrac{L}{1 + be^{-kt}}$ Carrying capacity Population size Time where as, L = curve maximum value y = output of function K = Logistic growth rate

7. **Reproductive fitness :** Reproductive fitness reflects the ability of individuals to pass on their genes to subsequent generations.

Example : Oyster lays 60-80 million eggs in a single spawn.

8. Population interactions :

Population Interactions

S. No.	Types of Interation	Species 1	Species 2	General Natue of Interaction	Examples
1	Amensalism	–	0	Biological interaction where one species causes harm to another organism without any cost or benefits to itself.	*Penicillum* killing bacteria
2	Mutualism	+	+	Interaction favourable to both species and obligatory.	Between fungus and algae (Lichen)
3	Commensalism	+	0	Relationship between two living organisams in which one organism benefits from the other without harming it.	Suker fish on shark
4	Competition	–	–	Direct inhibition of each species by the other.	Birds compete with squirrels for units and seeds
5	Parasitism	+	–	Symbiotic relationship where the parasite (smaller in size), lives on or inside the host (larger in size), causing it some harm.	*Ascaris* and tapeworm in human digestive tract
6	Predation	+	–	**Biological** interaction where **predator** (larger in size), kills and eats another organism, its prey (smaller in size).	Lion predatory on deet

('+' = benefitted; '–' = harm; '0' = neutral)

9. Types of Parasitism :

(a) Ecto- parasitism	(b) Endo- parasitism	(c) Brood parasites
Ectoparasitism is a form of parasitism where the parasite lives outside the body of the host. Ex : Lice and ticks	Endoparasites are parasites that live within the body of their host. Ex-hook worms	The brood parasite manipulates a host, either of the same or another species, to raise its young as if it were its own, using brood mimicry Ex-Cuckoo, mochokidae cat fish

Chapter 14. Ecosystem

1. An ecosystem is a geographical area that contains biotic or living parts, as well as abiotic factors or non-living parts together.

   The **four types of an** ecosystem are:

   (a) Terrestrial ecosystem
   (b) Aquatic ecosystem
   (c) Natural ecosystem
   (d) Man-made ecosystem

2. **Structure of the ecosystem :**

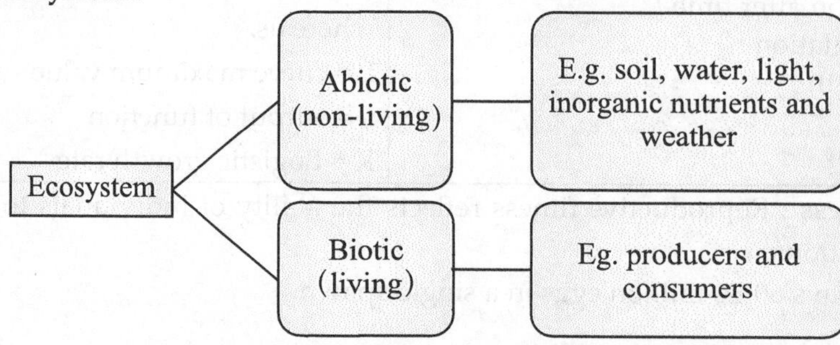

3. Ecological Adaption :

Morphological adaption	Physiological adaption
Morphological adaptation is a structural change that gives an organism a greater chance of survival in its habitat.	Internal and/or cellular features of an organism that enable them to survive in its environment.

4. Allen's rule is an eco-geographical rule which states that the limbs, ears and other appendages of animals living in colder climate are shorter than the animals of the same species living in warmer regions.

5. Pond ecosystem is a system of organisms that live together in a pond.

Functions of pond ecosystem :

(a) Recycle nutrients purify water

(b) Attenuate floods

(c) Recharge groundwater

(d) Provide habitats for wildlife.

6. Productivity : The rate of synthesis of energy-containing organic matter or biomass by a particular trophic level per unit area in a unit time.

Gross Primary Productivity	Net primary productivity	Secondary productivity
1. It is the amount of organic matter synthesised by the producers per unit of time. 2. It is the total productivity and includes the energy utilised for respiration by producers.	It is the amount of organic matter synthesised by the producers per unit area in unit time. It is equal to the organic matter synthesised by photosynthesis minus the rate of respiration and other losses.	It is the rate of synthesis of organic matter by consumers. It decreases with the rise in trophic level as some amount of productivity is lost through respiration at each tropic level.

7. Steps of Decomposition

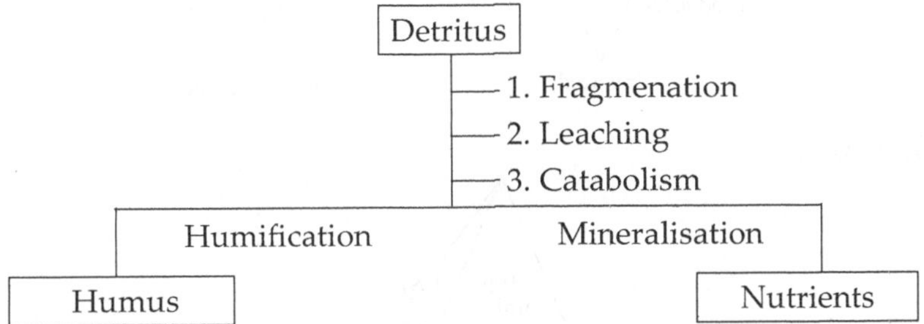

8. Factors affecting the Rate of Decomposition

(a) Climatic factors

(i) Temperature.

(ii) Soil moisture

(iii) Oxygen

(iv) Soil pH

(b) Chemical quality of detritus.

9. Various types of the food chain :

(a) Grazing food chain

(b) Detritus foor chain

10. Trophic levels :

(a) Producer — First tropic level (plants)

(b) Primary Consumer — Secondary tropic level (Herbivore)

(c) Secondary Consumer — Third tropic level (Carnivores)

(d) Tertiary Consumer — Fourth trophic level (Top carnivores)

11. An ecological pyramid is a graphical representation designed to show the biomass or bio productivity at each trophic level in a given ecosystem.

There are three types of Pyramid.

(a) Pyramid of Energy :

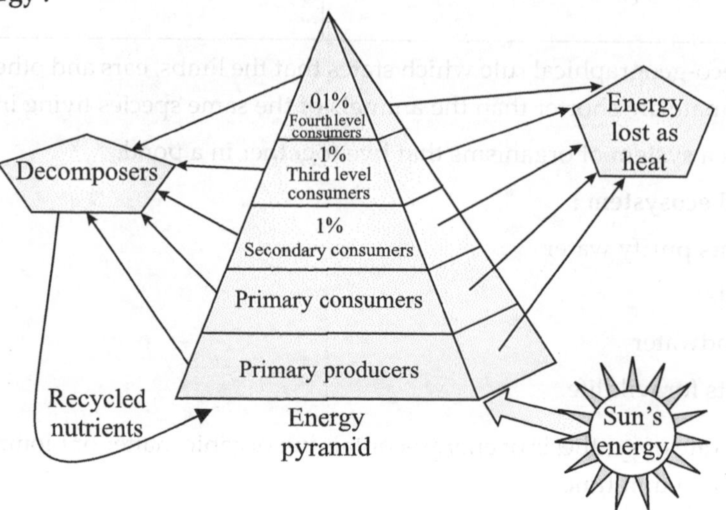

(b) Pyramid of Number :

(i) Inverted

Hyperparasites
Parasites
Herbivores
Producers
Inverted

(ii) Spindle Shaped/Urn Shaped

Carnivore
Herbivore
Producers
Spindle-shapped

(c) Pyramid of Biomass :

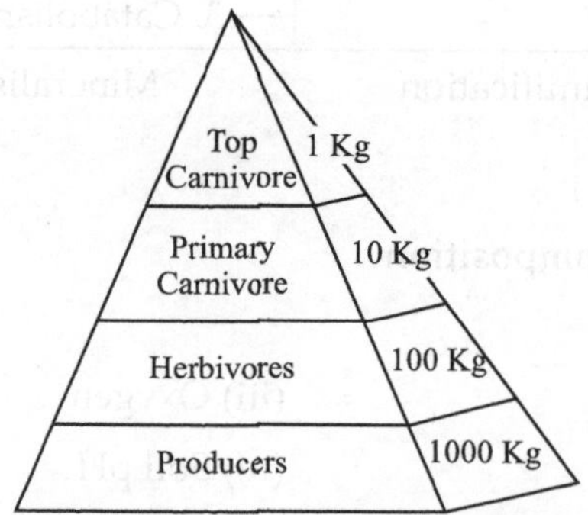

12. Natural cycles balance and regulate earth and its atmosphere. Example Carbon cycle, Phosphorus cycle, Nitrogen cycle, sulphur cycle which help in maintaing the nutritive value of earth.

13. PAR : Photosynthetically Active Radiation (PAR) is the amount of light available for photosynthesis, which is light in the 400 to 700-nanometer wavelength range.

14. Ten per cent Law : According to this law, during the transfer of organic food energy from one trophic level to the next higher level, only about ten percent of the transferred energy is stored as flesh.

15. Energy Flow Diagram :

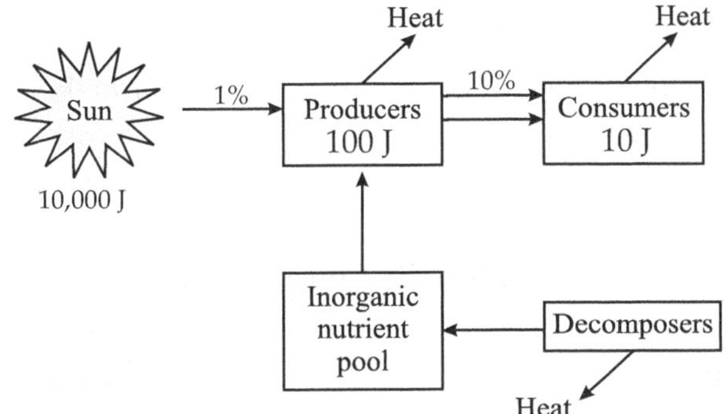

16. Succession : Ecological succession is the process of change in the species structure of an ecological community over time. Every succession leads to mesarch conditions.

17. Type of Succession

Primary Succession	Secondary Succession
1. Occurs in the area where no life existed earlier	Occurs in the area where life existed earlier but destroyed
2. Development of biotic community is very slow. Examples : bare rock, barren land, cooled lava.	Development of biotic community is fast. Examoles : Abandoned farm lands, burnt or cut forests.

Hydrarch	Xerarch
Plant succession starting in relatively shallow water, such as ponds and lakes, and culminating in a mature forest.	Xerarch succession is a form of plant succession starting in a bare ground that is very dry or is limited by the availability of water, and eventually culminating in a mature forest.

18. Types of Community :

Pioneer Community	Climax Community
1. It is the first species that invades a place where no life existed previously at the time of a biological succession.	They are the last species that comes into existence at the time of biological succession.
2. It is an unstable community.	It is a stable community.
3. They are poor competitors but good colonizers.	They are good competitors but poor colonizers.
4. The species can tolerate extremes of environmental conditions. Example are lichens, phytoplankton.	The species are sensitive to extremes environmental conditions. Example are forests, white spruce, etc.

19. Seral communitry :

(a) It is a transitional community that develops in an area during succession.

(b) This community is unstable and gets replaced by the subsequent community.

(c) The organisation of this community is simple as it has limited number of species.

(d) It has limited number of ecological niches.

Example : grass, herbaceous plants.

Chapter 15. Biodiversity and its Conservation

1. **Biodiversity :** It is defined as the totality of genes, species, and ecosystem in a particular environment, region, or entire world.

2. Biodiversity provides functioning ecosystems that supply oxygen, clean air and water, pollination of plants, pest control, wastewater treatment and many ecosystem services.
3. **Biodiversity loss** describes the decline in the number, genetic variability, and variety of species, and the biological communities in a given area.
4. **Biodiversity conservation :**

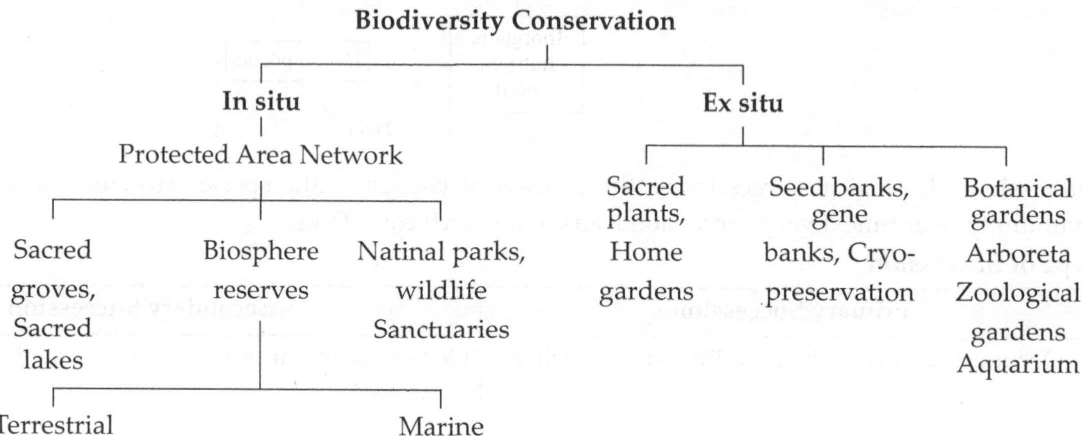

5. **Gene pool-** The total set of genes and alleles, present in a population.
6. **Genetic erosion-** It is a process whereby an already limited gene pool of an endangered species of plants or animal diminishes, even more, when an individual from the surviving population die off without getting a chance to meet and breed with others in their endangered low population.
7. **Hotspot :** A biogeographical region with significant levels of biodiversity that is threatened by human habitation.
8. **Endangered organisms :** A species that is very likely to become extinct shortly, either worldwide or in a particular political jurisdiction.
9. **Extinction :** Generally considered to be the death of the last individual of the species, although the capacity to breed and recover may have been lost before this point.
10. **Red data book :** Book contains a record of animals and plants which are known to be in danger and published by IUCN.
11. **Biosphere reserves :** Biosphere reserves are areas of terrestrial and coastal ecosystems promoting solutions to reconcile the conservation of biodiversity with its sustainable use.
12. **National parks and Sanctuary**

National Park	Sanctuary
1. They target the betterment of the whole wild life.	They target the betterment of only animals.
2. Human activities and human habitations are stricitly prohibited.	Limited human activities like collection of minor fore products, timber harvesting, etc are allowed.
3. They are maintained by the Central or State Legislation. Example : Jim Corbett National Park	They are formed by the order of State or Central Government. Private ownership is allowed. Example : Chilka Lake bird sanctuary.

13. **Ramsar sites :** The Ramsar sites, an international body forms the basis for the identification of wetlands.

Chapter 16. Environmental Issues

1. **Pollution :** The presence of harmful substances in air, land and water which contaminates the environment or give adverse effect to humans is known as pollution.

2. **Pollutant :** A pollutant is any substance that is present in the wrong amount, in the wrong place or at the wrong time.

3. **Pollutant Classification :**

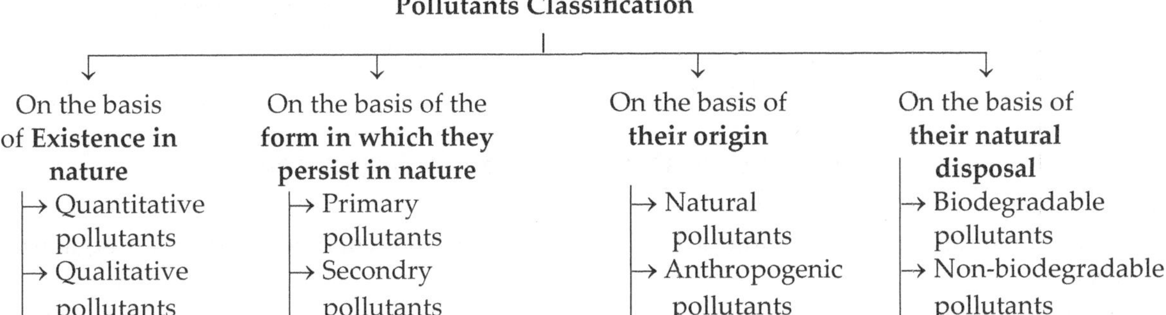

4. **Types of Pollution :**

Air pollution	Noise pollution	Water pollution	Thermal pollution	Soil pollution
It is a mixture of solid particles and gases in the air.	It is considered to be any unwanted or disturbing sound that affects the health and well being of human and other organism.	It is the pollution of bodies of water, such as lakes, rivers, sea as well as ground water.	It is defined as sudden increase of toxic chemicals in water.	It is defined as the presence of toxic chemicals in soil.

Solid waste management	Radioactive waste management
Waste management includes the activities and actions required to manage waste, from its inception to its final disposal.	Radioactive waste management refers to the safe treatment, storage and disposal of liquid, solid or gas discharge from nuclear industry operations to protect people and the environment

5. **Greenhouse effect** is the process by which radiations from the sun are absorbed by greenhouse gases and are not reflected back into space. This phenomenon insulates the surface of the earth of prevents it from freezing.

6. **Climate change** is a long term change in the average weather patterns that have come to define earth's local, regional and global climates.

7. **Ozone Layer Depletion-** When chlorine and bromine atoms come into contact with ozone in the stratosphere, they destroy ozone molecules. Its depletion increases the amount of UV-B that reaches the Earth's surface.

8. **Deforestation** is the removal of a forest or stands of trees from land that is then converted to a non-forest use.

9. **E-waste** is any electrical or electronic equipment that has been discarded. This includes working and broken items that are thrown in the garbage or donated to a charity reseller like Good will. Often, if the item goes unsold in the store, it will be thrown away.

❐

Multiple Choice Questions | Set 2 |

Chapter 1. Reproduction in Organisms

1. There are various types of reproduction. The type of reproduction adopted by an organism depends upon—

(a) The habitat and morphology of the organism
(b) Morphology of the organism
(c) Morphology and physiology of the organism
(d) The organism's habitat, physiology and genetic makeup

2. There is no natural death in single celled organisms like Amoeba and Bacteria because—

(a) They cannot reproduce sexually
(b) They reproduce by binary fission
(c) Parental body is distributed among the offspring
(d) They are microscopic

3. Identify the incorrect statement—

(a) In asexual reproduction, the offsprings produced are morphologically and genetically identical to the parent
(b) Zoospores are sexual reproductive structures
(c) In asexual reproduction, a single parent produces offspring with or without the formation of gametes
(d) Conidia are asexual structures in *Penicillium*

4. Which one of the following processes result in the formation of clone of bacteria ?

(a) Transformation (b) Transduction **(c) Binary fission** (d) Conjugation

5. Which one of the following is common to multicellular fungi, filamentous algae and protonema of mosses ?

(a) Diplontic life cycle (b) Members of kingdom Plantae
(c) Mode of nutrition **(d) Multiplication by fragmentation**

6. Transverse binary fission occur in—

(a) *Euglena* (b) *Amoeba* (c) *Hydra* **(d) *Paramecium***

7. Which of the following animal is having longitudinal binary fission ?

(a) *Euglena* (b) *Plasmodium* (c) *Planaria* (d) *Hydra*

8. Vegetative propagation in *Pistia* occurs by—

(a) Stolon **(b) Offset** (c) Runner (d) Sucker

9. Micropropagation is a technique—

(a) For the production of true to type plants (b) For production of haploid plants
(c) For production of somatic hybrid (d) For production of somaclonal plants

10. Example of Corm is—

(a) Ginger **(b) Colocasia** (c) Onion (d) Potato

11. Which part would be the most suitable for raising virus free plants for micropropagation ?

(a) Bark (b) Vascular tissue **(c) Meristem** (d) None of these

12. Banana is vegetatively propagated by—

(a) Tubers **(b) Rhizomes** (c) Bulbs (d) Suckers

13. A scion is grafted to stock. The quality of fruits produce will be determined by genotype of—
(a) Stock
(b) Scion
(c) Both stock and scion
(d) Neither stock nor scion
14. Vegetative propagation in Mint occurs by—
(a) Offset
(b) Rhizome
(c) Sucker
(d) Runner
15. When seeds are attached to parent plant, the type of germination in known as—
(a) Ovipary
(b) Epigeal
(c) Vivipary
(d) Hypogeal
16. Study of pollen grains is called—
(a) Ethnology
(b) Palynology
(c) Paleobotany
(d) Co-taxonomy
17. Sexual reproduction in fungi may occur by means of —
(a) Sporangiospore, oospore and ascospore
(b) Zoospore, oospore and ascospore
(c) Sporangiospore, ascospore and basidiospore
(d) Oospore, ascospore and basidiospore
18. External fertilisation occurs in majority in —
(a) Algae
(b) Fungi
(c) Liverworts
(d) Mosses
19. Which of the following statement support the view that elaborate sexual reproductive process appeared much later in the organic evolution ?
(i) Lower group of organisms have simpler body design
(ii) Asexual reproduction is common in lower groups of organisms
(iii) Asexual reproduction is common in higher groups of organisms
(iv) The high incidencė of sexual reproduction in angiosperms and vertebrates
Choose the *correct* answer from the options given below—
(a) (i) and (ii)
(b) (i) and (iii)
(c) (ii) and (iv)
(d) (ii) and (iii)
20. Offspring formed by sexual reproduction exhibit more variation than those formed by asexual reproduction because :
(a) Sexual reproduction is a lengthy process
(b) Gametes of parents have quantitatively or qualitatively different genetic composition
(c) Genetic material comes from parents of two different species
(d) Greater amount of DNA is involved in sexual reproduction
21. Choose the correct statement from amongst the following—
(a) Dioecious (hermaphrodite) organisms are found only in animals
(b) Dioecious organisms are found only in plants
(c) Dioecious organisms are found in both plants and animals
(d) Dioecious organisms are found only in vertebrates
22. What is common between negative reproduction and apomixis ?
(a) Both are applicable to dicot plants only
(b) Both bypass the flowering phase
(c) Both occur round the year
(d) Both produces progeny identical to the parent
23. In which of the following organism, self Fertilisation is seen ?
(a) Fish
(b) Roundworm
(c) Earthworm
(d) Liver fluke
24. Which of the following is hermaphrodite ?
(a) Ant
(b) Aphid
(c) Earthworm
(d) Cockroach

Chapter 2. Sexual Reproduction in Flowering Plants

1. Sexual reproduction in flowering plants was discovered by—
(a) Carmerarius
(b) Nawaschin
(c) Strasburger
(d) Maheshwari
2. Fertilisation involving carrying of male gametes by pollen tube is—
(a) porogamy
(b) siphonogamy
(c) chalazogamy
(d) syngonogamy

3. One of the most resistant biological material is—
   (a) lignin (b) cutin **(c) sporopollenin** (d) cellulose
4. Sporopollenin occurs in—
   (a) female gametophyte (b) male gametophyte
   (c) vegetative cells of pollen grain **(d) exine of pollen wall**
5. In angiosperms...........lead to the formation of a mature male gametophyte from a pollen mother cell.
   (a) two meiotic divisions **(b) one meiotic and two mitotic division**
   (c) one mitotic and two meiotic division (d) a single mitotic division
6. Usually pollen grains are shed at—
   (a) one-celled stage **(b) two-celled stage**
   (c) four-celled stage (d) five-celled stage
7. Meiosis occurs in—
   (a) endosperm cells (b) intercalary meristems
   (c) apical meristems **(d) spore mother cells**
8. Larger nucleus in a pollen grain is called.........
   (a) callus (b) generative nucleus
   (c) vegetative nucleus (d) none of these
9. Embryo sac occurs in—
   (a) embryo (b) axis part of embryo
   (c) ovule (d) stamen
10. Embryo sac is also called—
    (a) microspore (b) megaspore
    (c) megagametophyte (d) microgametophyte
11. Endosperm cell of an angiosperm has 36 chromosomes. The number of chromosomes in the gametes would be—
    (a) 11 **(b) 12** (c) 8 (d) 9
12. ....comprises the egg apparatus.
    (a) Polar nuclei (b) Antipodal cells
    (c) Egg cell and synergids (d) Male gametes
13. Starting from the innermost part, the correct sequence of parts in an ovule are—
    (a) egg, nucleus, embryo sac, integument (b) embryo sac, egg, nucellus, integument
    (c) egg, embryo sac, nucellus, integument (d) egg, embryo sac, integument, nucellus
14. Filiform apparatus is a characteristics feature of—
    (a) suspensor cells (b) egg **(c) synergids** (d) zygote
15. Synergids are—
    (a) Haploid (b) Diploid (c) Triploid (d) Tetraploid
16. In angiosperms, triple fusion is required in the formation of—
    (a) embryo **(b) endosperm** (c) suspensor (d) fruit wall
17. Double Fertilisation involve—
    (a) Fertilisation of the egg by two male gametes
    (b) Fertilisation of two eggs in the same embryo sac by two sperms brought by one pollen tube
    (c) Fertilisation of the egg and the central cell by two sperms brought by different pollen tubes
    (d) Fertilisation of egg and central cell by two sperms brought by the same pollen tube
18. Double Fertilisation is the process in plants that include—
    (a) Syngamy and triple fusion (b) Only triple fusio
    (c) Development of antipodal cell (d) None of the above
19. Micropyle occurs in—
    (a) ovary (b) ovule
    (c) seed **(d) Both (b) and (c)**

Multiple Choice Questions

20. The flowers which open their petals to expose their reproductive parts to allow pollination are called:*
(a) *Cleistogamous* (b) *Geitonogamous*
(c) *Chasmogamous* (d) *Autogamous*

21. Which among the following is triploid ?
(a) Megaspore (b) Embryo
(c) Endosperm (d) Microspore

22. Fragrant flowers with well developed nectaries are an adaptation for—
(a) Chiropterophily **(b) Entomophily**
(c) Anemophily (d) Hydrophily

23. The transfer of pollen to the stigma of another flower of the same plant is called—
(a) autogamy **(b) allogamy**
(c) geitonogamy (d) none of these

24. Which among the following exhibit Anemophily ?
(a) *Salvia* (b) *Vallisneria*
(c) Coconut (d) Bottle brush

25. Ornithophily refers to the pollination by which of the following ?
(a) Insect **(b) Birds** (c) Snails (d) Air

26. Both autogamy and geitonogamy are prevented in—
(a) Papaya (b) Cucumber (c) Castor (d) Maize

27. Secondary nucleus is formed by the fusion of—
(a) antipodal cells (b) egg apparatus
(c) two polar nuclei (d) synergids

28. ……is the phenomenon wherein, the ovary develops into a fruit without Fertilisation.
(a) Parthenocarpy (b) Apomixis
(c) Sexual reproduction (d) None of these

29. The 'eyes' of the potato tuber are………
(a) shoot buds **(b) axillary buds** (c) flower buds (d) root buds

30. Vegetative propagation in ginger is carried out by means of—
(a) rhizome (b) tuber (c) roots (d) bulb

31. Seedless banana is—
(a) Parthenocarpic fruit (b) Multiple fruit
(c) Drupe fruit (d) Tree fruit

Chapter 3. Human Reproduction

1. Which of the following is primary sex organ ?
(a) Scrotum (b) Penis **(c) Testis** (d) Prostate

2. Prostate gland is present—
(a) on ureter (b) on kidney (c) on testis **(d) around urethra**

3. The Leydig cells found in the human body are the secretory source of—
(a) Progesterone (b) Intestinal mucus
(c) Glucagon **(d) Androgens**

4. Clitoris in females is—
(a) homologous to penis (b) analogous to penis
(c) functional penis in female (d) non-functional

5. Site of Fertilisation in human beings is—
(a) ovary (b) uterus (c) vagina **(d) fallopian tube**

* Frequently asked Previous Years Board Exam Question

6. Identify the odd from the following—
(a) **Labia minora** (b) Fimbriae (c) Infundibulum (d) Isthmus

7. Spermatozoa are nourished during their development by—
(a) **Sertoli cells** (b) Interstitial cells
(c) Connective tissue cells (d) None of these

8. Location and secretion of leydig cells are—
(a) Liver-cholesterol (b) Ovary-estrogen
(c) **Testis-testosterone** (d) Pancreas-glucagon

9. Which of the following is immortal ?
(a) Somatic cell (b) Glomerular cell
(c) **Germ cell** (d) Cells of pituitary

10. Sertoli cells are regulated by the pituitary hormone known as—
(a) LH (b) **FSH** (c) GH (d) Prolactin

11. What is present in the middle piece of sperm ?
(a) Acrosome (b) **Mitochondria**
(c) Nucleus (d) Proximal centriole

12. Graafian follicles are found in—
(a) testis of mammal (b) ovary of frog
(c) ovary of cockroach (d) **ovary of mammals**

13. Egg is liberated from ovary in—
(a) **secondary oocyte stage** (b) primary oocyte stage
(c) oogonial stage (d) mature ovum stage

14. In 28 days of human ovarian cycle, ovulation occurs on—
(a) Day 1 (b) Day 5 (c) **Day 14/15** (d) Day 28

15. Corpus luteum develops from—
(a) Oocyte (b) Nephrostome
(c) **Graafian follicle** (d) None of these

16. Corpus luteum secretes—
(a) LH (b) Estrogen (c) **Progesterone** (d) FSH

17. In uterus, endometrium increases in thickness in response to—
(a) Oxytocin (b) **Estrogen** (c) LH (d) Relaxin

18. 1st polar body is formed at which stage of oogenesis—
(a) **1st meiosis** (b) 2nd mitosis (c) 1st mitosis (d) Differentiation

19. Which among the following has 23 chromosomes ?
(a) Spermatogonia (b) Zygote
(c) **Secondary oocyte** (d) Oogonia

20. About which day in a normal human menstrual cycle does rapid secretion of LH (popularly called LH surge) normally occurs ?
(a) **14th day** (b) 20th day (c) 5th day (d) 11th day

21. Sperms produce an enzymatic substance for dissolving egg membrane called—
(a) Hyaluronic acid (b) **Hyaluronidase** (c) Androgen (d) Estrogen

22. In the absence of acrosome, the sperm—
(a) Cannot get food (b) Cannot swim
(c) **Cannot penetrate the egg** (d) Cannot get energy

23. Role of placenta is to—
(a) **provide nutrition to developing embryo** (b) protect embryo from shock
(c) act as storage organ (d) convey nerve impulses

24. The embryo at 16 celled stage is known as —
(a) **Morula** (b) Gastrula (c) Blastula (d) Blastomere

25. A change in the amount of yolk and its distribution in the egg will affect—
    (a) Fertilisation
    (b) Formation of zygote
    (c) Pattern of cleavage
    (d) Number of blastomeres produced
26. Human embryo is protected by —
    (a) allantois
    (b) amniotic cavity
    (c) pleural cavity
    (d) peritoneal cavity
27. During cleavage, what is true about cells ?
    (a) nucleocytoplasmic ratio remains unchanged
    (b) size does not increase
    (c) less consumption of oxygen
    (d) division is like meiosis.
28. Villi of human placenta develop from —
    (a) chorion (b) allantois (c) yolk sac (d) amnion
29. Morula resembles a—
    (a) Mulberry fruit
    (b) Football
    (c) Hollow ball
    (d) None of these
30. 32-celled stage of human embryo is—
    (a) Larger than fertilised egg
    (b) Smaller than fertilised egg
    (c) Same size as fertilised egg
    (d) Four times larger than fertilised egg
31. Gastrulation involves the differentiation of —
    (a) ectoderm and endoderm
    (b) ectoderm, endoderm and mesoderm
    (c) ectoderm and mesoderm
    (d) mesoderm and endoderm
32. The mesoderm gives rise to all of the following structure in the fully developed animal except—
    (a) muscular system
    (b) nervous system
    (c) circulatory system
    (d) gonads
33. Extra embryonic membranes of mammalian embryo are derived from —
    (a) inner cells mass **(b) trophoblast** (c) formative cells (d) follicle cells
34. Signals for parturition originate from—
    (a) Both placenta as well as fully developed foetus
    (c) Placenta only
    (b) Oxytocin released from maternal pituitary
    (d) Fully developed foetus only
35. Hormone responsible for the secretion of milk after parturition is—
    (a) ICSH **(b) Prolactin** (c) ACTH (d) LH
36. Which layer develops first during embryonic development ?
    (a) Ectoderm (b) Mesoderm **(c) Endoderm** (d) Both (b) and (c)
37. Correct sequence in embryonic development is—
    (a) Fertilisation → Zygote → Cleavage → Morula → Blastula → Gastrula
    (b) Zygote → Fertilisation → Cleavage → Morula → Blastula → Gastrula
    (c) Morula → Fertilisation → Zygote → Cleavage → Blastula → Gastrula
    (d) None of these
38. Eyelids in human separate in :*
    (a) 14 weeks (b) 16 weeks **(c) 24 weeks** (d) 40 weeks

Chapter 4. Reproductive Health

1. The correct surgical procedure as a contraceptive method is—
    (a) Ovariectomy **(b) Hysterectomy** (c) Appendectomy (d) Castration
2. Diaphragms are contraceptive device used by females. Choose the correct option from the statement given below—
    (i) They are introduced into the uterus
    (ii) They are placed to cover the cervical region
    (iii) They act as physical barrier for sperm entry
    (iv) They act as spermicidal agents
    (a) (i) and (ii) (b) (i) and (iii) **(c) (ii) and (iii)** (d) (iii) and (iv)

* Frequently asked Previous Years Board Exam Question

3. Cu ions released from copper releasing Intra Uterine Device (IUDs)—
   (a) Prevent ovulation
   (b) Make uterus unsuitable for implantation
   (c) Increase phagocytosis sperms
   (d) Suppress sperm motility
4. Cu-T prevents pregnancy by preventing—
   (a) Fertilisation
   (b) Ovulation
   (c) Implantation of fertilised egg
   (d) None of these
5. Oral contraceptive pills help in the birth control by—
   (a) Killing sperms
   (b) Killing ova
   (c) Preventing ovulation
   (d) Forming barrier between sperm and ova
6. Which one of the following is the most widely accepted method of contraception in India at present ?
   (a) Cervical caps
   (b) Tubectomy
   (c) Diaphragms
   (d) IUD's (Intra Uterine Devices)
7. Which of the following birth control measures can be considered as the safest ?
   (a) The rhythm method
   (b) The use of physical barriers
   (c) Termination of unwanted pregnancy
   (d) Sterilization techniques
8. Which of the following is the component of oral pills ?
   (a) Progesterone
   (b) Oxytocin
   (c) Relaxin
   (d) None of these
9. Medical Termination of Pregnancy (MTP) is considered safe upto how many weeks to avoid pregnancy?
   (a) 8 weeks
   (b) 12 weeks
   (c) 18 weeks
   (d) 6 weeks
10. *In-vitro* Fertilisation is a technique that involves transfer of which one of the following into the fallopian tube ?
    (a) Zygote only
    (b) Embryo only, upto 8 cell stage
    (c) Either zygote or early embryo upto 8 cell stage
    (d) Embryo of 32 cell stage
11. The permissible use of the technique amniocentesis is for—
    (a) detecting any genetic abnormality
    (b) detecting sex of the unborn foetus
    (c) artificial insemination
    (d) transfer of embryo into the uterus of a surrogate mother
12. Amniocentesis is a process to—
    (a) determine any disease of heart
    (b) determine any hereditary disease of the embryo
    (c) know about the disease of brain
    (d) grow cell in culture medium
13. Full form of GIFT—
    (a) Gamete intrafallopian transfer
    (b) Gonad intrafallopian transfer
    (c) Both (a) and (b)
    (d) None of these
14. The technique called Gamete Intra Fallopian Transfer (GIFT) is recommended for those female—
    (a) Who cannot produce an ovum
    (b) Who cannot retain the foetus inside uterus
    (c) Whose cervical canal is too narrow to allow the passage for the sperms
    (d) Who cannot provide suitable environment for Fertilisation
15. The test tube baby programme employs which one of the following technique ?
    (a) Intra Cytoplasmic Sperm Injection (ICSI)
    (b) Intra Uterine Insemination (IUI)
    (c) Gamete Intra Fallopian Transfer (GIFT)
    (d) Zygote Intra Fallopian Transfer (ZIFT)
16. Which one of the following group include all sexually transmitted disease ?
    (a) AIDS, Syphillis, Cholera
    (b) HIV, Malaria, Trichomoniasis
    (c) Gonorrhea, Hepatitis-B, Chlamydiasis
    (d) Hepatitis-B, Haemophilia, AIDS

17. Amniocentesis is a technique used to—
    (a) Determine errors in amino acid metabolism in embryo
    (b) Pinpoint specific cardiac ailments in embryo
    (c) Determine any hereditary/genetic abnormality in embryo
    (d) All of the above

Chapter 5. Principles of Inheritance and Variation

1. G. J. Mendel was a—
   (a) British monk (b) Australian monk
   (c) Austrian monk (d) German scientist
2. The term genetics was proposed by—
   (a) Mendel **(b) Bateson** (c) Morgan (d) Johannsen
3. A cross is made between tall pea plant having green pods and dwarf pea plant having yellow pods. Tall plants in F_2 generation out of 16 would be—
   (a) 7 **(b) 12** (c) 13 (d) 15
4. A monhybrid cross produced tall and dwarf pea plants in ratio of 3 : 1. Their genotypes would be—
   (a) TT × Tt (b) Tt × Tt **(c) TT tt** (d) Tt × tt
5. The F_2 genotypic ratio of monohybrid cross is—
   (a) 1 : 1 **(b) 1 : 2 : 1** (c) 2 : 1 : 2 (d) 9 : 3 : 3 : 1
6. Genotypic ratio of a monohybrid cross is—
   (a) 5 : 1 (b) 3 : 1 **(c) 1 : 2 : 1** (d) 1 : 1
7. A character which is expressed in a hybrid is called—
   (a) recessive character **(b) dominant character**
   (c) codominant character (d) epistasis
8. When a dihybrid cross is fit into a punnett square with 10 boxes, the maximum number of different phenotypes available are—
   (a) 8 **(b) 4** (c) 2 (d) 2
9. The genetic ratio of 9 : 3 : 3 : 1 is due to—
   (a) Segregation of characters (b) Crossing over of character
   (c) Independent assortment of genes (d) Homologous pairing between chromosomes
10. In pea plants, yellow seeds are dominant to green. If a heterozygous yellow seeded plant is crossed with a green-seeded plant, what ratio of yellow and green seed plants would you expect in F_1 generation ?
    (a) 9 : 1 (b) 3 : 1 (c) 1 : 3 **(d) 50 : 50**
11. Which of the following genotypes show the heterozygous condition ?
    (a) Rr (b) RR (c) rr (d) None of these
12. Common test to find genotype of hybrid is by—
    (a) Studying sexual behaviour of F_1 progeny
    (b) Crossing F_1 individuals with recessive parents
    (c) Crossing one F_2 progeny with male parent
    (d) Crossing one F_2 progeny with female parent
13. Which one is a test cross ?
    (a) Ww × Ww **(b) Ww ww** (c) ww × ww (d) WW × Ww
14. Mating of an organism to a double recessive in order to determine whether it is homozygous or heterozygous for a character under consideration is called—
    (a) Reciprocal cross **(b) Test cross** (c) Dihybrid cross (d) Back cross
15. A dihybrid for qualitative trait is crossed with homozygous recessive individual of its type, the phenotype ratio is—
    (a) 1 : 2 : 1 (b) 3 : 1 **(c) 1 : 1 : 1 : 1** (d) 9 : 7

16. In order to find out the different types of gametes produced by a pea plant having genotype AaBb, it should be crossed with a plant with the genotype—
(a) AABB (b) AaBb (c) AABb **(d) aabb**

17. In *Mirabilis jalapa*, the number of F_2 red flowered plants in a cross of red flowered and white flowered would be—
(a) 1 (b) 2 (c) 8 (d) 6

18. Incomplete dominance is shown by—
(a) Primrose (b) *Mirabilis* (c) *Helianthus* (d) Pea

19. Mutual exchange of chromosome segments between a pair of homologous chromosomes is known as—
(a) mutation **(b) crossing over** (c) pleiotropism (d) translocation

20. Which of the following is best suited for codominance ?
(a) Both are recessive **(b) Both are dominant**
(c) One is recessive (d) One is dominant

21. Which one of the following is an example of polygenic inheritance ?
(a) Flower colour in *Mirabilis jalapa* (b) Production of male honeybee
(c) Pod-shape in garden pea **(d) Skin colour in humans**

22. Tt × tt is—
(a) Reciprocal cross (b) Hybridization
(c) Test cross (d) Back cross

23. The gene which controls many characters is called—
(a) Codominant gene (b) Polygene
(c) Pleiotropic gene (d) Multiple gene

24. Genes are situated on the—
(a) Chromosomes (b) DNA
(c) RNA (d) None of these

25. The forms of a single gene are referred as—
(a) alleles (b) loci (c) mutants (d) enzymes

26. The word gene for Medelian factor was introduced by ......in 1909.
(a) Watson and Crick (b) Sutton and Boveri
(c) Johannsen (d) Punnett

27. The parallelism between Mendelian factors and chromosomes led to the—
(a) Genic Balance Theory **(b) Chromosomal theory of inheritance**
(c) Law of independent assortment (d) Law of segregation

28. The co-inheritance of closely located genes on the same chromosome is referred as—
(a) linkage (b) chromosome mapping
(c) mutation (d) crossing over

29. All the genes which can inherit together from one generation to another form a—
(a) Locus **(b) Linkage group**
(c) Chromosome (d) Chromosome map

30. The chromosomal theory of inheritance was proposed by......in 1902.
(a) Mendel **(b) Sutton and Boveri**
(c) Bateson and Punnett (d) Watson and Crick

31. Coupling and repulsion hypothesis is concerned with........
(a) mutation **(b) linkage**
(c) Crossing over (d) pleiotropism

32. In......crossing over occurs at two places between the same non-sister chromatids
(a) Single crossing over **(b) Reciprocal crossing over**
(c) Multiple crossing over (d) Complementary crossing over

Multiple Choice Questions

33. The coupling and replusion hypothesis was given by......in 1906, in.......
(a) Bateson and Punnett, Pea plant (b) Morgan, Oenothera lamarkiana
(c) Morgan, *Drosophila* (d) Bateson and Punnett, *Drosophila*

34. ....carried out experiments to study linkage in *Drosophila*.
(a) Morgan (b) Correns
(c) Boveri and Brauer (d) Flemming

35. ....is the graphic representation of a linkage group.
(a) Gene map **(b) Chromosome map**
(c) Pedigree chart (d) Karyotype

36. ......are the chromosomes responsible for determining the sex of an individual.
(a) Autosomes **(b) Allosomes** (c) Both (a) and (b) (d) None of these

37. Sex chromosomes and chromosomal basis of inheritance was established by......
(a) McClung (b) Henking (c) Morgan **(d) Bateson**

38. .... in 1916, showed that Y-chromosome has no role in the determination of sex in Drosophila.
(a) Calvin Bridges (b) Morgan (c) McClung (d) Bateson

39. Drosophila exhibits......type of sex determination.
(a) XX-XY (b) XX-XO (c) ZZ-ZO (d) ZZ-ZW

40. Male transmits it X-chromosome to his daughters and female transmits its X-chromosome to both sons and daughters. This is called—
(a) Sex determination (b) Linkage
(c) Crossing over **(d) Criss-cross inheritance**

41. Generally in females, sex chromsomes are—
(a) Hmomorphic (b) Heteromorphic
(c) Both (a) and (b) (d) None of these

42. ....is/are an example/s of sex-linked inheritance in human beings.
(a) Colour blindness (b) Haemophilia
(c) Both (a) and (b) (d) None of these

43. In the given pedigree chart, the trait shown is :*

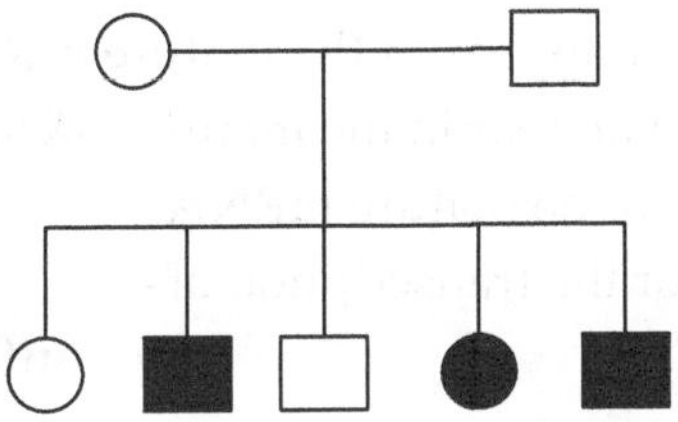

(a) Autosomal dominant **(b) Autosomal recessive**
(c) X-linked (d) Y-linked

44. Study the given monohybrid cross :*

P_1 TT × tt

F_1 Tt

A test cross for this F_1 will be :
(a) Tt × TT **(b) Tt tt** (c) Tt × Tt (d) TT × tt

Chapter 6. Molecular Basis of Inheritance

1. The first genetic material is—
(a) Protein (b) Carbohydrates **(c) DNA** (d) RNA

* Frequently asked Previous Years Board Exam Question

2. Length of the DNA with 23 base pairs is :*
   (a) 78·4 Å (b) 78·2 Å (c) 78 Å **(d) 74 8 Å**
3. In a DNA strand, the nucleotides are linked together by—
   (a) glycosidic bonds **(b) phosphodiester bonds**
   (c) peptide bonds (d) hydrogen bonds
4. Which of the following are the functions of RNA ?
   (a) It is carrier of genetic information from DNA to ribosome synthesising polypeptides
   (b) It carries amino acids to ribosomes
   (c) It is constituent component of ribosomes
   (d) All of the above
5. Khorana was awarded nobel prize for—
   (a) discovering DNA (b) discovering RNA
   (c) chemical synthesis of gene (d) discovering DNA polymerase
6. Who discovered DNA polymerase ?
   (a) Okazaki **(b) Kornberg**
   (c) Messelson and Stahl (d) Watson and Crick
7. Discontinuous synthesis of DNA occurs on one strand because—
   (a) DNA molecule being synthesized is very long
   (b) DNA dependent DNA polymerase catalyses polymerisation only in one direction (5′ → 3′)
   (c) It is a more efficient process
   (d) DNA ligase has to have some role
8. The promoter site and terminator site for transcription are located at—
   (a) 3′ (downstream) end and 5′ (upstream) end respectively of transcription unit
   (b) 5′ (upstream) end and 3′ (downstream) end respectively of transcription unit
   (c) 5′ (upstream) end
   (d) 3′ (downstream) end
9. With regard to mature mRNA in eukaryotes,which of the following is true ?
   (a) Exons and introns do not appear in the mature RNA
   (b) Exons appear but introns do not appear in the mature mRNA
   (c) Introns appear but exons do not appear in the mature mRNA
   (d) Both exons and introns appear in the mature mRNA
10. RNA polymerase II is responsible for the transcription of—
    (a) tRNA (b) rRNA **(c) hnRNA** (d) snRNA
11. RNA polymerase III transcribes—
    (a) hnRNA (b) tRNA (c) 5S rRNA **(d) Both (b) and (c)**
12. Removal of introns in a transcription unit is—
    (a) Transformation **(b) Splicing**
    (c) Tailing (d) Capping
13. Control of gene expression takes place at the level of—
    (a) DNA replication **(b) Transcription**
    (c) Translation (d) None of these
14. A specific nucleotide sequence to which RNA polymerase attaches to initiate transcription of mRNA from a gene called—
    (a) promoter gene (b) structural gene
    (c) operon (d) regulation gene
15. In eukaryotic cell, transcription, RNA splicing and RNA capping take place in—
    (a) Nucleus (b) Cytoplasm (c) Ribosomes (d) Golgi body
16. Peptide synthesis in a cell takes place in—
    (a) chloroplast **(b) ribosome** (c) mitochondria (d) golgi body

* **Frequently asked Previous Years Board Exam Question**

17. Which of the following play a role in protein synthesis ?
(a) Introns **(b) Exons**
(c) Both (a) and (b) (d) None of the above

18. The protein of DNA, which contains information for an entire polypeptide is called as—
(a) Cistron (b) Muton
(c) Recon (d) Operon

19. A gene of operon which forms the repressor protein is—
(a) Operator (b) Promoter **(c) Regulator** (d) Structural

20. In the absence of lactose, the operator gene of lac-operon is suppressed by genetic material—
(a) Structural gene (b) Repressor protein
(c) Regulator gene (d) Promoter gene

21. In *E-coli*, the lac operon gets switched on when—
(a) lactose is present and it binds to the repressor
(b) repressor binds to operator
(c) RNA polymerase binds to the operator
(d) lactose is present and it binds to RNA polymerase

22. In the Lac operon system, β-galactosidase is encoded by—
(a) a-gene (b) i-gene (c) y-gene **(d) z-gene**

23. Repressor protein is produced by—
(a) Regulator gene (b) Operator gene (c) Structural gene (d) Promoter gene

24. Triplet UUU codes for—
(a) leucine (b) methionine **(c) phenylalanine** (d) glycine

Chapter 7. Evolution

1. A biogenesis occurred about.................billion years ago.
(a) 1.2 (b) 1.5 (c) 2.5 **(d) 3.5**

2. According to Abiogenesis, life originated from :*
(a) Non-living matter (b) Pre-existing life
(c) Oxygen (d) Extra-terrestrial matter

3. Theory of chemical origin of life was prepared by—
(a) Miller and Fox **(b) Oparin and Haldane**
(c) Miller and Watson (d) Watson and Melvin

4. Coacervates were experimentally produced by—
(a) Oparin and Sidney Fox (b) Fischer and Huxley
(c) Jacob and Monod (d) Urey and Miller

5. Which of the following was likely to have been absent in a free molecule state, in the primitive atmosphere of the earth ?
(a) Carbon **(b) Oxygen** (c) Hydrocarbon (d) Nitrogen

6. Theory of spontaneous generation was proposed by—
(a) Spallanzani **(b) Aristotle** (c) F. Redi (d) Louis Pasteur

7. Evolutionary history of an organism is known as—
(a) Ontogeny **(b) Phylogeny** **(c) Ancestry** (d) Paleontology

8. A connecting link between reptiles and birds is—
(a) Dimetrodon (b) Dodo **(c) *Archaeopteryx*** (d) *Sphenodon*

9. A baby has a small tail. It is a case of—
(a) Mutation (b) Metamorphosis
(c) Atavism (d) Retrogressive evolution

10. Analogous organs arise due to—
(a) Divergent evolution (b) Artificial selection
(c) Genetic drift **(d) Convergent evolution**

* **Frequently asked Previous Years Board Exam Question**

11. Which of the following pairs of structures is homologous ?
(a) Wings of grasshopper and forelimbs of flying squirrel
(b) Tentacles of Hydra and arms of starfish
(c) Forelimbs of bat and forelegs of horse
(d) Wings of birds and wings of moth

12. Wings of pigeon, bat and mosquito exhibit the phenomenon called—
(a) Convergent evolution (b) Divergent evolution
(c) Atavism (d) All of these

13. Wings of bat and bird are—
(a) Homologous but not analogous (b) Neither homologous nor analogous
(c) Analogous but not homologous (d) Vestigial

14. Which one is linked to evolution ?
(a) Extinction (b) Competition **(c) Variation** (d) Reproduction

15. The theory of use and disuse of organ was proposed by—
(a) Darwin **(b) Lamarck** (c) de Vries (d) Hooker

16. Lamarck's theory of evolution is also called—
(a) Survival of the fittest (b) Special creation theory
(c) Inheritance of acquired characters (d) None of these

17. Birbal Sahni Institute of Palaeobotany is located at—
(a) Delhi **(b) Lucknow** (c) Dehradun (d) Kolkata

18. Who proposed the theory of origin of species by Natural Selection ?
(a) August Weismann (b) De Vries
(c) Charles Darwin (d) Charles Darwin and Alfred Wallace

19. The unit of natural selection is—
(a) an individual (b) a species (c) a germ (d) a population

20. Industrial melanism is related to—
(a) Skin darkening due to smoke (b) Drug resistance
(c) Defence against radiation **(d) Protective resemblance to surroundings**

21. Darwin's finches are good example of—
(a) Connecting link **(b) Adaptive radiation**
(c) Convergent evolution (d) Industrial melanism

22. Non-directional alteration in Hardy-Weinberg equilibrium is—
(a) Gene flow (b) Mutation
(c) Genetic drift (d) Gene recombination

23. Genetic drift operates only in—
(a) Island population **(b) Smaller population**
(c) Larger population (d) Mendelian population

24. Species occuring in different geographical areas are called—
(a) Sibling (b) Neopatric (c) Sympatric **(d) Allopatric**

25. Cause of speciation is—
(a) Random mating (b) Hybridisation
(c) Isolation (d) Migration

26. The largest unit in which gene flow is possible is:*
(a) Organism **(b) Population** **(c) Species** (d) Genes

27. The earliest hominids evolved about 4 million years ago were—
(a) *Homo erectus* (b) Neanderthal man
(c) Cro-magnon man **(d) *Australopithecus***

28. The golden age of reptiles is—
(a) Palaeozoic **(b) Mesozoic** (c) Cenozoic (d) Proterozoic

29. Hominids originated during—
(a) Pliocene (b) Paleocene **(c) Miocene** (d) Oligocene

30. Human ancestors who left cave paintings were—
(a) Neanderthal man
(b) Cro-magnon man
(c) Java apeman
(d) Peking man

Chapter 8. Human Health and Diseases

1. A non-communicable disease is—
(a) Diabetes (b) Measles (c) Influenza (d) Diphtheria
2. Which of the following is not a hereditary disease ?
(a) Cretinism (b) Cystic fibrosis (c) Thalassaemia (d) Haemophilia
3. Haemophilia is a condition where there is—
(a) no production of haemoglobin in the blood
(b) No production of melanin in the skin
(c) a failure of clotting of blood
(d) a delay in clotting of blood
4. The pathogen which causes Syphilis—*
(a) Neisseria
(b) Chlamydia
(c) Treponema
(d) Papilloma virus
5. The disease tetanus is also called—
(a) Bleeder's disease
(b) Lockjaw
(c) Whooping cough
(d) Gangrene
6. BCG vaccine is a preventive measure against—
(a) Tuberculosis (b) Typhoid (c) AIDS (d) Cholera
7. Passive immunity can be obtained by injecting—
(a) antibodies
(b) antigens
(c) antibiotics
(d) all of the above
8. In the cell mediated immune response, T-lymphocytes divide and secrete—
(a) antigens (b) plasmogens (c) collagens **(d) cytokines**
9. An organ related with immunity is—
(a) liver (b) kidney **(c) thymus** (d) heart
10. Memory cells are formed from—
(a) monocytes (b) eosinophils (c) neutrophils **(d) lymphocytes**
11. Antigens are present—
(a) inside the nucleus
(b) on the nuclear membrane
(c) on the cell surface
(d) inside the cytoplasm
12. Interferons are—
(a) anti-bacterial protein
(b) anti-viral protein
(c) complex protein
(d) anti-clotting protein
13. Colostrum provides the infant with—
(a) Autoimmunity
(b) Passive immunity
(c) Active immunity
(d) Innate immunity
14. This class of antibodies is made up of five immunoglobulin units—
(a) IgG **(b) IgM** (c) IgA (d) IgD
15. Aspirin is an—
(a) antibiotic **(b) antipyretic** (c) antiseptic (d) none of these
16. An auto immune disease is—
(a) AIDS
(b) Haemophilia
(c) Allergy
(d) Myasthenia gravis
17. In AIDS, which of these cells are most affected ?
(a) B-cells
(b) T-cells
(c) Monocytes
(d) Neutrophils
18. T- Lymphocyte first mature in—
(a) Thymus (b) Pancreas (c) Liver (d) Spleen

* Frequently asked Previous Years Board Exam Question

19. HIV that causes AIDS, first start destroying—
(a) **helper T - lymphocyte** (b) B- lymphocytes
(c) leucocytes (d) thrombocytes

20. Hashish and Ganja are got from—
(a) *Erythroxylon* (b) *Nicotiana* (c) *Papaver* **(d) *Cannabis***

21. The stimulant present in tea, coffee and cola drinks is —
(a) Cocaine (b) Tannin (c) Astringent **(d) Caffeine**

22. LSD is—
(a) Lysing diethyl amide (b) Lysergic acid
(c) Lysing dimethyl amide **(d) Lysergic acid dimethylamide**

23. Opium is obtained from :*
(a) *Papaver somniferum* (b) *Cannabis sativa*
(c) *Erythroxylum coca* (d) *Datura metel*

Chapter 9. Strategies for Enhancement in Food Production

1. Which one of the following species of bees are used for the commercial production of honey?
(a) *Apis dorsata* (b) *Apis indica* (c) *Apis florea* **(d) *Apis mellifera***

2. Which one of the following is a breed of cattle?
(a) Ayrshire (b) Ghagus
(c) Kadakanath (d) Scampi

3. 33% of India's gross domestic product comes from—
(a) Industry **(b) Agriculture**
(c) Export (d) Small scale cottage industry

4. Who gave the idea that every plant cell is totipotent ?
(a) P.R. White (b) E.C. Cocking
(c) F.C. Steward **(d) G. Haberlandt**

5. Root development is promoted by—
(a) Abscisic acid **(b) Auxin**
(c) Ethylene (d) Gibberellins

6. One of the commonly used plant growth hormone in tea plantation is —
(a) Ethylene (b) Abscisic acid
(c) Zeatin **(d) Indole-3- acetic acid**

7. The term 'totipotency' refers to the capacity of a—
(a) cell to generate whole plant (b) bud to generate whole plant
(c) seed to germinate (d) cell to enlarge in size

8. In the hexaploid wheat, the haploid (n) and basic (x) number of chromosomes are—
(a) $n = 21$ and $x = 21$ (b) $n = 21$ and $x = 14$
(c) $n = 21$ and $x = 7$ **(d) $n = 7$ and $x = 21$**

9. In hybridisation, the haploids combine the advantages of—
(a) Recombination (b) Segregation
(c) Fixation **(d) All of them**

10. Which one of the following substance is responsible for callus formation?
(a) 2, 4-D (b) NAA (c) BAP (d) PEG

11. The new varieties of plant are produced by—
(a) Introduction and selection (b) Mutation and selection
(c) Selection and hybridisation (d) Introduction and mutation

12. 'Norin gene' of dwarfness in wheat originated through spontaneous mutation in—
(a) India **(b) Japan** (c) Mexico (d) U.S.S.R.

* Frequently asked Previous Years Board Exam Question

Multiple Choice Questions

13. Pure line breed refer to—
    (a) Homozygosity and independent assortment
    (b) Homozygosity only
    (c) Heterozygosity
    (d) Heterozygosity and Linkage
14. The scientific process of which crop plant are enriched with certain desirable nutrient is called—
    (a) Crop protection
    (b) Breeding
    (c) Biofortification
    (d) Bioremediation
15. An explant is —
    (a) dead plant
    (b) part of the plant
    (c) part of the plant used in tissue culture
    (d) part of the plant that express a specific character
16. Micropropagation is a technique—
    (a) For production of thousand of plants
    (b) For production of haploid plants
    (c) For production of somatic hybrids
    (d) For production of somoclonal plants.
17. The technique of obtaining large number of plantlets by tissue culture method is—
    (a) Organ culture
    (b) Micro propagation
    (c) Macro propagation
    (d) Plantlet culture
18. Which of the following is paired incorrectly:*
    (a) Cyclosporin A - Trichoderma polysporum
    (c) Swiss Cheese - Propionibacterium
    (b) Streptokinase - Saccharomyces cerevisiae
    (d) Penicillin - Penicillium
19. 'Jaya' and 'Ratna' developed for green revolution in India are the varieties of—
    (a) Maize **(b) Rice** (c) Wheat (d) Bajra
20. Single cell protein refer to—
    (a) A specific protein extracted from pure culture of a single type of cell.
    (b) Sources of mixed proteins extracted from pure or mixed culture of organism from a single cell.
    (c) Protein extracted from a single cell
    (d) A specific protein extracted from a single cell

Chapter 10. Microbes in Human Welfare

1. Lactic Acid is formed by the process of —
    (a) Fermentation
    (b) Glycolysis
    (c) HMP pathway
    (d) None of these
2. Which of the following is maintained for optimum production of Vinegar?
    (a) Anaerobic condition
    (b) Temperature of 65°C
    (c) Aerobic condition
    (d) Microaerophilic condition
3. Ethanol is commercially produce through a particular species of—
    (a) *Saccharomyces* (b) *Clostridium* (c) *Trichoderma* (d) *Aspergillus*
4. Yeast is used in the production of—
    (a) Citric acid and Lactic acid
    (b) Lipase and Pectinase
    (c) Bread and Beer
    (d) Cheese and Butter
5. Lactic acid bacteria (LAB) grow in milk and convert it to curd and also improve its nutritional quality by increasing—
    (a) Vitamin A
    (b) Vitamin B_{12}
    (c) Vitamin B_6
    (d) Vitamin C and A
6. The sugar present in milk is—
    (a) Glucose **(b) Lactose** (c) Fructose (d) Sucrose

* Frequently asked Previous Years Board Exam Question

7. Bacteriophages kill—
   (a) Fungi (b) Parasites **(c) Bacteria** (d) Viruses
8. To speed up the malting process in brewing industry, the growth hormone used is—
   (a) Auxin (c) Kinetin
   (b) Gibberellic acid (d) Ethylene
9. Continuous addition of sugar in 'fed batch' fermentation is done to—
   (a) produce methane (b) obtain antibiotics
   (c) purify enzymes (d) degrade sewage
10. Which of the following antibiotic is active against fungus ?
   (a) Streptomycin **(b) Penicillin** (c) Tetracycline (d) Neomycin
11. The purpose of biological treatment of waste water is to—
   (a) reduce BOD (b) increase BOD
   (c) reduce sedimentation (d) increase sedimentation
12. Methanogens growing anaerobically on cellulosic material produce—
   (a) Methane (b) Methane and carbon dioxide
   (c) Methane and hydrogen (d) Methane, carbon dioxide and hydrogen
13. Methanogens do not produce—
   (a) oxygen (b) methane
   (c) hydrogen sulphide (d) carbon dioxide
14. Biogas production is carried out by —
   (a) thermoacidophiles **(b) methanogens**
   (c) halophyles (d) luminants
15. A common bio control agent for the control of plant disease is—
   (a) Trichoderma (b) Baculovirus
   (c) *Bacillus thuringiensis* (d) *Glomus*
16. Which one of the following is used as biological insecticides ?
   (a) Tiger beetle (b) Caterpillar (c) Silk moth (d) Majua poka
17. Which one of these is not used as a bioweapons ?
   (a) *Bacillus anthracis* (b) Botulinum toxin
   (c) *Bacillus thuringiensis* (d) Small pox
18. Rotenone is a —
   (a) Bioherbicides (b) Commonly used biofertilizer
   (c) Bioinsecticide (d) Juvenile hormone
19. Which one of the following is not a biofertilizer?
   (a) *Agrobacterium* (b) *Rhizobium* (c) *Nostoc* (d) *Mycorrhiza*
20. A good example of organic fertilizer which improves phosphorus uptake is —
   (a) Actinomycetes fungi (b) Rhizobium
   (c) Azospirillum (d) None of these
21. Nitrogen fixation in root nodules of Alnus is brought about by—
   (a) *Frankia* (b) *Azorhizobium* (c) *Bradyhizobium* (d) *Clostridium*
22. Mycorrhiza does not help the host plant in—
   (a) Enhancing its phosphous uptake capacity (b) Increasing its tolerance to drought
   (c) Enhancing its resistance to root pathogens **(d) Increasing its resistance to insect**
23. The common nitrogen fixer in Paddy field is—
   (a) *Frankia* (b) *Rhizobium* **(c) *Azospirillum*** (d) *Oscillatorlia*
24. The main source of biofertilisers are —
   (a) bacteria (b) cyanobacteria (c) fungi **(d) all of these**

Multiple Choice Questions

25. Nitrogen is absorbed by plant in the form of—
(a) NO^-_3 (b) NH_3 (c) NO^-_2 **(d) Both (a) and (c)**

26. Which one of the following help in the absorption of phosphorus from soil by plants?
(a) Glomus (b) Rhizobium (c) Frankia (d) Anabaena

Chapter 11. Biotechnology–Principles and Processes

1. The ability of a cell to grow into a complete plant is called—
(a) Protoplasmic fusion (b) Tissue culture
(c) Cellular totipotency (d) Somaclonal hybridisation

2. Plasmid is a—
(a) fungus (b) plastid
(c) extrachromosomal DNA (d) part of plasma membrane

3. Restriction endonuclease—
(a) Synthesises DNA
(b) Cuts the DNA molecule randomly
(c) Cuts on DNA molecule at specific sites
(d) Restricts the synthesis of DNA inside the nucleus

4. There is a restriction endonuclease called EcoRI. What does 'co' part in it stand for?
(a) Colon (b) Coelom (c) Coenzyme **(d) Coli**

5. Restriction enzyme Eco RI cut the DNA between bases G and A only when the sequence in DNA is—
(a) GATATC **(b) GAATTC** (c) GATTCC (d) GAACTT

6. Which among the following is regarded as molecular scissors ?
(a) Reverse transcriptase **(b) Restriction endonuclease**
(c) Exonucleases (d) Polymerases

7. Transfer of any gene into a completely different organism can be done through—
(a) genetic engineering (b) tissue culture
(c) transformation (d) none of these

8. The construction of the first recombinant DNA was done by using the native plasmid of—
(a) *E. coli* **(b) *Salmonella typhimurium***
(c) *B. thuringiensis* (d) Yeast

9. The most common bacterium used in genetic engineering is—
(a) *E. coli* (b) *Clostridium*
(c) *Bacillus thuringiensis* (d) *Salmonella*

10. Polyethylene glycol method is used for—
(a) Gene transfer without a vector (b) Biodiesel production
(c) Seedless fruit production (d) All of these

11. Viral genome corporated into host DNA is called—
(a) Phage **(b) Prophage**
(c) Bacteriophage (d) None of these

12. Significance of heat shock method in bacterial transformation facilitate—
(a) Binding of DNA to the cell wall
(b) Uptake of DNA through membrane transport protein.
(c) Uptake of DNA through transient pores in the bacterial cell wall
(d) Expression of antibiotic resistance genes.

13. For transformation, microparticles coated with DNA to be bombarded with gene gun are made up of—
(a) Silver or Platinum (b) Platinum or Zinc
(c) Silicon on Platinum **(d) Gold or Tungsten**

14. PCR is required for—
(a) DNA synthesis
(b) DNA amplification
(c) Protein synthesis
(d) Amino acid synthesis
15. Gel electrophoresis is used for—
(a) Construction of recombinant DNA by joining with cloning vectors
(b) Isolation of DNA molecule
(c) Cutting of DNA into fragments
(d) Separation of DNA fragments according to their size
16. Enzyme used in PCR technology is—
(a) Taq polymerase
(b) Polymerase
(c) Exonuclease
(d) Helicase
17. Which is used in molecular genetic engineering?
(a) Tomato (b) Tobacco (c) Carrot **(d) Arabidopsis**
18. Human Genome Project was initiated by—
(a) Francis Collins and Roderick
(b) Watson and Crick
(c) Beadle and Tatum
(d) Paul Berg and Waldmann
19. In Bt cotton, a transgentic plant, Bt refers to—
(a) Botanical
(b) Beta
(c) Biotechnology
(d) *Bacillus thuringiensis*
20. Bt cotton is resistant to—
(a) insects (b) herbicides (c) salt (d) drought
21. Bt toxin is—
(a) intracellular lipid
(b) polysaccharide
(c) intracellular crystalline protein
(d) extracellular crystalline protein
22. Transgenic plant was developed by—
(a) Cloned and genetically modified gene
(b) Introduction of foreign genes
(c) Genetic engineering
(d) Purified genes

Chapter 12. Biotechnology and its Applications

1. During gene cloning, what is called as 'gene taxi'?
(a) Vaccine **(b) Plasmid** (c) Bacterium (d) Protozoa
2. Terminator gene—
(a) Help in terminating flowering
(b) Help in terminating seed germination
(c) Used in hybridisation
(d) None of these
3. In transgenics, expression of transgene in Target tissue is determined by—
(a) Transgene **(b) Promoter** (c) Reporter (d) Enhancer
4. Transfer of DNA band from an agarose gel to a nitrocellulose or nylon membrane is referred to as—
(a) Southern transfer
(b) Western transfer
(c) Northern transfer
(d) Eastern transfer
5. *Bacillus thuringiensis* (Bt) strain have been used for designing novel—
(a) Bio-fertilizers
(b) Bio-metallurgical techniques
(c) Bio-mineralization processes
(d) Bio-insecticidal plants
6. Isolation of Bt gene from bacterium (*Bacillus thuringiensis*) was taken up in the year—
(a) 1977 **(b) 1980** (c) 1997 (d) 1990
7. What is true about Bt toxin?
(a) Bt toxin exist as active toxin in the *Bacillus*
(b) The activated toxin enter the ovaries of the pest to sterilize it and thus prevent its multiplication
(c) The concerned *Bacillus* has anti toxins.
(d) The inactive prtotoxin gets converted into active form in the insect gut.

8. Tobacco plant resistant to a nematodes have been developed by the introduction of DNA that produce in the host cells—
   (a) both sense and antisense RNA (b) a particular hormone
   (c) an antifeedant (d) a toxic protein
9. In RNA, genes are silenced using—
   (a) ss DNA (b) ds DNA (c) ds RNA **(d) ss RNA**
10. Golden rice is a promising transgenic crop when released for cultivation, it will help in—
   (a) producing a petrol like fuel from rice **(b) alleviation of vitamin A**
   (c) pest resistance (d) herbicide tolerance
11. Blindness is prevented by use of which crop in poor countries—
   (a) Golden rice (b) Wheat (c) Gram (d) Pea
12. A transgenic food crop which may help in solving the problem of night blindness in developing countries is—
   (a) Bt soyabean **(b) Golden rice**
   (c) Flavr savr tomatoes (d) Starlink maize
13. Silencing of a gene could be achieved through the use of—
   (a) Short interfering RNA (RNAi) (b) Antisense RNA
   (c) Both (a), (b) (d) None of the above
14. crg 1 endotoxin obtained from bacillus thuringiensis is effective against :
   (a) Nematodes (b) mosquidtoes
   (c) bollworm (d) Flies
15. Genetically engineered bacteria are being employed for production of—
   (a) Thyroxine **(b) Human insulin**
   (c) Cortisol (d) Epinephrine
16. The genetic defect Adenosine deaminase (ADA) deficiency may be cured permanently by—
   (a) Administering adenosine deaminase activators
   (b) Periodic infusion of genetically engineered lymphocytes having functional ADA cDNA
   (c) Enzyme replacement therapy
   (d) Introducing bone marrow cells producing ADA into cells at early embryonic stages
17. The first clinical gene therapy was done for the treatment of—
   (a) AIDS
   (b) Cancer
   (c) Cystic fibrosis
   (d) SCID (Severe Combined Immune Deficiency resulting from deficiency of ADA)
18. The first clinical gene therapy was given for treating—
   (a) Diabetes mellitus (b) Chicken pox
   (c) Rheumatoid arthritis **(d) Adenosine diaminase deficiency**
19. Name the drug used in cancer treatment produced by using biotechnology—
   (a) Terramycin (b) HGH (c) TSH **(d) Interferon**
20. Amplification of gene of interest by using DNA polymerase may go upto—
   (a) 0.1 million times (b) 1.0 million times
   (c) 1.0 billion times (d) 1.0 trillion times
21. Product of biotechnology is—
   (a) Humulin (b) Bt cotton (c) Biofertilisers **(d) All of these**
22. Human insulin is being commercially produced from a transgenic species of—
   (a) *Rhizobium* (b) *Saccharomyces* **(c) *Escherichia*** (d) *Mycobacterium*

Chapter 13. Organisms and Populations

1. Niche overlap indicates—
   (a) mutualism between two species
   (b) active cooperation between two species

(c) two different parasites on the same host
(d) sharing of one or more resources between the two species

2. Geometric representation of age structure is a characteristic of—
(a) landscape
(b) ecosystem
(c) biotic community
(d) population

3. Study of human population growth is—
(a) Anthropology
(b) Geography
(c) Demography
(d) Biology

4. The primary reason for increase in human population is—
(a) The increase in agricultural production
(b) The increase in birth rate and decrease in death rate
(c) The improvement in medical technology
(d) All of the above

5. If 10 individuals in a laboratory population of 100 mice die during the period of one year, the death rate in this population will be—*
(a) 110
(b) 0.01
(c) 0.1
(d) 90

6. Maximum growth rate occur in—
(a) senescent phase
(b) lag phase
(c) exponential phase
(d) stationary phase

7. Exponential population growth is—
(a) dN/dt = rN
(b) dt/dN = rN
(c) dN/rN = dt
(d) rN/dN = dt

8. Population density for a particular area can be calculated by—
(a) $D = \dfrac{N}{S}$
(b) $S = N \times D$
(c) $D = S \times N$
(d) $N = \dfrac{S}{D}$

9. The equation $\dfrac{N_n}{N_i}$ = B represents which of the following ?
(a) natality
(b) growth rate
(c) mortality
(d) all of these

10. When the number of individuals added to the population exceeds the number of individuals lost, the population experiences—
(a) zero population growth
(b) exponential population growth
(c) decrease in biotic potential
(d) decrease in carrying capacity

11. In the sigmoid growth curve, the upper asymptote represent the period of—
(a) establishment
(b) positive acceleration
(c) negative acceleration
(d) equilibrium

12. A population has more young individuals compared to the older individuals. What would be the status of the population after some years—
(a) It will decline
(b) It will stabilise
(c) It will increase
(d) It will first decline and then stabilise

13. Which one of the following is the characteristics of a biological community?
(a) Sex-ratio
(b) Stratification
(c) Natality
(d) Mortality

14. A protozoan reproduces by binary fission. What will be the number of profozoans in its population after six generations?
(a) 128
(b) 24
(c) 64
(d) 32

15. Which one of the following organism reproduce sexually only once in its life time?
(a) Banana plant
(b) Mango
(c) Tomato
(d) Eucalyptus

* **Frequently asked Previous Years Board Exam Question**

16. Lichens are well known combination of an algae and fungus where fungus is—
    (a) an epiphytic relationship between algae
    (b) a parasitic relationship with the algae
    (c) a symbiotic relationship with the algae
    (d) a saprophytic relationship with algae
17. Annual migration does not occur in case of—
    (a) Arctic tern
    (b) Salmon
    (c) Siberian fox
    (d) Salamander
18. The physiological capacity of a species to produce their progeny in absence of predators, diseases and other inhibiting factors is called—
    (a) carrying capacity
    (b) biotic potential
    (c) environmental resistance
    (d) population explosion

Chapter 14. Ecosystem

1. Which one of the following is not a functional unit of an ecosystem?
   (a) Energy flow (b) Decomposition (c) Productivity **(d) Stratification**
2. Term ecosystem was given by—
   (a) Odum
   (b) Haeckel
   (c) Transley
   (d) Mobius and forbes
3. Maximum absorption of rainfall water is done by—
   (a) Tropical deciduous forest
   (b) Tropical everygreen forest
   (c) Tropical savannah
   (d) Scrub forest
4. According to Allen's rule, mammal in cold region have ...... to conserve body heat—
   (a) Small extremities (legs, tails, and ears)
   (b) Longer extremities
   (c) Longer body mass
   (d) Smaller body mass
5. Animals from colder climate generally have smaller limbs. This is called—
   (a) Niche rule **(b) Allen's jule** (c) Ehrlich rule (d) None of these
6. Animals undergo inactive stage during winter is known as—
   (a) aestivation **(b) hibernation** (c) adaptation (d) acclimatisation
7. Pneumatophores are present/common in—
   (a) xerophytes (b) hygrophytes **(c) mesophytes** (d) halophytes
8. Identify the possible link "A" in the following food chain Plant → Insect → Frog → "A→ Eagle.
   (a) Rabbit (b) Wolf **(c) Cobra** (d) Parrot
9. Which one of the following animals may occupy more than one trophic levels in the same ecosystem at the same time?
   (a) sparrow (b) lion (c) goat (d) frog
10. Identify the likely organism (1), (2), (3) and (4) in the food web shown below—

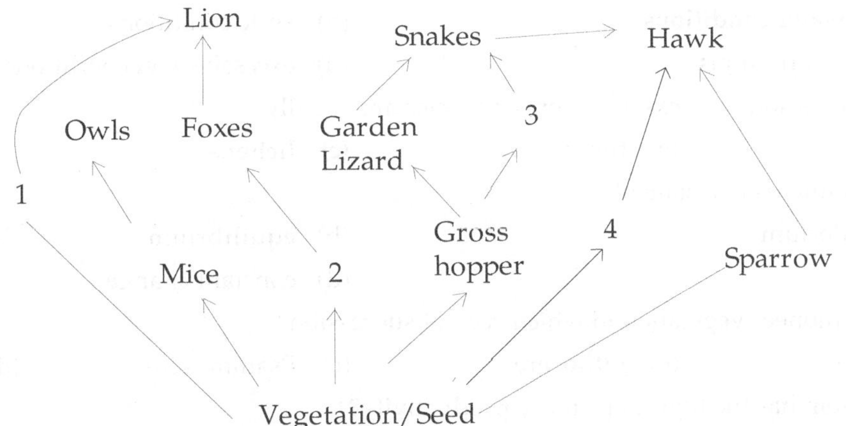

	(1)	(2)	(3)	(4)
(a)	**deer**	**rabbit**	**frog**	**rat**
(b)	dog	Squirrel	bat	deer
(c)	rat	dog	tortoise	crow
(d)	squirrel	cat	rat	pigeon

11. Decomposers are—

(a) autotrophs (b) photoheterotrophs

(c) organotrophs (d) heterotrophs

12. Energy transferred from one trophic level to another is—

(a) 5% **(b) 10%** (c) 15% (d) 20%

13. The '10 percent law' is related to—

(a) Mendelian genetics

(b) Non-Mendelian genetics

(c) Energy transfer from lower trophic level to higher trophic level

(d) Energy consumption during photosynthesis in C_4 plants

14. Photosynthetically active radiation (PAR) represent the following range of wavelength—

(a) 450–950 nm (b) 340 450 nm **(c) 400–700 nm** (d) 500–600 nm

15. Which of the following representation shows the pyramid of number in a forest ecosystem ?

(A) (B) (C) (D)

(a) D (b) A **(c) B** (d) C

16. The upright pyramid of number is absent in—

(a) Pond **(b) Forest** (c) Lake (d) Grassland

17. Pyramid of energy in the river ecosystem is—

(a) always upright (b) always inverted (c) constant (d) declining

18. Which one of the following is not a gaseous biogeochemical cycle in ecosystem?

(a) Sulphur cycle **(b) Phosphorus cycle**

(c) Nitrogen cycle (d) Carbon cycle

19. Among the following biogeochemical cycles which one does not have losses due to respiration?

(a) Phosphorus (b) Nitrogen (c) Sulphur **(d) All of the above**

20. Both hydrarch and Xerarch successions lead to—

(a) medium water conditions (b) xeric conditions

(c) highly dry conditions (d) excessive wet conditions

21. In Primary succession on rocks, the pioneer species are usually—

(a) algae (b) fungi **(c) lichens** (d) bryophytes

22. Climax community is in a state of—

(a) non-equilibrium **(b) equilibrium**

(c) disorder (d) constant change

23. Lichens is the pioneer vegetation in which type of succession?

(a) hydrosere **(b) Lithosere** (c) Psammosere (d) Xerosore

24. Which ecosystem has the highest primary productivity?

(a) Pond (b) Ocean (c) Desert **(d) Forest**

25. Zone of atmosphere near the earth surface is—

(a) Stratosphere (b) Mesosphere **(c) Troposphere** (d) Thermosphere

26. Xerophytes are mostly—

(a) succulents (b) water related
(c) mesophytes (d) none of these

Chapter 15. Biodiversity and its Conservation

1. The world biodiversity day is celebrated annually on–

(a) 5th June **(b) 22nd May** (c) 22nd April (d) 16th September

2. Quinine is obtained from—

(a) bark of cinchona (b) root of cinchonas
(c) wood of cinchona (d) leaves of cinchona

3. Extinction of a Species in a food chain is compensated by—

(a) food chain (b) ecological pyramid
(c) food web (d) none of these

4. If the Bengal tiger becomes extinct—

(a) Hyaenas and wolves will become scarce
(b) The wild areas will be safe for man and domestic animals
(c) Its gene pool will be lost for ever
(d) The population of beautiful animals like deers will get stabilized

5. Which is mainly responsible for extinction of wildlife ?

(a) Hunting of flesh (b) Destruction of habitats
(c) Pollution of air and water **(d) All of these**

6. Wildlife is conserved—

(a) In-situ (b) Ex-situ
(c) Both (a) and (b) (d) Selective hunting of predators

7. New approach to conservation is establishment of—

(a) Reserve forests (b) Sanctuaries
(c) Biosphere reserves (d) National parks

8. Biosphere reserves are being threatened with—

(a) Population growth (b) Rains
(c) Pollution (d) All of these

9. First National Park developed in India is—

(a) Gir (b) Kaziranga **(c) Jim Corbett** (d) None of these

10. Which one of the following is not included under in-situ conservation?

(a) National park (b) Sanctuary
(c) Botanical garden (d) Biosphere reserve

11. Manas sanctuary is located at—

(a) Rajasthan **(b) Assam** (c) Bihar (d) Gujarat

12. Endangered plant species are conserved through—

(a) Herbarium **(b) Gene library**
(c) Gene bank (d) Reducing Pollution

13. An ex-situ method of conservation of endangered species is—

(a) Cryo-preservation (b) Wildlife sanctuary
(c) Biosphere reserves (d) National park

14. Gene library consists of—

(a) All fragments of a genome **(b) Frozen germplasm**
(c) Tissue culture (d) Living bacteria having clones of genes

15. Organisation responsible for maintaining Red Data Book/Red List is—
(a) **IUCN** (b) WWF
(c) CITES (d) IBWL

16. Which one is endangered species of India ?
(a) Horse (b) Fox (c) **Indian wild ass** (d) Elephant

17. Red list contains information about—
(a) Red coloured flowers (b) Red coloured fishes
(c) Red coloured insects (d) **Endangered plants and animals**

18. WWF-N has logo for conservation of—
(a) Tiger (b) **Giant panda** (c) Red panda (d) Polar bear

19. World Summit on sustainable development (2002) was held in—
(a) Argentina (b) **South Africa** (c) Brazil (d) Sweden

20. Chipko movement was launched for the protection of—
(a) Wetlands (b) Grasslands (c) **Forest** (d) Livestock

Chapter 16. Environmental Issues

1. Effect of pollution is observed first on—
(a) **micro-organism** (b) food crop
(c) green vegetable (d) herbivores

2. Which of the following is secondary pollutant?
(a) NO (b) NO_2 (c) SO_2 (d) **PAN**

3. The blue baby syndrome results from—
(a) Excess of TDS (Total dissolved solid) (b) Excess of chlorides
(c) **methaemoglobin** (d) excess of dissolved oxygen

4. In 1984, the Bhopal gas tragedy took place because methyl isocyanate—
(a) Reacted with ammonia (b) Reacted with CO_2
(c) **Reacted with water** (d) Reacted with DDT

5. Which is always present in photochemical smog ?
(a) Ozone (b) CO_2 (c) **SO_2** (d) CH_4

6. Photochemical smog pollution does not contain—
(a) PAN (peroxy acyl nitrate) (b) Ozone
(c) Nitrogen dioxide (d) **Carbon dioxide**

7. Which of the following is biogas?
(a) CO_2 (b) N_2O (c) **CH_4** (d) N_2

8. In gobar gas, the maximum amount is that of—
(a) butane (b) **methane**
(c) propane (d) carbon dioxide

9. The intensity level of whispering noise is—
(a) 10–15 dB (b) **20–30 dB** (c) 45–50 dB (d) 50–55 dB

10. Common indicator organism of water pollution is—
(a) *Eichhornia crassipes* (b) ***Escherchia coli***
(c) *Entamoeba histolytica* (d) *Lemna paucicostata*

11. Which of the following are the indicators of pollution?
(a) **Lichen** (b) Fungi (c) Algae (d) None of these

12. Biochemical oxygen demand (BOD) in a river water—
(a) has no relationship with concentration of oxygen in the water
(b) **gives a measures of Salmonella in the water**
(c) increase when sewage gets mixed with river water
(d) remains unchanged when algal bloom occur

Multiple Choice Questions

13. When domestic sewage mixes with water.....
(a) Small animal likes rats will die after drinking river water
(b) The increased microbial activity releases micronutrients such as iron
(c) The increased microbial activity uses up dissolved oxygen
(d) The river water is still suitable for drinking as impurities are only about 1.1%

14. In which one of the following the BOD (Biochemical Oxygen Demand) of sewage (s), distillery effluent (DE), paper mill effluent (PE) and sugar mill effluent (SE) have been arranged in ascending order ?
(a) SE < PE < S < DE
(b) PE < S < SE < DE
(c) S < DE < PE < SE
(d) SE < S < PE < DE

15. The term 'Bio-magnification' refers to the—
(a) Growth of organism due to food consumption
(b) Increase in population size
(c) Blowing up of Environmental Issue by man
(d) Increase in conc. of non-degradable pollutants as they pass through food chain.

16. Lead Concentration blood is considered alarming if it is—
(a) 30 μg/100 ml
(b) 4–6 μg / 100 ml
(c) 10 μg/100ml
(d) 20 μg/100ml

17. Which of the following metal causes harmful effects?
(a) Lead
(b) Cobalt
(c) Uranium
(d) All of these

18. Mercury pollution causes—
(a) black foot disease
(b) itai–itai disease
(c) blue baby syndrome
(d) minamata disease

19. Which element is the cause of itai-itai disease?
(a) Hg
(b) Pb
(c) Cd
(d) As

20. The World Environment Day is celebrated on—
(a) 6th of June
(b) 5th of June
(c) 6th of August
(d) 5th of May

21. Checking of re-radiating heat by atmospheric dust O_3, CO_2, and water vapour is—
(a) green house effect
(b) solar effect
(c) ozone layer effect
(d) radioactive effect

22. Which of the following is not a green house gas?
(a) CO_2
(b) CFC
(c) O_2
(d) SO_2

23. Freon gas causing stratospheric ozone depletion is mainly released from—
(a) refrigerator
(b) automobile
(c) thermal power plant
(d) steel industry

24. Ultravoilet radiation from sunlight causes a reaction that produces—
(a) Fluorides
(b) Carbon monoxide
(c) Sulphur dioxide
(d) Ozone

25. Ozone hole result in—
(a) UV radiation reaching the earth
(b) Cataract
(c) Increase in skin cancer
(d) All of the above

26. Greenhouse effect is due to—
(a) CO_2
(b) CO
(c) NO
(d) PO_4

27. Montreal Protocol aims at :*
(a) Reduction of ozone depleting substances
(b) Biodiversity conservation
(c) Control of water pollution
(d) Control of CO_2 emission.

* Frequently asked Previous Years Board Exam Question

Technical Terms

Set 3

Chapter 1. Reproduction in Organisms

1. Give one word answer for the following :
   (i) The technical term used for a bisexual condition.
   (ii) The technical term used to denote unisexual condition.
   (iii) The technique used to improve the varieties of many plants or horticultural importance.
   (iv) One plant which propagates by the means of its leaf.
   (v) The phenomenon of splitting of a parent cell into numerous daughter cell each of which grows into an adult.
   (vi) One organism where chain budding is found.
   (vii) The organism where cell division is itself a mode of reproduction.
   (viii) The process of transfer of pollen grains from anther of one flower to the stigma of another flower.
   (ix) The group of organism that produce non-motile gametes.
   (x) The phenomenon where the female gamete directly develops into a new organism.
   (xi) The period from birth to natural death of an organism.
   (xii) The phenomenon of fusion of male and female gamete.
   (xiii) The phase that all organism have to pass through before they can reproduce sexually.
   (xiv) The outermost protective covering of fruit.
   (xv) The most critical event in sexual reproduction.
   (xvi) Terminal stage of ageing in the life cycle of plants.

Ans. (i) Homothallic and monoecious
(ii) Heterothallic and dioecious
(iii) Grafting
(iv) *Bryophyllum*
(v) Multiple fission
(vi) Yeast
(vii) Protists/Amoeba
(viii) Pollination
(ix) Angiosperms
(x) Parthenogenesis
(xi) Life span
(xii) Fertilisation
(xiii) Juvenile/growth phase
(xiv) Pericarp
(xv) Fertilisation
(xvi) Senescence

2. Name the relationship between the first two words and suggest a suitable word.
   (i) Doob grass : Runner : : potato : ....
   (ii) Endogamy : Self fertilisation :: Exogamy : ....

Ans. (i) Tuber
(ii) Cross-Fertilisation

Chapter 2. Sexual Reproduction in Flowering Plants

1. Give One word answer to the following :
   (i) The type of reproduction which is uniparental.
   (ii) Units of vegetative propagation.
   (iii) The method of artificial vegetative propagation where stock from one plant is bound firmly to the scion from another plant.

(iv) The arrangement of more than one flower on a floral axis.
(v) The type of floral arrangement where the main axis is elongated and does not terminate into a flower.
(vi) Male reproductive whorl of a flower.
(vii) Another name for pollen grains.
(viii) An anther with two lobes.
(ix) Pollination type aided by bats.
(x) Pollination type facilitated by snails.
(xi) Fertilisation in seed plants which is aided by pollen tube carrying the male gametes.
(xii) The plant part which after fertilisation develops into a fruit.
(xiii) Gynoecium with free carpels.
(xiv) The process of formation of haploid megaspores from diploid megaspores mother cell.
(xv) Gynoecium with fused carpels.
(xvi) The arrangement of ovules in the chambers of the ovary.
(xvii) The condition where flowers have different heights of stamens and styles to prevent self-pollination.
(xviii) The tissue on which ovules develops.
(xix) Flowers tend to be unisexual, so that self pollination is not possible.
(xx) The phenomenon of development of more than one embryo in a seed.
(xxi) Seeds which have presence of endosperm.
(xxii) The three cells at the chalazal end of the ovule.
(xxiii) Conical protective sheath around plumule in a monocot seed.
(xxiv) The layer that provides nourishment to the growing sporogenous cells.
(xxv) Single large cotyledon of monocot seeds.

Ans. (i) Asexual reproduction
(ii) Vegetative propagules
(iii) Grafting
(iv) Inflorescence
(v) Racemose inflorescence
(vi) Androecium
(vii) Male gametophyte
(viii) Dithecous
(ix) Chiropterophily
(x) Malacophily
(xi) Siphonogamy
(xii) Ovary
(xiii) Apocarpous
(xiv) Megasporogenesis
(xv) Syncarpous
(xvi) Placentation
(xvii) Heterostyly
(xviii) Placenta
(xix) Unisexuality/Dicliny
(xx) Polyembryony
(xxi) Endospermous or Albuminous seeds
(xxii) Antipodal cells
(xxiii) Coleoptile
(xxiv) Tapetum
(xxv) Scutellum

2. Name the following :
(i) Vegetative propagation by roots.
(ii) Monoecious plant.
(iii) Female reproductive part of the flower.
(iv) The plant which has vegetative propagation by leaves.
(v) An inhibitory substance which causes seed dormancy.
(vi) Bird pollinated flower.
(vii) The tissue on which ovules develop.
(viii) The plant part which develops into a fruit.
(ix) A non-endospermic seed.

Ans. (i) Sweet potato
(ii) Cucurbits

(iii) Gynocecium
(iv) Bryophyllum
(v) Abscisic acid (ABA)
(vi) Red silk cotton
(vii) Placenta
(viii) Ovary
(ix) Pea

3. Name the following :

(i) Two plants where the vegetative propagation occurs by leaves.
(ii) Layer of pollen grain is comparable to cellulosic wall of any cells.
(iii) Region of pollen grain through which the pollen tube emerges.
(iv) A plant which flowers every twelve years.

Ans. (i) (a) *Bryophyllum* (b) *Kalanchoe*
(ii) Intine (an inner thin layer)
(iii) Germ Pore
(iv) Kurunji/Neelakurunji (*Strobilanthes kunthianus*).

Chapter 3. Human Reproduction

1. Give One word answer to the following :

(i) A milky viscous mixture of sperms and secretions of the accessory sex glands.
(ii) Seminal vesicles, Cowper's glands, prostate gland, urethra and penis of male reproductive system.
(iii) Cells synthesizing and secreting male hormones.
(iv) Cells providing attachment and nutrition to developing sperms.
(v) Development and differentiation of a spermatozoan from a spermatid.
(vi) Detachment of fully mature spermatozoa from Sertoli cells.
(vii) Fusion of male and female gametes.
(viii) Site of fertilisation.
(ix) The structure which provides vascular connection between foetus and uterus.
(x) Process of formation, development and maturation of haploid female gamete.
(xi) The follicle bearing egg at secondary oocyte stage.
(xii) Pregnancy hormone.
(xiii) Innermost vascular lining of uterus.
(xiv) Beginning of menstruation in human females.
(xv) Permanent cessation of menstruation.
(xvi) Activation of sperms for fertilizing the egg.
(xvii) The hormone that induces ovulation.
(xviii) Hollow, muscular, inverted pear-shaped organ also called womb.

Ans. (i) Semen
(ii) Accessory sex organs
(iii) Interstitial cells/Leydig cells
(iv) Sertoli cells/Nurse cells
(v) Spermiogenesis
(vi) Spermiation
(vii) Fertilisation
(viii) Ampulla-isthmus junction
(ix) Umbilical cord
(x) Oogenesis
(xi) Graafian follicle
(xii) Progesterone
(xiii) Endometrium
(xiv) Menarche
(xv) Menopause
(xvi) Capacitation
(xvii) Luteinizing Hormone (LH)
(xviii) Uterus

2. Write the technical term for the following :

(i) Funnel lying close to the ovary
(ii) The period of endometrial repair and regeneration
(iii) Copulatory organ in human female

Ans. (i) Infundibulum
(ii) Proliferative phase
(iii) Vagina

3. Give the scientific term for the following :
(i) Onset of menstrual cycle in female
(ii) The structures that pick up ova from body cavity

Ans. (i) Menarche
(ii) Fimbrae.

4. Give scientific term for each of the following :
(i) Process of transformation of spermatids to spermatozoa.
(ii) Transparent non-cellular layer of ovum.
(iii) Activation/preparation of human sperm to fertilise the ovum.
(iv) The enzyme that separates the cells of corona radiatia layer over the egg.
(v) Type of cleavage in fertilised egg of humans.
(vi) Process of synthesis of yolk in the oocyte.
(vii) Formation of various organs and organ systems from germinal layers.
(viii) Period between fertilisation and birth of young one.
(ix) Term for animals which give birth to young ones.
(x) Stage at which mammalian embryo is implanted in the uterus.
(xi) The embryonic stage represented or characterised by blastocoel cavity.
(xii) Embryonic stage with three germ layers.

Ans. (i) Spermiogenesis
(ii) Zona Pellucida
(iii) Capacitation
(iv) Hyaluronidase
(v) Holoblastic cleavage
(vi) Vitellogenesis
(vii) Organogenesis
(viii) Gestation period
(ix) Viviparous
(x) Blastocyst
(xi) Blastula
(xii) Gastrula

Chapter 4. Reproductive Health

1. Give one word answer of the following :
(i) The tremendous increase in size and growth rate of population.
(ii) Inability to produce children in spite of unprotected sexual cohabitation.
(iii) International or voluntary termination of pregnancy before full term.
(iv) A foetal sex-determination test based on the chromosome pattern in the cells found in Amniotic fluid surrounding the developing embryo.
(v) The surgical method of contraceptions in males.
(vi) The surgical method of contraceptions in females.
(vii) The procedure to form an embryo in the laboratory by directly injecting the sperm into an ovum.
(viii) Surgical method to block gamete transport and thereby present conception.
(ix) The transfer of the embryo with more than 8 blastomers into the uterus.
(x) The transfer of the zygote or embryo with upto 8 blastomers into the fallopian tube.
(xi) The virus which is the causative agent of AIDS.
(xii) Sexually transmitted disease caused by bacteria.
(xiii) The diseases that are transmitted through sexual intercourse.

Ans. (i) Population explosion
(ii) Infertility
(iii) MTP (Medical Termination of pregancy)
(iv) Amniocentesis

Technical Terms

(v) Vasectomy
(vi) Tubectomy
(vii) ICSI-Intracytoplasmic Sperm Injection
(viii) Sterilisation
(ix) Intra Uterine Transfer (IUT)
(x) Zygote Intra Fallopian Transfer Technique (ZIFT)
(xi) Human Immunodeficiency Virus (HIV)
(xii) Syphilis
(xiii) STD's (Sexually Transmitted Diseases)

Chapter 5. Principles of Inheritance and Variations

1. Give One word answer to the following :
(i) Passing traits from parents to offspring.
(ii) A trait that disappeared in the 1st generation and reappeared in the 2nd generation in Mendel's experiments.
(iii) Traits from two parents are placed together to determine the traits of their offspring.
(iv) Different forms of physical characteristics.
(v) A trait that appeared in the 1st and 2nd generation in Mendel's experiments.
(vi) The scientific study of heredity.

Ans. (i) Heredity
(ii) Recessive traits
(iii) Genetic inheritence
(iv) Phenotypic Trait
(v) Dominant traits
(vi) Genetics

2. Write the term that best fits the definition or description below :
(i) Outward appearance
(ii) A tool used to show gene combination and predict the probability of a trait for a set of parents.
(iii) Gene combination inside an organism.
(iv) The mathematical chance that something will happen.
(v) Different form of a gene.

Ans. (i) Phenotype
(ii) Punnett square
(iii) Genotype
(iv) Probability
(v) Allele.

3. Give One word answer to the following :
(i) The phenomenon of inheritance of closely located genes together to the next generation retaining parental combination.
(ii) The genes located in the same chromosome and getting inherited together.
(iii) The type of linkage exhibiting only parental combinations in offspring.
(iv) The physical process of exchange of corresponding parts between the chromatids of a pair of homoglous chromosomes.
(v) The point of contact between the non-sister chromatids when the paired chromosomes starts separating.
(vi) One map unit distance on a chromosome map.
(vii) Chromosomes that carry genes for somatic characters.
(viii) Organisms having homorphic sex chromosomes.
(ix) Organism having heteromorphic sex chromosomes.
(x) The characters present in both sexes but expressed only in one sex.

Ans. (i) Linkage
(ii) Linked genes

(iii) Complete linkage
(iv) Crossing over
(v) Chiasmata
(vi) Centimorgan or map unit (m.u.)
(vii) Autosomes
(viii) Homogametic
(ix) Heterogametic
(x) Sex linked characters

Chapter 6. Molecular Basis of Inheritance

1. Give One word answer to the following :

(i) Polynucleotide chain of deoxyribonucleotides.
(ii) Single stranded nucleic acid found in all living cells.
(iii) Mode of DNA replication in which one daughter strand retains the nucleotides of parent strand.
(iv) Region located downstream, towards 3′end of the non-template or coding strand.
(v) DNA $\xrightarrow{\text{Transcription}}$ RNA $\xrightarrow{\text{Translation}}$ Protein
(vi) Process of copying the genetic message coded in DNA into mRNA molecule.
(vii) Process of removal of unwanted ribonucleotides or introns from mRNA primary transcript.
(viii) Genes that transcribe mRNA coding for more than one type of polypeptide chains.
(ix) Process used to check the accuracy of DNA replication as it occurs and to replace the mispaired bases.
(x) Addition of poly A residues to the 3′ end of newly formed hnRNA.
(xi) Process involving change in language from nucleotide order in a mRNA molecule to amino acid sequence in a polypeptide.
(xii) Triplet sequence of nitrogenous bases in mRNA.
(xiii) UAA, UAG, UGA

Ans. (i) DNA
(ii) RNA
(iii) Semi-conservative replication
(iv) Terminator region
(v) Central Dogma
(vi) Transcription
(vii) Splicing
(viii) Polycistronic genes
(ix) Proofreading
(x) Tailing/polyadenylate
(xi) Translation
(xii) Codons
(xiii) Stop codons

2. Give One word answer to the following :

(i) Single circular extrachromosomal DNA molecule.
(ii) Viral genome incorporated in the DNA molecule of a bacterium.
(iii) Genes that code for polypeptides.
(iv) Genes that activate or inactivate other genes.
(v) Switching on of specific gene to produce RNA for the synthesis of required enzyme.
(vi) Substances that may suppress the activity of specific genes to synthesize enzymes.
(vii) Genes that controls transcription of mRNA from structural genes.
(viii) Site of an operon where RNA polymerase binds.
(ix) It is an operon which normally remains switched off.
(x) The operon that normally remains switched on.

Ans. (i) Plasmid
(ii) Prophage
(iii) Structural genes
(iv) Regulatory genes
(v) Induction
(vi) Repressors
(vii) Operator gene
(viii) Promoter gene
(ix) Inducible operon
(x) Repressible operon.

Chapter 7. Evolution

1. Substitute the following statements with one or few words -
   (i) All the living animals of a specific area.
   (ii) Ontogeny repeats phylogeny.
   (iii) Reduced organ without any function in the organism.
   (iv) Remains of ancient plants and animals.
   (v) Java ape-man was discovered by.
   (vi) Man has evolved from.
   (vii) Zoological name of Java man is.
   (viii) Common ancestors of ape and man.
   (ix) The first man who lived in caves was.
   (x) The oldest fossil was recovered from.
   (xi) The fossil recovered from Shivalik hills of India is.
   (xii) Ancestor of man who first walked erect was.
   (xiii) First social life is seen in.
   (xiv) The diploid chromosome number in Homo sapiens is

Ans. (i) Fauna
   (ii) Biogenetic law or Recapitulation Theory
   (iii) Vestigial organ
   (iv) Fossils
   (v) Dubois
   (vi) Ape like ancestors
   (vii) *Homo erectus* or *Pithecanthrops erectus*
   (viii) Dryopithecus or Ramapithecus
   (ix) Neanderthal man
   (x) Africa
   (xi) Ramapithecus
   (xii) Australopithecus
   (xiii) Neanderthal man
   (xiv) 46.

Chapter 8. Human Health and Diseases

1. Give One word answer to the following :
   (i) Typhoid causing organism
   (ii) Chemicals which control allergic reactions.
   (iii) Enzyme which destroys cell wall of many bacteria.
   (iv) Leukocytes which are chief cells of immune system.
   (v) Drugs used to suppress the rejection of transplanted organs.
   (vi) Cells that stimulate B-cells to form more antibodies.
   (vii) Proteins released by the cells making them less susceptible to viral infections.
   (viii) Diagnostic Test for AIDS

Ans. (i) *Salmonella typhi*
   (ii) Antihistamines
   (iii) Lysozyme
   (iv) Lymphocytes
   (v) Immunosuppressants
   (vi) Helper T-cells
   (vii) Interferons
   (viii) ELISA

Chapter 9. Strategies for the Enhancement in Food Production

1. Give one word answer of the following :
   (i) The process of management of animals for milk and their product for human consumption.
   (ii) The most common species of honey bee in India.
   (iii) The strategy used to increase homozygosity in cattle for desired traits.
   (iv) The chemical which induce polyploidy.
   (v) The phenomenon of the continuous inbreeding especially close inbreeding may led to the reduced fertility and productivity.

(vi) An invitro technique of regeneration of a whole plant from any part of a plant by growing it on culture medium under aseptic conditions.
(vii) The capacity of a cell explants to grow into a whole plant.
(viii) The part of plant taken for tissue culture.
(ix) The method of growing or producing thousands of plant through tissue culture.
(x) The plants produce from tissue culture are genetically identical to the original plant from which they are grown.
(xi) The merging of protoplasts obtained from two different cells forms a hybrid Protoplast.
(xii) Situation in which a crop needs more of a given nutrient yet show no deficiency symptoms.
(xiii) The scientific name of micro organism which produce high quality of protein.
(xiv) The method for developing crops with higher levels of vitamins, mineral, proteins, and heal their fat to improve public health.

Ans. (i) Dairying
(ii) *Apis indica*
(iii) Inbreeding
(iv) Colchicine
(v) Inbreeding depression
(vi) Tissue culture
(vii) Totipotency
(viii) Explant
(ix) Micropropagation
(x) Somaclones
(xi) Protoplast fusion
(xii) Hidden hunger
(xiii) *Methylophilus methylotrophus*
(xiv) Biofortification

Chapter 10. Microbes in Human Welfare

1. Give one word answer of the following :
(i) Traditional drink of southern India which is prepared by the yeast fermentation of coconut water.
(ii) The organism commercially used for the production of single cell protein.
(iii) The first organic acid produced by microbial fermentation.
(iv) The process of controlling plant diseases and pests using biological method.
(v) The Municipal waste containing large quantities of human excreta.
(vi) The cyanobacterium that can fix atmospheric nitrogen.
(vii) Free living bacteria that can fix nitrogen in the soil.
(viii) Viruses used in biological control of insects, pests and other arthopods.
(ix) The category of microbes occurring naturally in sewage and making it less polluted during the treatment.
(x) A super bug of Pseudomonas with multiple plasmids.

Ans. (i) Toddy
(ii) Spirulina
(iii) Lactic acid
(iv) Biocontrol
(v) Sewage
(vi) *Oscillatoria*
(vii) *Azospirillum*
(viii) Baculoviruses
(ix) Aerobic microbe
(x) Chakrabarty bug

2. Give the biological names of the following :*
(i) The mould from which penicillin is obtained.
(ii) Baker's yeast.
(iii) The microbe used to control insect larvae growing on cotton.
(iv) The microbe used to produce Swiss cheese.
(v) The fungus that is being developed as a bio-control agent.
(vi) A symbiotic nitrogen fixing bacterium found in root nodules.

Ans. (i) *Penicillium notatum*
(ii) *Saccharomyces cerevisiae*
(iii) *Bacillus thuringiensis*
(iv) *Propionibacterium sharmanii*
(v) *Trichoderma polysporum*
(vi) *Rhizobium leguminosarum*

* **Frequently asked previous years Board Exam Question.**

Chapter 11. Biotechnology: Principles and Processes

1. Give One word answer to the following :

(i) Another name for recombinant DNA technology.
(ii) Process of separation of DNA fragments according to their size.
(iii) Technology used for amplifying DNA.
(iv) Enzyme aiding in joining of DNA fragments.
(v) Another name for restriction endonucleases.
(vi) Carriers vehicles of desired DNA fragments which can undergo independent replication ?
(vii) A circular extrachromosomal DNA in bacteria.
(viii) The polymerase used in PCR.
(ix) Compound used for staining DNA in gel.

Ans. (i) Genetic engineering
(ii) Gel electrophoresis
(iii) Polymerase Chain Reaction
(iv) DNA ligase
(v) Molecular scissors
(vi) Cloning vectors
(vii) Plasmid
(viii) Taq Polymerase
(ix) Ethidium bromide.

Chapter 12. Biotechnology and its Application

1. Give one word answer of the following :

(i) A molecular diagnostic technique to detect the presence of a pathogen in its early stage.
(ii) A method of cellular defence which works in all eukaryotic organism.
(iii) The first transgenic cow.
(iv) The non-compensated use of bioresources by organisations.
(v) The resultant plant after successful insertion of the desired gene.
(vi) A multinational company outside India tried to sell new varieties of turmeric without proper rights. What is such an act referred to ?
(vii) Animal whose DNA is manipulated to possess and express an extra (Foreign) gene.
(viii) The process involved in the production of nematode-resistant tobacco plant, using genetic engineering.
(ix) Genetically engineered variety of tomato.
(x) The soil bacterium that produces a protein/chemical that is toxic to insect pests.
(xi) The deficiency for which first clinical gene therapy was given.
(xii) Replacement of defective gene by a normal healthy and functional gene.
(xiii) The organism having genes from other organism through the technique of genetic engineering.
(xiv) Crops which posses and express on or more genes obtained from others through genetic engineering.
(xv) Patenting and exploitation of bioresources of other nations without authorisation or agreement.

Ans. (i) ELISA
(ii) RNA interference
(iii) Rosie
(iv) Biopiracy
(v) BT corn
(vi) Biopiracy
(vii) Transgenic animals
(viii) RNA Interference
(ix) Flavr-Savr
(x) *Bacillus thuringiensis*
(xi) Adenosine Deaminase Deficiency
(xii) Gene Therapy
(xiii) Transgenic
(xiv) Genetically modified crops
(xv) Biopiracy

Technical Terms

2. (i) Name the enzyme which cuts the DNA molecule into fragments with sticky ends.
   (ii) What are VNTRs ?
   (iii) Name two benefits that can be derived by genetically modified potato in the future.
   (iv) Name two drugs manufactured via recombinant micro-organisms.
   (v) What technique is used to identify a specific mRNA from the total RNA isolated from a tissue ?
   (vi) Which human gene has been introduced into the farm animals in order to modify milk ?
   (vii) What type of cells were used in cloning 'Dolly sheep' ?
   (ix) Enzyme EcoRI is derived from ?
   (x) Name the enzyme derived from *Bacillus amyloliquefaciens*.
   (xi) Which animal was the first animal clone ?
   (xii) Who carried out this first animal cloning ?
   (xiii) What is the full form of VNTR ?
   (xiv) Who produced the first transgenic sheep ?
   (xv) Which bacterium produces proteins, that are fatal to insect larvae ?

Ans. (i) Restriction enzymes
   (ii) Short repetitive nucleotide sequences which are specific for a person
   (iii) Increased production and increased starch contents
   (iv) Insulin, Growth hormone
   (v) PCR
   (vi) Human anti-hemophilic factor IX gene
   (vii) Mammary cells into enucleated egg cells
   (ix) *Escherichia coli*
   (x) BamHI
   (xi) Dolly, the Sheep
   (xii) Ian Wilmut
   (xiii) Variable Number Tandem Repeats
   (xiv) J.P. Simons and Ian Wilmut
   (xv) *Bacillus thuringiensis*

Chapter 14. Ecosystem

1. Give one word answer of the following :
   (i) The amount of biomass or organic matter produce per unit area over a certain time period by plant during photosynthesis.
   (ii) The process in which the humus is degraded by certain microbes and thus inorganic nutrients are released.
   (iii) The amount of energy or biomass remaining in a producer after meeting the cost of its respiration.
   (iv) The role of formation of new organic matter by consumers.
   (v) The process of formation of humus from detritus or organic matter.
   (vi) Detritus food chain begins with.
   (vii) The dominant producer in a deep aquatic ecosystem.
   (viii) Dead organic matter or remains of plant such as leaves, bark, flower and dead remains of animal including faecal matter.
   (ix) Each energy step or levels in a food chain.
   (x) The transfer of energy (food) from the producer through a series of consumers.
   (xi) The herbivores which depends on plant for their food.

(xii) The species that invade a bare area.

(xiii) The community which is established at the terminal stage of succession and is in equilibrium with the environment.

(xiv) The total amount of nutrient like carbon, phosphorus, calcium etc. present in soil at any time.

Ans. (i) Primary production

(ii) Mineralisation

(iii) Net primary productivity

(iv) Secondary productivity

(v) Humification

(vi) Dead and organic matter

(vii) Phytoplanktons

(viii) Detritus

(ix) Trophic level

(x) Food chain

(xi) Primary carnivores

(xii) Pioneer species

(xiii) Climax community

(xiv) Standing state

Chapter 15. Biodiversity and its Conservation

1. Give one word answer for the following :

(i) Exploring molecular genetics, species level diversity for product of economic importance.

(ii) The Amazon rain forest being so huge produces 20% of the oxygen in our planet's atmosphere.

(iii) Sixth episode of extinction is in progress at hundred to thousand times faster rate.

Ans. (i) Bioprospecting

(ii) Lungs of the planet

(iii) Holocene or Anthropocene extinction

2. Give one word answer for the following :

(i) Variety of living organisms inhabiting an ecosystem.

(ii) Management of the earth's resources in a way which restores and maintains the balance between human requirements and other species.

(iii) Species that are introduced into the ecosystem from outside.

(iv) Maintenance of genetic diversity either in botanical gardens, zoological parks or in gene banks.

(v) Institution where valuable genetic material is stored either in the form of seeds or vegetative material.

(vi) Total set of genes and alleles present in a population.

(vii) Maintenance of genetic resources in wild places and nature reserves.

(viii) Large areas of scenic and natural beauty maintained for scientific, educational and recreational use.

(ix) Species found in localized and specific regions.

(x) Forest patches around places of worship being protected by tribal communities.

(xi) Preservation of organs and tissues at a very low temperature.

(xii) Priority areas of conservation, extremely rich in species but under threat of extinction.

Ans. (i) Biodiversity

(ii) Conservation

(iii) Exotic species

(iv) Ex-situ conservation

(v) Gene/Germplasm bank

(vi) Gene pool

(vii) In-situ conservation

(viii) National parks

(ix) Endemic species

(x) Sacred groves

(xi) Cryopreservation

(xii) Hot spots

Chapter 16. Environmental Issues

1. Give one word answer of the following :

(i) The city in which the entire public road transport runs on CNG.

(ii) The gases released during Bhopal gas tragedy in December 1984.

(iii) An industry which can cause air pollution, thermal pollution and eutrophication.

(iv) Fine powder of recycled modified plastic.

(v) The major water pollutant from farm run-off and bathroom water.

(vi) The excessive growth of algae (Free floating) that causes colouration of water bodies.

(vii) The organism that is reffered to as the, Terror of Bengal.

(viii) The other named given to Jhum cultivation.

(ix) The process of restoring a forest that once existed but was removed at same point of time in the past.

(x) One gas is most abundant out of four commonest green house gases.

(xi) The gas holds up ultraviolet rays.

(xii) The unit used for measuring ozone thickness.

(xiii) The type of UV radiation can be lethal to organism.

(xiv) The year was the Air prevention and control of pollution Act amended to include noise as air pollution.

Ans. (i) Delhi

(ii) Methyl isocyanate (MIC)

(iii) Fertilizer factory

(iv) Polyblend

(v) Phosphates and nitrates

(vi) Algal bloom

(vii) Water hyacinth (*Eicchornia crassipes*)

(viii) Slash and burn agriculture

(ix) Reforestation

(x) CO_2

(xi) Ozone

(xii) Dobson unit

(xiii) UV-B radiations

(xiv) 1981

Expand the Term

Set 4

Chapter 3. Human Reproduction

LH	Luteinising Hormone
GnRH	Gonadotropin Releasing Hormone
FSH	Follicle Stimulating Hormone
hPL	human Placental Lactogen
hCG	human Chorionic Gonadotropin

Chapter 4. Reproductive Health

STDs	Sexually Transmitted Diseases
RCH	Reproductive and Child Healthcare
MMR	Maternal Mortality Rate
IMR	Infant Mortality Rate
IUD's	Intra Uterine Devices
CDRI	Central Drug Research Institute
MTP	Medical Termination of Pregnancy
VD	Venereal Disease
PID	Pelvic Inflammatory Disease
HIV	Human Immunodeficiency Virus
ART	Assisted Reproductive Technology
IVT	In Vitro Fertilisation
ZIFT	Zygote Intra Fallopian Transfer
IUT	Intra Uterine Transfer
AIDS	Acquired Immuno Deficiency Syndrome
RTI	Reproductive Tract Infection
ICSI	Intra Cytoplasmic Sperm Injection
ET	Embryo Transfer
AI	Artificial Insemination
IUI*	Intra Uterine Insemination
IUCD	Intrauterine Contraceptive Device

Chapter 5. Principles of Inheritance and Variation

F_1 generation	First Filial Generation
F_2 generation	Second Filial Generation
n	Haploid
2n	Diploid

Chapter 6. Molecular Basis of Inheritance

RNA	Ribonucleic Acid
DNA	Deoxyribonucleic Acid
^{32}P	Radioactive Phosphorus
^{35}S	Radioactive Sulphur
mRNA	messenger Ribonucleic Acid
tRNA	transfer Ribonucleic Acid
rRNA	ribosomal Ribonucleic Acid
ESTs	Expressed Sequence Tags
SNPs	Single Nucleotide Polymorphisms
VNTRs	Variable Number Tandem Repeats
ELSI	Ethical, Legal and Social Issues
UTRs	Untranslated Regions
ATP	Adenosine Triphosphate
AMP	Adenosine Monophosphate
AA	Amino Acid

Chapter 8. Human Health And Diseases

AMI	Antibody Mediated Immunity
HIR	Humoral Immune Response
CMIR	Cell Mediated Immune Response
CMI	Cell Mediated Immunity

* **Frequently asked previous years Board Exam Question.**

MALT Mucosa Associated Lymphoid Tissue
ELISA Enzyme Linked Immuno Sorbent Assay
C-Onc Cellular Oncogenes
WHO World Health Organisation
NACO National AIDS Control Organisation
NGOs Non Governmental Organisations
LSD Lysergic Acid Diethylamide
DPT Diphtheria, Pertussis and Tetanus
AIDS Acquired Immunodeficiency Syndrome
HCLV III Human Cell Leukemia Virus III
HIV Human Immunodeficiency Virus
Ig Immunoglobulin
kT cells Killer T-cells
hT cells Helper T-cells
sT cells Suppressor T-cells
SCID Severe Combined Immuno Deficiency
ARC AIDS Related Complex
AMIS Antibody Mediated Immune System
CMIS Cell Mediated Immune System

Chapter 9. Strategies For Enhancement in Food Production

SCP Single Cell Protein
MOET Multiple Ovulation Embryo Transfer

Chapter 10. Microbes in Human Welfare

LAB Lactic Acid Bacteria
STPs Sewage Treatment Plants
BOD Biological Oxygen Demand
COD Chemical Oxygen Demand
IARI Indian Agricultural Research Institute
KVIC Khadi and Village Industries Commission
Bt *Bacillus thuringiensis*
IPM Integrated Pest Management

Chapter 11. Biotechnology : Principles and Processes

Ori Origin of Replication
UV Radiation Ultra Violet Radiation
EFB European Federation of Biotechnology
PCR Polymerase Chain Reaction
rDNA recombinant Deoxyribonucleic Acid
BAC Bacterial Artifical Chromosome
YAC Yeast Artifical Chromosome

Chapter 12. Biotechnology and its Applications

IPB Indian Patent Bill
SCT Stem Cell Transplantation
GEAC Genetic Engineering Approval Committee
GMOs Genetically Modified Organisms
SCID Severe Combined Immunodeficiency Disease
ADA* Adenosine Deaminase
HGH Human Growth Hormone
HGP Human Genome Project

Chapter 14. Ecosystem

PAR Photosynthetically Active Radiation
GPP Gross Primary Productivity
NPP Net Primary Productivity
CAM Crassulacean Acid Metabolism

* Frequently asked previous years Board Exam Question.

Chapter 15. Biodiversity and its Conservation

IUCN	International Union for Conservation of Nature and Natural Resources
MAB	Man and Biodiversity Programme
WWF	The World Wildlife Fund for Nature
IBWL	Indian Board for Wildlife
IBP	International Biological Programme
UNDP	United Nation Development Programme

Chapter 16. Environmental Issues

ESP	Electrostatic Precipitator	**JFM**	Joint Forest Management
CNG	Compressed Natural Gas	**TBT**	Tributyltin
DDT	Dichloro Diphenyl Trichloroethane	**BOD**	Biological Oxygen Demand
CFCs	Chloro Fluoro Carbons		

Definitions

Set 5

Chapter 1. Reproduction in Organisms

1. Define (i) Juvenile phase (ii) Reproductive phase (iii) Senescent phase.

Ans. (i) **Juvenile phase** — It is the pre-reproductive phase in which all organism require a certain growth and maturity in the life before reproducing sexually.

(ii) **Reproductive phase** — It is the phase in the life cycle where an organism posses all the capacity and potential to reproduce sexually. It is the end of juvenile phase or vegetative phase.

(iii) **Senescent phase**—It is the post-reproductive phase in the life cycle where an organism slowly looses the rate of metabolism, reproductive potential and show deterioration of the physiological activity of the body.

2. What is asexual mode of reproduction ?

Ans. The mode of reproduction in which formation and fusion of gamete does not takes place is called asexual mode of reproduction.

3. What is a clone ?*

Ans. Clone is a group of individuals of the same species that are morphologically and genetically identical.

4. Define gemmules.

Ans. Internal asexual reproductive unit, as bud in sponges used in asexual reproduction are called gemmules.

5. What is conidia ?

Ans. The asexual, non-motile spore produced externally/exogenously in some fungi is called conidia.

6. What is budding ?

Ans. It is a mode of asexual reproduction in which one or more outgrowth (bud) protrude out, which later gets detached from the parent body to grow into a new individual e.g. yeast.

7. What is isogamy ?

Ans. It is a type of sexual reproduction in which the fusion gametes are similar. These cannot be categorized into male and female and are produced in equal number. It occurs in primitive aquatic organisms.

8. What is zygote ?

Ans. Zygote is a single diploid cell formed by fusion of male and female gamete.

9. What are antherozoids ?

Ans. The male gamete or male reproductive unit is called antherozoids or spermatozoa.

10. What do you mean by staminate ?

Ans. The unisexual male flower bearing only stamen is called staminate.

11. What is meant by pistillate ?

Ans. The unisexual female flower bearing only pistil is called pistillate.

12. What is gametogenesis ?

Ans. The process of formation of gamete in male and female organs is called gametogenesis.

13. What is meant by meiocyte ?

Ans. In diploid organisms, gamete mother cell which undergoes meiosis to produce four genetically different haploid gamete is called meiocyte.

---

* **Frequently asked previous years Board Exam Question.**

14. What is syngamy ?

Ans. The fusion of male and female gamete to form a zygote is called syngamy.

15. Define external fertilisation.

Ans. The fusion of compatible gamete outside the body of an organism is called external fertilisation.

16. What is bisexual flower ?

Ans. A flower which posseses both the male and female reproductive structures *i.e.* both stamen and carpel is called a bisexual flower.

17. What is parthenogenesis ?

Ans. The development of a new individual from an unfertilised egg in called parthenogenesis.

18. What is complete parthenogensis ?

Ans. Certain insects have no sexual phase and no males. They depends exclusively on the parthenogenesis for the self-reproduction. This type of parthenogenesis is known as the complete parthenogenesis.

19. What is Arrhenotoky ?

Ans. If only males are produced parthenogenitically then the process is called arrhenotoky.

20. What is Thelotoky ?

Ans. It is the process of production of only female individuals parthenogenetically.

21. What is amphitoky ?

Ans. In this process parthenogenetic eggs may develop into any sex.

22. What is strobilation ?

Ans. It is a special type of asexual reproduction in which several transverse fissions occur simultaneously giving rise to a number of organisms which often do not separate e.g., *Aurelia*.

23. What do you understand by oestrous cycle ?

Ans. The reproductive cycle in non-primate females is called oestrous cycle e.g. cows, sheep.

24. What is menstrual cycle ?

Ans. The reproductive cycle in primate females is called menstrual cycles.

25. What are seasonal breeders ? Give Example.

Ans. Some animals exhibit reproductive cycle only during favourable season. They are called seasonal breeders e.g. frog, many birds.

26. Define continuous breeders.

Ans. Some animals are reproductively active throughout their reproductive phase. Such animals are called continuous breeders e.g. cattle and rabbit.

27. What is neoteny ?

Ans. When larva develops the gonads and reproductive parts without undergoing metamorphosis, the process is called neoteny.

28. Define the term hermaphrodite and name few hermaphrodite animals.

Ans. The organisms which has both the male and female sex organs in a single individual is referred as hermaphrodite animals. e.g., Earthworm, Leech, Tapeworn.

29. Define the following :

(i) Plasmotomy
(ii) Gemmation
(iii) Regeneration
(iv) Exogamy
(v) Paedogamy
(vi) Automixis
(vii) Parthenogenesis.

Ans. (i) Plasmotomy : It is a kind of asexual reproduction occurs in multinucleate protozoans like *Opalina* in which the cytoplasm divides but nuclei does not. Later each daughter cell regains the normal number of nuclei by nuclear division unfertilised.

(ii) **Gemmation :** Gemmation or budding occurs in certain lower metazoans like sponge (*Sycon*). During budding the body of concerned animal gives out a small outgrowth called bud, which ultimately develops into a new individual.

(iii) **Regeneration :** It is a process by which the organism develop or regenerate their lost or worn out parts. It is a means of asexual reproduction in certain protozoans, sponges, etc.

(iv) **Exogamy :** In exogamy, the male and female gametes produced by different parents unite to form a zygote.

(v) **Paedogamy :** It is a kind of syngamy in which sexual union of young individual produced immediately after the division of the adult parent cell by mitosis.

(vi) **Automixis (or autogamy) :** When the gamete nuclei of same cell unite together to form new individuals, this process is called automixis *e.g., Paramecium.*

(vii) **Parthenogenesis :** (Partheno — *virgin,* genesis—*reproduction*). It is a special type of sexual reproduction in which the egg of individual develops into young ones without fertilisation of eggs by sperms. In honey bee colony queen bees and worker bees develop from fertilised eggs and drones (sterile males) are produced from unfertilised ova. It may be natural or artificial.

30. What is gametes ?

Ans. Gamete is the sexual reproductive unit. They are usually haploid cells.

Chapter 2. Sexual Reproduction in Flowering Plants

1. What is hypanthodium ?

Ans. In this main axis forms a cup shaped receptacle with a small opening. Flowers are enclosed within the cup in cymose clusters e.g. Banyan.

2. What is mixed inflorescence ?

Ans. Inflorescence, in which the axis starts as a racemose inflorescence and shows branching in a cymose fashion is called mixed inflorescence e.g., Panicle of spikelets in oat and rice, Corymb of capituala in *Ageratum*, mixed spadix in banana.

3. Explain the following terms :

(i) Hypogynous
(ii) Epigynous
(iii) Perigynous
(iv) Monoecious
(v) Dioecious.

Ans. (i) Hypogynous : It is a condition where ovary is situated above all the other floral parts of the flower. In such condition, the ovary is said to be superior *e.g.,* mustard and tomato.

(ii) Epigynous : A condition in which the thalamus is expanded and grows over the ovary. It is fused with the ovary. All the other floral parts arise on the top of the ovary. Such an ovary is said to be inferior and the flower is termed epigynous. E.g., Cucumber, Apple.

(iii) Perigynous : In this condition, the thalamus is a cup-shaped, disc-shaped or flask-shaped structure around the gynoecium and other floral parts like sepals, petals and stamens are borne from the aim of the thalamus. e.g., Pea, Plum.

(iv) Monoecious : Plants that bear flowers of both sexes, e.g., male and female flower are called monoecious. e.g., Cucurbits, Maize.

(v) Dioecious : The plants which produce exclusively staminate or pistillate flowers are called dioecious *e.g.,* Date palm, mulberry and *Coccinia grandis.*

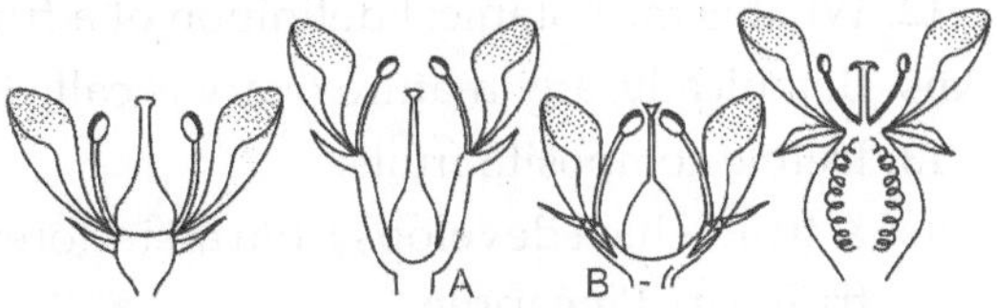

A B

Hypogyny Perigyny A and B Epigyny

4. Define the following :

(i) Actinomorphic
(ii) Zygomorphic Flower
(iii) Trimerous Flower
(iv) Tetramerous Flower
(v) Pentamerous Flower

Ans. **(i)** **Actinomorphic Flower :** A flower is termed as actinomorphic when its symmetry remains undisturbed if one cuts the flower into two halves through any vertical plane passing through the axis e.g., *Petunia.*

(ii) **Zygomorphic Flower :** A flower in which the symmetry remains intact if is cut into two equal halves only through one particular plane passing through the centre is termed as zygomorphic flower.

(iii) **Trimerous Flower :** A flower having three or multiple of three floral leaves in each whorl is called trimerous flower.

(iv) **Tetramerous Flower :** A flower having four or multiple of four floral leaves in each whorl is called tetramerous flower.

(v) **Pentamerous Flower :** A flower having five or multiple of five floral leaves in each whorl is called pentamerous flower.

5. Define the following :

(i) Cruciform corolla

(ii) Vexillary corolla or Papilionaceous corolla.

(iii) Tetradynamous stamens

(iv) Didynamous stamens

(v) Synandrous

Ans. **(i)** **Cruciform corolla :** Presence of four free unguiculate petals arranged in the form of a cross. Each petal consists of a lower narrow portion called claw and the upper broad portion limb eg., *Brassica.*

(ii) **Papilionaceous corolla :** It is irregular corolla consisting of five petals, the largest posterior standard or vexillum, two lateral, free petals called wings and two anterior fused petals called keel or carina *e.g.,* Pea.

(iii) **Tetradynamous :** There are six stamens in which four inner stamens are large and two outer are shorter in length *e.g., Brassica.*

(iv) **Didynamous :** There are four stamens out of which two stamens are larger and two are shorter in length *e.g., Ocimum.*

(v) **Synandrous :** When both filaments and anthers of stamens are fused completely. *e.g.,* Cucurbita.

6. What is sporopollenin ?

Ans. It is a chemically stable substance which forms the exine layer of the pollen grain.

7. What is Hilum ?

Ans. Hilum is a scar on the seed surface, formed due to the detachment of the seed from funiculus.

8. What is hypocotyl ?

Ans. The region of the embryonal axis between radicle and point of attachment to cotyledons is the hypocotyl.

9. What is epicotyl ?

Ans. The region of the embryonal axis between the plumule and point of attachment to the cotyledons is the epicotyl.

10. What is coleoptile ?

Ans. It is the covering sheath of plumule in monocot embryo.

11. What is coleorhiza ?

Ans. The root cap protects the radicle in monocot grain. It is in turn protected by a sheath called coleorhiza.

12. What is the botanical definition of a fruit ?

Ans. Botanically, any mature ovary is called a fruit.

13. Define composite fruits.

Ans. A fruit which develops from an inflorescence by the fusion of flower and their parts is called a composite fruit e.g., Pineapple.

14. What are monocarpic plants ?

Ans. The plants which produces fruits only once in their life are called monocarpic plants e.g., *Agave* (*i.e.,* century plant).

15. What are geocarpic fruits ?

Ans. The fruits which develop under the ground are called geocarpic fruits e.g., Ground nut.

16. What is perisperm ?

Ans. During the formation of seed, nucellus of ovule is used up and disappears. However in some cases it persists in the form of food storing thin layer which is called perisperm.

17. Define the term 'Proembryo' ?

Ans. Proembryo is the embroynic stage of development of embryo the in angiosperms represented by a filament of a few cells before the embryo formation.

18. What is placentation ?

Ans. Placentation is the arrangement of ovules in the chambers or locules of the ovary.

19. What is cleistogamy ?

Ans. Sometimes bisexual flowers remain closed and never open; such flowers are called cleistogamous. In such flowers, the anthers and stigma lie close to each other. When anthers dehisce, the pollen grains come in contact with stigma and pollination takes place. Such a type of pollination occurring in closed flowers is referred as cleistogamy, *e.g., Arachis hypogea* (Groundnut).

20. What is emasculation ?*

Ans. Emasculation is the removal of anthers from bisexual flower before maturing for cross fertilisation.

21. Define the following :

(i) Parthenocarpy
(ii) Aril
(iii) Chalaza
(iv) Pomology
(v) Agamospermy
(vi) Intersexual flower
(vii) Geigtonogamy
(viii) Parthenogenesis

Ans. (i) **Parthenocarpy :** The formation of fruits without fertilisation is called parthenocarpy. E.g. Banana, Grapes fruits produced are seedless.

(ii) **Aril :** It is the fleshy, juicy edible covering on the seed that arises as an outgrowth from the ovule stalk.

(iii) **Chalaza :** Chalaza is the part of the nucellus from where integument arise.

(iv) **Pomology :** Pomology is the branch of the horticulture which deals with the study of fruits and its cultivation.

(v) **Agamospermy :** It is the mode of apomixis in which seeds are formed but are asexual in nature as the embryo develops directly without gamete fusion.

(vi) **Intersexual flower :** A flower that contains both stamen and pistil is called intersexual flower.

(vii) **Geitonogamy :** It is a type of self pollination in which pollen of one flower get deposited on the stigma of other flower belonging to the same plant.

(viii) **Parthenogenesis :** It is the development of a new individual from a single gamete without fusion with another gamete.

Chapter 3. Human Reproduction

1. What is reproduction ?

Ans. It is a means of perpetuation of species. It is not essential to the life of any organism but helps in maintaining the continuity of the species. All living organisms divide by two methods, asexual and sexual.

2. What are secondary sex organs ?

Ans. The organs which perform important functions in reproduction but they do not secrete hormones or produce gametes are called secondary sex organs.

* **Frequently asked previous years Board Exam Question.**

3. What is gametogenesis ?

Ans. It is the process of formation of gametes in sexual reproduction.

4. What is nebenkern ?

Ans. It is a spiral ring formed from mitochondria around the distal centriole and proximal part of axoneme.

5. What is manchette ?

Ans. Manchette is a temporary thin sheath of cytoplasm around the head of the sperm, guiding the shape of developing head. It also covers head, neck, middle piece and tail and help in protein transport along the length of sperm.

6. What is azoospermia ?

Ans. Complete absence of sperms in semen is called azoospermia.

7. Define semen.

Ans. Semen is the fluid ejaculated at the time of coitus containing sperms and secretion of accessory glands.

8. What is endometrium ?

Ans. Endometrium is the innermost layer of uterus which is formed as a preparation to support pregnancy in case fertilisation occurs. It is highly vascularized and rich in glands.

9. What is atresia ?

Ans. Degenerative process of ovarian follicles (*i.e.* oogonia) which do not ovulate during the menstrual cycle is called atresia.

10. What is fourchette ?

Ans. Labia minora fuses posteriorly to form a membranous fold called fourchette.

11. What is corona radiata ?

Ans. These are radially elongated follicle cells outside the zona pellucida layer of the ovum.

12. What is corpus albicans ?

Ans. Degenerating corpus luteum is called corpus albicans.

13. What is gestation period ?

Ans. It is a period from fertilisation till birth of child.

14. What is umbilical cord ?

Ans. It is a vascular connection between foetus and its placenta connecting to mother's uterus.

15. What is embryonic development ?

Ans. It is the structural and physiological changes in the zygote leading to the formation of adult individual.

16. What is the Transitory three cell stage ? How is it formed ?

Ans. The morula consisting of three blastomeres is known as transitory three celled stage. Immediately after the formation of first two blastomeres, one of these usually divides a little sooner than the other resulting in transitory three celled stage.

17. What are foetal membranes ? Give their name.

Ans. Foetal membranes are extra-embryonic membranes that provide protection and nourishment to the developing foetus. These are : Amnion, Allantois, Chorion, Yolk sac.

18. Define menarche.

Ans. It refers to the time when menstruation first begins. During the menarche period, menstruation may be irregular and unpredictable. It is also known as female puberty.

19. What is parturition ?

Ans. Parturition is the process of giving birth to a baby in viviparous animals.

20. What is Menopause ? Explain briefly.

Ans. It is the climacteric period when ovulation and menstrual cycle stop in human female. The average period of menopause is between 45 to 55 years. The main cause of menopause is decline in the estrogen level.

21. Define the following :
   (i) Cleavage
   (ii) Insemination
   (iii) Capacitation
   (iv) Copulation
   (v) Morphogenesis
   (vi) Differentiation

Ans. (i) **Cleavage :** It is the repeated mitotic divisions in the zygote which transforms a single celled zygote (fertilised egg) into a hollow, spherical and multicellular blastocyst/ blastula.

(ii) **Insemination :** It is the deliberate introduction of sperm into a female animal or plant for the purpose of fertilizing the egg in female for sexual reproduction.

(iii) **Capacitation :** The changes in a mammalian sperm which prepares it to fertilise the ovum is termed as capacitation.

(iv) **Copulation :** The sexual union of two individuals, resulting in insemination or deposition of male gametes in close proximity of female gamete is termed copulation.

(v) **Morphogenesis :** The process in which embryo gradually assumes the specific shape and morphology. Thus the size and volume of embryo increases.

(vi) **Differentiation :** The process in which the cells composing the different tissues and organs become different from each other structurally and functionally. The differentiation can be either morphological, physiological or chemical.

22. Define castration ?

Ans. Removal of testis in male is called castration.

Chapter 4. Reproductive Health

1. What do you mean by reproductive health ?

Ans. Reproductive health means total well being in all aspects of reproduction *i.e.* physical emotional, behavioural, social and physiological.

2. What do you mean by contraception ?

Ans. The birth control method which prevent conception are known contraception.

3. What do you mean by infertility ?

Ans. It is the inability to produce children inspite of unprotected sexual cohabitation.

4. What is meant by artificial insemination ?

Ans. It is a technique by which semen collect from the husband or a healthy donor is artificially introduce either into the vagina or into the uterus of the female.

5. What is amniocentesis ?

Ans. Amniocentesis is a prenatal test in which small amount of amniotic fluid is removed from the sac surrounding the fetus for testing. Amniotic fluid helps in developing baby to move in the womb, which allows for proper bone growth.

6. What is sterilisation ?

Ans. It is a surgical method to block gamete transport and thereby prevent conception.

7. Define tubectomy.

Ans. In female a small part of the fallopian tube is removed and tied up through a small incision in the abdomen. This is called tubectomy.

8. Define vasectomy.

Ans. In males a small part of the vas deferens is removed and tied up through a small incision in the scrotum. This is called vasectomy.

9. Define hysterectomy ?

Ans. Surgical removal of uterus is called hysterectomy.

10. What do you mean by STDs?

Ans. Infection or disease that are transmitted through sexual intercourse are collectively called sexually transmitted disease (STDs).

11. Define the following :

(a) Reproductive health (b) Amenorrhea

Ans. (a) Reproductive health is a state of complete physical, mental and social well-being in all matters relating to the reproductive system. Also refers to the ability of the individual to produce healthy offspring.

(b) Lactational amenorrhea is the temporary postnatal infertility that occurs when the female is amenorrheic (not menstruating) and fully breast-feeding. Breastfeeding prevents the resumption of normal ovarian cycles by disrupting the pattern of pulsatile release of GnRH from the hypothalamus and hence LH from the pituitary.

Chapter 5. Principles of Inheritance and Variation

1. Define heredity. Give example of inherited traits.

Ans. The transmission of characters from parents to offspring from one generation to another is called heredity. E.g. Height, skin colour, ability to roll tongue, haemophilia.

2. What are alleles ? Give an example.

Ans. Alternative forms of a gene are called as alleles. E.g. : The alleles for length of stem are T and t.

3. What are multiple alleles ?

Ans. If a gene has more than two alleles then they are said to be multiple alleles. e.g. Blood groups are a best example for multiple allelism.

4. What are contrasting traits? Explain with an example.

Ans. Contrasting traits are the two alternative forms of a character. For example, tall and dwarf length of the stem in *Pisum sativum* are the contrasting traits of the character height of the plant. Mendel selected pure lines for his experimentation.

5. What is crossing over ?

Ans. The phenomenon of exchange of chromosomal segments between the homologous chromosomes during meiosis is called crossing over.

6. Define pure line.

Ans. A pure line is the one that produces the identical progeny on self pollination or self fertilisation. On self pollination, a pure tall plant produces all tall plants and a dwarf plant produces only dwarf plants.

7. What are isoagglutinins ?

Ans. Antibodies produced by one individual that causes agglutination of cell of other individual of same species. The antibodies anti A and anti B are called isoagglutinins. These antibodies cause agglutination reactions with antigen.

8. Define the following terms :

(i) Linkage
(ii) Sex Linkage
(iii) Complete Linkage
(iv) Sex Determination
(v) Allele Frequency
(vi) Chromosome
(vii) Haemophilia

Ans. (i) Linkage : The tendency of genes located on the same chromosome, to stay together in hereditary transmission, is known as linkage. The genes located on the same chromosome are called linked genes.

(ii) Sex Linkage : Sex linkage is the phenotypic expression of an allele that is dependent on the sex of the individual, and is directly related to the genes on the sex chromosomes.

(iii) Complete Linkage : The genes closely located on the chromosome show complete linkage as they have no chance of separation by crossing over and are always transmitted together to the same gamete and the same offspring. Thus, the parental combination of traits is inherited as such by the young one.

(iv) **Sex Determination :** It is a biological system that determines the development of sexual characteristics in an organism.

(v) **Allele Frequency :** Allele frequency is a measure of the relative frequency of an allele on a genetic locus in a population.

(vi) **Chromosome :** The DNA which carries genetic information in cells is normally packaged in the form of one or more large strucrtures macromolecules called chromosomes.

(vii) **Haemophilia :** Haemophilia or 'Bleeder's disease' is the name given to a genetic disorder that impairs the body's ability of blood clotting. It is a sex linked disease.

9. What do you mean by chromosomal sex determination?

Ans. Determination of sex in an individual by types of allosomes is known as chromosomal sex determination.

10. What is sex-linked inheritance ?

Ans. The inheritance of a trait (phenotype) that is determined by a gene present on sex chromosomes is sex linked inheritance.

11. What is criss cross inheritance ?

Ans. It is the transmission of gene from mother to son or father to daughter. Criss cross inheritance is also known as skip generation inheritance, in which a character is inherited to the second generation through the carrier of first generation. *E.g.,* X-linked recessive genes are transmitted from a male parent to female child and to a male grandchild.

12. Define mutation.

Ans. A sudden, discrete and discontinuous change in the genome of an organism which is heritable is known as mutation.

13. Define homologous chromosomes ?*

Ans. A couple of homologous chromosomes, are a set of one maternal and one corresponding chromosomes that pair up with each other inside a cell during meiosis.

14. Define Aneuploidy.*

Ans. Aneuploidy refers to the chromosomal variation due to a loss or a gain of one or more chromosomes deviating from normal genome number for that species, due to non-disjunction of the homologous chromosome.

Chapter 6. Molecular Basis of Inheriance

1. What is bacteriophage ?

Ans. Viruses which infect bacterial cells are known as bacteriophage.

2. Define the term capsid.

Ans. The protein coat that envelopes viral genetic material is known as a capsid.

3. What is primer?

Ans. A primer is a short, single strand of RNA that provide starting point for DNA synthesis.

4. What is a replication fork ?

Ans. Unwinding of DNA molecule into two strands results in the formation of a Y-shaped structure called replication fork.

5. What are the Okazaki fragments ?

Ans. Okazaki fragments are short fragments of DNA created on the lagging strand during DNA replication.

6. What are polycistronic mRNA ?

Ans. When the genes in an operon are transcribed, a single mRNA is produced for all the genes in that operon. This mRNA is said to be polycistronic because it carries the information for more than one type of protein.

7. What is translation ?

Ans. Translation is the process by which the linear sequence of codons in a molecule of mRNA directs the specific linear sequence of amino acids.

---

* **Frequently asked previous years Board Exam Question.**

8. What is genetic code ?

Ans. A genetic code comprises a triplet of nitrogen bases (A, G, C and T, U) which could translate the genetic information from DNA to a sequence of amino acids found in the protein of an organisms.

9. What do you mean by lac operator ?

Ans. The operator is a short region of DNA that lies partially within the promoter and that interacts with a regulatory protein that controls the transcription of the operon.

10. What are operators ?

Ans. These are the segments of DNA that regulate the activity of the structural genes of the operon.

Chapter 7. Evolution

1. Define :

(i) Ontogeny
(ii) Phylogeny
(iii) Geological time scale
(iv) Mass extinction
(v) Fossil
(vi) Molecular homology
(vii) Mutations
(viii) Hybridisation
(ix) Genetic drift
(x) Isolation
(xi) Neo-Lamarckism
(xii) Cultural evolution
(xiii) Variations
(xiv) Differential reproduction

Ans. (i) Ontogeny : Ontogeny is the study of process of individual development (both physical and psychological) throughout his life.

(ii) Phylogeny : Phylogeny is the study of evolutionary history and relationships among or within group of organisms.

(iii) Geological time scale : It is a tabulated form showing the sequence and duration of the eras and the periods with their dominant form of life, since the beginning of life on the earth.

(iv) Mass extinction : When plants and animals disappear and become extinct on a large scale, over a relatively short span of time, it is known as mass extinction.

(v) Fossil : A fossil may be defined as any trace or remains of impression of a plant or animal of past geologic ages, preserved in the earth's crust.

(vi) Molecular homology : Homology seen at the level of molecules in organisms is termed molecular homology. For example, the proteins present in man and apes are similar.

(vii) Mutations : They are discontinuous variations which develop due to permanent changes in genotype.

(viii) Hybridisation : It is the crossing over of organisms which are genetically different in one or more traits. It helps in intermingling of genes of different groups of the same variety, species or different species.

(ix) Genetic drift : It is the elimination of the genes of certain characters when a section of population migrates or dies because of natural calamity. It changes gene frequency of remaining population.

(x) Isolation : It is the segregation or separation of population by some barriers which prevents interbreeding. The reproductive isolation between the population due to certain barriers leads to the formation of new species.

(xi) Neo–Lamarckism : This theory considers the change in environment produces modifications in species directly due to its effect on germ cells or indirectly through somatic cells which have the diploid number of chromosomes..

(xii) Cultural evolution : Changes in the culture of society marked with the changes in ideas, traditions, customs, etc., as a result of scientific and technological development is termed as cultural evolution.

(xiii) **Variations :** The difference in form and structure which occur naturally among individuals within the same species and which may result from genetic changes such as mutations or from differences in such factors as nutrition or the density of the population.

(xiv) **Differential reproduction :** All individuals reproduce to produce offsprings. It has been observed that some of the offsprings are able to survive to reach adulthood. Those which become adult and reproduce with varying degrees of success, produce many offsprings while some only a few. This is called differential reproduction.

2. Define the term biota, fauna and flora.

Ans. Biota : The animals living and plants growing in a particular geographical region or geographic period constitutes biota.

Fauna : The animals living in a particular region or environment or period constitute fauna.

Flora : The plants growing in any geographical region or period constitute flora.

3. Define biogenesis.

Ans. Biogenesis : Biogenesis refers to production of life from pre-existing organisms or matter. Biogenesis is observed at all patterns of life including like when bacteria divides, production of seeds from plants etc.

4. Define protobionts.

Ans. Protobionts are considered to be the evolutionary precursors of prokaryotic cells which resemble very simple cells. They are thought to have originated as an array of microspheres of diverse organic and inorganic compounds enclosed by lipid membranes.

5. Define :

(i) Somatic mutations

(ii) Heritable mutation

(iii) Spontaneous mutations

Ans. (i) Somatic mutations : Mutations which occurs in somatic cells are called as Somatic mutations.

(ii) Heritable mutations : Mutations which transmitted from one generation to another are called as Heritable Mutations.

(iii) Spontaneous mutations : Mutations which occur spontaneously in our body.

Chapter 8. Human Health and Diseases

1. Define epidemiology.

Ans. The branch of biology which deals with the study of mode or mechanism of infection or diseases is called epidemiology.

2. What is disease agent ?

Ans. A disease agent is a substance or agent which in excessive presence causes a disease.

3. What are pathogens ?

Ans. These are living organisms which cause diseases.

4. What is a "reservoir of infection" ?

Ans. The term "reservoir of infection" refers to the place where the pathogen normally resides but does not cause any disease.

5. What is pus ?

Ans. Pus is collection of dead cells, body fluids and bacteria (dead) and often develops in infected wounds.

6. What are lymphoid organs ?

Ans. Lymphoid organs are those organs where proliferation and maturation of lymphocytes take place.

7. Define anaphylaxis.

Ans. The sudden, violent and fatal reaction produced by an allergen in a sensitive individual is called anaphylaxis.

8. What is Rh factor ?

Ans. A specific protein on the surface of RBC which was first discovered in Rhesus monkey is called Rh-factor or Rhesus factor.

9. What is opsonization ?

Ans. Opsonization is the coating of bacteria or other pathogens to facilitate their subsequent phagocytosis.

10. What are interferons ?

Ans. Interferons are proteins released by cells in response to a viral infection which they help to combat.

11 What are oncogenes ?

Ans. These are active cancer causing genes.

12. What is Cri-du-chat syndrome ?

Ans. In Cri-du-chat syndrome, there is deduction of short arm of chromosome 5. It is characterised by microcephaly, severe mental degradation, cry of child resembles that of a cat. Also known as cat cry syndrome.

13. What are congenital diseases ?

Ans. Congenital diseases are those diseases present since birth due to genetic or metabolic disorder.

14. What is Albinism ?

Ans. Albinism is a congenital disorder characterised in humans by the complete or partial absence of pigment (melanin) in their skin, eyes, or hairs.

15. What is holandric interitance ?

Ans. It is a type of sex-linked inheritance. The genes which are present on Y-chromosome can be transmitted by a father to their sons only and not to the daughters. These genes are called holandric genes and the inheritance is called holandric inheritance. It is because a male can pass its Y-chromosome to sons only.

16. What do you understand by organ transplants and immunosuppression ?

Ans. **Organ transplant :** Replacement of damaged organ by artificial or doner organ is known as organ transplant.

e.g., Kidney transplant, small piece of skin from other part of the same body.

Immunosuppression : Process to suppress the immune system and stop production of antibodies against the transplant organ is known as immunosuppression.

17. What is erythroblastosis fetalis ?

Ans. It is a haemolytic disease of the new born in the second and afterwards preganancy due to destruction of Rh^+ RBC in the foetus by anti-Rh antigen present in the mother's blood formed during first pregnancy which has filtered through the placenta.

18. Define :

(i) Allergy
(ii) Allergens
(iii) Virulence
(iv) Invasiveness
(v) Pyrogens
(vi) Autoimmunity
(vii) Immunology
(viii) Acquired immunity
(ix) Infection
(x) Incubation period
(xi) Toxigenicity
(xii) Epidemic
(xiii) Communicable disease
(xiv) Carrier
(xv) Immunosuppressants
(xvi) Inflammation
(xvii) Vaccination

Ans. (i) Allergy : Allergy is the hypersensitivity of the body against foreign particles, e.g., some environmental factors or substances.

(ii) Allergens : The allergy causing weak antigens are called allergens.

(iii) Virulence : The relative ability of pathogens to damage or harm the host is known as virulence.

(iv) **Invasiveness :** The ability of organisms (pathogen) to invade the host and grow in its tissue is termed as invasiveness.

(v) **Pyrogens :** These are the chemical substances released by white blood cells in order to set the body's thermostat at a higher temperature.

(vi) **Autoimmunity :** It is an abnormality which sometimes occur in the immune system and instead of destroying foreign molecules it attacks the body's cells itself or the antibodies are produced against person's own tissues.

(vii) **Immunology :** It is the science concerned with the mechanism of body protection against pathogenic micro-organisms and foreign cells.

(viii) **Acquired immunity :** It is that immunity which is acquired by an organism due to vaccinations or inoculations during its lifetime.

(ix) **Infection :** When the body's defence is won over by pathogens and symptoms of disease appear, it is called infection.

(x) **Incubation period :** The period between entry of pathogen and appearance of symptoms is called incubation period.

(xi) **Toxigenicity :** The ability of pathogens to produce toxins or poisons which affect the host cell and destroy them.

(xii) **Epidemic :** The diseases which spread to various persons through different agents suddenly in a large scale in a large area.

(xiii) **Communicable disease :** These are diseases which are transmitted from one man to another. These disease are caused by viruses, bacteria, protozoa, fungi and worms.

(xiv) **Carrier :** It is an animal or organisms that harbours the pathogenic micro–organisms and carries it to animal which is then invaded by pathogen.

(xv) **Immunosuppressants :** The drugs to suppress the immune system and stop production of antibodies against transplanted organs are called immunosuppressants.

(xvi) **Inflammation :** A tissue reaction resulting from irritation by a foreign material causing migration of leukocytes and increased flow of blood to the area produced swelling, reddening, heat, pain.

(xvii) **Vaccination :** The inoculation with a biological preparation (a vaccine) to produce immunity is termed vaccination.

Chapter 9. Strategies for Enhancement in Food Production

1. What is dairying ?

Ans. The management of animal including breeding and raising for milk and their products for human consumption.

2. What is somatic embryo ?

Ans. The e

3. What is protoplast fusion ?

Ans. The merging of protoplast obtained from two different cells to form a hybrid protoplast is called protoplast fusion.

4. What is meant by germplasm collection ?

Ans. The collection of all the diverse alleles of all the gene of a crop plant is called germplasm collection.

5. What do you meant by outbreeding ?

Ans. The breeding between different breeds of animals called outbreeding.

6. Define heterosis.

Ans. The cross between two unrelated individuals or species or genus may develop an improved hybrid with improved or increased function quality called heterosis.

7. Define inbreeding depression.

Ans. The continuous inbreeding especially close inbreeding may lead to the reduced fertility and productivity. This is called inbreeding depression.

8. What is inbreeding in plants ?

Ans. The process of mating more closely related plants within same breed is called inbreeding.

9. What is micropropagation ?

Ans. Micropropagation is the production of healthy plantlet by rapid vegetative multiplication under aseptic and controlled conditions.

10. What is single cell protein ?

Ans. The biomass obtained from micro-organism can be treated or processed in industry to be used as food and is called single cell protein.

11. Define the following terms :

(i) Somatic hybrid (ii) Somaclones (iii) Tissue culture.

Ans. (i) Somatic hybrid — Somatic hybrid is a product obtained by fusion of somatic cell protoplast, obtained from two different varieties or species of plant cultured a suitable nutrient medium under sterile condition.

(iii) Somaclones — The genetically identical plants developed from any part of the plant by tissue culture or micro propagation are called somaclones.

(iv) Tissue culture — The technique to generate whole plant from an explants with in vitro condition on a suitable synthetic medium is called tissue culture.

Chapter 10. Microbes in Human Welfare

1. What is 'Toddy' ?

Ans. Toddy is a traditional drink of Southern India which is prepared by the yeast fermentation of coconut water.

2. What are flocs ?

Ans. Flocs are masses of bacteria associated with fungal hyphen, which form mesh-like structure during secondary treatment of sewage.

3. What are biofertilizers ?

Ans. Biofertilizers are organisms that enrich the nutrients in the soil.

4. What are methanogens ?

Ans. Methanogens are micro-organisms that produce methane as a metabolic byproduct in anaerobic conditions.

5. What is biogas ?

Ans. The gas produced by the anaerobic action of bacteria on biomass is known as biogas.

6. What are biopesticides ?

Ans. Biopesticides are living organism which are able to kill or repel specific pests. Nowadays bio pesticides are used in place of chemical fertilizers.

7. What are the baculoviruses ?

Ans. Virus used in biological control of insects, pests and other arthopods.

8. What are broad spectrum antibiotics ?

Ans. A broad spectrum antibiotic is one which can inhibit the growth of both gram positive and gram negative bacteria. For example : Tetracycline.

Chapter 11. Biotechnology : Principles and Processes

1. What are plasmids ?

Ans. Plasmids are DNA molecules, usually circular, that are independent of the chromosomal DNA.

2. What are vectors ?

Ans. Plasmids used in genetic engineering are known as 'vectors'.

3. What is the enzyme that digests bacterial cell wall in genetic engineering?

Ans. Enzyme called lysozyme digests the cell wall.

4. What is the origin of replication ?

Ans. The site from where DNA starts to replicate is called as the Origin of replication or ori site.

5. What is recombinant DNA molecule ?

Ans. The DNA molecule created in the laboratory by ligating two different pieces of DNA from two different organisms is called as Recombinant DNA or rDNA.

Chapter 12. Biotechnology and its Applications

1. What is Bt Cotton ?

Ans. It is a transgenic variety of cotton which contains a foreign gene obtained from bacterium *Bacillus thuringiensis*.

2. What is "Flavr savr" ?

Ans. It is a transgenic tomato variety which has blocked production of the polygalacturonase.

3. What is chimeric DNA ?

Ans. DNA molecule produced by inserting a foreign segment of DNA into the DNA molecule of a vector is called chimeric DNA.

4. What is plasmid ?

Ans. Plasmid is a circular extra–chromosomal DNA molecule present in bacterial cell which replicates autonomously independent of bacterial chromosomal DNA.

5. What are stem cells ?

Ans. Stem cells are undifferentiated biological cell. These can differentiate into specialised cells and can divide to produce more stem cells.

6. What are transgenic animals ?

Ans. Animals that have had their DNA manipulated to posses and express an extra (foreign) gene are known as transgenic animals.

7. What are Genetically Modified Organism ?

Ans. Plant, fungi, bacteria and animals that have had their DNA manipulated to posses and express and extra (Foreign) gene are known as transgenic or genetically modified organism.

8. What is therapeutic cloning ?

Ans. Therapeutic cloning involves creating a cloned embryo for the sole purpose of producing embryonic stem cells with the same DNA as the donor cell. These stem cells can be used in experiments aimed at understanding disease and developing new treatments for disease. To date, there is no evidence that human embryos have been produced for therapeutic cloning.

9. What is patent ?

Ans. A patent is a set of exclusive rights granted by state (national government) to an inventor or their assignee for a limited period of time in exchange for a public disclosure of an invention.

10. Define Biopiracy.

Ans. The term biopiracy also known as biological theft is illegal or unethical exploitation of the natural occurring biochemical or genetic resources of a country without providing prior information or compensation to the country.

Chapter 13. Organisms and Populations

1. Define demography.

Ans. The study of statistics of human population covering all aspects and parameters is referred as demography. It takes into account factors like natality, mortality, sex-ratio, age groups, etc.

2. Define population and community.

Ans. Population— Population is a group of individuals of same species, which can reproduce among themselves and occupy a particular area in a given time.

Community– It is an assemblage of several population in a particular area and time and exhibit interaction and inter dependence through trophic relationship.

3. What is habitat ?

Ans. The place where an organism lives and reproduce is called its habitat.

4. Define population explosion.

Ans. An enormous increase in population in a short span of time is called population explosion.

5. What is migration ?

Ans. It is defined as the rate at which the individuals move out of the population.

6. What is population density ?

Ans. Population density indicates the size of population. It is the number of individuals of a population found in a unit area or space at a given time.

$$D = \frac{N}{S}$$

N = Number of individuals of the given area

D = Density of population

S = Area occupied by individual

7. Define carrying capacity.

Ans. Carrying capacity measures the feeding capacity of an environment or ecosystem for a population under given set of conditions to support the maximum number of the species in a population.

8. Define each of the following :

(i) Zero population growth
(ii) Birth control
(iii) Carrying capacity
(iv) Biotic potential

Ans. (i) Zero population growth : When the number of individuals added to the population becomes equal to the number of individuals lost from the population then the size of population will remain constant. This is known as zero population growth.

(ii) Birth control : It is an attempt to control the number of offsprings produced.

(iii) Carrying capacity : It is a measure of the feeding capacity of an environment or ecosystem for a population of a maximum number of species under a given set of conditions. The population generally stabilizes around carrying capacity. All environments can sustain only a limited size of population. This limit of constant (K) is called carrying capacity.

(iv) Biotic potential : It is defined as the physiological capacity of organisms to produce their offspring under natural and environmentally optimum conditions.

9. What is meant by sex ratio ?

Ans. Sex ratio is depicted as number of females per 1000 males.

10. What is MTP ?

Ans. Medical Termination of Pregnancy is the legalized termination of pregnancy before the foetus becomes viable. It is usually done to abort genetically abnormal or deformed foetus.

11. What is a test tube baby ?

Ans. When a child is developed in the uterus by implanting the embryo from fertilisation outside the body then the child is known as test tube baby.

12. What is adaptation ?

Ans. Adaptation is the quality of the organism (morphological, physiological, behavioural) that enables the organism to survive and reproduce in its habitat. It adapts organism to live in different type of habits.

13. What is amensalism ?

Ans. The mode of interaction between two species where one is harmed and the other is unaffected is called amensalism.

14. What is mycorrhiza ?

Ans. Mycorrhiza is a symbiotic association between a fungus and the roots of higher plants.

15. What is meant by 'Competitive release' ?

Ans. Competitive release is a phenomenon in which a species whose distribution is restricted to a small geographical areas because of the presence of a competitively superior species, is found to expand drastically when the competing species is removed.

16. What is Gause's competitive exclusion principle ?

Ans. Gause's competitive exclusion principle state that two closely related species competing for the same resource cannot co-exist indefinitely and the competitively inferior one will be eliminated eventually by the superior one.

Chapter 14. Ecosystem

1. Define phenotypic adaptation.

Ans. Non-genetic, temporary and inheritable changes in physical or morphological characteristics behaviour and physiology, which develop in response to changes in environment and are useful to organism in adjustment to prevailing environmental condition are called phenotype.

2. What is aerenchyma ?

Ans. Aerenchyma is a modified parenchyma, where the cells are arranged with a regular air spaces or air chambers to facilitate diffusion of gases in order to provide buoyancy to aquatic plants like lotus and water hyacinth.

3. What is Allen's rule ?

Ans. According to Allen's rule, mammals in colder climate have shorter ears and shorter limbs to minimize heat loss.

4. Define food chain.

Ans. It is referred to the transfer of energy (food) from the producers through a series of consumers.

5. Define trophic level.

Ans. Each energy step or level in a food chain is a called a trophic level.

6. Define food web.

Ans. A number of food chain interconnected with each other forming a web-like pattern is called food web.

7. What is detritus?

Ans. Dead organic matter or remain of plant such as a leaves, bark, flower and dead remain of animals including faecal matter constitute detritus.

8. What are primary carnivores ?

Ans. The herbivores (primary consumer) which depend on plant for their food are called primary carnivores. They are also called secondary consumer.

9. What is meant by humification ?

Ans. The

10. Define mineralization.

Ans. It is process in which the humus is degraded by certain microbes and thus inorganic nutrients are released.

11. What is secondary productivity ?

Ans. Secondary productivity is defined as the rate of formation of new organic matter by consumers.

12. What is net primary productivity ?

Ans. The amount of energy or biomass remaining in a producer after meeting the cost of its respiration is called the net primary productivity.

13. What is meant by 10% law ?

Ans. It states that only 10% of every of a trophic level is transferred to the next higher trophic level.

14. Define climax community.

Ans. The community which is established at the terminal stage of succession and is in equilibrium with the environment is called climax community.

15. Define pioneer species.

Ans. The species that invades a base area are called pioneer species.

16. What are stenothermals ?

Ans. The organism whose survival capacity is restricted to a narrow range of temperature are called stenothermals.

17. What do you mean by the term eurythermal ?

Ans. The organism which can tolerate and thrive in a wide range of temperature is called eurythermal.

18. What are ectotherms ?

Ans. Ectotherms are those animals whose body temperature changes and matches with that of the environment in which they are living. They are also called cold blooded animals.

19. Define standing state.

Ans. The total amount of nutrient like carbon, phosphorus, calcium etc. present in soil at any time is called standing state.

20. Define primary production.

Ans. It is defined as the amount of biomass or organic matter produce per unit area over a certain time period by plants during photosynthesis.

Chapter 15. Biodiversity and its Conservation

1. What is biodiversity ?

Ans. Biodiversity is the variety of life on earth and includes variation at all levels of biological organization from genes to species to ecosystems.

2. What is meant by species diversity ? Give the approximate number of plant and animal species recorded in India.

Ans. Species diversity is the variety of species and their relative abundance present within a region. The two are combined to obtain a mathematical expression that aids in ensuring species diversity. The diversity includes the whole range of organisms found on earth. The number of identified species worldwide is between 1·7 to 1·8 million. However, the estimates of total known species may be 50 million. A large number of plant and animals species are yet to be identified. There are many more species to be present in the tropics.

3. What are exotic species ?

Ans. Species that have been introduced from another geographical region to an area outside its natural range called exotic species.

4. What is genetic diversity ?

Ans. It is the measure of variety in genetic information contained in the organism.

5. What is fragmented habitat ?

Ans. When large sized habitats are broken into smaller parts due to human activities they are called fragmented habitat and it leads to population decline.

6. What is mass extinction ?

Ans. Due to natural calamities like volcanic eruptions, prolonged drought, heavy rains, earthquakes, asteroid collision etc a large number of species become extinct which is called mass extinction.

7. What is national park ?

Ans. A national park is a protected area reserved for wildlife where human activities are not permitted.

8. What do you mean by gene bank ?

Ans. Gene banks are institutions that store genetic complements of as many living forms as possible in the form of seeds, pollens, ovules, tissue culture and frozen germplasm as ova and sperm.

9. What is cryopreservation ? What are its applications ?

Ans. Cryopreservation is a method of preserving biological tissue or cells at extremely low temperatures. Although there are many applications of cryopreservation, the most common uses of the process are to preserve sperm, embryos, eggs and plant seeds.

10. What do you mean by orthodox seed ?

Ans. Orthodox seeds are those seeds that can be stored in a viable condition for many years by reducing the moisture to 5 percent and lowering the temperature to –10 °C to –20 °C.

11. What are sacred groves ? What is their role in conservation ?

Ans. Sacred groves are tracts of forest which are regenerated around places of worship. Sacred groves are found in Rajasthan, Western Ghats of Karnataka and Maharashtra, Meghalaya, and Madhya Pradesh. Sacred groves help in the protection of many rare, threatened, and endemic species of plants and animals found in an area. The process of deforestation is strictly prohibited in this region by tribals. Hence, the sacred grove biodiversity is a rich area.

12. What is a hotspot ?

Ans. A hotspot is an area having endangered endemic species with very high levels of species richness.

13. What are Ramsar sites ?

Ans. Ramsar sites are conserved wetlands which are of international importance under Ramsar Convention.

14. What are endangered species ?

Ans. The species which are in danger of extinction and whose number have been reduced to a critical level are called endangered species.

15. What is the Red Data Book ?

Ans. The Red Data book is a compilation of data on species threatened with extinction (endangered) and is maintained by IUCN.

Chapter 16. Environmental Issues

1. What is reforestation ?

Ans. Reforestation is the process of restoring a forest that once existed but was removed at some point of time in the past.

2. What is hybrid vehicle technology ?

Ans. Vehicle running on dual mode like petrol and CNG are hybrid vehicle. As CNG is a green fuel there is conservation of fossil fuel and reduction in the environmental pollution.

3. What is eutrophication ?

Ans. Eutrophication is excessive growth of algae, plant and animals in water bodies due to the nutrient enrichment particularly with nitrogen and phosphorus.

4. What is an algal bloom ?

Ans. The excessive growth of algae (Free-floating) that causes colouration of water bodies is called algae bloom.

5. What is biological magnification ?

Ans. Biological magnification refers to increase in concentration of toxic materials like heavy metals and pesticides at each trophic level of food chain due to their accumulation.

6. Define the following :

(i) Biodiversity (ii) Eutrophication (iii) PAR

Ans. (i) 'Biological diversity' refers to variation among living organisms from all sources like, inter alia, terrestrial, marine and other aquatic eco-systems and the ecological complexes; this comprises of diversity within species, between species and of ecosystems.

(ii) Eutrophication or more precisely hyper-trophication, refers to enrichment of a water body with nutrients, usually with an excess amount of nutrients. This process induces growth of plants and algae and, due to the biomass load, may result in oxygen depletion of the water body. For example algal bloom.

(iii) Photosynthetically Active Radiation (PAR) is the amount of light available for photosynthesis with wavelengths between 400 and 700 nm. PAR changes seasonally and varies depending on the latitude and time of day.

7. What is BOD ?

Ans. BOD stand for biochemical oxygen demand which represent the amount of dissolved oxygen that would be consumed if all the organic matter in one litre of water were oxidized by micro organism. More BOD means the water sample is polluted by organic matter.

8. What is EI Nino effect?

Ans. EI Nino effect refers to the odd climatic change due to rise in temperature. It lead to increased melting of polar ice caps as well as other places like the Himalayan snow caps. This will result in a rise in sea level that can submerge many coastal areas and leads to loss of biodiversity.

9. What is polyblend ?

Ans. Polyblend is a fine powder of recycled modified plastic. Polyblend was produced to recycle plastic waste.

10. What is meant by ozone hole ?

Ans. Depletion of ozone widely in the stratosphere, particularly over the antarctic region result in formation of thinned layer called ozone hole.

11. What is ozone shield ?

Ans. The thin layer of ozone around the atmosphere that prevent entry of harmful UV rays in earth's atmosphere is called ozone shield.

12. What is polar vortex ?

Ans. Antarctic air circulating over the polar region is called polar vortex.

13. What is meant by Jhum cultivation ?

Ans. In Jhum cultivation farmer clear the trees of the forest and burn the plant remains. The ash of the burnt vegetation contains minerals and is used as fertilizers. The land is then used for farming or cattle grazing. After cultivation, farmers move to another area and this process is repeated.

❐

Very Short Answers

| Set 6 |

Chapter 1. Reproduction in Organisms

1. On what factor does the type of reproduction adopted by an organism depends ?

Ans. The organism's habitual, physiology and genetic make-up determines the types of reproduction adopted by it .

2. Define life span. Give the life span of an elephant.*

Ans. Life span measures the length of time an individual person lives from birth to the natural death. It refers to the longest period over which the life of any organism or species may extend. Elephants can live up to 20-90 years in the wild.

3. Name the type of fission with examples.

Ans. (i) Binary fission – *Amoeba*

(ii) Transverse fission – *Paramecium*

(iii) Multiple fission – *Plasmodium*

4. Name one organism where chain budding is formed.

Ans. Yeast.

5. How does *penicillium* reproduces asexually?

Ans. *Penicillium* reproduces asexually by spore formation.

6. Mention two inherent characteristic of Amoeba and yeast that enables them to reproduce asexually?

Ans. (i) They are unicellular organism

(ii) They have a very simple body structure

7. Mention a characteristic feature and a function of zoospore in some algae.

Ans. Zoospore is a microscopic motile structure. These are special structure by which algae reproduces asexually.

8. List three advantages of using plant tissue culture for propagation.

Ans. (i) We can produce several plants.

(ii) Genetic constitution of all plants is same.

(iii) Disease free plants can be produced in large number.

9. Meiosis is an essential event in the sexual life cycle of any organism. Give two reasons.

Ans. (i) Meiosis helps in formation of gamete by reductional division and maintains constant number of chromosome.

(ii) Meiosis brings variation in offsprings by recombination of genes.

10. The cell division involved in gamete formation is not of the same type in different organism. Justify.

Ans. The parent may be haploid or diploid but the gamete always have to be haploid. Diploid parent undergoes meiosis to produce haploid gametes whereas haploid parents undergo mitosis to produce haploid gametes.

11. With which type of reproduction do we associate reduction division ? Analyse the reason for it .

Ans. With sexual reproduction because diploid organism undergoes meiosis to form haploid gametes.

12. In a developing embryo, analyse the consequences if cell divisions are not followed after differentiation.

Ans. If differentiation does not follow division, embryo will not develop and it will not develop into a new organism.

13. What is the major difference you observe in the offsprings produced by asexual reproduction and in the progeny produced by sexual reproduction ?

Ans. Offspring produced through asexual reproduction are genetically identical (clone) to each other as well to their parents, while sexually produced offspring shows genetic variation leading to evolution.

---

* **Frequently asked previous years Board Exam Question.**

14. If the chromosome, numbers in meiocytes of human beings, rat, elephant, rice, butterfly and onion all 46, 42, 56, 24, 380 and 32 respectively. What will be the chromosome number in gametes of these species ?

Ans. The chromosome numbers in the gamete of these species will be :
Human beings = 23, rat = 21, elephant = 28, rice = 12, butterfly = 190 and onion = 16.

15. Name two organisms which reproduce asexually by multiple fission.

Ans. The two organisms which reproduce asexually by multiple fission are—*Amoeba* and *Plasmodium*. The only difference is that amoeba undergoes multiple fission during unfavourable condition whereas plasmodium undergoes multiple fission in favourable conditions.

16. Name the animals in which the following type of reproduction takes place :
(i) Multiple fission
(ii) Exogenous budding
(iii) Fragmentation
(iv) Binary fission
(v) Isogamy
(vi) Anisogamy
(vii) Parthenogenesis.

Ans. (i) Multiple fission—Malarial parasite (*Plasmodium*)
(ii) Exogenous budding—*Hydra*
(iii) Fragmentation—Star fish, *Planaria*
(iv) Binary fission—*Amoeba, Paramecium*
(v) Isogamy—*Monocystis*
(vi) Anisogamy—Human beings
(vii) Parthenogenesis—Drones of honey bee.

17. Which type of division is involved in the asexual reproduction ?

Ans. Only mitotic divisions occur during asexual reproduction.

18. Name one plant which propagate by the means of its leaf.

Ans. *Bryophyllum.*

19. Name the vegetative propagates in the following :
(i) Agave
(ii) *Bryophyllum*

Ans. (i) Agave – Bulbils
(ii) *Bryophyllum* – Leaf buds or adventitious roots.

20. Mention a unique feature with respect to flowering and fruiting in bamboo species.

Ans. Bamboo flowers once in its life time generally after 150-100 yrs of vegetative growth. It produces large number of fruit and dies.

21. Name the unit of vegetative propagation in water hyacinth. Explain giving reason why it has become the most invasive aquatic weed.

Ans. Offset. Since the formation of hyacinth, offset does not involves two parents, this process is asexual therefore they spread quickly.

22. Name a plant having vegetative propagation by bulbs.

Ans. A plant having vegetative propagation by bulbs is *Agave* or century plant, where the flower bud develops into bulbs, which when falls on the ground forms a new plant.

23. How does potato multiply ?

Ans. Potato multiplies by buds (eyes) of the tuber (vegetative propagation).

24. Name two organisms which shows complete parthenogenesis.

Ans. Indian snake : *Typhlina brahmina*
Caucasian rode lizard : *Lacerta saxicola armeniaca.*

25. Name the phenomenon and the cell responsible for the development of new individuals without fertilisation as seen in honey bees.

Ans. The phenomenon is called parthenogenesis and the cell responsible for it is female gamete ovum.

26. Name the phenomenon and one bird where the female gamete directly develops into a new organism.

Ans. The phenomenon in called parthenogenesis and the bird is Turkey.

27. Pollen grains are usually shed individually. What term is used when all the pollen grains of anther locule are shed as one unit ? Name a plant in which this condition is observed.

Ans. In *Calotropis procera* all the pollen grains are shed as one unit this is called pollinium.

28. Name the group of organisms that produce non-motile gametes. How do they reach the female gamete for fertilisation ?

Ans. Angiosperms produce non-motile gametes. They reach the female gametes with the help of air or water.

29. Cucurbits and papaya plant bears staminate and pistillate flowers. Mention the categories they are put under separately on the basis of the type of flowers they bear.

Ans. Cucurbit is a monoecious plant having staminate and pistillate flowers on the same plant, papaya has staminate and pistillate flowers on separate plants and hence it is dioecious.

30. Name two plants having haploid body.

Ans. Algae and Bryophytes.

31. How many chromosomes are present in male gamete of a rat?*

Ans. 21 chromosomes .

32. Name one plant and animal in which internal fertilisation takes place.

Ans. Plant - Algae

Animal - Fish

33. Name the vegetative propagule of Bryophyllum.*

Ans. Adventitious buds in the notches of a leaf.

34. Name two organisms which shows neoteny.

Ans. *Necturus* (mud puppy) and *Ambystoma* (tiger salamander).

35. Mention the site where syngamy occurs in amphibians and reptiles respectively.

Ans. In amphibians, external fertilisation occurs hence, syngamy occurs in the medium of water. In reptiles, internal fertilisation occurs hence, syngamy occurs within the body of female parent.

36. Name one bisexual and one unisexual animal.

Ans. Bisexual – Earthworm

Unisexual – Cockroach

37. Name one oviparous and viviparous animal.

Ans. Oviparous animal – Reptile

Viviparous animal – Mammals

38. Write two advantages of vivipary.

Ans. Two advantages of vivipary :

(i) Proper embryonic care and protection of the young one.

(ii) The chances of survival of young one is greater.

Chapter 2. Sexual Reproduction in Flowering Plants

1. What is chemical nature of exine ?

Ans. Exine is formed of a chemical called sporopollenin.

2. Mention the location and function of tapetum.

Ans. Tapetum is the innermost layer of cells in a microsporangium. It nourishes the developing pollen grains (micropores) and secretes both enzymes and hormones.

3. What is placenta ?

Ans. Placenta is the vascular tissue found in the ovary over which ovules are borne by means of their stalks or funicles.

4. What is the function of synergids ?

Ans. Synergids have following functions :

* Frequently asked previous years Board Exam Question.

(i) Absorption of nourishment from nucellus.
(ii) Secretion of chemotaxically active substances from its filiform apparatus for guiding pollen tube.
(iii) Act as shock absorber for ingrowing pollen tube.
(iv) Forms seat for pollen tube discharge.

5. What is the function of suspensor ?

Ans. It pushes the embryo deep into the endosperm to enable it to derive nutrition from the endosperm.

6. Suppose haploid number of chromosomes in a flowering plant is 12. What will be the ploidy in the cells of integuments, nucellus, Antipodals, endosperm and embryo of that plant.

Ans.

Name of the Part		No. of Chromosomes
(i) Integument	(2n)	12 × 2 = 24
(ii) Nucellus	(2n)	12 × 2 = 24
(iii) Antipodals	(1n)	12 × 1 = 12
(iv) Endosperm	(3n)	12 × 3 = 36
(v) Embryo	(2n)	12 × 2 = 24

7. What do the following parts form in a seed and what is their ploidy level ?

(i) Integuments
(ii) Nucellus
(iii) Secondary nucleus
(iv) Zygote
(v) Primary endosperm nucleus
(vi) Micropyle

Ans.

	Part of seed	Ploidy level
(i) Integuments	Seed coats	Diploid (2n)
(ii) Nucellus	Perisperm	Diploid (2n)
(iii) Secondary nucleus	Endosperm	Triploid (3n)
(iv) Zygote	Embryo	Diploid (2n)
(v) Primary endosperm nucleus	Endosperm	Triploid (3n)
(vi) Micropyle	Micropyle	It is not a cellular structure.

8. Mention the function of micropyle of ovule.

Ans. Micropyle is a pore-like structure of ovule which facilitates the entry of pollen tube towards egg cell of embryo sac.

9. (i) What are monocarpic plants ?
(ii) Which plants are more in nature, monocarpic or polycarpic ?

Ans. (i) Monocarpic plants are annual plants which bear flowers and fruits once in their life cycle. Example of Monocarpic Plant—*Triticum aestivum* (Wheat).
(ii) Polycarpic plants are more common in nature and produce fruits every year. Example *Mangifera indica* (Mango).

10. What are cleistogamous flowers ?

Ans. Cleistogamous flowers are the flowers which are bisexual and never opens.

11. What are the advantages of self pollination ?

Ans. Advantages of Self Pollination :
(i) Parental characters can be preserved.
(ii) Self-pollination helps in maintaining pure lines.
(iii) No extra energy is required, thus it is an economical process.
(iv) It ensures seed production.

12. What are the disadvantages of self pollination ?

Ans. Disadvantages of Self Pollination :
(i) There is a gradual decrease in the vigour of the plants produced through self-pollination.
(ii) The yield is decreased.
(iii) The size and number of seeds also goes on decreasing generation after generation.
(iv) The plants became susceptible to diseases.

13. What are the different agencies which aid in cross pollination ? Give example.

Ans.

(i) Wind	Anemophily	Date palm, Maize, *Cannabis*
(ii) Water	Hydrophily	*Ceratophyllum, Vallisneria*
(iii) Insect	Entomophily	*Salvia*, Pea
(iv) Bat	Chiropterophily	*Kigelia pinnata*, (sausage tree)
(v) Bird	Ornithophily	Rose apple (*Syzygium*).

14. Name the agencies of pollination that occurs in the following plants :
Vallisneria, Salvia, Cannabis, Kigelia pinnata, Bottle brush and Sunflower.

Ans. *Vallisneria* — Water
Salvia — Insects
Cannabis — Air
Kigelia pinnata — Bat
Bottle brush — Birds
Sunflower — Bees.

15. Give two examples for each of the following :
(i) Insect-pollinated (ii) Water pollinated (iii) Wind pollinated flower.

Ans. (i) Insect pollinated flowers are *Salvia, Calotropis*; (ii) Wind pollinated flowers are Coconut, Date palm; (iii) Water pollinated flowers are *Vallisneria, Ceratophyllum*.

16. How many nuclei are present in mature pollen tube in angiospermic plant just before fertilisation ?

Ans. Two male nuclei and a vegetative nucleus.

17. Name the parts into which the pericarp is differentiated in a succulent fruit.

Ans. Pericarp is differentiated into Epicarp, Mesocarp and Endocarp in succulent fruits.

18. What is siphonogamy ?

Ans. The fertilisation in angiosperms is facilitated by the pollen tube which carries the male gametes to the female gamete; and is referred as siphonogamy.

19. What are the various types of seeds in which food is stored ?

Ans. In Endospermic seeds or Albuminous seeds food is stored in the endosperm while in the non-endospermic or Ex-albuminous seeds food is stored in cotyledons. This food whether it is stored in endosperm or in cotyledons is utilized during seed germination.

20. What are the following parts in a fruit formed from ?

(i) Pericarp (iii) Tegmen (v) Endosperm
(ii) Testa (iv) Embryo (vi) Seed

Ans. (i) Pericarp — Ovary wall
(ii) Testa — Outer Integument
(iii) Tegmen — Inner Integument
(iv) Embryo — Zygote
(v) Endosperm — Secondary Nucleus
(vi) Seed — Ovule

21. In how many broad categories fruits are divided ?

Ans. Into three categories :
(i) Simple fruits
(ii) Aggregate fruits
(iii) Composite or Multiple fruits

22. In which type of germination cotyledons are carried above the surface of soil ?

Ans. In case of epigeal germination (*e.g.*, Castor, Bean) the cotyledons are carried above the surface of soil.

23. Above the surface of soil, which part of seed is responsible for giving rise to caruncle ?

Ans. The outer integument of seed is responsible for giving rise to caruncle.

24. What is fate of following in a mature fruit ?

(i) Integuments

(ii) Ovary wall

(iii) Egg cells

(iv) Secondary nucleus

(v) Antipodal cells

Ans. In mature fruit :

(i) Integuments forms testa and tegmen.

(ii) Ovary walls forms fruit wall.

(iii) Egg cells forms embryo after fertilisation.

(iv) Secondary nucleus forms endosperm.

(v) Antipodals cells degenerate.

25. What is syngamy ?

Ans. Fusion of egg or oosphere with male nucleus forming diploid zygote is called syngamy.

26. What is the mode of seed dispersal in the following plants :

Coconut, *Calotropis*, Waterlily, Poppy, Balsam, Banyan tree.

Ans. Plant		**Mode of seed dispersal**
Coconut	—	By water, as outer coat of fruit is fibrous.
Calotropis	—	By wind, as seeds are hairy
Waterlily	—	By water, since seeds are light and small
Poppy	—	By wind, since seeds have censor mechanism
Balsam	—	By explosion of fruit
Banyan Tree	—	By animals since seeds are sticky

27. (i) Name the two major kinds of reproduction.

(ii) What is the name given to the sheath covering the plumule ?

(iii) Give an example of plant exhibiting hypogeal germination.

(iv) Name the tallest angiosperm.

(v) List any two monocarpic plants.

(vi) What is the name given to the axis which bears flowers ?

(vii) Which technique would you see for propagation of improved varieties of mango and rose ?

(viii) Name four whorls of flower.

(ix) Name the term for the portion which forms upper part of the graft.

(x) List the different parts of carpel.

(xi) Name abiotic agencies of pollination.

(xii) Name biotic agencies of pollination.

(xiii) Name the seed in which endosperm is present.

(xiv) Name the two substances which causes seed dormancy.

(xv) Name any plant hormone which induces parthenocarpy.

(xvi) Name a cultivated plant in which neither fruit nor seeds are formed.

(xvii) Name the essential whorls of a flower.

(xviii) Who discovered double fertilisation for the first time ?

(xix) What technical term is given to a population of genetically identical plants derived from an individual ?

(xx) What term is given to the part which usually becomes the supporting portion *i.e.,* root ?

(xxi) Name the two processes involved in double fertilisation.

(xxii) Name the parts of stamens.

(xxiii) Name the layer which is physiologically and metabolically active in anthers.

(xxiv) Name the common term used for the mature microspore in flowering plants.

(xxv) Name the edible part of the fruit of apple and coconut.

(xxvi) Which plant exhibits vivipary ?

(xxvii) Give example of smallest seed and largest seed.

(xxviii) Name some ornamental plants, which are commercially multiplied by micro-propagation. (any three).

(xxix) Name the special kind of inflorescence in which the main axis grows in a racemose pattern bearing pairs of opposite cymes forming false whorls around the nodes of the main axis.

(xxx) What is the shield-shaped single cotyledons of monocots called ?

(xxxi) Name a plant in which the pollen grains are not covered by exine.

(xxxii) Name the products of coconut tree homologous to the pulp of mango.

Ans. (i) Asexual and sexual reproduction.

(ii) Coleoptile.

(iii) *Pisum sativum.*

(iv) *Eucalyptus regnans*

(v) (a) Agave and (b) Bamboo.

(vi) Peduncle.

(vii) Grafting for mango and cutting for rose.

(viii) The four whorls of flower are : (a) Calyx (b) Corolla (c) Androecium and (d) Gynoecium.

(ix) Scion.

(x) Carpel consists of three parts viz., ovary, style and stigma.

(xi) (a) Wind (b) Water.

(xii) (a) Insects (b) Birds (c) Bats.

(xiii) Endospermic seed or albuminous seed *e.g.,* Maize.

(xiv) (a) Abscisic acid (ABA) (b) Coumarin.

(xv) Gibberelins.

(xvi) Potato (*Solanum tuberosum*) and sugarcane (*Saccharum officinarum*).

(xvii) Androecium (male) and gynoecium (female) are two essential whorls of flower.

(xviii) S. G. Nawaschin (1897) discovered double fertilisation for the first time in species of *Lilium* and *Fritillaria.*

(xix) Clone.

(xx) Stock.

(xxi) Syngamy and Triple fusion.

(xxii) Filament, connective and anther.

(xxiii) Tapetum.

(xxiv) Pollen grains.

(xxv) Apple—Fleshy receptacle or Thalamus, Coconut—Endosperm.

(xxvi) *Rhizophora.*

(xxvii) Smallest seed—Orchids, largest seed — Coconut.

(xxviii) Orchids, *Gladiolus carnations, Chrysanthemum.*

(xxix) Verticillaster.

(xxx) Scutellum.

(xxxi) *Zostera marina.*

(xxxii) Coir, which is derived from dry fibrous mesocarp.

28. Which plant has the smallest flower and which has the largest flower ?

Ans. *Wolffia globosa,* the duckweed has the smallest flower about 0.1 mm in diameter. *Rafflesia arnoldii,* root parasite found in forests of Malaysia has the largest flower about one metre in diameter.

Chapter 3. Human Reproduction

1. Name the glands found in the human male reproductive system.

Ans. Prostate, Seminal vesicles and Cowper's glands.

2. Name various organs of female reproductive system.

Ans. Two ovaries, oviduct, uterus, cervix and vagina.

3. What are the main functions of testes and ovaries ?

Ans. **Testes :** Produce sperms and male hormone testosterone.

Ovaries : Produce ova and female hormone like estrogen and progesterone.

4. What is gonad ? Name the male and female gonads of human body.

Ans. The reproductive organ that produces special sex cells, gametes is called gonad.

Male gonad : Testes produces male gametes called sperms.

Female gonad : Ovary produces female gametes called ova.

5. What are the signs of puberty in human male and female ?

Ans. **In Boys :** (i) Enlargement of testes and starting of spermatogenesis.

(ii) Low pitched, deep and hoarse voice.

In Girls : (i) Menstruation and ovulation starts.

(ii) Widening of pelvis, and development of breasts.

6. The number of chromosomes in an animal body is 46. What would be in its sperm cell ?

Ans. 23 Chromosomes in sperm cell as sperms being the male gametes are haploid.

7. How many spermatozoa are produced from a secondary spermatocytes and how many ovum are produced from a primary oocyte ?

Ans. Two spermatozoa are produced from a secondary spermatocyte and one ovum is produced from a primary oocyte.

8. What is the function of middle piece of sperm ?

Ans. It is known as "power-house" of sperm because it generates energy needed for movement of sperm.

9. Name the stages of spermatogenesis.

Ans. The stages of spermatogenesis are :

(i) Formation of spermatids. This stage includes three phases-

(a) Multiplication phase. (b) Growth phase. (c) Maturation phase.

(ii) Metamorphosis of spermatids or spermiogenesis.

10. How many sperms will be produced from 100 primary spermatocytes and how many eggs will be produced from 50 primary oocytes ?

Ans. 400 sperms will be produced from 100 primary spermatocytes and 50 eggs will be produced from 50 primary oocytes.

11. Give one example of sperm lysin.

Ans. Hyaluroniclose enzyme.

12. What is fertilizability period of sperm and ovum in human ?

Ans. 5 days for sperm inside female and 12-24 hours for ovum.

13. What is the significance of fertilisation ?

Ans. (i) Fertilisation signifies the fusion of gametes.

(ii) The diploid number of chromosomes characteristic of the species is restored by fertilisation.

(iii) It introduces genetic variation in the species.

14. What are the two basic objectives of fertilisation ?

Ans. (i) The fusion of male and female gametes to restore diploid number of chromosomes.

(ii) To stimulate the completion of second meiotic division of the secondary oocyte and begins a series of changes leading to cleavage of the zygote.

15. Define spermiogenesis.*

Ans. Spermiogenesis is the final stage of spermatogenesis, which includes the maturation of spermatids into mature, motile spermatozoa. The transformation of spermatid into sperms takes place in sertoli cells.

16. At what time does menstruation starts in females ?

Ans. At the age of puberty usually around 12 years of age.

17. Which process is called weeping of uterus for the lost ovum ?

Ans. Menstruation is also called as weeping of uterus for the lost ovum.

18. The corona radiata on an ovum acts as a barrier for the penetration of sperms. How do sperms overcome this barrier ?

Ans. The follicular cells forming corona radiata are glued together by hyaluronic acid. To dissolve the acid, the acrosome part of the sperm contains hyaluronidase enzyme which causes the lysis of hyaluronic acid and thus cells of corona radiata.

19. Name the enzyme present in the acrosome of sperm.

Ans. Sperm contains Lysin or Hyalurodinase enzyme and neuraminidase.

20. Name the development stage of mammals which gets implanted in the wall of uterus.

Ans. Blastocyst (Blastodermic Vesicle) formed as a result of cleavage of zygote, gets implanted in the uterus.

21. If endoderm does not differentiate into archenteron, what abnormality you consider to occur ?

Ans. As archenteron represents the primitive gut so digestive system and other associated structures will not develop.

22. What is the function of human Chorionic Gonadotropin ?

Ans. Human Chorionic Gonadotropin (hCG) is secreted by chorionic cells of placenta soon after the blastocyst gets implanted in the uterine wall. hCG functions to maintain corpus luteum and thus maintains pregnancy. It stimulates the secretion of progesterone which maintains endometrial lining of uterus throughout pregnancy. hCG also increases secretion of mucus in cervix, forming protective plug during pregnancy.

23. What is the characteristic feature of gastrula ?

Ans. Presence of three primary germinal layers ectoderm, mesoderm and endoderm.

24. Name the two female sex hormones which control the menstrual cycle.

Ans. Estrogen and Progestrone.

25. (i) What is fertilisation ?
(ii) Name the organs which secretes Testosterone.
(iii) Name the cells that provides nourishment to the developing sperms.
(iv) How many eggs are released from the ovary at the time of ovulation ?
(v) Name the hormone secreted by the corpus luteum.
(vi) Which hormone of the anterior pituitary stimulates the process of ovulation ?
(vii) Where does fertilisation occurs in human female ?
(viii) What is the site for implantation ?
(ix) Name the process during which spermatozoa are produced.
(x) Name the process during which ova are produced.
(xi) Name the loose skin over glans penis.
(xii) Name the method of surgical operation in male in which the vas deferens is cut for family planning.
(xiii) What is the full form of AIH ?
(xiv) Give the typical terms for birth canal and fertilisation canal.
(xv) Name the method of surgical removal of testis in male.
(xvi) Name the process of release of spermatozoa from sertoli cells.
(xvii) Name the method of surgical operation in female where fallopian tube is cut and tied for family planning.

---

* **Frequently asked previous years Board Exam Question.**

(xviii) Name the structure where sperms are temporarily stored and matured.

(xix) What is the gestation period in human female ?

(xx) What are the three parts of male urethra ?

(xxi) Give an example of an animal that reproduce both asexually and sexually.

(xxii) How many sperms are ejaculated by a male during one mating into vagina of the female ?

(xxiii) Name the different parts of mammalian sperm.

(xxiv) Name the developmental stage in humans, which gets implanted in the uterine wall ?

(xxv) Which hormone is supposed to promote motility to spermatozoa ?

(xxvi) How much semen is discharged per ejaculation in man ?

(xxvii) For how many days : (a) Proliferative phase (b) Secretory phase extends ?

(xxviii) Which glands of female are homologous to Cowper's glands of male ?

(xxix) Which structure in female is homologous to penis of males ?

(xxx) Name the skin fold surrounding the vestibule in females.

(xxxi) Name three parts of fallopian tubes.

(xxxii) Which hormone accomplishes the menstrual cycle in human females ?

Ans. (i) Fusion of sperm and ovum.

(ii) Testes.

(iii) Cells of Sertoli.

(iv) Only one ovum.

(v) Progesterone.

(vi) Luteinizing hormone (LH).

(vii) In the fallopian tube.

(viii) Endometrial wall of uterus.

(ix) Spermatogenesis.

(x) Oogenesis.

(xi) Prepuce.

(xii) Vasectomy.

(xiii) Artificial Insemination by husband

(xiv) Vagina and fallopian tubes respectively.

(xv) Orchiectomy.

(xvi) Spermiation.

(xvii) Tubectomy.

(xviii) Epididymis.

(xix) Around 280 days (Nine months).

(xx) Prostatic, membranous and spongy.

(xxi) Hydra.

(xxii) About 350 million sperms.

(xxiii) Head, middle piece and tail.

(xxiv) Blastocyst stage (blastula).

(xxv) Testosterone hormone.

(xxvi) 3 to 5 ml.

(xxvii) (a) 10-12 days, (b) 6-10 days.

(xxviii) Bartholin glands.

(xxix) Clitoris.

(xxx) Labia minora and labia majora.

(xxxi) Ampulla, isthmus, Infundibulum.

(xxxii) Progesterone and estrogen.

26. Mention functions of the following :

(i) Sertoli cells

(ii) Acrosome

(iii) Leydig cells

(iv) Epididymis

(v) Golgi body (in formation of sperm)

(vi) Vas deferens

(vii) Fallopian tubes

(viii) Vagina

(ix) Uterus

(x) Corpus luteum

(xi) Seminiferous tubules

(xii) Scrotum

(xiii) Graafian follicles

(xiv) Bartholin glands

(xv) Seminal vesicles

(xvi) Proximal centriole.

Ans. (i) **Sertoli cells :** The main function of Sertoli cells is to provide nourishment to spermatids or developing sperms.

(ii) **Acrosome :** It facilitates fertilisation by rupturing vitelline membrane as it contains hyalourinidase enzymes which helps the sperm to penetrate into the ovum.

(iii) **Leydig cells :** They are found in the connective tissue around the seminiferous tubules of testes. They secrete testosterone into the blood.

(iv) **Epididymis :** It is a convoluted tube that leads into vas deferens. In the epididymis the sperms are temporarily stored till they become motile.

(v) **Golgi body :** It results in the formation of acrosome in the sperm.

(vi) **Vas deferens :** It is a straight muscular tube carrying sperms from epididymis to urethra.

(vii) **Fallopian tubes :** It receives the ovum from the ovary and the ovum is fertilised inside the fallopian tube.

(viii) **Vagina :** It receives the semen from the male during mating. It serves as the lower portion of the birth canal and helps in menstrual flow.

(ix) **Uterus :** It is a pear-shaped organ specialized for anchoring and nourishing the developing foetus.

(x) **Corpus luteum :** It secretes progesterone hormone for maintaining pregnancy.

(xi) **Seminiferous tubules :** They form the spermatozoa and secretes testosterone.

(xii) **Scrotum :** This keeps the testes at a cooler temperature than the body temperature to produce viable spermatozoa.

(xiii) **Graafian follicles :** These are rounded enclosure for developing ova and produce female sex hormone oestrogens.

(xiv) **Bartholin glands :** Bartholin glands secrete mucus fluid which facilitates the entry of penis during copulation.

(xv) **Seminal vesicles :** It secretes the viscous fluid which is rich in fructose to nourish and activate the spermatozoa.

(xvi) **Proximal centriole :** The proximal centriole passes into the ovum during fertilisation and helps in first division of the fertilised ovum.

Chapter 4. Reproductive Health

1. Reproductive health refers only to a healthy reproductive functions. Comment.

Ans. Reproductive health refers to the total well-being in all aspects of reproduction *i.e.* physical, behavioural, psychological and social.

2. The present population growth refer to healthy reproductive functions.

Ans. According to WHO, reproductive health means, a total well being in all aspect of reproduction, i.e., physical, emotional, behavioural and social. Therefore, a society with people having functional reproductive organs (physical) and emotional and behavioural interactions among them in all sex-related aspects are considered as reproductively healthy.

3. Comment on the RCH programme of the government to improve the reproductive health of the people.

Ans. The basic aims of the RCH programmes are creating public awareness regarding reproduction related aspects and providing facilities to build up in healthy society with added emphasis on the health of mother and child.

4. Give two examples of natural method of contraception

Ans (i) Periodic abstinence

(ii) Coitus interruptus

5. Mention any two characteristics that an ideal contraceptive should have.

Ans. An ideal contraceptive should :

(i) Be easily available.

(ii) Be effective and easily removable with least or no side effects.

6. What are the commonly used barrier methods of contraceptions?

Ans. Condoms, diaphragms, cervical caps and vaults are the commonly used barrier methods of contraception.

7. Mention of any two events that are inhibited by the intake of oral contraceptive pills to prevent pregnancy in humans.

Ans. Two events that are inhibited by the intake of oral contraceptive pills to prevent pregnancy in humans are ovulation and implantation.

8. Write the role of hormones in contraception ?

Ans. Progestogens or progestin– estrogen combination plays an important role in contraception.

9. Name an oral pill used as contraceptive of human females. Explain how does it prevents pregnancy.

Ans. 'Saheli' is an oral pill used as a contraceptive by females. Oral pills inhibits ovulation and implantation, as well as, after the quality of cervical mucus to prevent and retard entry of sperms.

10. How can pregnancy due to unprotected sex be prevented?

Ans. Progestin-estrogen combination and IUDs administered within 72 hours of intercourse can prevent pregnancy.

11. Describe the chemical methods of birth control.

Ans. Chemical methods of birth control are as follows :

(i) Spermicidal creams, jellies and foams are introduced in the vagina just before coitus that kills sperms.

(ii) Pills are taken as oral contraceptive that inhibits ovulation and thus implantation.

12. Explain how do the following acts as contraceptives

(i) CuT (ii) Saheli

Ans. (i) CuT is an ideal contraceptive for humans in males the Cu ions released thus supresses the motility of the sperm and lowers the fertilising capacity of the sperm.

(ii) Saheli contains a oral drug that blocks estrogen receptors in the uterus. It alters the lining of the uterus which prevents a fertilised egg from implanting.

13. How do copper and hormone releasing IUD's act as contraceptive ? Explain.

Ans. The copper releasing IUD's release Cu ions, which suppress sperms motility and the fertilising capacity of sperms. The hormone releasing IUD's makes the uterus unsuitable for implantation and the cervix hostile to the sperms.

14. Define and give the role of amniocentesis.*

Ans. Amniocentesis is a procedure in which amniotic fluid is removed from the uterus for testing or treatment. Amniotic fluid is the fluid that surrounds and protects a baby during pregnancy. This fluid contains foetal cells and various proteins. It helps in developing baby to move in the womb, which allows for proper bone growth. It is done to find genetic abnormalities in the foetus, prenatal diagnosis of genetic defects and sex of the foetus. Thus, this technique is misused in the present time.

15. Classify the following contraceptive measures into different methods of birth control.

(i) Saheli
(ii) Tubectomy
(iii) Vasectomy
(iv) Condoms
(v) Diaphragms
(vi) Cervical caps

Ans. (i) Oral pills
(ii) Surgical method
(iii) Surgical method
(iv) Barrier method
(v) Barrier method
(vi) Barrier method

16. A couple where both husband and wife are producing functional gametes, but the wife is still unable to conceive, is seeking medical aid. Describe any one that you can suggest to this couple to become happy parents.

Ans. Methods : IVF/ZIFT/AI

IVF : Ova from wife and sperm from the husband is collected. It is induced to formed zygote under laboratory conditions.

* **Frequently asked previous years Board Exam Question.**

Very Short Answers

17. Describe the different methods of embryo transfer.

Ans. Embryo is transferred into the uterus or fallopian tube in the following ways :

(i) The zygote or embryo with upto 8 blastomers is transferred into the fallopian tube. This is called zygote intra fallopian transfer (ZIFT)

(ii) Embryo with more than 8 blastomers are transferred into the uterus. This is called intra uterine transfer (IUT)

18. In case of an infertile couple, the male partner can inseminate normally but the mobility of sperms is below 40 percent. Judge which kind of ART is suited in this situation to form an embryo in the laboratory, without involving a donor .

Ans. Intra Cytoplasmic sperm injection.

19. In GIFT, gametes are transferred to the fallopian tube. Can gametes be transferred to the uterus to achieve the same result ? Explain.

Ans. The uterine environment is not congenital for the survival of the gamete. If is directly transferred to the uterus, they will undergo degeneration or could be phagocytosed and hence viable zygote would not be formed.

20. Name two sexually transmitted disease caused by bacteria.

Ans. Syphilis and gonorrhea.

21. Mention early symptoms of STD's.

Ans. Early symptoms of STD's includes itching, fluid discharge, slight pain and swelling in the genital region.

22. Name the causative agent and give any one symptom of Gonorrhoea.

Ans. Causative agent—*Neisseria gonorrhoeae*

Symptoms: itching in genital region, fluid discharge, pain, swelling, sterility, burning sensation during urinating, vaginal bleeding.

23. Name three incurable sexually transmitted diseases and their causative organism.

Ans.

Sexually transmitted disease	Causal agent
(i) Hepatitis –B	Hepatitis – B virus
(ii) Genital herpes	Herpes simplex virus
(iii) AIDS	HIV (Human Immunodeficiency Virus)

Chapter 5. Principles of Inheritance and Variation

1. Who coined the term genetics and in which year ?

Ans. William Bateson, in 1906.

2. Study the following and suggest which traits are dominant-constricted pods, yellow cotyledons, wrinkled seeds, green pods, round seeds, axial flowers, grey seed coat.

Ans. Yellow cotyledons, Green pods, Round seeds, Axial flowers, Grey seed coat are the dominant traits.

3. What are the possible genetic make up of tall pea plants, that Mendel worked with ?

Ans. TT or Tt.

4. A Mendelian experiment consisted of breeding tall pea plants bearing violet flowers with short pea plants bearing white flowers. The progeny all bore violet flowers, but almost half of them were short. This suggests that the genetic make up of the tall parent can be depicted as :

(i) TTWW (ii) TTww (iii) TtWW (iv) TtWW

Ans. (iii) TtWW.

5. Assuming a typical monohybrid cross in which one allele is completely dominant to the other, what ratio is expected if the F_1's are crossed?

Ans. 3:1

6. In which cross would you obtain a 9:3:3:1 ratio for F_2 Progeny. What is the test cross ratio of the same cross ?

Ans. Dihybrid cross. Test cross ratio is 1 : 1 : 1 : 1.

7. Give the ratio of the following :

(i) Monohybrid cross
(ii) Dihybrid cross
(iii) F_2 phenotypic ratio of monohybrid cross
(iv) F_2 phenotypic ratio of dihybrid cross
(v) F_2 genotypic ratio of monohybrid cross
(vi) F_2 genotypic ratio of dihybrid cross

Ans. (i) Monohybrid cross ratio = 3 : 1

(ii) Dihybrid cross ratio = 9 : 3 : 3 : 1

(iii) F_2 phenotypic ratio of monohybrid cross = 3 : 1

(iv) F_2 phenotypic ratio of dihybrid cross = 9 : 3 : 3 : 1

(v) F_2 genotypic ratio of monohybrid cross = 1 : 2 : 1

(vi) F_2 genotypic ratio of dihybrid cross = 1 : 2 : 2 : 4 : 1 : 2 : 1 : 2 : 1

8. What proportion of the plants in the F_2 progeny were tall, in Mendel's cross between a tall and a short pea plant ?

Ans. Three-fourth of the plants.

9. What phenotypic ratios are likely to occur in dihybrid crosses when dealing with two independently segregating gene pairs, when both parents are heterozygous ?

Ans. 9 : 3 : 3 : 1

10. Under what conditions does one expect a 9 : 3 : 3 : 1 ratio ?

Ans. Dihybrid cross (F_2) with independently assorting, completely dominant genes.

11. Define reciprocal cross. Give an example.

Ans. The reciprocal cross involves two crosses concerning the same character but with reversed sexes.

e.g., If in first cross, TT is male parent and tt is female parent then in second cross, TT is female parent and tt is male parent.

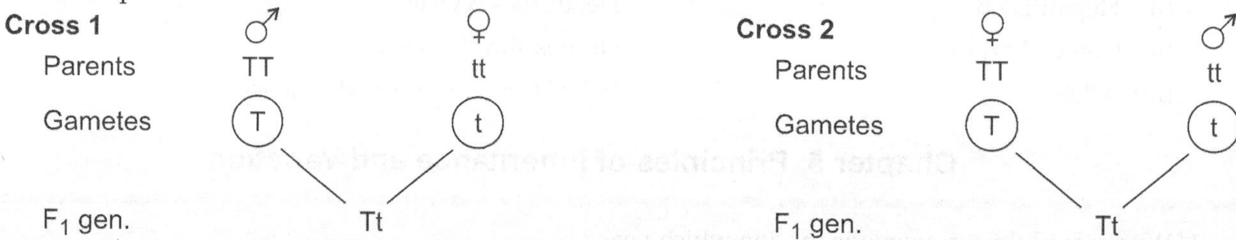

12. Name two factors responsible for genetic variation in an organism.

Ans. Recombination and crossing over.

13. When a tall pea plant was self pollinated one-fourth of the progeny were dwarf. Give the genotype of the parent and dwarf progenies.

Ans. Genotype of parent (Tt).

Genotype of dwarf progenies (tt).

14. Mention any two contrasting pairs of inherited traits in human beings.

Ans. (i) Free and attached ear lobes.

(ii) Blue and brown eye colour.

15. Under what conditions does one expect a 1 : 1 : 1 : 1 ratio ?

Ans. This occurs in a cross involving doubly heterozygous individuals crossed to completely recessive individuals. The genes involved in this cross, assort independently of each other.

16. What is the genetic basis of blood types in ABO system in man ?

Ans. Genetic basis for blood types is due to the presence of three types of alleles I^A, I^B, I^O. It was explained by Bernstein.

Very Short Answers

17. Name the inheritance where both the genotype and phenotype ratios are same.

Ans. Incomplete dominance is the inheritance pattern where both the genotype and phenotype characters are same *i.e.,* 1 : 2 : 1.

18. Give the genotypic ratios that may appear among the progeny of simple crosses and the genotypes of the parents that may give rise to each ratio.

Ans.

Genotypic ratio	Parental genotype
1 : 2 : 1	Aa × Aa
1 : 1	Aa × aa
	AA × AA
	aa × aa
	AA × aa
	AA × Aa

19. Which is the Universal donor? What is the condition required to be a universal donor ?

Ans. Blood group O is considered as Universal donor. The condition required is that the blood group is without any antigens on RBC to react with antibodies of plasma of other blood groups. O group is without any antigens on RBC, so it is considered as universal donor.

20. Which is the Universal recipient ? What is the condition required to be a Universal recipient ?

Ans. Blood group AB is considered as Universal recipient. Any blood group without antibodies in the plasma to react with the antigens of the donor is considered as universal recipient.

AB group is without any antibodies in plasma. Hence, AB is considered as universal recipient.

21. What are the antigens causing ABO blood grouping? Where are these antigens present?

Ans. Isoagglutinogen A(antigen A) and Isoagglutinogen B (antigen B) are the antigens responsible for ABO blood grouping. These antigens are present on the surface of red blood cells.

22. What are the antibodies of ABO blood grouping? Where are they present ?

Ans. Isoagglutinin A (anti A) and Isoagglutinin B (anti B) are the antibodies of ABO blood grouping. These antibodies are present in the blood plasma.

23. What is the blending theory of inheritance ?

Ans. An old theory that says that offspring shows traits intermediate between those of the parents.

24. Why is haemophilia uncommon in females?*

Ans. Haemophilia is due to the presence of a recessive sex-linked gene 'h', carried by the 'X' chromosome. A female becomes haemophilic only when both X chromosomes carry the gene (X^hX^h). A female having only one allele for haemophilia (XX^h) appears normal. Thus, the possibility of a human female becoming haemophilic is extremely rare because she has to be homozygous recessive for the trait.

25. Who carried out his genetical experiments on *Drosophila* ?

Ans. Thomas Hunt Morgan.

26. Name a plant where cytoplasmic inheritance is seen.

Ans. *Mirabilis jalapa.*

27. What is crossing over ?

Ans. Mutual exchange of chromosomes segments between non-sister chromatids during production of gametes is called crossing over.

28. What are the different types of crossing over ?

Ans. Types of crossing over :

(i) Single crossing over

(ii) Double crossing over (reciprocal and complimentary)

(iii) Multiple crossing over

29. What is linkage ?

Ans. Tendency of two or more genes of the same chromosome to remain together during the process of inheritance.

---

* **Frequently asked previous years Board Exam Question.**

30. What is the main significance of linkage ?

Ans. The phenomenon of linkage has great significance as it reduces the possibility of variability in gametes and retains parental combination unless crossing over separates the linked genes.

31. Who first studied cytoplasmic inheritance in four O′ clock plant ?

Ans. C. Correns.

32. What are sex limited characters ?

Ans. Sex limited characters are expressed in only one sex. The genes are present on autosomes in both males and females and their phenotypic expression is limited to one sex only due to internal hormonal environment. *e.g.,* Secondary sexual characters in male and females.

33. Who discovered sex linked inheritance ? How is it discovered ?

Ans. Sex linkage was discovered by Thomas H. Morgan. He discovered it by conducting experiments on eye character in fruit fly, *Drosophila.*

34. What type of inheritance do the XY linked genes follow ?

Ans. The XY linked genes exhibit incomplete sex linkage.

35. Concerning the sex chromosomes of the XY system, which type of gamete do the male and the female individuals respectively produce ?

Ans. The individual of the male sex is XY so he forms gametes containing either the X chromosome or the Y chromosome in a 1 : 1 proportion. The individual of the female sex is XX and thus she forms only gametes containing an X chromosome.

36. What is hemizygous condition ? Explain with one example.

Ans. It is a condition in which an individual has only one member of a chromosome pair or segment rather than the usual two; refers in particular to X-linked genes in males who under usual circumstances have only 1X-chromosome.

e.g., X-linked genes in males and sex-limiting genes in males.

37. Enlist any two sex linked recessive characters in human beings.

Ans. The two sex linked recessive characters are :

(i) Haemophilia

(ii) Colour-blindness

(iii) Duchenne Muscular Dystrophy

38. What is sex reversal ?

Ans. Sex reversal was reported by Crew in fowls. In fowls the fertile hen may become fertile cock due to the damage to the ovary or after the natural cessation of egg laying.

39. Compare the importance of Y chromosome in Human beings and *Drosophila.*

Ans. In Human beings, a single gene (SRY) on Y chromosome develops maleness. But in *Drosophila,* Y chromosome lacks male determining factors, but contains only genetic information essential to male fertility. This concept was explained by Calvin Bridges.

40. What are sex influenced characters ?

Ans. Sex influenced characters are autosomal traits that are influenced by sex of an organism.They may appear in both sexes, but act as dominant in one sex and recessive in other.

e.g., baldness in human beings and horns in sheep.

41. Explain any sex influenced character in Human being.

Ans. Sex influenced character in human being can be explained with an example of Baldness.

Baldness causing gene, B is dominant in males but recessive in females. This is more expressive in males than in females. In male both homozygous dominant (BB) and heterozygous condition (Bb) causes baldness. But homozygous dominant (BB) only causes baldness in female.

42. How many alleles of genes are required to condition X-linked traits in female and male individuals respectively ?

Ans. For each correspondent gene to an X-linked trait women present two alleles since they have two X chromosomes. Men present only one allele of genes related to X-linked traits since they have one X chromosome.

43. Define polygenic inheritance.*

Ans. Polygenic inheritance, also known as quantitative inheritance, refers to a single inherited phenotypic trait that is controlled by two or more different genes.

44. (i) Which term is given to character that an organism pass to offsprings through genes ?
(ii) Who proposed that every part of body produced very small invisible particles called pangenes ?
(iii) Who coined the term genetics ?
(iv) Which name given to the offspring of many generations that have the same traits ?
(v) Name the number that describes how likely it is that an event will occur.
(vi) Which chart shows all the possible combinations of alleles that can result from a genetic cross ?
(vii) In which condition neither of two alleles of a gene is dominant or recessive.
(viii) What name is given to the position of a particular gene on a specific chromosome ?
(ix) What is the name given to sum of all the genes shared by a population ?
(x) Name the proportion of the phenotypic variation that is explained by genetic factors.
(xi) Name the traits which can be expressed in terms of units of length, weight, time, number etc.
(xii) Which term is used for net effect of the trait depends upon the number of genes ?
(xiii) Name the trait influenced by multiple factor.
(xiv) An example of polygenic trait.
(xv) How many chromosomes are present in meiocytes of a fruit fly?*
(xvi) If the haploid number of chromosomes in a plant species is 20, how many chromosomes will be present in the cells of the shoot tip ?*

Ans. (i) Trait
(ii) Darwin
(iii) William Bateson in 1906
(iv) Pure breed
(v) Probability
(vi) Punnett Square
(vii) Codominance
(viii) Locus
(ix) Gene pool
(x) Heritability
(xi) Quantitative traits
(xii) Polygenic Inheritance
(xiii) Polygenic
(xiv) Human skin colour
(xv) 8 or 4 pairs
(xvi) 40 chromosomes

45. (i) Who indicated the involvement of chromosome in the mechanism of inheritance ?
(ii) How many linkage groups are present in *Drosophila melanogaster* ?
(iii) Which term is given to the distance between the linked genes on the same chromosomes ?
(iv) Who coined the term Linkage value ?
(v) At which point homologous chromosomes are held together ?
(vi) Who discovered the relationship between the cross over frequency and the distance between loci ?
(vii) Who discovered the X-chromosome ?
(viii) Who observed the sex determination in grasshopper ?
(ix) What we called the highly condensed structure of DNA ?

Ans. (i) Boveri and Sutton
(ii) 4
(iii) Centimorgan
(iv) T.H. Morgan
(v) Chiasmata
(vi) H. Sturtevant
(vii) H. Henking
(viii) C.E. McClung
(ix) Chromatin

46. (i) Name those scientists who rediscovered the Mendelian's principle.
(ii) Who have worked on *Drosophila* ?
(iii) By which process anthers are removed from flower before maturity.
(iv) Who named factors as genes ?
(v) Which law is also known as the law of splitting of hybrids ?
(vi) Who devised checker board or Punnett's Square ?
(vii) Who introduced the term genotype or phenotype ?
(viii) Who gave the terms homozygous and heterozygous ?

* **Frequently asked previous years Board Exam Question.**

(ix) Who discovered complementary gene inheritance ?
(x) Who studied the inheritance of human skin colour ?

Ans. (i) De Vries, Correns, Tschermak.
(ii) Thomas Morgan.
(iii) Emasculation.
(iv) W. Johannsen.
(v) Law of Segregation.
(vi) R. C. Punnett.
(vii) W. L. Johannsen.
(viii) Bateson and Saunders.
(ix) W. Bateson and R. C. Punnett.
(x) Davenport and Davenport.

47. How many types of chromosomes are found in an organism ?

Ans. Two types of chromosomes-autosomes and sex-chromosomes are present in an organism.

48. Name few human disorders caused by recessive gene present on X chromosome.

Ans. Hemophilia, colour blindness, muscular dystrophy.

49. Name few human disorders caused by recessive gene present on autosomes.

Ans. Albinism, phenylketonuria, tay-sachs or Infantile Amaurotic Idiocy

50. Name a human disorder caused by variations on chromosome.

Ans. Down Syndrome (Trisomy of Chr-21)

51. What is the location of gene for hemophilia ?

Ans. On X chromosome.

52. Is Albinism a deficiency disease ?

Ans. No, it is a autosomal recessibe disorder.

53. Name few human disorders caused by mutant gene or dominant gene disorder.

Ans. Brachydacyly, polydactyly, huntingtion's disease.

54. Name few human disorders caused due to structural variation of chromosome.

Ans. Cri-du-chat syndrome, Chronic Granulocytic Leukemia.

Chapter 6. Molecular Basis of Inheritance

1. Which experiment proved that DNA replication in living cells follows the semiconservative scheme ?

Ans. The Meselson-Stahl experiment proved that DNA replication in living cells follows the semiconservative scheme proposed by Watson and Crick.

2. Which mode of the replication is present in the living cells ?

Ans. Semi conservative mode of replication is present in the living cells.

3. Who figured out semiconservative model of replication was correct ?

Ans. Meselson and Stahl figured out that semiconservative model of replication was correct.

4. Which base pairing of the two nucleotides on the opposite complementary DNA strands connected via hydrogen bonds is correct ?

Ans. G-C, T-A or A-U base pairing of the two nucleotides on the opposite complementary DNA strands connected via hydrogen bonds is correct.

5. What was the objective of Hershey and Chase experiment ?

Ans. The objective of Hershey and Chase experiment was "Which viral component is responsible for replication, protein or DNA ?".

6. On which organism Hershey and Chase conducted their experiments ?

Ans. Hershey and Chase conducted their experiments on the T2 phage, a virus.

7. Who demonstrated that DNA, is the carrier of genetic information ?

Ans. In 1944 the Avery-MacLeod-McCarty experiments demonstrated that DNA, rather than RNA, is the carrier of genetic information.

8. Who gave the concept of central dogma of molecular biology ?

Ans. Concept of central dogma of molecular biology was given by Francis Crick in 1958.

9. Which direction of the replication is possible in case of all the natural DNA polymerases ?

Ans. 5′ → 3′ direction of the replication is possible in case of all the natural DNA polymerases

10. Which are the different phases of replication ?

Ans. Initiation, elongation, termination are the different phases of replication.

11. During which phase of the cell cycle DNA replication occurs ?

Ans. During S phase of the cell cycle DNA replication occurs.

12. Where does transcription occurs ?

Ans. Transcription occurs in the cell nucleus where DNA is housed.

13. Amino acids are the building blocks of proteins. What are the building blocks of amino acids ?

Ans. The building blocks of amino acids are the triplet of nitrogen bases referred as codon, each coding for a specific amino acid.

14. Give significant contribution of G. Gamow.

Ans. **George Gamow** : A physicist suggested that the genetic code should be made up of three nucleotides. He stated that since there are only four bases and if they have to code for 20 amino acids, the code should constitute a combination of bases.

15. Name the chain termination codons.

Ans. UAA, UAG, UGA.

16. There are three codons that do not specify an amino acid. What is the function of these codons ?

Ans. These three codons are stop codons which signals for the termination of translation.

17. Which two processes are involved in protein synthesis ?

Ans. Transcription and translation are the two processes involved in protein synthesis.

18. Identify the codon that would be the first codon translated into a protein during translation in the following sequence:

5′—A C G U A G G U C A U G C C G C G U A C G G G—3′

Ans. The first codon is always an AUG or GUG.

5′—A C G U A G G U C **A U G** C C G C G U A C G G G—3′

19. Who discovered bacteriophages ?

Ans. Bacteriophages were discovered by Twort and d'Herelle in 1915 and 1917.

20. Which term is given to the viruses that kill their infected host cell ?

Ans. Viruses that kill their infected host cell are called virulent.

21. Name any two phages which attack E.coli ?

Ans. T_4-Bacteriophage and phage lambda.

22. Which phage causes lytic infection in E.coli ?

Ans. T_4-Bacteriophage causes lytic infection in E.coli.

23. Which phage causes lysogenic infection in E.coli ?

Ans. Lysogenic infections are caused by the phage lambda.

24. Who is the one who first discovered the tobacco mosaic virus (TMV) ?

Ans. Dimitri Ivanovski gave the first concrete evidence for the existence of TMV in 1892.

25. Who discovered reverse transcriptase ?

Ans. Howard Temin and David Baltimore, discovered reverse transcriptase in 1972.

26. Name the phenomenon by which interchange of genetic material can take place in bacteria.

Ans. Conjugation, transduction and transformation.

27. Which was the first discovered example of a gene regulation? Who discovered this model ?

Asn. The first discovered example of a gene regulation system was the lac operon, discovered by Jacques Monod.

28. What was the observation of Jacques Monod Experiment ?

Ans. The observation of Jacques Monod Experiment is that the proteins involved in lactose metabolism are expressed by E.coli only in the presence of lactose and absence of glucose.

29. What is function of the lac Regulatory Gene ?

Ans. The regulatory gene lac I bind to the operator gene of the lac operon and check synthesis of mRNA.

30. Which energy source is preferred by *E.coli* ?

Ans. Glucose is the preferred and most frequently available energy source for E. coli. The enzymes to metabolize glucose are made constantly by *E. coli*.

31. What is the effect of Glucose and Lactose on the lac Operon ?

Ans. When E.coli is exposed to both lactose and glucose, the organism first metabolise glucose and then temporarily cease to grow until the genes of lac operon become included to provide the ability to metabolise lactose.

32. What is the effect of Glucose and Cyclic AMP on the lac Operon ?

Ans. The presence or absence of glucose affects the lac operon by affecting the concentration of cyclic AMP. The concentration of cyclic AMP in *E. coli* is inversely proportional to the concentration of glucose: as the concentration of glucose decreases, the concentration of cyclic AMP increases.

33. (i) What was determined by chemical analysis w.r.t. structure of DNA ?

(ii) What is Transformation ?

(iii) Who supported the semi conservative mode of DNA replication ?

(iv) Who discovered Okazaki fragments ?

(v) When was Okazaki fragments discovered ?

(vi) Who discovered that all Okazaki fragments are subsequently joined together by DNA ligase to form a long continuous DNA strand ?

(vii) Who renamed nucleic acid ?

(viii) Who demonstrated that RNA is the genetic material in RNA containing viruses like TMV ?

(ix) What term is given to Location of DNA split, Y-shaped region where DNA is unwound ?

(x) Name the enzyme that corrects overwinding ahead of replication fork.

(xi) Length of Okazaki fragments in bacteria.

(xii) Length of Okazaki fragements in eukaryotes.

Ans. (i) DNA was a weak acid rich in phosphorous.

(ii) The phenomenon by which DNA isolated form one type of cell, when iontroduced into another type, is able to bestow some of its properties of the former to the latter is referred to as transformation.

(iii) Meselson and Stahl

(iv) Reiji and Tsuneko Okazaki

(v) In 1968

(vi) Anderson and De Pamphilis ,1979; Alberts 2003

(vii) Altman

(viii) H. Fraenkel

(ix) Replication fork

(x) Topoisomerase.

(xi) 1000-2000 nucleotides

(xii) 100-200 nucleotides

34. (i) What is a Virion ?

(ii) Who discovered TMV and in which year ?

(iii) How many capsomeres occur in a TMV in protein coat ?

(iv) Name the scientists who first demonstrated the sexuality in bacteria ?

(v) What is transduction ?

(vi) Who first demonstrated the phenomenon of bacterial conjugation ?

(vii) Name the scientist who discovered the phenomenon of transduction. Name the organism also.

(viii) Who discovered the phenomenon of RNA splicing ?

(ix) Who proposed the gene battery model to explain the gene expression in eukaryotes ?
(x) Who coined the term 'gene' ?
(xi) Who coined the term nucleosomes ?
(xii) Who coined the term cistron, recon and muton ?
(xiii) About how much of the DNA in a eukaryotic cell is expresses at any time.

Ans. (i) A simple virus particle
(ii) D. Iwanowski, in 1892
(iii) Approximately 2130
(iv) Joshua Lederberg and Edward L. Tatum (1946) in *E. coli*
(v) Transfer of genetic material from one cell to another by a bacteriophage
(vi) Joshua Lederberg and Edward L. Tatum in 1946
(vii) N. D. Zinder and J. Lederberg (1952); *Salmonella typhimurium*
(viii) Phillip Sharp Richard Roberts.
(ix) R. J. Britten and N. Davidson in 1969
(x) W. Johannsen, in 1909
(xi) P. Oudet in 1975
(xii) Seymour Benzer (1957)
(xiii) 1%

35. In which two categories genes are classified with respect to their expression and what are their characteristics ?

Ans. Classification of gene with respect to their expression :

Constitutive (housekeeping) genes : (i) Expressed at a fixed rate, irrespective to the cell condition. (ii) Their structure is simpler.

Controllable genes : (i) Are expressed only as needed. Their amount may increase or decrease with respect to their basal level in different condition. (ii) Their structure is relatively complicated with some response elements.

Chapter 7. Evolution

1. What is anthropology ?

Ans. The study of interrelations of biological, cultural, geographical and historical aspects of man is called Anthropology.

2. What is meant by the term fossil record ?

Ans. The sequential arrangement of fossil as determined by their age and the rock strata in which they are found.

3. What is pangaea ?

Ans. Pangaea is a single land mass supposed to exist millions of years ago that later broke into land masses known as continents at present.

4. What is meant by the term extinction ?

Ans. Elimination of a species is termed as extinction.

5. What is a species ?

Ans. Species are a group of plants and animals that have high degree of similarity and generally can inbreed only among themselves.

6. What is geographical isolation ?

Ans. When a population of a species is separated from similar members due to geographical barriers, is called geographical isolation.

7. What do you mean by speciation ? Define the following terms :

(i) Sympatric speciation (ii) Allopatric speciation.

Ans. **Speciation :** Speciation refers to the origin of new species. It occurs by mutation, isolation and selection.

(i) **Sympatric speciation :** Origin of new species in the populations occupying the same geographical area.

(ii) **Allopatric speciation :** Formation of new species in geographically isolated population.

8. Which geological era was dominated by reptiles?*

Ans. Mesozoic Era.

Chapter 8. Human Health and Diseases

1. Name different disease causing agents.

Ans. Physical agents, Chemical agents, Nutrient agent, Biological agent, Mechanical agents.

2. Name a few diseases for which the agents are not yet identified.

Ans. Peptic ulcer, Hypertension and Coronary Heart Disease (C.H.D.) are the diseases for which the agents are yet to be identified.

3. List the direct mode of transmission of diseases.

Ans. (i) Direct contact
(ii) Droplet infection
(iii) Contact with soil
(iv) Biting by an infected animal
(v) Transplacental transmission.

4. List the indirect modes of transmission of diseases.

Ans. (i) By vector
(ii) Vehicle borne
(iii) Airborne
(iv) Fomite borne

5. What are the factors on which the degree of virulence depend ?

Ans. The degree of virulence depends upon :

(i) Invasiveness
(ii) Ability to produce poisons
(iii) Effect on the host cells which leads to destruction (toxigenicity).

6. What are the factors which influence infection ?

Ans. (i) Tissue affinity
(ii) Hypersensitivity
(iii) Ineffective dose
(iv) Portal of entry

7. Name the causative organism of following diseases :

(i) Typhoid
(ii) Whooping cough
(iii) Gonorrhoea
(iv) Ascariasis
(v) Filariasis
(vi) Ancylostomiasis
(vii) Amoebiasis
(viii) Benign malaria
(ix) Measles
(x) Leprosy
(xi) Diphtheria

Ans. (i) *Salmonella typhi*
(ii) *Bacillus pertussis*
(iii) *Neisseria gonorrhoeae*
(iv) *Ascaris lumbricoides*
(v) *Wuchereria bancrofti*
(vi) *Ancyclostoma duodenale**
(vii) *Entamoeba histolytica*
(viii) *Plasmodium*
(ix) *Rubella virus*
(x) *Mycobacterium leprae*
(xi) *Coroynebacterium dipntheriae*

* Frequently asked previous years Board Exam Question.

Very Short Answers

8. What is the incubation period of *Corynebacterium diphtheriae* ?

Ans. 24 hours.

9. Which of the following diseases are caused by Virus—TB, Cholera, Influenza, Trachoma, Mumps and Filaria.

Ans. Influenza, Mumps, Trachoma are caused by viruses.

10. List the different species of plasmodium.

Ans. *Plasmodium vivax, P. ovale, P. falciparum and P. malariae.*

11. Explain two characteristics of cancer cells.*

Ans. Characteristics of cancer cells:

(i) Cancer cells show uncontrolled mitotic divisions which result in unorganised growth.

(ii) Cancer cells lose their property of differentiation and proliferation.

(iii) Cancer cells also fail in Apoptosis or programmed cell death.

(iv) Cancer cells acquire the capacity to draw out their supply of blood and blood vessels (Angiogenesis).

12. What care should be taken in case of a dog bite ?

Ans. Wash the wound with soap and clean water and with an antiseptic solution. Consult a doctor for anti–rabies vaccine without any delay.

13. Name the two components of the immune system.

Ans. The two components of the immune system are : (i) Humoral Immune System (ii) Cell Mediated Immune System.

14. What are the two main defence mechanisms of the body ?

Ans. (i) A non-specific defence mechanism

(ii) A specific defence mechanism.

15. What is the rate of production of antibodies by B-cells ?

Ans. 2000 molecules per second.

16. What is the function of helper T-cells ?

Ans. Helper T-cells stimulate antibody production by B-cells.

17. What is the function of suppressor T-cells ?

Ans. These cells suppress the total immune reaction preventing it from attacking the body's own cells.

18. What is the function of lymph nodes ?

Ans. Lymph nodes are small solid structures located at intervals along the lymphatic system. These lymph nodes entrap any micro-organism or antigen entering the lymphatic fluid. The trapped antigen activates the lymphocytes initiating an immune response, resulting in the enlargement of lymph glands.

19. What is the most important characteristic of the immune system ?

Ans. Immune system is able to distinguish between self and non-self organs cells.

20. What blood group should you have to become a universal donor or universal recipient ?

Ans. The blood group for universal donor should be "O" where as blood group should be 'AB' for a universal recipient.

21. On the basis of location in the human body name the four different types of cancer.

Ans. (i) Sarcoma

(ii) Carcinoma

(iii) Leukemia

(iv) Lymphoma.

22. Name two major forms of cancer found in Indian men and women.

Ans. In India, major form of cancer in men is the mouth and throat cancer and in women it is the uterine cancer (cancer of the cervix).

23. What is the role of histamine in inflammatory response ?

Ans. Histamine is released by lymphocytes which causes blood vessels to dilate (vasodilation).

---

* **Frequently asked previous years Board Exam Question.**

24. Name the technique that promises cure for diseases like jaundice and liver cancer.

Ans. Recombinant DNA technique by which a successful vaccine has been produced against Hepatitis B virus

25. What is D.P.T. ?

Ans. It is a vaccine which can immunize the infants against Diphtheria, Pertussis (whooping cough) and Tetanus.

26. Which vaccine is administered to prevent the following diseases ?

(i) Diphtheria
(ii) Whooping cough

Ans. (i) Diphtheria-D.P.T. vaccine
(ii) Whooping cough-D.P.T. vaccine

27. What are the symptoms for AIDS related complex (ARC) ?

Ans. Swollen Lymph nodes, fever, night sweats and weight loss.

28. Name the scientists associated with each of the following :

(i) Who discovered passive immunity ?
(ii) Who coined the terms vaccine and vaccination ?
(iii) Cowpox vaccine.
(iv) Father of immunology.
(v) Vaccine for anthrax, rabies, chicken pox.
(vi) Vaccine against polio.
(vii) First vaccine against measles.
(viii) BCG vaccine against tuberculosis.
(ix) Isolation of bacteria that cause T.B.
(x) Treatment of Rabies.
(xi) Who studied Blood Groups ?
(xii) Sex Chromatin in Animals.
(xiii) Discovered number of chromosomes in human somatic cells.
(xiv) Discovered Rh-factor.
(xv) Who noted Barr body ?
(xvi) Germ theory of disease.
(xvii) Malarial parasite.

Ans. (i) Von Behring .
(ii) Edward Jenner.
(iii) Edward Jenner.
(iv) Edward Jenner.
(v) Louis Pasteur.
(vi) Salk.
(vii) John F Enders.
(viii) Albert Calmette and Camille Guerin.
(ix) Robert Koch.
(x) Louis Pasteur.
(xi) Landsteiner.
(xii) Murray Barr.
(xiii) Tijo and Levan.
(xiv) Landsteiner and A. L. Weiner.
(xv) Murrary Barr.
(xvi) Louis Pasteur.
(xvii) Ronald Ross.

29. Mention one significant function or role of each of the following :

(i) Suppresser cells
(ii) Stem cells
(iii) Complement system
(iv) Tears
(v) Saliva
(vi) Memory cells
(vii) Lysozyme
(viii) Syndrome

Ans. (i) **Suppresser Cells :** Prevents immune system from attacking body's own cells.

(ii) **Stem Cells :** Both kinds of lymphocytes B-cells and T-cells originate from stem cells in the bone marrow.

(iii) **Complement System :** Is composed of a normal thermolabile protein constituents of blood serum that participates in antigen-antibody reactions. They assign body's defence mechanism.

Very Short Answers

(iv) **Tears :** Tears are secreted by tear glands and regularly washes the exposed surface of the eyes. The tears contains an enzyme Lysozyme which prevents eye infections.

(v) **Saliva :** Secreted by salivary glands contains Lysozyme which attacks the bacteria present in food and water and destroys many of them.

(vi) **Memory Cells :** These specialised cells that remain in the lymph nodes even after the antigen has disappeared. They help to fight infection when attacked by same antigen again.

(vii) **Lysozyme :** It is an enzyme present in sweat, saliva and destroys the cell wall of many bacteria.

(viii) **Syndrome :** The sign or symptoms associated with any process and constituting whole picture of disease.

30. (i) Name any two congenital disease.

(ii) Name two kinds of toxins.

(iii) Where are the B-cells and T-cells are formed ?

(iv) Name two basic types of immunity ?

(v) Who is called father of medicine ?

(vi) Who is called father of microbiology ?

(vii) Who discovered the Lepra virus ?

(viii) What is the expanded form of ATS ?

(ix) Which material does small pox vaccine contain ?

(x) List five major groups or immunoglobulins.

(xi) Name the types of cancer of adipose tissues.

(xii) The retrovirus causing AIDS.

(xiii) The inborn resistance to diseases of other organisms.

(xiv) The ability of an animal or plant to resist infection by parasitic organisms.

(xv) The branch of medical science that deals with the nature of the spread of communicable diseases.

(xvi) A state of complete physical, mental and social well being and not mere absence of disease or infirmity.

(xvii) The foreign molecules that gain access to the body.

(xviii) The condition when the muscles cells are the targets, the muscles are destroyed and severe weakness results.

(xix) The word antigen is an abbreviated form of which term ?

(xx) In which year plague was brought in India ?

(xxi) Which disease is responsible for Black Death in Europe in 1348 ?

(xxii) Who is called father of antiseptic surgery ?

(xxiii) Name three diseases caused by bacteria.

(xxiv) Name two diseases caused by protozoa and helminth.

(xxv) Name the most important types of phagocytes.

(xxvi) Name the antibody which is most effective in allergies.*

(xxvii) Disease spreading suddenly on a large scale in a large area.

Ans. (i) Haemophilia, Phenylketonuria.

(ii) Endotoxin and Exotoxin.

(iii) From stem cells in bone marrow.

(iv) Innate immunity and acquired immunity.

(v) Hippocrats.

(vi) Louis Pasteur.

(vii) Hensen (1874).

(viii) Anti Tetanus Serum.

(ix) Attenuated virus.

(x) IgM, IgG, IgA, IgD and IgE.

---

* **Frequently asked previous years Board Exam Question.**

(xi) Lipoma.
(xii) HIV (Human Immunodeficiency Virus)
(xiii) Natural Immunity.
(xiv) Immunity.
(xv) Epidemiology.
(xvi) Health.
(xvii) Antigen.
(xviii) Myasthenia gravis.
(xix) Antibody generating.
(xx) In 1895 from Hong-Kong.
(xxi) Plague.
(xxii) Joseph Leister.
(xxiii) Tuberculosis, Diphtheria and Whooping cough (Pertussis).
(xxiv) Protozoan diseases : Amoebic dysentry, Malaria.
Helminths diseases : Filariasis, Ascariasis.
(xxv) Macrophages and Neutrophils.
(xxvi) IgE
(xxvii) Epidemic

Chapter 9. Strategies for Enhancement in Food Production

1. Name any two disease of poultry.

Ans. Ranikhet and Bird Flue.

2. Write the name of the following.
   (i) The most common species of bees suitable for apiculture.
   (ii) An improved breed of chicken.

Ans. (i) *Apis indica/ Apis dorsate*
(ii) Leghorn/Rhode island red/Minorca.

3. What is the purpose of animal breeding ?

Ans. Purpose of animal breeding :
(i) increase in yield (ii) improving quality of product

4. Name two factors on which the yield of milk depends.

Ans. (i) Quality of the breed (ii) Type of food (fodder)

5. List any three important component of poultry farm management.

Ans. Three important components of poultry farm management are :
(i) Selection of disease – free and suitable breeds.
(ii) Proper and safe farm conditions.
(iii) Proper feed and water.

6. State the disadvantage of inbreeding among cattle. How it can be overcome?

Ans. Continuous inbreeding reduce fertility and productivity. This is called inbreeding depression. It can be overcomed by outbreeding. Selected animal should be mated with unrelated superior animal.

7. Success rate of artificial inseminations in cattle is fairly low. Identify any other means to improve the successful production of hybrids. State the advantage of this technique.

Ans. Multiple Ovulation Embryo Transfer (MOET) technology can be used to improve production of hybrids. This techniques produce 8-10 eggs at a time and the genetic mother is only available for super ovulation. By this technique, head size can be increased in a short time.

8. Describe the technology that has successfully increased the herd size of cattle in a short time to meet the increasing demand of growing demands of growing human population.

Ans. Multiple ovulation embryo transfer (MOET) technology has successfully increased the herd size of cattle. The cow is administered with FSH to induce follicular maturation and super ovulation to produce 6 to

8 eggs. It is then mated or artificially inseminated. The fertilised eggs are recovered non-surgically and transferred to surrogate mother where they develop into an improved variety.

9. Name any two common Indian millet crops. State one characteristic of millets that has been improved as a result of hybrid breeding so as to produce high yield millets crops.

Ans. Maize, jowar, bajra, sorghum

Resistance to water stress has been improved.

10. What is outbreeding? How is it useful in animal breeding?*

Ans. Out-breeding is the breeding of the unrelated animals, which may be between individuals of the same breed but having no common ancestors for 4-6 generations (out-crossing) or between different breeds (cross-breeding) or different species (interspecific hybridisation).

Significance of outbreeding in animals:

(i) Outcrossing is the best breeding method for animals that are below average in productivity in milk production, growth rate in beef cattle, etc.

(ii) A single outcross often helps to overcome inbreeding depression.

(iii) Cross-breeding allows the desirable qualities of two different breeds to be combined. The progeny hybrid animals may themselves be used for commercial production.

11. Give one example of disease caused each by fungi, bacteria and viruses in crop-plant.

Ans. Fungal disease-Red rot of sugarcane, brown rust of wheat; Bacterial disease-citrus canker, black rot of crucifier, Viral disease-Turnip mosaic, tobacco mosaic.

12. Write the scientific names of Sugarcane grown in North India and South India, respectively. Mention their characteristic features.

Ans. Sugarcane species grown in North India is *Saccharum barberi*. It has low sugar content and yield. Sugarcane grown in South India is *Saccharum officinarum*. It failed to grow in north India but is rich in sugarcane and possess thick stem with high yield.

13. Suggest two features of plant that will prevent insect and pest infestation.

Ans. (i) Increasing hair growth on aerials part of plants

(ii) Rendering the flower nectar less

(iii) Enabling plant to secrete toxin to kill insect.

14. How are evaluation and testing of new crop variety carried out ?

Ans. Evaluation is done by growing these plants in research fields and recording their performance under ideal fertilizer application, irrigation and other crop management practices.

15. How can healthy potato plants be obtained from a desired potato variety which is virus infected? Explain.

Ans. Even though a plant is infected by virus, the meristem (apical or axillary) is free of virus so the meristem can be removed and grown in vitro in sterile condition and special nutrient/culture medium.

16. What can be used as the reference material for comparison of any new improved variety ?

Ans. The best available local cultivar.

17. Mention any two methods used for breeding for disease resistance apart from conventional method for hybridization.

Ans. (i) Selection among somaclonal variants (ii) Genetic engineering.

18. What is inbreeding depression and how is it caused in organism ? Write any two advantages of inbreeding.

Ans. Continued inbreeding especially close inbreeding, usually reduces fertility and even productivity. This is called inbreeding depression.

Advantages of inbreeding :

(i) It exposes harmful recessive genes that are eliminated by selection

(ii) It help in developing a pure line in any animal and accomulation of superiar genes

19. Write the importance of bagging of unisexual flowers in crop improvement programme.

Ans. The emasculated flowers are covered with a bag of butter paper to prevent contamination of stigma with unwanted pollen. This process is called bagging. When this stigma attains receptivity, mature pollen grains are dusted on the stigma and the flowers are rebagged to allow the fruits to develop.

---

* **Frequently asked previous years Board Exam Question.**

20. State any one significance of interspecific hybridization in plants.

Ans. It is important for breeding disease-resistant plant varieties.

21. Give one significant contribution of M.S. Swaminathan.

Ans. M.S. Swaminathan was an Indian geneticist, known for developing short duration high-yielding varieties of rice. He brought green revolution in India.

22. Genetic variability is the root of any breeding programme. How does this variation occur ?

Ans. (i) All the wild varieties and relatives of the cultivated crops and collected and preserved

(ii) For their characteristics, evaluation of these genetic collection are done.

23. Find out what the various component of the medium used for propagation of an explants in vitro are ?

Ans. The major component of the medium for *in vitro* propagation are :

(i) Water
(ii) Agar Agar
(iii) Sucrose
(iv) Inorganic salts
(v) Vitamins
(vi) Amino acids
(vii) Growth hormones like auxins, cytokinins.

24. How can pollen grains of wheat and rice which tend to lose viability within 30 minutes of their release be made available months later for breeding programmes ?

Ans. The pollen grains can be stored in liquid nitrogen (–196°C). Such stored pollen grains do not lose their viability for years and can be used in breeding programmes.

25. Name the disease resistant variety of cowpea developed by plant breeding technique.*

Ans. Pusa Komal variety of Cowpea.

Chapter 10. Microbes in Human Welfare

1. Milk starts to coagulate when Lactic Acid Bacteria (LAB) is added to warm milk as a starter. Mention any other two benefit that LAB provides.

Ans. (i) It increases nutritional quality of curd by increasing vitamin B_{12} content.

(ii) It checks the growth of disease causing organism in the gut.

2. LAB in producing curd.

Ans. *Lactobacillus lactis* and *Streptococcus lactis* bacteria produce acid which cause coagulation of milk protein called casein and changes milk into curd.

3. Consuming curd keeps the gastro intestinal tract in form. Give reason.

Ans. Curd contain lactic acid bacteria which checks the growth of disease-causing microbes and protects the gastro intestinal tract.

4. What is the role of baker's yeast in the production of bread ?

Ans. *Sacchromyces cerevisaie* (Baker's yeast) causes the alcoholic fermentation of sugars in wheat flour and produce ethanol and CO_2. CO_2 raises the bread when baked and make it soft.

5. What is puffed up appearance of Idli/Dosa due to which metabolic pathway is involved in it ?

Ans. Puffed up appearance of dough is due to production of CO_2 as result of fermentation.

6. Name two liquid household product-obtained by microbial activity.

Ans. The products are vinegar and toddy.

7. How is cheese prepared ?

Ans. Cheese is partially degraded flavoured milk curd formed by the activity of micro-organism. Processed cheese is formed by blending and emulsifying different types of cheese with or without cream.

8. Name any two varieties of cheese and mention the names of microbes used.

Ans. Swiss cheese – *Propionibacterium shermanii*

Roquefort cheese – *Penicillium roqueforti.*

* **Frequently asked previous years Board Exam Question.**

Very Short Answers

9. Name the bacterium responsible for large holes seen in swiss "cheese". What are these holes due to ?

Ans. *Propionibacterium shermanii*

The holes are because of production of large amount of CO_2.

10. Name the pest that Ladybird beetle and Dragonflies help to get rid of.

Ans. Ladybird beetle and Dragonflies control mosquito and aphids.

11. Name the source of statin and state its action on the human body.

Ans. Statin is produced by the yeast *Monascus purpureus*. It acts as a blood cholesterol lowering agent by competitively inhibiting the enzyme responsible for synthesis of cholesterol.

12. Give the scientific name of the microbes from which cyclosporine A and statin are obtained. Write one medical use of each one of these drugs.

Ans. Cyclosporin that is used as an immuno-suppressive agent during organ transplantation in patient, is produced by the fungus *Trichoderma polysporum*. Statin produced by yeast *Monascus pupureus* have been commercialised as blood cholesterol lowering agents.

13. What is chakravarthy bug ? Give its scientific name and its application.

Ans. Chakravarthy bug is a super bug of *Psedomonas* with multiple plasmids. Its scientific name is *Pseudomonas putida* they are helpful in removing oils spills.

14. The excreta of cattle do not contain any cellulose but human excreta may contain cellulose. Comment.

Ans. The rumen of cattle contains methenogens which keep in the digestion of cellulose but these bacteria are not present in human stomach so cellulose is not digested.

15. Mention the product and it use by each of the microbes listed below :

(i) *Streptococcus* (ii) *Lactobacillus* (iii) *Saccharomyces cerevisiae.*

Ans.

S.No.	Microbes	Products	Uses
(i)	*Streptococcus*	Streptokinase	Bioactive molecule used to remove blood clot (thromboytic)
(ii)	*Lactobacillus*	Lactic acid	Coagulates and partially digest milk protein and form curd
(iii)	*Sacchoromyces*	Ethanol and CO_2	For bread-making fermenting maltose cereals and fruit juices.

16. Mention the three enzymes of industrial importance.

Ans. Three enzyme of industrial importance are proteases, pectinase and lipase.

17. Bottled fruit juices are clearer to those made at home. Explain.

Ans. Bottled fruit juices contain enzyme pectinase and protease for making them clear.

18. Name the group of organisms and the substrate they out on to produce biogas.

Ans. Group of organism - methanogens

Substrate – cellulosic material/cow dung/agriculture waste.

19. Write the constituents of biogas.

Ans. Biogas comprises a mixture of methane, carbon dioxide, hydrogen sulphide and ammonia.

20. Describe how biogas is generated from activated sludge ? List the component of biogas.

Ans. The activated sludge is pumped into a large tank called anaerobic sludge digest. Here bacteria grows anaerobically and digest the aerobic bacteria and the fungi in the sludge. During this digestion bacteria produce a mixture of gases which forms biogas, is made up of methane, hydrogen sulphide and carbon dioxide.

21. What is the chemical nature of biogas ? Name an organism which is known to be employed in biogas.

Ans. The biogas contain methane CO_2 and H_2S. Methane bacterium a type of methanogen which is employed for biogas production.

22. What function do methanogens perform in the rumen of cattle ?

Ans. Methanogens help in the digestion of cellulose in the rumen of cattle.

23. Name the genus to where baculovirus belong. Describe the role in the integrated pest management programme.

Ans. Baculoviruses belong to the genus *Nucleopolyhedrovirus.* They control only species-specific pest, do not effect non-target organism and beneficial insects are conserved, they aid in IPM problems and there is no negative impact on plants or other animals.

24. How has the bacterium *Bacillus thuringiensis* helped us in controlling caterpillars of insect pests ?

Ans. *Bacillus thuringiensis* produce endotoxin which when ingested and released in the gut of the larvae of insect pest, disrupts the insect gut lining thereby killing earthen. It is used as a Bio-insecticide.

25. Mention the role of cyanobacteria as a biofertilizer.

Ans. It is a biological organism that fixes atmospheric nitrogen.

26. Give the significance of biofertilizers.

Ans. Biofertilizers increase the soil fertility and reduce the use of chemical fertilizer and in turn reduce the changes of environmental pollution.

27. How is presence of cyanobacteria in the paddy fields beneficial to rice crops ?

Ans. Cyanobacteria can fix atmospheric nitrogen. In paddy fields, the Cyanobacteria act as an important fertilizer. They also add organic matter to soil and increase its fertility.

28. How do mycorrhizal fungus helps the plant harbouring them ?

Ans. The mycorrhizal fungi absorbs phosphorus from the soil and transfer them to the host cells. They also impart resistance to host plants against root pathogens. They also help plant to tolerate salinity and drought.

29. Who won the Nobel Prize for the discovery of penicillin ?

Ans. Chain, Florey and Fleming won the Nobel prize for the discovery of penicillin in 1945.

30. Who discovered the first antibiotic ? Name the antibiotic.

Ans. Alexander Fleming discovered the first antibiotic named penicillin.

31. What are the properties of antibiotics ?

Ans. The properties of an antibiotics are as follows :

(i) It should not kill useful micro organism found in alimentary canal of man

(ii) It should not cause side effect in the host.

32. What would have happened if antibiotics were not discovered ?

Ans. If antibiotics were not discovered, bacterial and fungal disease would not have been controllable.

33. Name any two properties of fungus which are used in the production of antibiotics.

Ans. Two species of fungus, used in the production of antibiotics :

(i) *Penicillium notatum* (For penicillin production)

(ii) *Aspergillus fumigatus* (for fumagillin production)

34. What does the Ganga action plan tend to achieve ?

Ans. Ganga action plan tend to save the major rivers from pollution. Under these plans, a large number of storage treatment plant were built so that only treated sewage is discharged in the river.

Chapter 11. Biotechnology : Principles and Processes

1. Who invented DNA Fingerprinting ?

Ans. Sir Alec Jefferies invented DNA fingerprinting.

2. What is the use of DNA fingerprinting ?

Ans. More than 99.9% of the base sequence in DNA is same in all humans. Less than 0.1% is unique to each individual. Scientists make use of this difference to generate DNA fingerprints which are specific to different individuals.

3. How accurate is DNA Fingerprinting ?

Ans. In organism, 99.9% of the genetic code is exactly same in each individual. Only 0.1% of it differs in every individual, Hence, if the technique of DNA fingerprinting is followed in reliable manner, without any manipulation, the chance of wrong result in highly unlikely.

4. Name the source organism from which Ti plasmid is isolated. Explain use of this plasmid in biotechnology.

Ans. Ti plasmid is isolated from *Agrobacterium tumifaciens*. Ti plasmid of *Agrobacterium tumifaciens* has been modified into a cloning vector which is not pathogenic to plant but still is able to use the mechanisms to deliver genes of interest into plants.

5. How can bacterial DNA be released from the bacterial cell for biotechnology experiments ?

Ans. The bacterial cell wall is digested by the enzyme lysozyme to release DNA from the cell.

6. (i) State the role of DNA ligase in biotechnology.

(ii) What happens when *Meliodogyne incognita* consumes cells with RNA i gene ?

Ans. (i) DNA ligase joins the DNA fragments with same sticky 'ends', links okazaki fragments or dis continuously synthesized fragment, link desired gene with plasmid to form recombinant DNA.

(ii) The specific mRNA of the nematode is silenced and the parasite dies.

7. What are the advantages of recombinant DNA technology ?

Ans. Advantages of Recombinant technology :

(i) Provide substantial quantity

(ii) No need for natural or organic factors

(iii) Controlled Tailormade product

(iv) Unlimited utilizations

(v) Cheap

(vi) Resistant to natural inhibitors

8. From which bacterium gene barnase and barstar has been derived ?

Ans. Gene barnase (encoding a ribonuclease) and barstar (encoding a ribonuclease inhibitor) derived from *Bacillus amyloliquefaciens*.

9. What is the advantage of using multiple restriction enzymes to cut the DNA during DNA fingerprinting ?

Ans. Restriction enzymes cuts DNA at specific site, where complementary DNA binds. Multiple restriction enzyme increases the specificity of DNA binding sites, giving more reliable results.

10. Who discovered restriction endonuclease ?

Ans. These were discovered by Nathans.

11. What is the use of restriction endonucleases ?

Ans. Restriction endonucleases cut DNA molecules at specific places into many smaller fragments called restriction fragments.

12. State the role of transposons in silencing of mRNA in Eukaryotic cells.

Ans. Transposons or mobile genetic elements in viruses are the source of the complementary dsRNA which in turn bind to specific mRNA and cause RNA interference of the parasite.

13. Which two vectors are used for gene transfer in plants ?

Ans. Ti plasmid of *Agrobacterium tumefaciens* and Ri plasmid of *A. rhizogenes* have been effectively used as vectors for gene transfer to plant cells.

14. What are the commonly used vectors for genetic engineering ?

Ans. The commonly used vectors for genetic engineering are :

(i) pBR322(Plasmid)

(ii) pUC19(Plasmid)

(iii) Bacteriophage M13

(iv) Cosmids

(v) Yeast Artificial Chromosome (YAC)

15. What are the commonly used vectors for transformation in plant cells ?

Ans. The commonly used vectors for transformation in plant cells are :

(i) *Agrobacterium tumefaciens*

(ii) *E. coli*

(iii) Cauliflower mosaic viruses (CaMV)

(iv) Brome mosaic viruses (BMV)

16. What is the function of DNA ligase enzyme ?

Ans. The joining or ligation of DNA fragments is achieved by the use of DNA ligase enzyme. DNA ligase is a mediate in annealing, sealing or joining of DNA fragments used in genetic engineering for the production of hybrid DNA.

17. What is a gene gun used for ?

Ans. Gene gun is the most effective method used for gene transfer and creation of transgenic plants. This method is also known as Biolistics or Particle bombardment method.

18. What are the non-viral methods of introducing foreign DNA into host cell ?

Ans. Non-viral methods include microinjection, gene gun, electroporation, hydrostatic pressure, etc.

19. From a group of genes of the same lengths, how the gene with desired sequence is separated ?

Ans. From a group of genes of the same lengths, the gene with desired sequence is separated by Southern Blotting technique.

20. Name the machine where gene cloning is carried out.

Ans. Gene cloning is carried out in a computerised machine known as Thermocycler that uses the method called Polymerase Chain Reaction.

21. What characteristic about the DNA molecule fragments allows them to separate when an electric current is applied to gel in the electrophoresis box ?

Ans. DNA has a partial negative charge that is attracted to the positive end of the electrical current. Hence, DNA fragments move towards the positive terminal, separating on the basis of their size.

22. How DNA fragments are separated ?

Ans. DNA fragments of different sizes can be separated by using gel electrophoresis.

23. When did human genome project begin ?

Ans. The Human Genome Project formally began in 1990 and was completed in 2003.

Chapter 12. Biotechnology and its Applications

1. Give significant contribution of Frederick Sanger ?

Ans. He made contribution to genetics and molecular biology. He was awarded Nobel Prize in Chemistry for structure of insulin protein and protein sequencing.

2. Give significant contribution of H.Boyer ?

Ans. Herber W.Boyer, Stanley, Cohen and Paul Berg together worked on first Artificial recombinant DNA. They also discovered the restriction enzyme endonuclease.

3. What is the best method of achieving sustainable agriculture ?

Ans. Yields of almost all crops are significantly lower in developing countries than developed countries, because of poor soil and pest infestation. Some people argue that the use of organic farming methods, integrated pest management and mixed cropping would be a more appropriate solution than the use of GM technology.

4. What is a GM crop ?

Ans. Genetically modified crop are the crops whose genome has been engineered/altered, to favour the expression of desired traits.

5. Mention the source organism of the gene cry I_{ac} and its target pest.

Ans. Source organism – *Bacillus thuringienses*

Target pest – Cotton bollworm.

6. Bt cotton is resistant to pest, such as lepidopteran, dipterans and coleopterans. Is Bt cotton resistant to other pest as well ?

Ans. Bt cotton is made resistant to only certain specific texa of pests. It is quite likely that in future some other pest may infest that Bt cotton plants. It is similar to immunisation against small-pox which does not provide immunity against other pathogens like those that cause cholera, typhoid, etc.

7. For which variety of Indian rice, patent was filed by a USA company ?

Ans. Indian Basmati was crossed with semi-dwarf variety and was claimed as a new variety for which patent was filed by a USA company.

8. What is golden rice ?

Ans. Golden rice is genetically engineered rice variety which is rich in vitamin A and the genes of Beta carotene are also present in it which is the precursor of Vitamin A.

9. Mention the problems when we consume GM food.

Ans. When we consume GM food, following problems arise :

(i) GM food may cause toxicity

(ii) The bacteria present in the alimentary canal of human could take up the antibiotic resistant gene which is present in GM food, that will cause problems.

10. How transgenic tomatoes have been produced and which name is given to that ?

Ans. In tomato using antisense RNA technology, transgenic plants have been produced which are either 'bruise resistant' (suitable for transport and storage) or exhibit 'delayed ripening' giving more time for ripening on the plant, thus permitting more time for sugar accumulation and also giving higher shelf life. Such trangenic tomato were called 'Flavr Savr'.

11. Are foods from genetically engineered plants less nutritious than comparable foods ?

Ans. Nutritional assessments for foods from genetically engineered plants that have been evaluated by FDA through the consultation process have shown that such foods are generally as nutritious as foods from comparable traditionally bred plants.

12. What kind of GM foods are on the market internationally ?

Ans. All GM crops available on the international market today have been designed using one of three basic traits : resistance to insect damage; resistance to viral infections; and tolerance towards certain herbicides.

13. Given scientific term used for preservation of germplasm at very low temperature.

Ans. Cryopreservation.

14. What was the speciality of the milk produced by the transgenic cow Roise ?

Ans. The first transgenic cow, Rosie produced milk with human alpha-lactalbumin which was nutritionally, more balanced product for human babies than natural cow milk.

15. How transgenic animals, that produces useful products, can be created ?

Ans. Transgenic animals that produce useful biological products can be created by introduction of the portion of DNA (for genes) which codes for a desired product.

16. Name any four transgenic animals commonly produced. Which animal constitutes a major proportion among transgenic animals.

Ans. Four commonly produced transgenic animals are mice, pigs, cow and sheep. 25% of the transgenic animal are mice.

17. Name the biological product made in transgenic animal to treat emphysema. Explain.

Ans. Human protein α-1 antitrypsin is produced to treat emphysema. The gene for this protein is isolated and introduced into a mouse. The transgenic mouse produces the protein, which is to be isolated purified and used on human beings after further clinical trails.

18. How have Transgenic animals proved to be beneficial in :

(i) Production of biological products. (ii) Chemical safety testing.

Ans. (i) **Rosie :** The transgenic cow, produced human proteins containing human α- lactalbumin. Transgenic animals have been made to produce α-1 anti-trypsin used to treat emphysema.

(ii) **Toxicity testing :** Transgenic animals are more sensitive to toxic substance, so the results are obtained in less time.

19. How are two short polypeptide chains of insulin linked together ?

Ans. Two shorts polypeptide chain of insulin are linked together by disulphide bridges.

20. How many amino acids are arranged into two chains of insulin ?

Ans. A molecule of insulin consist of two short polypeptide chain A and B. The chain A has 21 amino acid and chain B has 30 amino acids.

21. State the role of C peptide in human insulin.

Ans. C-peptide (extra stretch of polypeptide) makes the insulin inactive.

22. How is mature, functional insulin hormone different from its pro-hormone form ?

Ans. Nature functional insulin is obtained by processing of pro-hormone which contains extra peptide called C- peptide. This c- peptide is removed during maturation of pro-insulin to insulin.

23. Name a recombinant vaccine that is currently being used in vaccination programs.

Ans. Hepatitis B recombinant vaccine, Engerix-B is used for vaccination of hepatitis's virus.

24. A boy has been diagnosed with ADA deficiency. Suggest any one possible treatment.

Ans. Bone marrow transplant/enzyme replacement therapy/ gene therapy.

25. How transgenic fish are produced by micro-injection technique ?

Ans. In micro-injection technique, Transgene is directly injected into the male pronuclei of fertilised egg, which developed into a transgenic gish.

26. State the cause of adenosine deaminase enzyme deficiency.

Ans. Deletion of gene coding for adenosine deaminase.

27. Why is patent given ?

Ans. Patent is given for producing new produces or inventions or for modifiying and improving earlier inventions.

28. What is legally wrong in the US patent law ? Mention the common items of biopiracy.

Ans. US patent law does not recognize technologies and methods in use in other countries as 'prior past. The common items of biopiracy are soil micro-organisms, plants, animals and their genetic materials.

29. Write the two uses of PCR technique in diagnosis.

Ans. Two uses of PCR technique :

(i) It is used to detect HIV in suspected AIDS patients.

(ii) It is used to detect mutation in gene, in suspected cancer patients.

30. ELISA technique is based on the principles of antigen and antibody interaction. Can this technique be used in the molecular diagnoses of a genetic disorder, such as phenylketonuria ?

Ans. Yes, one can use antibody against the enzyme that is responsible for the metabolism of phenylalanine to develop ELISA-based diagnostic technique. The patient in which the enzyme protein is present would give negative results in ELISA when compared to normal individual.

Chapter 13. Organisms and Populations

1. Who is responsible for the sex of a child-mother or father ?

Ans. Father.

2. At what time the sex is determined ?

Ans. The sex is determined at the time of fertilisation and formation of zygote.

3. What are the two factors which have raised the life expectancy in the developing countries ?

Ans. The two factors which have raised the life expectancy in the developing countries is due to better health care facilities and improved hygiene, diet and nutrition.

4. What is the relationship between population and poverty ?

Ans. Poverty is directly related to population. With increase in population, there develops a scarcity of resources leading to poverty.

5. How does man destroy his own species ?

Ans. Man destroys his own species through deprivation, exploitation, subjugation, violence and wars.

6. Give examples of positive checks on population.

Ans. Hunger, war, spread of epidemic diseases, other natural calamities like floods, earthquake and drought.

7. What demographers study ?

Ans. They study birth-rate, mortality, migrations, survival time and family size.

8. Who put forward the theory of positive checks of population ?

Ans. T. R. Malthus.

9. How is the population growth determined ?

Ans. Population growth is the increase in the size of the population over a period of time. It is designated as N_{t+1} and can be predicted by the formula :

$N_{t+1} = N_t$ + (Natality + Immigration) – (Mortality + emigration)

$N_{t+1} = N_t + (B + I) - (D + E)$

10. Name the two types of population curves.

Ans. J and S shaped curves.

11. In which organisms is the J-shaped curve obtained ?

Ans. Unicellular organisms.

12. Which growth curve is represented by a small population ?

Ans. 'J' shaped growth curve.

13. When do population growth curves assume 'J' shape and sigmoid of 'S' shape ?

Ans. When the food supply becomes insufficient in comparison with population, mass starvation death is result. The graph of population growth will become 'J' shaped.

When there is no increase in population and environment can sustain a limited population, growth curve will become 'S' shaped.

14. How does the number of organisms in a population increase ?

Ans. By birth and immigration.

15. Define Brood parasitism.*

Ans. Brood parasitism, also called social parasitism, is the unique population interaction in which the parasitic bird deposits its eggs in the nests or broods of another individual (host) and let them incubate. For example- a cuckoo lays its eggs in the nest of a crow.

Chapter 14. Ecosystem

1. Mention any two significant roles predation play in nature.

Ans. Predation play following two important roles in nature :

(i) They keep the prey population under control.

(ii) They act as conductors of energy transfer across trophic levels.

2. Name the four functional aspects of an ecosystem.

Ans. Functional aspects of an ecosystems are :

(i) Productivity

(ii) Decomposition

(iii) Energy flow

(iv) Nutrients cycle

3. What is approximate value of net primary productivity of the biosphere ?

Ans. The approximate value of net primary productivity of the biosphere is 170 billion tonnes of (dry weight) organic matter.

4. What is primary productivity ? In what units could you express productivity ?

Ans. The rate at which solar radiation is trapped by producers for the synthesis of organic compound through photosynthesis is called primary productivity. It is expressed as g C m^{-2} Y^{-1} or K cal m^{-2} $year^{-1}$.

5. List two factors that determine the vegetation and soil type of ecosystem.

Ans. Two factors that determine the vegetation and soil type of an ecosystem one :

(i) Precipitation

(ii) Temperature

6. Name the basic requirement of any ecosystem to function and sustain properly.

Ans. A constant input of Solar energy is the ultimate science of all energy and requirement of any ecosystem to function and sustain properly.

7. What is biogas? Name any two main constituents of biogas.*

Ans. Biogas is the mixture of gases produced by the anaerobic breakdown of organic matter such as agricultural waste, manure, municipal waste, plant material, sewage, green waste or food waste by bacteria.

The two main constituents of biogas are methane and carbon dioxide.

8. What is a detritus food chain made up of ? How do they meet their energy and nutritional requirements ?

* **Frequently asked previous years Board Exam Question.**

Ans. Dead organic remains of the plants and animals their feacal matter. They obtain energy by decomposing the dead materials.

9. What is common to earthworm, mushroom, soilmites and dung beetle in an ecosystem ?

Ans. They are all detritivores *i.e.* decomposing organisms, which feed on dead remains of plants and animals.

10. "Decomposition is an oxygen requiring process". Comment.

Ans. Detritus is rich in nitrogen and sugar. For oxidation of nitrogen and sugar, oxygen is required by a class of aerobic microbes.

11. Name the dominant producers in a deep aquatic ecosystems. What other name could you give to a primary consumer ?

Ans. Phytoplanktons are the dominant produces in aquatic ecosystem. Primary consumers are also known as herbivores.

12. What is the starting point of grazing food chain and detritus food chain, respectively ?

Ans. The grazing food chain starts with producers while the detritus food chain starts with decomposers.

13. Man can be primary as well as a secondary consumer. Justify this statement.

Ans. Man has a varied diet. When on a vegetarian diet, they are primary consumers and when on a non-vegetarian diet, they are secondary consumers.

14. Name the trophic level occupied by :

(i) Secondary consumers (ii) Tertiary consumers (iii) Quaternary consumers

Ans. (i) 3rd trophic level (ii) 4th trophic level (iii) 5th trophic level

15. Name one example of inverted pyramid in number.

Ans. Inverted Pyramid of number is observed in tree ecosystem.

16. In terrestrial ecosystem, DFC and GFC are interlinked at certain level. Justify the statement.

Ans. Some of the organism of DFC are prey to the animal of GFC, hence they are interlinked.

17. What are the shortcomings of ecological pyramids in the study of ecosystem ?

Ans. The ecological pyramid assumes in simple food chain and does not accommodate food webs. Thereby it does not take into account the fact that species may belong to two or more trophic levels at a time. Also saprophytes despite their vital role in ecosystem are given no place in the ecological pyramids.

18. How is stratification represented in a forest ecosystem ?

Ans. Stratification in the vertical distribution of species at different levels. Trees occupy vertical strata, shrub the second layer and herbs/ grasses occupy the bottom layers.

19. Name the two forms of reserviors of carbon that regulate the ecosystem carbon cycle.

Ans. Dissolved carbon in oceans and carbon in rocks, sediments and fossils are the two forms of carbon reservoirs.

20. Explain the function of reservoir in a nutrient cycle. List the two types of nutrient cycle in nature.

Ans. The function of the reservior is to meet the deficit which occurs due to imbalance in the rate of influx and efflux. The two types of nutrient cycles are :

(i) Gaseous (ii) Sedimentary

21. Mention the role of pioneer species in primary species in primary succession on rocks.

Ans. The pioneer species invade a bare area and paveway for other species.

22. Mention one similarity between hydrarch and xerarch successions.

Ans. Both hydrarch and xerarch succession result in mesarch (medium water) conditions.

23. How much of carbon is dissolved in the oceans ?

Ans. About 71% of carbon is dissolved in the oceans.

24. Under what condition would a particular stage in the process of succession revert back to an earlier stage?

Ans. Natural or human induced disturbances like fire, deforestation, etc.

25. State what does standing crop of a trophic level represent.

Ans. Standing crop represents the mass of living material (biomass) at a particular time.

26. List any two ways of measuring the organic standing crop of a trophic level.

Ans. Standing crop is measured as the organic biomass or the number in a unit area.

27. What are the contribution of R.Mishra ?

Ans. R. Mishra is the Father of ecology in India. He laid the foundation for understanding of tropical communities and their succession, environmental responses of plants population and productivity and nutrient cycling in tropical forest and grassland ecosystem.

28. Which organism constitute the last trophic level ?

Ans. Tertiary Consumer.

Chapter 15. Biodiversity and its Conservation

1. Name the different types of biodiversity conservation techniques.

Ans. In situ conservation and Ex situ conservation.

2. Name the three important components of biodiversity.

Ans. (i) Genetic diversity (ii) Species diversity (iii) Ecosystem diversity

3. Name three products of forest insects.

Ans. Silk, Honey and Lac.

4. How much land should be under forest in a country?

Ans. 1/3 land should be under forest in a country.

5. How much area is covered by forests in India ?

Ans. 75 million hectare (23% of total land area).

6. Who coined the term Wildlife ?

Ans. William Hornaday coined the term Wildlife.

7. Give the scientific importance of wildlife.

Ans. Wildlife acts as gene bank and provides research materials to produce new varieties of species.

8. Which is the most serious threat to the wildlife ?

Ans. Destruction of habitats is the most serious threat to wildlife.

9. Which animal from India become extinct due to excess hunting ?

Ans. Cheetah, *Acinonyx jubatus*.

10. What % of the living species of animal is threatened ?

Ans. 10% of the living species of animal is threatened.

11. The accelerated rate of species extinction that the world is facing now is largely due to human activities. Group such activities under four major heads and explain.

Ans. Habitat loss and fragmentation; overexploitation; introduction or invasion of alien species; co-extinction.

12. Write the advantages of In situ conservation.

Ans. (i) It aids in maintaining gene banks.

(ii) It helps in preservation of genetic variability.

(iii) It is vital for maintenance of species and check its extinction.

13. Which species should be paid special attention in conservation ?

Ans. Those species who are the sole representative of their genus should be paid special attention in conservation.

14. Categorize the following into *in-situ* and *ex-situ* approaches of biodiversity conservation.

(i) Botanical gardens; (ii) Wild life sanctuaries; (iii) Gene bank; (iv) Biosphere reserves; (v) Sacred forests/lakes; (vi) Pollen banks; (vii) Tissue culture; (viii) Cryo-preservation.

Ans. Ex-situ : Botanical gardens, Gene Bank, Pollen banks tissue culture, Cryopreservation.

In-situ : Wildlife Sanctuaries, Biospheres reserves, Sacred forest/Lakes.

15. Give two soap substitutes provided by forest.

Ans. Ritha, Shikakai.

16. What is the use of field gene bank in conservation ?

Ans. Field gene banks used for collecting Cocoa, Coffee, Rubber, Banana, Coconut, Sweet potato, Mango. These are used to store plants for a short period.

17. What is the drawback of recalcitrant seed?

Ans. These seeds are killed by drying and freezing temperature during ex-situ conversation.

18. Give an example of recalcitrant seed.

Ans. Jackfruit is an example of recalcitrant seed.

19. (i) Name the large areas of scenic and natural beauty maintained for scientific, educational and recreational use.

(ii) Give significant contribution of E.Wilson.

Ans. (i) National parks

(ii) Coined the term Biodiversity

20. Write the names of some major seed banks for agricultural plants :

Ans. Major seed banks for agricultural plants are :

Wheat, barley	–	US small grains collection
Maize	–	CIMMYT, International centre for maize and wheat improvement, Mexico
Rice	–	IRRI, International rice research Institute, Philippines
Potato	–	CIP, International potato centre, Lima and Peru

21. Name the bird which has recently become extinct.

Ans. Dodo of Mauritius.

22. Give two examples of Indian endangered species.

Ans. Great Indian Bustard, Kashmir stag.

23. What for the abbreviation ZSI and IBWL stand ?

Ans. Zoological Survey of India, Indian Board of Wildlife.

24. What is the importance of Red Data Book ?

Ans. The IUCN produces Red Data Book on species at risk (endangered) including vascular plants, mammals. BSI brings out an inventory of endangered plants in the form of Red Data Book. Similarly a Red Data Book for endangered animals has been completed by Zoological Survey of India.

Chapter 16. Environmental Issues

1. Name the main source of air pollution.

Ans. Pollutants emitted by automobiles are the main source of air pollution.

2. What is biomagnification? Write two effects of biomagnification.*

Ans. Biomagnification refers to the accumulation of non-biodegradable toxic substances in the food chain. Following are the effects of biomagnification on living organisms and the environment:

(i) Biomagnification makes humans more prone to cancer, kidney problems, liver failure, birth defects, respiratory disorders, and heart diseases.

(ii) The pesticides and chemicals such as DDT and mercury released into the lakes and rivers are ingested by the aquatic organisms. These get accumulated in their body tissues and are transferred to other organisms that feed on them.

3. Mention the six harmful effects of noise on human health.

Ans. Six harmful effects of noise on human health are :

(i) Stress
(ii) Sleeplessness
(iii) Increased rate of heat beat
(iv) Altered breathing patterns.
(v) Damage of eardrum if sound levels are very high
(vi) Damage of heaving abilities at low level noise

* Frequently asked previous years Board Exam Question.

4. List two advantages of the use of unleaded petrol in automobile on fuel.

Ans. (i) Allow the catalytic convertor to remain active.

(ii) Reduce air pollution.

5. It is a common practice to plant tree and shrubs mean the boundary walls of building. What purpose do they serve ?

Ans. The plant growing near the boundary will act as barrier for sound pollution and act as dust catchers.

6. A Factory drains its waste water into the nearby lake. It has caused algal bloom.

(i) How was the algal bloom caused ?

(ii) What would be the consequences ?

(iii) Name the phenomenon that caused.

Ans. (i) Algal bloom is caused due to large amounts of nutrients present in the waste water.

(ii) It causes deterioration of water quality and increased fish mortality rate. Some bloom forming algae are toxic to human beings and animals.

(iii) Accelerated eutrophication.

7. Mention the information that the health workers derive by measuring BOD of water body.

Ans. By measuring BOD of a water body, health workers find the amount of dissolved oxygen in water. The lesser the amount of dissolved oxygen, the more polluted the water body will be.

8. How do algal bloom affect the life in the water bodies ?

Ans. (i) Algal bloom pollutes water and deteriorates the water quality by depleting the oxygen content. This makes the water toxic and increase its BOD.

(ii) Algal blooms lead to death of aquatic organism due to oxygen depletion.

9. State the cause of accelerated eutrophication.

Ans. Pollutants from human activities, effluents from industries effluents from home, sewage and agricultural chemical waste radically accelerated the aging process.

10. Explain accelerated eutrophication. Mention any two consequences of this phenomenon.

Ans. Accelerated eutrophication is the nutrient enrichment of water bodies due to human activities like passage of sewage. Consequences are :

(i) Large amount of nutrients in waters causes excessive growth of planktonic algae (called algal bloom) which impart characteristic colour to water bodies.

(ii) Depletion of oxygen content of water leading to the death of aquatic organism.

11. Explain the process of secondary treatment given to the primary effluent up to the point it shows significant change in the level of biological oxygen demand (BOD).

Ans. The primary effluent is passed into large aeration tank where it is constantly agitated. Air is pumped into it mechanically. This allow vigorous growth of useful aerobic microbes into floes. These microbes consume the major part of organic matter in the effluent (this significantly reduces the BOD of the effluent).

12. BOD of two samples of water A and B were 120 mg/L and 400 mg/L respectively, which sample is more polluted ?

Ans. Sample B is more polluted than sample A.

13. What can be the effect of discharging hot (thermal) water into a water body with organisms in it ?

Ans. (i) Number of organism will die that are sensitive to high temperature.

(ii) Hot water will reduce oxygen content of water and thus increased BOD will affect the aquatic animals.

14. Name four harmful/ toxic compound found in defunct ships.

Ans. Defunct ship contains :

(i) Asbestos

(ii) Poly chlorinated biphenyl (PCB)

(iii) Tributylin (TBT)

(iv) Mercury

15. Mention how e-waste is produced and dispersed off. Write the solution for its treatment.

Ans. Irreparable computers and other electronic good are known as electric wastes (e-wastes) E-wastes are buried in land fills or incnerated. Recycling is the only solution for the treatment of e-wastes.

16. Name any two sources of e-wastes and write two different ways for their disposal.

Ans. Sources : Irreparable computers and electronic goods.

Methods of disposal : burying in landfills, incineration, recycling.

17. Write a short note on the contribution of Ahmed Khan of Bangalore.*

Ans. Ahmed Khan, a plastic sack manufacturer in Bengaluru was successful in finding a problem of plastic waste accumulation :

(i) In collaboration with R. V. College of Engineering and the Bangalore City Corporation, he developed polyblend, a powdered structure manufactured from waste plastic which when mixed with bitumen (an important ingredient in constructing roads), increased the water repelling property of roads that increased the average lifetime of roads three times.

(ii) The raw material for creating polyblend is any plastic film waste. So, against the price of ₹ 0.40 per kg that rag pickers had been getting for plastic waste, Khan now offers ₹ 6. So, he was successful in finding a problem with plastic waste accumulation.

18. What is Dobson unit?*

Ans. Dobson unit is used for measuring the thickness of ozone in a column of air from ground to top of the atmosphere. One Dobson Unit is the equivalent of 2.69×10^{16} molecules of ozone per square cm.

19. State the effect of UV-B on human eye.

Ans. UV-B is absorbed by human eye and at high does it causes inflammation of cornea. This is called snow blindness cataract.

20. Mention the effect of UV rays on DNA and proteins in living organisms.

Ans. The high energy of UV rays breaks the chemical bonds within DNA and protein molecules.

21. Refrigerants are considered to be a necessity in modern living, but are said of be responsible for ozone holes detached in Antarctica. Justify.

Ans. The widely used refrigerants are CFC's or chlorofluoro carbons. CFC's discharged in the lower part of atmosphere move upward to the stratosphere. Here the UV rays act on them and release chlorine atoms. These free chlorine atoms react with ozone to release molecular oxygen. Chlorine atoms are not consumed in this reaction and hence these continuously degrade ozone and have resulted in ozone hole.

$$ClF_2C - Cl \xrightarrow{h\nu} ClF_2C^{\bullet} + Cl^{\bullet}$$

$$Cl^{\bullet} + O_3 \longrightarrow ClO^{\bullet} + O_2$$

$$ClO^{\bullet} + O_3 \longrightarrow Cl^{\bullet} + 2O_2$$

22. State the purpose of signing the Montreal protocol.

Ans. Montreal protocol was signed at Montreal, in 1987 to curb the emission of ozone depleting substances.

23. (i) What is Amrita Devi Bishnoi wildlife protection award given for?

(ii) Where did the Chipko Movement start?

Ans. (i) It is given to individual or community from rural areas that has shown extra ordinary courage and dedication for protecting wildlife.

(ii) It started in Garhwal.

❒

* **Frequently asked previous years Board Exam Question.**

Short Answers

Set 7

Chapter 1. Reproduction in Organisms

1. Do all the gametes found from a parent organism have the same genetic composition.

Ans. Gametes do not have the same genetic composition formed from the parent organism by meiosis. Gametes formed by mitotic division have same genetic composition.

2. How is the community of species maintained generation after generation ?

Ans. Community of species is maintained generation after generation by the process of reproduction.

3. In yeast and amoeba, the parent cell divides to give rise to the new individual cell. How does the cell division differ in these two organisms ?

Ans. In Amoeba, binary fission takes place whereas in yeast cell division occurs by budding.

4. Give the basic features of asexual reproduction.

Ans. Basic features of asexual reproduction :

(i) Only one parent is involved.
(ii) All the cell divisions are mitotic.
(iii) All the offsprings are genetically similar to the parent.
(iv) The reproductive unit is a fragment or specialised part of the parent.

5. Name various methods of asexual reproduction in animal with example.

Ans. Asexual reproduction in animals can be of following types :

(i) Binary fission *e.g., Amoeba.*
(ii) Budding *e.g., Hydra.*
(iii) Fragmentation *e.g.,* Planarian flatworms.
(iv) Plasmotomy *e.g., Opalina.*
(v) Gemmule formation *e.g.,* Freshwater sponge.
(vi) Regeneration *e.g.,* Sponges, Coelenterates, etc.

6. List two advantages and two disadvantages of asexual reproduction.

Ans. Advantages of asexual reproduction :

(i) It is resulted by simple process of divisions amitosis or mitosis.
(ii) Large number of offsprings are produced from single parent which are dispersed to far off places.

Disadvantages of asexual reproduction :

(i) The offsprings so produced do not possess much capacity to adapt rapidly to environment.
(ii) As there is no fusion of gametes, hence genetic recombination, interaction, etc., do not occur, so there is no variation caused in offsprings.

7. Give examples for each of the following :

(i) Vegetative propagation by stems.
(ii) Vegetative propagation by roots.
(iii) Vegetative propagation by leaves .

Ans. Vegetative propagation by stems : Onion, *Colocasia*, Potato.

Vegetative propagation by roots : *Murraya sp. Albizia lebbeck, Dalbergia sissoo.*

Vegetative propagation by leaves : *Bryophyllum, Begonia, Saintpaulia.*

8. Give two examples of each in which the following is practised—Cutting, Layering and Grafting.

Ans. Cutting — Sugarcanes, Grapes, Rose, *Bougainvillia*
Layering — Jasmine, Citrus
Grafting — Rubber plant, Apple, Mango, Guava

9. Give examples of plants which are propagated vegetatively from underground stems, creeping stems and aerial shoots.

Ans. Underground Stems : Mint, *Chrysanthemum,* Banana, Turmeric, Ginger, *Aspidium, Adiantum*

Creeping Stems : Runners (Mint grass), stolons (strawberry) and offset (*Eicchornia*)

Aerial Shoots : *Opuntia,* Sugarcane.

10. Explain the common method of vegetative propagation in rose and sugarcane.

Ans. In rose and sugarcane, cutting is the most common method of vegetative propagation. In this method, stems are cut obliquely into long pieces with minimum two nodes and one internode. Each cutting is planted in moist soil, which then develops adventitious roots and shoots and grows into new plants.

11. Define Grafting.

Ans. Grafting is a process/method, where root portion called stock is taken from one plant and the stem part called scion is taken from another plant. The scion is attached to the stock and becomes an integrated part of the rooted plant.

12. Enumerate the merits of vegatative propagation (only three).

Ans. Merits of Vegatative Propagation :

(i) It makes possible propagation of plants which cannot reproduce sexually *e.g.,* Banana, Grapes, Oranges, Rose and Jasmine.

(ii) This method is more rapid and less expensive.

(iii) All plants produced will have same characters and hereditary potential.

13. What do you understand by mound layering ?

Ans. In mound layering the lower branch of a plant is bent down close to soil and covered with moist soil in such a way that its growing tip remains above the soil surface. After a few days the covered portion of stem produces adventitious roots. The rooted branch is then cut and grown as in independent plant, *e.g.,* Lemon, Jasmine.

14. List two advantages and two disadvantages of sexual reproduction.

Ans. Advantages of sexual reproduction :

(i) Fusion of gametes leads to genetic combination and causes variations in offsprings.

(ii) It plays an important role in evolution since organisms so produced adapt quickly to environment.

Disadvantages of sexual reproduction :

(i) Two parents of opposite sexes are needed since it is biparental reproduction.

(ii) There are fifty percent chances of conception after fusion.

15. List various methods of sexual reproduction.

Ans. Various methods of sexual reproduction are :

(i) **Syngamy :** The fusion of gametes takes place completely and permanently. It is of following kinds :

(a) Autogamy
(b) Exogamy
(c) Hologamy
(d) Paedogamy
(e) Mesogamy
(f) Isogamy
(g) Anisogamy
(h) Macrogamy
(i) Microgamy

(ii) **Conjugation** is temporary union of the two individuals of same species. *E.g.,* Ciliates.

(iii) **Automixis** (Autogamy)

16. Mention two important characteristics in the sexual reproduction of frog and bony fishes.

Ans. Both of them release the mature gamete simultaneously. They also release a large number of gamete in the water to enhance the chances of syngamy.

17. Which is the most critical event in sexual reproduction ?

Ans. Fertilisation of gametes is the most critical event in sexual reproduction.

18. Write short notes on :

(i) Oogenesis
(ii) Gametogenesis
(iii) Advantages of Artificial insemination
(iv) Gemmulation
(v) Precocious puberty
(vi) Menopause
(vii) Menstrual cycle.

Ans. (i) **Oogenesis :** The process of formation of egg or ovum from egg mother cell by meiotic cell division in ovary is known as oogenesis. Oogenesis takes place in the graafian follicles in ovary.

(ii) **Gametogenesis :** Gametogenesis is the process of formation and development of gametes both sperm and ova in reproductive organs *i.e.,* testis and ovary respectively. Gametes are always haploid hence meiosis occur in the germinal epithelium of testis and ovary.

(iii) The advantages of artificial insemination are :

(a) Healthy progeny may be produced.

(b) It prevents the spread of certain disease and sterility.

(c) New characters may be brought in the progeny.

(d) It is economical.

(e) It may be used at the proper time of the reproductive cycle.

(iv) **Gemmulation :** Poriferans (*e.g., Spongilla*) reproduce by formation of special internal buds called archaeocytes. They remain covered by hard covering. During unfavourable periods the gemmules are released. As the favourable period returns those gemmules germinate into adult sponges.

(v) **Precocious puberty :** Normally a women's fertile life starts from the age of puberty (about 13 years), but under some abnormal conditions like high level of sex hormone (LH and FSH), menstruation starts at an early age than the normal, it is called precocious puberty.

(vi) **Menopause :** The natural physiological stoppage of menstruation or the arrest of reproductive capacity at the age of 45 to 50 is called menopause. At this stage, women is unable to bear the children.

(vii) **Menstrual cycle :** Menstrual cycle is defined as the cyclic changes occurring in ovaries and female reproductive tract of female primates during the reproductive phase.

It is meant to prepare them for fertilisation and pregnancy. The cycle takes about 28 days. The menstrual cycle is characterized by vaginal bleeding which includes the unfertilised egg, mucus and blood and is called menstrual flow/menstruation which lasts for 4-5 days.

Chapter 2. Sexual Reproduction in Flowering Plants

1. What is the cyathium type of inflorescence ?

Ans. Cyathium is a special type of inflorescence. It is modified cymose inflorescence, consisting of involucre of bracts which forms a cup. Single female flower arises in the middle surrounded by a large number of male flowers. Female flower is pedicellate consisting of pistil only, whereas male flowers are in the form of stamens. *E.g., Euphorbia.*

2. Write short notes on the following :

(i) Verticillaster (ii) Hypanthodium (iii) Cyathium

Ans. (i) **Verticillaster :** In this inflorescence, the condensed form of biparous cyme with a cluster of sessile flowers arise in the axil of a leaf at the node. The first axis ends in flower. Two branches arise below it bearing branches in an alternate manner *E.g., Ocimum.* Due to overcrowding, the two opposite clusters form an apparent whorl or verticil.

(ii) **Hypanthodium :** In this inflorescence, the main axis form a cup-shaped receptacle with an apical opening. Flowers are enclosed with the inner lining of cup. The flowers are unisexual. The female flowers develop at the base of the cavity and male flowers towards the apical pore, *E.g., Ficus.* Each male flower represented by a single stamen and each female flower by a single pistil.

(iii) **Cyathium :** It is a special type of condensed cymose inflorescence which appears like a single flower having a conical receptacle and a green or coloured involucre of 4-5 bracts with nectariferous glands. The receptacle bears a single female flower in the centre and a large number of centrifugally arranged male flowers around it. These flowers have single essential organ; pistil in female and stamen in male flowers. Cyathium inflorescence is characteristic of *Euphorbia, poinsettia.*

3. Define inflorescence. What are the types ?

Ans. The arrangement of flowers on the floral axis is called inflorescence.

Depending on whether the apex gets converted into a flower or continues to grow, two major types of inflorescences are :

(i) Racemose (ii) Cymose

In racemose type, the main axis continues to grow, the flowers are borne laterally in an acropetal order.

In cymose type, the main axis terminates in a flower, hence is limited in growth. The flowers are borne in a basipetal order.

4. What are the functions of Calyx, Corolla and Thalamus ?

Ans. Functions of Calyx :

(i) The sepals protect the inner floral parts during bud condition.

(ii) They support inner parts like petals, stamens and carpels at the time of pollination.

(iii) Sometimes they attract the pollinating agents as in *Garden nasturtium.*

(iv) In case of *Sonchus* and other members of family Compositae, they get modified and help in dispersal of seeds.

Functions of Corolla :

(i) Being beautifully coloured, they attract the insects for pollination.

(ii) The petals protect the inner floral parts.

Function of Thalamus : Thalamus supports the arrangement of floral parts on it; but it may become fleshy and edible part of some fruits (*e.g.,* Apple).

5. Classify fruits according to the nature of Pericarp and condition of Gynoceium.

Ans. (i) On the basis of nature of pericarp, the fruits may be dry (dry pericarp) or succulent (fleshy pericarp).

(ii) On the basis of the condition of gynoecium, the fruits are classified as—

(a) **Simple fruits :** Simple fruits develop from the monocarpellary (single) ovary or multicarpellary syncarpous ovary. E.*g.,* Maize, Mango.

(b) **Aggregate fruits :** These fruits develop in groups, each of which is formed from free carpel of apocarpous their number corresponds to the number of carpels in a flower. *E.g.,* Buttercup, Custard apple.

(c) **Composite fruits :** These are complex fruits formed from the whole inflorescence. *E.g.,* Fig, Pinapple.

6. Describe the process of double fertilisation in angiosperms. What is its significance? *

Ans. Double fertilisation in angiosperms involves a fusion of a female gametophyte with two male gametes. The haploid sperm-egg combines to form a diploid zygote. The other sperm nucleus fuses with the other two haploid polar nuclei of the megagametophyte to form a triploid nucleus, which develops into an endosperm and is called the primary endosperm nucleus. This entire phenomenon of fertilisation that involves the fusion of the egg and one male gamete and the fusion of the other male gamete with the secondary or the polar nuclei is called double fertilisation.

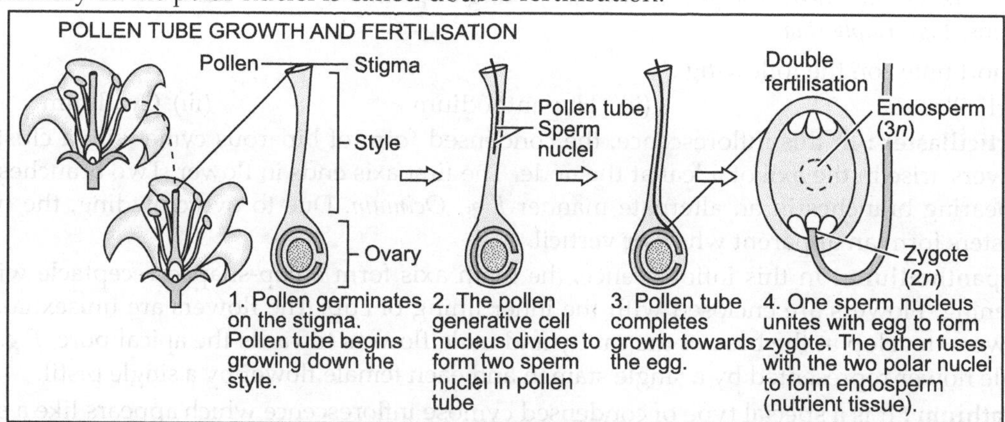

Significance: Double fertilisation leads to the development of triploid endosperm which provides sufficient food material to the developing healthy seeds. It introduces genetic variations in the species. It restores the diploid condition by fusion of haploid male and female gametes.

7. What are the advantages of cross-pollination ?

Ans. Advantages of Cross-Pollination :

(i) It increases variability and certain new and useful characters appear in the offsprings.

(ii) Sometimes harmful characters are eliminated.

(iii) It leads to production of new varieties.

* **Frequently asked previous years Board Exam Question.**

(iv) Several disease resistant varieties of important crop plants have been produced as a result of cross pollination.
(v) It increases the power of adaptation among the offspring.
(vi) The number of seed production can be increased among useful cereal crops.

8. State four features of flowers pollinated by insects.*

Ans. Characteristics of flowers pollinated by insects are :
(i) These flowers are often attractive, with bright petals and are fragrant to attract the insect visitors to them.
(ii) They often have small, compact and sticky stigmas to allow the insect to perch on it.
(iii) Pollen grains are sticky and spiny so that they can stick easily to insect's body parts.
(iv) Many of the insect-pollinated flowers also secrete nectar which attracts bees, butterflies or other similar insects to the flowers.

9. What is the significance of dispersal of seeds ? Give any two points ?

Ans. Dispersal of seeds away from the main plant by a number of modes is very important. These modes may include one or a combination of gravity, wind, water or animals. Seed dispersal has several advantages for the plant such as :
(i) Protection from predation and avoids intraspecific competition.
(ii) Distribution of seeds avoids the problem of competition for basic requirements and results in the healthy growing plant. Helps in rapid evolution.

10. What are the features of bird pollinated flower ?

Ans. Features of bird pollinated flowers are :
(i) Large-sized, scanted flowers.
(ii) Flowers have bright colours to attract birds from a long distance. Usually yellowish- orange, reddish or bluish in colour.
(iii) Corolla is usually funnel-shaped and floral parts are strong and leathery.
(iv) These flowers produce large amount of nectar. *E.g., Bigonia*, Bottle brush.

11. What is double fertilisation ? Write its significance.

Ans. Double fertilisation involves fusion of egg and one male gamete forming diploid zygote and fusion of second male gamete with diploid secondary nucleus forming triploid primary endosperm nucleus.

Significance : By double fertilisation, endosperm is formed from primary endosperm nucleus. This endosperm is required for developing embryo.

12. What is triple fusion ?

Ans. During formation of primary endosperm nucleus, two polar nuclei fuses to form secondary nucleus which again fuses with one of the male nucleus forming primary endosperm nucleus. This process is called triple fusion, as it involves fusion of three nuclei.

13. What are the main functions of fruits ?

Ans. Function of fruits :
(i) Protection of seeds in the early stages till they become fully mature.
(ii) Dispersal of seeds.
(iii) Their decay provides raw material for the benefit of new seedling.
(iv) Many fruits are consumed by human being as food and it is source of glucose and pectin.

14. How seed dormancy can be broken ?

Ans. Seed dormancy can be broken by :
(i) Weakening of the seed coat in the alimentary canal of seed eating birds and animals.
(ii) Microbes of soil also weakens the seed coat.
(iii) Treatment of seed with concentrated H_2SO_4 also helps in breaking seed dormancy.
(iv) The inhibitors may be activated by heat or cold.
(v) Growth hormones like gibberellins can overcome dormancy.

15. Give significant contribution of scientist—P. Maheshwari ?

Ans. Panchanan Maheshwari was an Indian botanist known for his invention of the technique of test fertilisation of angiosperms or plant tissue culture.

* **Frequently asked previous years Board Exam Question.**

Chapter 3. Human Reproduction

1. How do primary sex organs control the growth, function and maintenance of secondary sex organs ?

Ans. Primary sex organs produce gametes and sex hormones whereas secondary sex organs neither produce gametes nor secrete sex hormones but perform important functions in reproduction. The activities of secondary organs are ultimately controlled by the primary sex organs. The Leydig cells of testes secrete testosterone to control the functioning, growth and maintenance of secondary sex organs. Similarly, the Graafian follicles of ovary secrete oestrogens and progesterone to control the growth and functions of female secondary sex organs. Thus, primary sex organs control the growth, function and maintenance of secondary sex organs.

2. List various changes occurring during spermiogenesis.

Ans. The following changes occur during spermiogenesis :

(i) The developing spermatozoa increase in their length.

(ii) The weight of the gamete is reduced alongwith development of locomotory structures.

(iii) Loss of water from the nucleus results in its compaction, wherein it forms the major part of head of spermatozoa. DNA, RNA and nucleolus become concentrated.

(iv) Golgi complex of spermatid gives rise to acrosome on the head of the sperm.

(v) The two centrioles (proximal and distal) of the spermatid get arranged one after the other, behind the nucleus in the neck region.

(vi) Distal centriole forms the axial filament of the spermatozoa.

(vii) Mitochondria from different parts of the spermatid get arranged in the middle piece around axial filament and form nebenkern.

(viii) Much of the cytoplasm of a spermatid is lost and forms a thin layer called manchette around middle piece.

3. State the role of hormones, acting during menstrual cycle.

Ans. Role of hormones in the menstrual cycle :

(i) **FSH (Follicle Stimulating Hormone) :**

(a) This hormone is secreted by anterior pituitary, it stimulates the growth of selected ovarian follicles and maturation of the primary oocyte.

(b) It also stimulates the follicular cells of Graafian follicle to secrete a female sex hormone named Oestradiol (Oestrogen)

(ii) **Role of Luteinizing Hormone (LH) :** It induces the mature Graafian follicle to burst and eject its egg into fallopian tube and stimulates the formation of corpus luteum.

(iii) **Role of Progesterone :**

(a) It maintains the hypertrophy of endometrial lining in the uterus and fallopian tube.

(b) It is required for proper implantation of foetus in the wall of uterus.

(c) It stimulates the endometrial glands to secrete a nutrient fluid for foetus.

(d) It inhibits the release of FSH.

4. What is ovulation ? What happens to graafian follicle after ovulation ?

Ans. The release of eggs (at secondary oocyte stage) after rupturing of graafian follicle is called ovulation. After the ovulation, the granulosa cell as well as the stroma cells form theca interna rapidly multiply to fill the cavity of graaffian follicle which becomes corpus luteum. If fertilisation occurs, corpus luteum grows further and secrete hormone progesterone. If fertilisation does not take place, corpus luteum regresses and forms a white body called corptus albicans.

5. (i) Name the various foetal membranes. (ii) Mention any four functions of placenta.

Ans. (i) The name of various foetal membranes are :

(a) Amnion (b) Chorion (c) Allantois (d) Yolk sac

Short Answers

(ii) Functions of Placenta are :

(a) **Nutrition :** The mother supplies all nutrients like minerals, amino acids, sugars, lipids, vitamins etc., required by the foetus through the placenta.

(b) **Exchange of gases :** Placenta helps in exchange of respiratory gases (O_2 from mother and CO_2 from foetus) between foetus and mother.

(c) **Excretion :** The nitrogenous waste generated by the foetus is eliminated through placenta.

(d) **Antibodies :** Placenta is permeable to antibodies of the mother which enter the foetal blood and provide immunity against several diseases.

(e) **Hormones :** Placenta is an important endocrine gland that produces a number of hormones.

6. Write short notes on :

(i) Placenta (ii) Implantation

Ans. (i) Placenta : Placenta is a link between foetus (embryo) and mother's uterine wall formed by the union of foetal and uterine tissue. It helps embryo in nutrition, respiration and excretion by either taking nutrients, O_2 from mother or transporting excretory products, CO_2 to mother.

The placenta serves primarily as an organ that permits the interchange of material carried in the blood of mother and foetus. The main functions of placenta are :

(a) **Nutrition :** Supply of nutrient material to foetus.

(b) **Respiration :** Supply of O_2 to foetus and receives CO_2 back from it.

(c) **Excretion :** Fluid nitrogenous waste products escape through the placenta.

(d) **Barrier :** The placenta is a barrier like semi-permeable membrane.

(e) **Storage :** The placenta stores fat, glycogen and iron for the embryo before the formation of liver.

(f) **Hormonal function :** The placenta secretes extra ovarian hormones-estrogen and progesterone in female during pregnancy that serves to maintain foetus.

(ii) Implantation : It is embedding of the blastocyst into endometrium of uterus. The blastocyst comes in contact with the endometrium in the region of embroynal knob. The surface cells of trophoblast secrete lytic enzymes which cause corrosion of endometrial lining. These give rise to finger-like projection called chorionic villi, which helps in fixation and absorption of nutrients. The blastocyst sinks into a pit formed in the endometrium and then got completely buried in it and covered by cells of endometrium. This is implantation, which begins about 7th day after fertilisation of ovum. It takes 3 days to complete.

7. Briefly describe the secretory phase of the menstrual cycle.

Ans. During the secretory phase, progesterone hormone is produced by the ovaries. Progesterone and estrogen are secreted by the corpus luteum that develops from the Graafian follicle. Luteinizing hormone (LH) induces the formation of a corpus luteum (which degenerates if pregnancy does not occur). Uterine movement or contraction decreases.

Progesterone stimulates the further build-up of the cells in the endometrium of the uterus and stimulates the glands in the uterus to secrete substances that maintain the endometrium and prevents it from breaking down. The glands become enlarged and if implantation does not occur the endometrium sheds after 28 days. Therefore, this phase of the menstrual cycle is called the secretory phase (Days 14-28 of the menstrual cycle). Decline of progesterone level starts menstrual phase.

8. What is cleavage ? Give difference between holoblastic and meroblastic cleavage.

Ans. Cleavage : It is the segmentation which divides the fertilised egg by repeated mitotic division into a hollow multicellular blastula with increasing number of blastomeres and of progressively decreasing size.

Holoblastic Cleavage : In this cleavage, the furrows divide the entire fertilised egg completely. It may be equal or unequal type *e.g.*, in frog eggs in human beings.	**Meroblastic Cleavage :** In this cleavage, the furrows do not bisect the entire egg completely. The furrow remains confined to the small amount of active cytoplasm at the animal pole. It may be discoidal or superficial. *e.g.*, in birds.

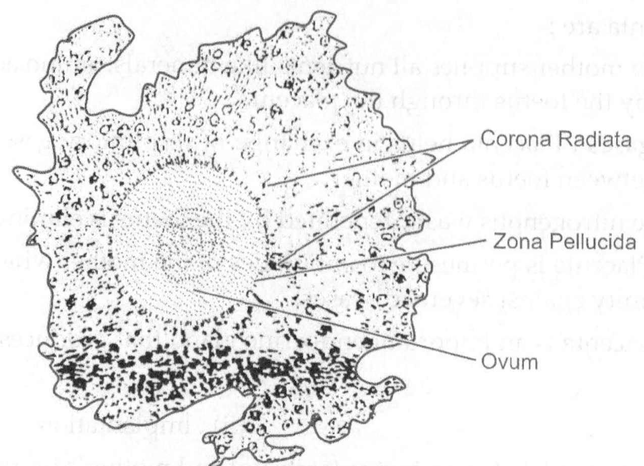

Fig. : Structure of human ovum

9. What is Lactation and which two hormones are responsible for it ?

Ans. Lactation is the secretion of milk from the mammary glands, soon after child birth. In humans, the process of feeding milk is called breastfeeding or nursing. A principal hormone in promoting lactation is prolactin released from the anterior pituitary gland. It is released during response to, prolactin releasing hormone (PRH) secreted by hypothalamus. The principal stimulus in maintaining prolactin secretion during lactation is the sucking action of the infant. Sucking initiates nerve impulses from receptors in the nipples to the hypothalamus. The impulse decrease release of prolactin inhibiting hormone (PIH) and/ or decrease release of prolactin releasing hormone (PRH), so more prolactin is released by the anterior pituitary gland. The sucking action also initiates nerve impulses to the posterior pituitary gland via the hypothalamus. These impulses stimulate the release of the hormone oxytocin from the posterior pituitary gland. Oxytocin induces the ejection of milk by compressing the alveoli.

10. Give an account of hormonal control of oogenesis.*

Ans. In females, the ovarian cycle is regulated by hormones secreted by the hypothalamus, pituitary, and ovaries. The hypothalamic hormone GnRH (Gonadotropin-Releasing Hormone) causes the release of FSH (Follicle-Stimulating Hormone) and LH (Luteinizing Hormone) from the anterior pituitary. Simultaneously, estrogen and progesterone are released from the developing follicles present in ovaries. FSH stimulates the development of ova, which further develop in structures called follicles. Follicle cells in turn produce inhibin, which inhibits FSH production. LH also plays a role in the development of ova, as well as in the induction of ovulation and stimulation of estradiol and progesterone production by the ovaries. Estradiol and progesterone are steroid hormones that prepare the body for pregnancy. Estradiol is the reproductive hormone in females that assists in endometrial regrowth, ovulation, and calcium absorption; it is also responsible for the secondary sexual characteristics of females. These include breast development, flaring of the hips, and a shorter period necessary for bone maturation. Progesterone assists in endometrial regrowth and inhibition of FSH and LH release.

Chapter 4. Reproductive Health

1. Define reproductive health ?

Ans. Reproductive health is a state of complete physical, mental and social well-being and not merely the absence of disease relating to the reproductive system and its functions and process.

2. What is reproductive fitness ? Explain it with the help of an example.*

Ans. Reproductive fitness is the individuals reproductive success (equal to overage contribution to gene pool). For example : Ascaris lays 2 crore eggs.

3. Give four causes of infertility in males.*

Ans. Causes of infertility in males are :

(i) Y chromosome deletion and other genetic problem leads to infertility in males.

(ii) Blockage in the Vas deferens, acidity in urethra causes infertility.

* **Frequently asked previous years Board Exam Question.**

(iii) Deformities in the shape and size of the sperm.
(iv) Age, senility and injury to reproductive part are some other causes of infertility.

4. Describe the lactational amenorrhea method of birth control.

OR

Define Amenorrhea.

Ans. Lactational amenorrhea is based on the principle that during the period of intense lactation after parturition, stoppage of mensuration occurs or the ovulation does not takes place.

5. Mention any four characteristics that an ideal contraceptive should have.

Ans. An ideal contraceptive should :
(i) Be easily available.
(ii) Be effective and reversible with least or on side effect
(iii) Not interfere with the sexual drive/desire
(iv) Be user-friendly.

6. What is meant by artificial insemination ?

Ans. It is a technique by which semen collected from the husband or a healthy donor is artificially introduced either into the vagina or into the uterus of the female.

7. Write an account on any two sexually transmitted diseases.

Ans.

Disease	Transmission	Symptoms	Prevention and Cure
(i) Syphilis *(Treponema palladium).*	Entry of bacteria through direct contact, through cuts or cracks in the skin, sexual intercourse with an infected person.	Lesions in the mucous membrane of urinogenital tract and oral cavity.	(a) Hygienic conditions (b) To avoid going to prostitutes. (c) Use of antibiotics. (d) Avoid promiscuity and use of condom
(ii) Gonorrhoea *(Neisseria gonorrhoae)*	Transmission during sexual acts from one person to another.	Inflammation of mucous membrane of urinogenital tract results in burning sensation during urination.	(a) Avoid direct or indirect contact with the suffering persons. (b) Use of antibiotics. (c) Personal hygiene is very important. (d) Suffering persons must go in for proper treatment.

Chapter 5. Principles of Inheritance and Variations

1. Explain the concept of probability in relation to genetics.

Ans. In genetics, the probability concepts decide the extent of agreement of the experimental results with a theoretically expected model, to predict the results of a hybridization programme.

Definition : Number of a possible outcomes (or happening) of an event divided by the total number of possible outcomes of that event. Probability is the study of the likelihood of the occurrence of a particular event or offspring. The chances or probability that an event will take place can be expressed as a fraction (1/4), ratio (1 : 4) or % (25%).

Example : In monohybrid cross heterozygous F_1 hybrid (Tt) can produce two kinds of gametes (T) and (t). The probability of gamete that carries the (T) allele is 1/2. Similarly the probability of gamete that carries the (t) allele is 1/2.

2. Define gene. What are its important characteristics ?

Ans. Gene is the unit of inheritance. Important characteristics are as follows :
(i) Genes are found on chromosomes at fixed positions.
(ii) Each gene is responsible for a providing particular trait to an organism.
(iii) Chemically gene is a segment of a large polynucleotide molecule called DNA (deoxy ribonucleic acid) and RNA (ribonucleic acid).

3. What is the principle of segregation ? Why is it important ?

Ans. The principle of segregation states that an organism possesses two alleles for any one particular trait and that these alleles separate during the formation of gametes. In other words, one allele goes into each gamete. The principle of segregation is important because it explains how the genotypic ratios in the haploid gametes are produced.

4. Assume that in a series of experiments, plants with round seeds were crossed with plants with wrinkled seeds and the following offspring were obtained : 220 round and 180 wrinkled.

(i) What is the most probable genotype of each parent ?

(ii) What genotypic and phenotypic ratios are expected ?

(iii) Based on the information provided in part (b) above, what are the expected (theoretical) numbers of progeny (400 total) of each phenotypic class ?

Ans. (i) Assuming that round (W) is dominant to wrinkled (w) : Ww × ww

(ii) 1 : 1 - both genotypic and phenotypic ratio.

(iii) 200

5. How do Mendel's experiments show that traits may be dominant or recessive ?

Ans. Mendel took one tall pea (TT) plant and one short pea (tt) plant and produced progeny from them. The plants grown from F_1 seeds represent first filial (F_1) generation. All F_1 plants were tall. Then Mendel self-pollinated F_1 plants and found that all plants obtained in F_2 progeny were not tall. Instead 75% were tall (dominant) while remaining 25% were dwarf (recessive traits).

From the experiment, he concluded that F_1 progeny is not a true breed but carries both traits. He concluded one trait is dominant over the other, that is why plant appear tall. It is also called the law of segregation.

6. How garden pea (*Pisum sativum*) proved to be a suitable plant for Mendel's work ?

Ans. Garden pea (*Pisum sativum*) proved a suitable plant for Mendel's work due to the following reasons :

(i) These plants are easy to cultivate and complete their life cycle in one season only.

(ii) They have sharply distinct heritable differences, but Mendel selected only seven characters.

(iii) Method of pollination in these plants is self pollination. Therefore they are pure lines.

(iv) The hybrids, produced are fertile.

(v) It is easy to conduct cross pollination in these plants, if required.

(vi) Large number of offspring will be obtained so experimental error can be minimized.

7. How do Mendel's experiments show that traits are inherited independently ?

Ans. When a pea plant of round green (RRyy) seeds is crossed with a pea plant having wrinkled yellow seeds (rrYY), plants in F_1 progeny have round and yellow seeds. However in F_2 generation, all traits appear independently. In F_2 generation, Mendel got the following result (called dihybrid ratio) :

Round Yellow—9

Round Green—3

Wrinkled Yellow—3

Wrinkled Green—1

He concluded that alleles controlling different traits are not linked and are inherited independently. It is also known as Law of Independent Assortment.

8. When a cross is made between tall plants with yellow seeds (TtYy) and tall plant with green seed (Ttyy). What proportions of following phenotypes in the offspring could be expected to be :

(i) Tall and green (ii) Dwarf and green.

Ans. A cross between tall plant with yellow seeds and tall plant with green seeds will produce :

(i) three tall and green plants (ii) one dwarf and green plant

Parents: Tall, yellow seed plant TtYy × Tall, green seed plant Ttyy

Gametes: TY, Ty, ty, tY ↓ Ty, ty

	Ty	ty
TY	TTYy Tall yellow	TtYy Tall yellow

Ty	TTyy Tall green	Ttyy Tall green
ty	Ttyy Tall green	ttyy Dwarf green
tY	TtYy Tall yellow	ttYy Dwarf yellow

Phenotypes : (Tall and green = 3, Dwarf and green = 1)

10. What is back cross ?

Ans. A back cross is the crossing of an F_1 progeny to a homozygous individual (either dominant or recessive). Backcross method is the one whose main objective is to incorporate desired gene, either dominant or recessive, to the highly productive, commercially successful variety which lacks that specific gene. For example, if F_1 heterozygous tall plant is crossed with homozygous dominant parent, then all offsring produced are tall.

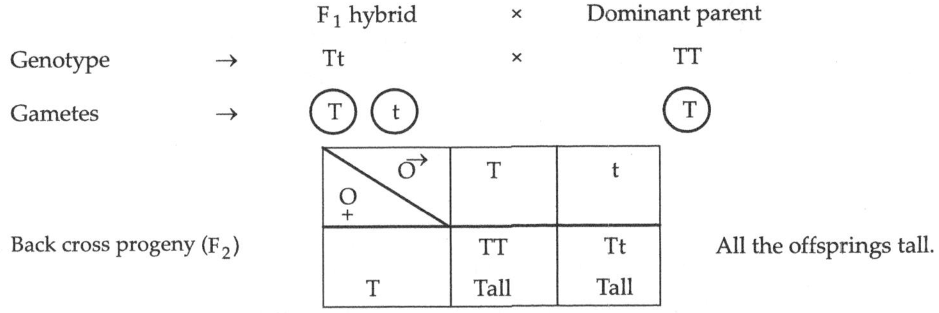

Fig. : Back Cross

11. What is the principle of independent assortment ? How is it related to the principle of segregation ?

Ans. According to the principle of independent assortment, genes for different characteristics that are at different loci segregate independently of one another. The principle of segregation indicates that the two alleles at locus separate; the principle of independent assortment indicates that the separation of alleles at one locus is independent of the separation of alleles at other loci.

12. What is the importance of blood groups ?

Ans. Importance of blood groups : A knowledge of blood groups is necessary :

(i) For successful blood transfusion.

(ii) To solve disputed parentage.

(iii) To identify the criminals.

(iv) For understanding the human population.

13. If mother is with O Blood group and father with A Blood group (homozygous), what is the genetic probability of their children being O or A for their blood group ?

Ans.

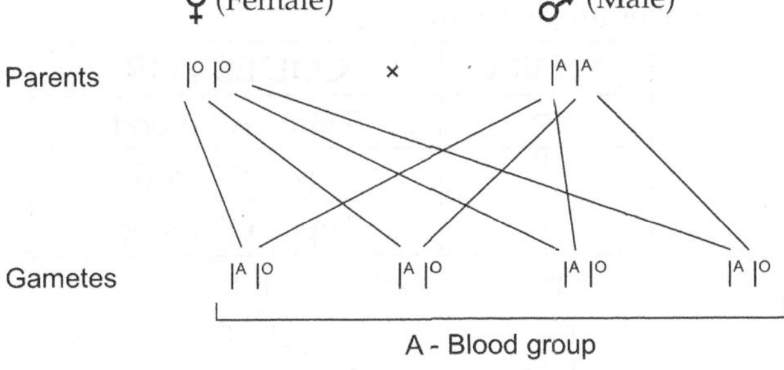

14. A Rh positive woman, whose father is Rh negative, marries a Rh negative man. Is there any possibility that their child be a Rh negative ?

Ans. Yes, there is a possibility of having Rh negative child.

The woman is Rh positive (Rh +ve), but her father is (Rh –ve). So the genotype of the woman is heterozygous Rh +ve (+ –). Man is Rh –ve (– –). If we make a marriage and observe their progeny there is 50% chance of having Rh –ve child.

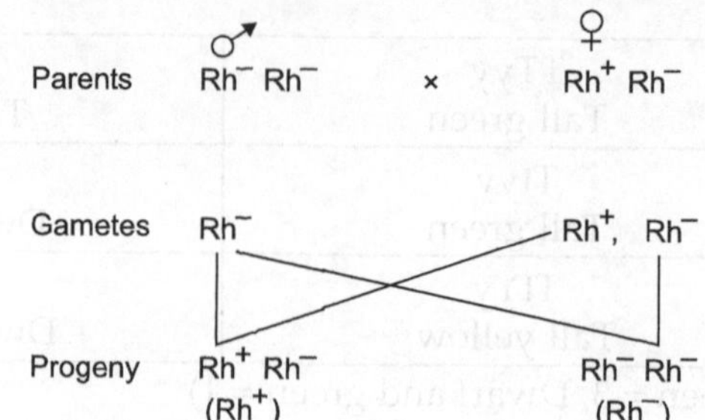

15. A woman with type O blood and a man who is type AB have are expecting a child. What are the possible blood types of the kid ?

Ans. Blood group of woman = O

Blood group of man = AB

If woman is O and man is AB the possible blood types of their child are : A and B blood groups

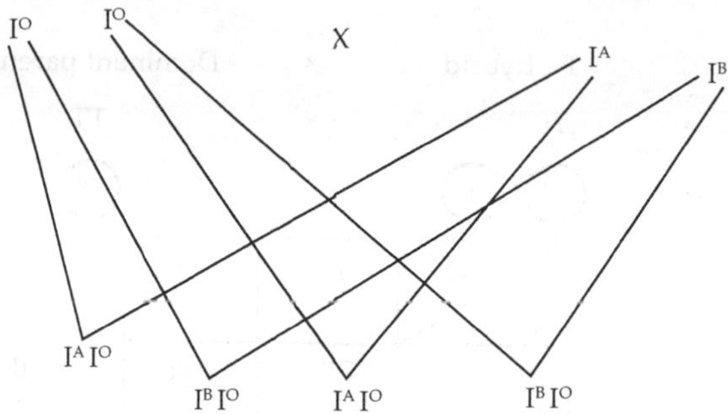

Fig. : Diagrammatic representation of possible blood types

The child blood group will be either A or B but not be O.

16. What are the possible blood types of a child whose parents are both heterozygous for B blood type?

Ans.

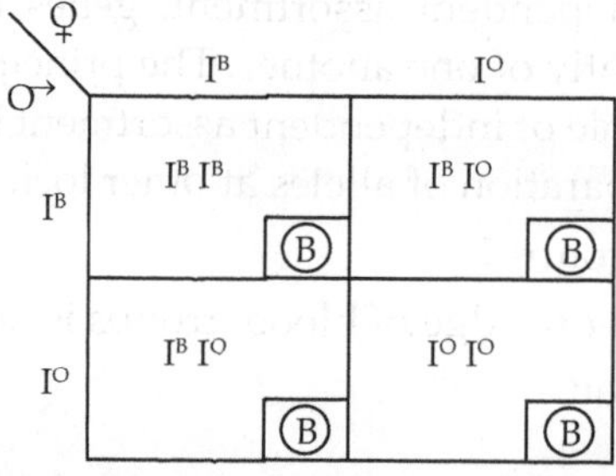

Fig. : Diagrammatic representation of possible blood types

The possible blood types of a child are blood group B and blood group O.

17. A woman with blood group O married a man with blood group AB. Show the possible blood groups of the progeny. List the alleles involved in this inheritance.*

Ans. Blood groups exists as four possible phenotypes : A, B, AB and O. There are 3 alleles for the gene that determines blood type. The alleles are as follows :

ALLELE	CODES FOR
I^A	Type "A" Blood
I^B	Type "B" Blood
i	Type "O" Blood

$I^o I^o$ × $I^A I^B$

I^o I^A, I^B

I^A, I^o I^B I^o

(Blood Group A) (Blood Group B)

Possible Blood groups of the progeny are A and B

* **Frequently asked previous years Board Exam Question.**

18. If the mother is a carrier of colour blindness and the father is normal, show the possible genotype and phenotype of the offspring of the next generation, with the help of a punnet square.*

Ans. If a woman has only one colour blind 'gene' she is known as a 'carrier' but she won't be colour blind. When she has a child she will give one of her X chromosomes to the child. If she gives the X chromosome with the faulty 'gene' to her son he will be colour blind, but if he receives the 'good' chromosome he won't be colour blind.

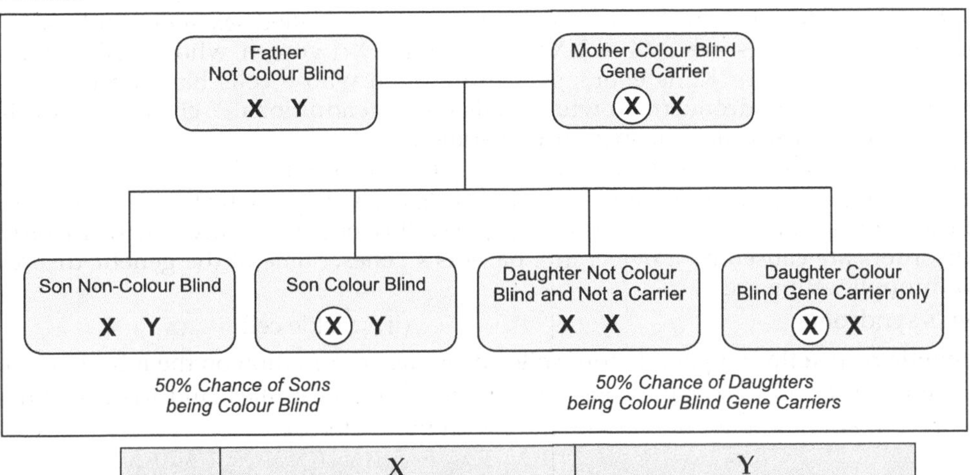

	X	Y
X^C	$X X^C$ Carrier daughter	X^CY Colour blind son
X	XX Normal daughter	XY Normal son

19. Explain how each of the following appears to disrupt Mendelian ratios : Incomplete Dominance, Codominance, Pleiotropy, Epistasis.

Ans. Incomplete dominance : The heterozygote's phenotype is intermediate between those of the two homozygotes. This goes against the idea that two alleles should produce only two phenotypes, with one allele dominant over the other. Instead of a 3:1 phenotypic ratio, the ratio is 1 : 2 : 1.

Codominance : The heterozygote fully expresses two different and equally expressed alleles. This goes against the idea that two alleles should produce only two phenotypes, with one allele dominant over the other. Instead of a 3 : 1 phenotypic ratio, the ratio is 1 : 2 : 1.

Pleiotropy : One gene has multiple phenotypic expressions. Mendel's laws imply that each gene controls only one trait. One allele can therefore change the phenotype in multiple ways.

Epistasis : One gene affects the expression of another gene. Entire classes of phenotypes corresponding to one gene can seem to disappear if the allele of the other gene changes.

20. Describe the relationships among chromosomes, DNA, genes and alleles.

Ans. Chromosomes contain tightly packed DNA and associated proteins. DNA are the strands of genetic material that contains genes, sequences of nucleotides that code for amino acids. Those genes come in varieties called alleles. Alleles are the alternative forms of characters or genes.

21. What do you mean by linkage group ? Explain with example.

Ans. Linkage Group : All the genes which are inherited together from one generation to another generation constitute a linkage group. Number of linkage group in an organism is equal to 'n' number of chromosome. *e.g.*, it is 12 in rice, in *Pisum* the number of linkage group are 7, while in human beings, the number of linkage group is 23 in human female and 24 in human male and in Drosophila it is 4.

22. What are the main characteristics of X-linked traits ?

Ans. Characteristics of X-linked traits :

(i) X-linked genes are never passed from father to son. The Y-chromosome is the only sex chromosome that passes from father to son.

(ii) Males are never carriers– if they have a single mutated gene on the X-chromosome, it will be definitely expressed.

(iii) Males are termed hemizygous for genes on X-chromosome, females are either homozygous or heterozygous.

23. What is the role of the Y-chromosome in human sex determination ?

Ans. The Y chromosome contains the SRY gene that acts as a switch for other sex determining genes. This gene activates in the embryo so that it develops as a male and dismantles all female embryonic structures.

24. What is Sex-influenced dominance ?

Ans. Sex-influenced dominance is the phenomenon in which the manifestation of a phenotype of a gene in heterozygosity depends on the sex of the individual. For example, hereditary baldness is a dominant phenotypical form if the individual is male and it is a recessive form if the individual is female.

25. What are the main diseases caused by errors of the number of sex chromosomes in the cells of an individual ?

Ans. Diseases caused by abnormal number of sex chromosomes are called sex aneuploidies.

The main sex aneuploidies are : 44 + XXX, or trisomy X (women whose cells have an addition X chromosome); 44 + XXY, or Klinefelter's syndrome (men whose cells have an extra X chromosome); 44 + XYY, or double Y syndrome (men whose cells have an additional Y chromosome); 44 + X, Turner's syndrome (women whose cells lack an X chromosome).

26. What is genetic counselling and what is the role of genetic counsellor ?

Ans. Genetic counselling is a process where medical experts carry out genetic tests and look at family history and medical records. Their aim is to identify the possibility of passing any genetic disorders to the baby. Genetic disorders are caused by a defect in a parents's genes. Some of the genetic disorders that can be identified through testing are :

(i) Down's syndrome (ii) Sickle cell

Role of genetic counsellor : A genetic counsellor provides information on the inheritance of illnesses and their recurrence risks; addresses the concern of patients, their families, and their health care providers; and supports patients and their families dealing with these illnesses.

Chapter 6. Molecular Basis of Inheritance

1. Describe the structure of a nucleosome with the help of a well-labelled diagram.*

Ans. A nucleosome is the basic structural unit of DNA packaging in eukaryotes. The negatively charged DNA is wrapped around positively charged histone octamer to form nucleosome. Each nucleosome has a core histone octamer and a linker DNA. The core particle consists of 8 histones proteins which occur in pairs : H_2A, H_2B, H_3 and H_4. About 200 bases of DNA are involved with each nucleosome. DNA winds around the core (1¾ or 1.65 turns involving approximately 146 base pairs) and two successive nucleosomes are joined together by linker DNA H_1 histone (approximately 80 bp) which holds the entire complex. There are repeating units in chromatin fibre like beads on string.

Structure of a nucleosome :

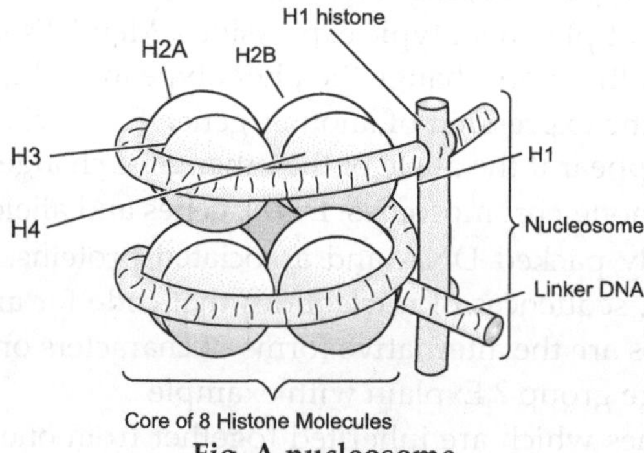

Fig. A nucleosome

2. Which indirect evidences suggested that DNA is the genetic material ?

Ans. Indirect evidences for DNA as the genetic material, includes :

(i) The quantity of DNA in somatic cells is constant, while gametes have half of the quantity.

(ii) DNA is located in the chromosomes.

(iii) Mutagens that affect DNA molecules cause increased phenotypic mutation.

(iv) The wavelengths of UV radiation which cause the most mutations are also those which are strongly absorbed by pure DNA.

(v) All cells in one organism have the same type of DNA, as shown by Chargaff's work, but the DNA varies between different species.

(vi) In non-dividing tissue such as nervous tissue, the DNA is not broken down, but other cellular molecules are constantly being recycled and replaced.

(vii) DNA is copied exactly during each cell cycle and is therefore consistent and self-perpetuating.

* **Frequently asked previous years Board Exam Question.**

3. Write in brief about semi-conservative mode of replication.

Ans. In the replication process, each original strand is used as a template to synthesize the new chain by pairing C with G and A with T. Thus, when replication is complete, there are two DNA double strands with exactly the same base sequence as the original double-stranded DNA. This replication method is termed semi conservative replication, in which one strand is the original that served as a template parent strand and the other being a newly created strand or daughter strand.

4. How are errors that occurred during DNA replication repaired ?

Ans. The DNA polymerase I enzyme carries out the DNA replication and participates in the repair and proofreading of DNA by catalysing the addition of mononucleotide units. A normal DNA polymerase I catalyse the repair activity and remove the nucleotides residue of primer RNA. Many enzymes are also involved in this repair mechanism where the endonucleases cut the defective part of DNA and then the gap is filled with DNA polymerase I and finally the DNA ligase fills that part. This is known as the proof-reading of DNA.

5. What is Reverse transcription ? Mention the term used for the group of viruses which show this kind of transcription.

Ans. Reverse transcription is a method of synthesizing DNA from a template of RNA. This process is performed by certain retroviruses whose genetic code is made up of single-stranded RNA molecules.

6. What are the functions of a gene ?

Ans. Functions of Genes :

(i) Genes controls the function of DNA and RNA.

(ii) They carry out all the information necessary for the proper and efficient functioning of the cell.

(iii) They replicate and inherited by the progeny faithfully, transferring the parental characterstics to further generations.

(iv) Genes provide instructions for coding amino acids, which further form proteins, which act as the building blocks of each cell in body.

7. Write in brief about the characteristics of genetic code ?

Ans. Characteristics of the genetic code :

(i) It is a triplet code. Each three-nucleotide codon in the mRNA specifies one amino acid.

(ii) It is comma free and mRNA read three bases at a time without skipping any bases.

(iii) It is non-overlapping/non-ambiguous. Each nucleotide is a part of only one codon and is read only once during translation.

(iv) It is almost universal. In nearly all organisms, most codons have the same amino acid meaning.

(v) It is degenerate, means among 20 amino acids, 18 are encoded by 2 or more codons.

(vi) The code has start and stop signals. AUG is the usual start signal and defines the open reading frame.

(vii) Stop signals are codons with no corresponding tRNA. Stop signals are also known as nonsense or chain-terminating codons. There are generally three stop codons : UAG, UAA, and UGA.

8. List the three amino acid sequences that could result from the following base sequence :

5′—A U A G C A G G A C U U—3′

Ans. There are three possible reading frames for this sequence because it can be in the 5′ to 3′ direction. If start with the first base and read by threes it gives : AUA, GCA, GGA, and CUU. This sequence codes for the amino acids 5′- isoleucine-alanine- glycine-leucine – 3′. If reading starts with the second base from the 5′ end and read in three, it gives : UAG, CAG, GAC. The first codon codes for a stop codon, the second for glutamine, and the third for aspartate. If reading begin at the third base from the 5′ end, it gives AGC, AGG, ACU. This sequence codes for the amino acids 5′—serine-arginine-threonine—3′.

9. What was the modification of one gene-one enzyme hypothesis ?

Ans. Two ideas were given to modify the one gene-one enzyme hypothesis, in which the first idea was that all enzymes are proteins, all proteins are not enzymes. Secondly, many protein are made up of sub-units called polypeptides, with each distinct polypeptide under the control of a gene.

Example : The adult haemoglobin consists of 4 polypeptides : 2α and 2β. A separate gene codes each type. This led to the re-evaluation of Beadle and Tatum's hypothesis into "one gene one polypeptide".

10. Explain the following in connection with the concept of lac operon :

(i) Structural Genes

(ii) An operator

(iii) A promoter

(iv) A regulatory gene

Ans. (i) Structural Gene : In the operon model, structural gene are those genes that control the production of a protein that performs a structural role in cell. In lac operon, structural gene code for the enzyme that regulate lactose metabolism.

(ii) An operator : The operator is the region of DNA of the operon that is the binding site for the regulatory protein.

(iii) A promoter : The promoter is the DNA sequence of the operon recognized by DNA-dependent RNA polymerase. The initiation site for RNA synthesis is immediately downstream of the promoter. The gene for DNA-dependent RNA polymerase is not part of the operon, since the RNA polymerase enzyme transcribes all bacterial operons.

(iv) A regulatory gene : The regulatory gene codes for a regulatory protein. The lac repressor, encoded by the lac gene, is the regulatory protein of the lac operon.

11. Explain about the negative regulation of the lac operon.

Ans. Immediately upstream of the three lac genes is a fourth gene which is responsible for negative regulation of the operon. This gene expresses a protein called the lac repressor. The repressor can bind to a short sequence of the lac promoter. When repressor binds to that sequence it physically blocks access by RNA polymerase to the DNA. As a result, as long as it is bound no lac mRNA can be initiated. The site at which the repressor binds is called an "operator".

Chapter 7. Evolution

1. Write short notes on :

(i) Coacervates (ii) Abiogenesis (iii) Primordial soup

Ans. (i) Coacervates : The cluster of molecular aggregates in colloidal form which are bounded by a membrane, grow by absorbing molecules from environment and divide by budding are termed as coacervates. According to the "Theory of chemosynthesis", the life on this earth originated as a result of chemical combination of various inorganic materials occurring in the atmosphere.

Formation of Coacervates : The earth was very hot at first and was like a ball of fire. No life could be possible in such conditions. The atmosphere of earth consisted of gases *i.e.,* hot gases including ammonia, carbon monoxide and water vapours. These substances combined with each other to form simpler organic substances such as CH_4, amino acids, carbohydrates and lipids, etc. As the earth cooled, water vapours were condensed and rains started. The amino acid molecules got linked to form polypeptide chains, they united with each other to form proteins. These proteins joined with nucleic acids and formed nucleoproteins. The fluid sediments, an aggregate of molecules, rich in substances formed globules called coacervates. They got enveloped by lipid membrane.

(ii) Abiogenesis : The theory of spontaneous generation or abiogenesis states that life originated spontaneously from non-living matter biogenetically (*i.e.,* without the involvement of living things). This view was held by Aristotle, Van Helmont and many others. Aristotle, believed that worms and larvae of insects developed from dew, decay and manure. Similarly it was believed that fish and dogs were produced from mud, maggots from decaying meat, insects from plant juices and micro organisms from air or water.

French microbiologist Louis Pasteur went well against the theory and disapproved it. He demonstrated that life originates from pre-existing life.

(iii) Primordial soup : In the second phase of chemical evolution, origin of organic compounds took place. The chemical reactions resulting into the formation of carbohydrates, fatty acids, amino acids and synthesis of other complex organic substances probably occurred in sea which has been termed as hot dilute soup or primordial soup by Haldane in 1920.

2. What is oxygen revolution ? What are its effects ?

Ans. The liberation of oxygen into atmosphere, produced by blue-green algae like prokaryotes due to photosynthesis is called oxygen revolution in the early history of earth. The oxygen brought about a number of versatile changes is primitive atmosphere. The free oxygen oxidised CH_4 and NH_3 forming CO_2, N_2 and water. Free oxygen also formed ozone layer above the earth surface. Due to oxygen the reducing atmosphere changed into oxidising atmosphere which checked the further chemical evolution.

3. (i) Write two differences between homologous organs and analogous organs.*

(ii) Give one example of homologous organs and one example of analogous organs found in plants.*

Ans. (i)

Homologous Organs	Analogous Organs
Those organs which are similar in their structure, morphology and anatomy but different in their functions are known as homologous organs.	Those organs which are different in their structure, morphology and anatomy but similar in their functions are called analogous organs.
Inherited from common ancestors.	They are inherited from different ancestors.
They have a similar developmental pattern.	Dissimilar developmental pattern.

(ii) Homologous organs: Leaves of Pitcher plant and Cactus are homologous in their functions.
Analogous organs: Thorns of Bougainvillea and tendrils of Cucurbita are both modifications of a stem arising from axillary bud but perform different functions like protection and climbing, respectively.

4. Give a brief account of Oparin Haldane theory of origin of life.

Ans. The theory was put forward by Russian biochemist A.I. Oparin (1923) and English biologist J. B. S. Haldane (1928). They believed that the origin of life was a gradual process of changes thought to have occurred about 4000-3600 millions years ago. The main steps of the theory are as follows :

(i) Free atoms (Hydrogen, Carbon, Nitrogen)

↓

(ii) Inorganic molecules (H_2, H_2O, CH_4, NH_3, CO_2, HCN)

↓

(iii) Simple organic molecules (Sugars, Fatty acids, Glycerol, Amino Acids, Nitrogen bases *i.e.*, Purines, Pyrimidines)

↓

(iv) Complex organic molecules (Polysaccharides, Proteins, Lipids, Nucleotides, Nucleic acids)

↓

(v) Coacervates (Cluster of complex organic molecules bounded by fatty acid membrane, division by budding)

↓

(vi) Eobionts (Nucleic acid controlled coacervates)

↓

(vii) Prokaryotes (Without organised nucleus)

↓

(viii) Eukaryotes (With well organised nucleus)

5. How did Stanley Miller test Oparin Haldane's theory on the origin of life ?

Ans. Stanley Miller and H.C. Urey built an apparatus of glass tubes and flasks and created an atmosphere rich in H_2, NH_3, CH_4 and water vapour in one chamber and allowed condensed liquids to accumulate in another chamber. They supplied energy by heating the liquid containing chamber as well as by electric sparks from electrodes in gaseous chamber. After running the experiment for a week, they found that liquid in the apparatus had many organic compounds including amino acids. This proved abiotic synthesis of organic compounds.

6. Explain any three molecular (genetic) evidences in favour of organic evolution.

Ans. The evidences in favour of organic evolution are as follows :

(i) Evidence from Paleontology and Metabolic Process : Paleontology is the science which studies past life with the help of fossils. It provided various valuable reasons for organic evolution because it is based on study of fossils which are formed over time. Fossils which are formed by preserving under rocks or at sea level provide the information about developmental changes occurred from simple to complex and more developed forms. Same biochemical reactions are found from bacteria to humans.

(ii) Evidence from Genetics : Genetics is the branch of science which deals with the study of heredity. Mutation caused in any organism can lead to new species. Some important mutations include

***** **Frequently asked previous years Board Exam Question.**

Ancon Sheep, Double Toed Cats, Hornless Cattle, Red Sunflower, Large-sized Banana, etc. Light coloured moth *Biston betularia* underwent a mutation to produce a dark from *Biston carbonaria.* The latter is more suitable for industrial areas and has survived while the white parent form is now restricted to small unpolluted pockets.

(iii) Evidence from embryology and cytology : Embryology is the study of early growth and development of organisms. Monocots and dicots are quite different from each other but show similar embryonic development pattern which indicates that monocots have evolved from dicots. Basic cell structure of cell is same in all organisms.

7. Taking the example of heart and gill slits, explain how embryological evidence support the theory of organic evolution.

Ans. The heart develops a two chambered structure in embryos of fishes and this condition is retained by their adult also. Heart of all other vertebrates in early embryonic stage is two chambered but later on it become three chambered in most reptiles and amphibians while four chambered in bird and mammals. The early embryo of all vertebrates have gill-slits and gill-pauches which retained as such in adults of fishes but these are replaced by lungs in amphibians, reptiles, birds and mammals. These characters give clear evidences that all other vertebrates are evolved from fish like ancestors.

8. Give the contribution of Alfred Russel Wallace ?

Ans. Alfred Russel Wallace was a British naturalist, who independently gave the theory of evolution through natural selection. He also created the basis for the zoogeorgraphic regions, which are still in use. He discussed all the factors then known to influence the current and past geographic distribution of animals with each geographic region.

9. Taking the example of gill clefts, mention how the embryological evidence supports the theory of organic evolution.

Ans. The science which deals with the development of multicellular organism from a fertilized egg (zygote) to an adult is called Embryology. When the zygote develops into an adult by repeated mitotic cell division, it passes through a series of stages. If these stages of embryos belonging to different forms are closely examined, they are found to resemble each other in the beginning but gradually the characteristic or features of the species appear due to evolution. This observation justifies the statement given by Haeckel *i.e.,* Ontogeny *repeats phylogeny.* On examining the development of embryo in some vertebrates *e.g.,* shark, lizard and mammal, all the embryos show gill clefts, tail and yolk-sac at some stage of differentiation. However these are gradually replaced in the adults by their specific structures. It suggests that all these organisms originated from a common ancestor during evolution.

10. Mention any four ways in which fossil records help in studying organic evolution.

Ans. (i) The fossil record can help in building a broad historical sequence of biological evolution.

(ii) It can also help in reconstruction of evolutionary history of an organism.

(iii) The well-preserved fossil can give information about the form, structure, habits and behaviour of extinct species, so it reconstructs the entire habitat of an organism.

(iv) The fossil record is also useful in indicating a connecting link between two groups of organisms *e.g., Archaeopteryx.*

11. What is a connecting link ? Give examples.

Ans. A living organism showing characters of two related groups is known as a connecting link *e.g., Protopterus, Peripatus.*

12. Write a brief note on fossil parks.

Ans. Protected areas which have rich deposits of fossils are called fossil parks. Important fossil parks of India are :

(i) 260 million years old coal forming forests of Odissa.

(ii) 100 million years old fossil forests of Raj Mahal hills in Bihar.

(iii) 50 million years old fossil forests at Mandla district in Madhya Pradesh.

13. How can the age of fossils can be determined ?

Ans. The approximate age of fossils can be determined by using radioactive carbon dating and radioactive potassium dating method. The age of fossil is determined by analysing the radioactive material present in the rock from which fossil is recovered. Since unstable uranium converts into stable nuclei through radioactive emission of α and β particles. Hence, by estimating the amount of lead in a rock and comparing to available data of conversion of uranium to lead one can determine the approximate age of rock.

14. Write short notes on :
(i) Fossilization
(ii) Pseudo fossil

Ans. (i) Fossilization : It is the process of the formation of fossil, during which entire organism or its parts in form or impression often get buried in sand or mud. The best place for fossilization is ocean because salty water checks the decay of organisms. The harder parts of the organism after their deads settle to the bottom and are covered by sediments. The process continues and older fossils are preserved deeper and recent ones are preserved in the upper layers.

(ii) Pseudo fossil : Sometimes rocks present structures resembling to plant remains. Such structures present in the rocks are normally products of minerals in rock crevices. Actually mineral substances in rock crystallize and develop into patterns resembling outlines of plants, which are called pseudofossils.

15. On the basis of Lamarck's theory, explain the evolutionary trend in the development of giraffe's neck.

Ans. One of the examples cited by Lamarck in support of his theory of organic evolution is stretching of the giraffe's neck. He believed that giraffe's neck became longer through the generations as the giraffe stretches it's neck to reach the leaves on tall trees. The information about neck stretching was transmitted in some way to the gametes. Thus, the long neck became a permanent trait, in giraffe's lifetime and passed from generation to generation by hereditary factors.

16. How does Natural Selection work ?

Ans. In the process of natural selection, individuals in a population who are well-adapted to a particular set of environmental conditions have an advantage over those who are not so well adapted. The advantage comes in the form of survival and reproductive success. For example, those individuals who are better able to find and use a food resource well, on average, live longer and produce more offspring than those who are less successful at finding food. Inherited traits that increase individuals' fitness are then passed to their offspring, thus giving the offspring the same advantages.

17. What are the drawbacks of Darwin's theory ?

Ans. Darwin's Theory had the following drawbacks :

(i) **Arrival of the Fittest :** Darwinism can explain the survival of fittest but is unable to account for arrival of the fittest. It is difficult to suppose that small variations could have formed organs like wings and mimicry colouration which are useful only in their fully developed state but what about their presence in early stages.

(ii) **Over Specializations :** Tusks of elephants and the antlers of some deer have overgrown because of their usefulness.

(iii) **Vestigial Organs :** A number of vestigial organs persisting generation after generation has no explanation according to Darwinism.

(iv) **Unchanged forms :** Pteridophyte (*Selaginella*) is continuing on this earth for the last 200 million years without any change, whereas, Darwin proposed addition of useful variation with every generation.

(v) Darwin did not differentiate between Somatic and Germinal variations.

(vi) Darwin also failed to explain how an animal would evolve from aquatic animals.

18. Explain industrial melanism.*

Ans. Industrial melanism was responsible for increase in the population of dark-coloured moths along with the loss of lichens during industrialisation in England.

The tree trunks became darker (after industrialisation) with deposits of soot and smoke; hence the number of dark moths increased due to the added advantage of camouflage as they were not easily visible to their predators while the white-winged ones were easily picked up by the predators. Thus, dark ones were selected by nature (natural selection) and light ones failed to survive.

19. What is Neo-Darwinism ? Explain.

Ans. The modification of Darwin's theory of Natural Selection is called Neo-Darwinism. Neo-Darwinism is supported by Wright, Stebbins, Simpson and Huxley. According to Neo-Darwinism, the whole breeding population of species is considered instead of individuals. The population has a gene pool which shows

* **Frequently asked previous years Board Exam Question.**

various combinations and recombination. New variations arise because of mutation. Thus, there is slow mutation or variation when a breeding population is isolated. A new species is formed on account of variation.

20. What is sickle cell anaemia ? Why has natural selection not eliminated sickle cell anaemia ?

Ans. Sickle cell anaemia is a genetic disorder caused by a variant form of haemoglobin in red blood cells Individual homozygous for this disorder die at an early stage. In heterozygotes also the red blood cells with variant form of haemoglobin become sickle-shaped and were unable to bind oxygen efficiently Even then the natural selection could not eliminate this abnormal allele from the population. This variant is often found in tropical Africa where malaria is common. Since sickle-shaped RBC effectively kills the malaria parasite, the individuals heterozygous for this variant are able to cope with malaria infections much better than normal persons. The natural selection thus maintains this variant form of haemoglobin along with the normal haemoglobin in malaria prone areas.

21. What is mutation ? What are its different types ?

Ans. A mutation is a change in a DNA sequence. Mutations can result from DNA copying mistakes made during cell division, exposure to inoizing radiation, exposure to chemicals called mutagens, or infection by viruses. Germ line mutations occur in the eggs and sperm and can be passed on to offspring while somatic mutations occur in body cells and are not passed on. Mutations may occur spontaneously in nature or may be artificially induced. Different types of mutation are :

(i) Gene Mutation : It is a permanent change in the DNA sequence that makes up a gene. It is also known as point mutation. These are the changes involving only one base pair of the DNA. Very small mutations or mutation involving two or three base pairs are generally also considered as point mutations. The alteration in the base sequence may be due to base substitution and frame shift mutation.

(ii) Chromosome Mutation : It is a mutation involving a long segment of DNA, it is a any change in the structure or arrangement of the chromosomes. These mutations can involve deletions, insertions, or inversions of sections of DNA.

(iii) Genomatic mutation : They involve variation in chromosome number of a whole genome. These variations are of two types : aneuploidy and euploidy.

22. List the different ways which alter the sequence of DNA.

Ans. Gene Mutation : It is a permanent change in the DNA sequence that makes up a gene. It is also known as point mutation. The alteration in the base sequence may be due to

(i) Base substitutions involve the replacement of one base (and its complementary partner) by one of the other bases (and its complementary partner). This replacement may takes place during the replication of DNA. This base pair replacement is of two type :

(a) Transition mutations involve replacing a purine with another purine (and its partnering pyrimidine with another pyrimidine), or replacing a pyrimidine with a pyrimidine (and its partnering purine with another purine). For example, if an A-T pair is replaced by a G-C pair, this is a transition.

(b) Transversion mutations involve replacing a purine with a pyrimidine (and its partnering pyrimidine with a purine), or replacing a pyrimidine with a purine (and its partnering purine with a pyrimidine). For example, if an A-T pair is replaced by either a T-A pair or a C-G pair, this is a transversion.

(ii) Frame shifts are created when one or a few bases are inserted or deleted, thus altering the ultimate reading frame of the ribosome. Protein-coding DNA is divided into codons which are three bases long, insertions and deletions in these codons can completely change a gene so its message cannot be decoded correctly. Such mutations are called frame shift mutations.

23. What are the chromosomal and genomatic mutations ?

Ans. Chromosome Mutation : It is a mutation involving a long segment of DNA. It is a change in the structure or arrangement of the chromosomes. These mutations can involve deletions, insertions, or inversions of sections of DNA. In some cases, deleted sections may attach to other chromosomes, and disrupts both the chromosomes the one that loses the DNA and the one that gains it.

(i) Translocations : It is the transfer of a chromosomal segment onto a non-homologous chromosome.

(ii) Insertion : Insertions are mutations in which extra base pairs are inserted into a new place in the DNA.

(iii) Deletions : Deletions are mutations in which a section of DNA is lost, or deleted.

(iv) Duplication : It is the doubling of one or several chromosomal fragments.

Genomatic Mutation : They involve variation in chromosome number of a whole genome. These variations are of two types: aneuploidy and euploidy.

(i) **Aneuploidy :** The abnormal conditions where one or more chromosomes of a normal set of chromosomes are missing or present in more than their usual number of copies. This type of variation involves one or few chromosomes but not the entire set.

(ii) **Euploidy :** Variation that involve entire sets of chromosomes are known as euploidy.

24. What are the different mutagens which induces mutations ?

Ans. Any physical or chemical agent that causes mutations are called mutagens.

(i) **Radiations :** α-, β- and λ-radiations of radioactive substances, X-rays, nutrons and protons induce mutations by ionising the matter. Non-ionising UV-rays also induce mutations but not to the extent asionising radiations. The natural radiations like cosmic rays, are responsible for spontaneous mutation in living beings. Besides gene mutation, X-rays and other high energy radiations also produce all types of chromosomal aberrations.

(ii) **Chemical :** Several chemical substances also induce mutations. Example 5-bromouracil, 2-amino purine; hydroxylamine, mustard gas, nitrous acid, phenol formaldehyde, ferrous and manganous salts, and acridine dyes. Chemical substances can produce both gene mutations and chromosomal alterations. Besides, colchinine induces polyploidy by inhibiting the formation of spindle in cell division.

(iii) **Temperature :** A rise in temperature may sometimes lead to disturbance in genes which cause mutation.

25. Explain the evolution of long neck of giraffe according to Charles Darwin.*

Ans. Darwinian Theory of Evolution explains that evolution is the result of natural selection and natural selection is based on the inherited characters of organisms. Adaptive ability of organisms is the one which helps organisms in evolution through natural selection. Initially, there were two varieties, short necked and long necked giraffe. Giraffe's long neck is the result of their adaptation to their nature since the ground vegetation disappeared and is replaced by high trees. The elongated necks are the outcome of their attempt to feed leaves on tall trees. Long neck could feed easily and they survived whereas short necked could not feed easily and they could not survive. This character then passed on from generation to generation leading to evolution.

26. Explain the term "Reproductive Isolation". Give one example.

Ans. It refers to the process of preserving the integrity of a species by preventing hybridization with individuals of another species due to some inbuilt genetic mechanism.

For example : Two individuals belonging to *Brassica campestris* growing at far off distances can easily interbreed producing fertile progeny. But, an individual of *Brassica campestris* will not interbreed with an individual of *Brassica nigra* though they might be growing closely in the same field. Thus, we can say that *Brassica campestris* and *Brassica nigra* are reproductively isolated.

27. Illustrate with example of wheat, how species originate by polyploidy ?

Ans. Polyploidy is a genetic mechanism, which also lead to the formation of new species. It means the multiplication of basic sets of chromosomes. For instance, Triploid (3n), Tetraploid (4n), Pentaloid (5n), Hexaploid (6n) are referred to as polyploids. The process is common in plants. Different varieties of wheat have been obtained by polyploidy. A species of wheat *Triticum monococcum* (2n = 14) is selectively breed with wild grass *Aegilops speltoides* (2n = 14) to produce another species of wheat *Triticum durum* (2n = 28).

28. Write a short note on evolution of horse.

Ans. Evolution of modern horse started in Eocene epoch. The first fossil horse, called Eohippus was discovered in North America whose height was 40 cm with a short neck and head. Forelimbs had four fingers and one splint of first finger while the hind feet had three toes and two splint toes of first and fifth toe. Teeth had complete cement while the molar teeth had no serrations.

Mesohippus : Was an intermediate horse during oligocene epoch. It was about 60 cm high in shoulders. Fore feet had three fingers and one splint finger while hind feet had three toes. Molar teeth had serrations.

Merychippus : The ruminating horse arose in Miocene epoch, which was about 100 cm high with a longer neck three fingers and three toes. Molar teeth with well developed serrations.

* **Frequently asked previous years Board Exam Question.**

Pliohippus : Arose in pliocene epoch was 120 cm high. Each fore and hind foot had one complete finger and toe and two splints. It is also called first one toed horse.

Equus : The modern horse arose in Pleistocene epoch in North America about 150 cm high, long head and neck. The fore foot and hind foot each with one finger and one toe and two splints. Molar teeth with enamelled ridges suitable for grinding.

29. Write briefly about :

(i) Ramapithecus man

(ii) Neanderthal man

Ans. (i) Ramapithecus man : Fossil of this ancestor was excavated by E. Lewis (1930) from Shivalik in India and it stood erect on its hind feet, was ape like and lived on the tree tops. Small canines and large molars suggested that Ramapithecus ate hard nuts and seeds like modern man.

(ii) Neanderthal man : *Homosapiens neanderthalensis* was the true man, including man of today. Fossils of Neanderthal were obtained from Neander Valley in Germany by C. Fuhlrott (1856). They were legendary cave dwellers and had heavy brow ridges and humped backs.

30. Give two characteristic features of each of the following :*

(a) Ramapithecus

(b) Cro-Magnon man

Ans. (a) The fossil evidence clearly indicates that **Ramapithecus,** lived in Africa and Asia (about 10-15 million years ago) were the forerunners of Homonids. They had Chimpanzee-like feet, flexible wrists and small canines, short face, thickly enameled teeth, small brain and cranial capacity of around 650 cc.

(b) Following distinctive features are regarded as common feature of the **Cro-Magnon race** : Long headed skull, straight forehead, the slightly projecting brow ridges, the cranial vault noticeably flattened, and the occipital bone (at the back of the head) projects backward. The cranial capacity is large, about 100 cubic inches. Although the skull is relatively long and narrow, the face appears quite short and wide.

31. Give three characters of *Australopithecus*.

Ans. (i) They are considered as the first ape-man. They were short statured average almost four feet.

(ii) The cranial capacity was of 600 cm^3.

(iii) The vertebral column had a distinct lumbar curve with pelvis broad and basin like, indicating an adaptation to bipedal posture. They walked almost erect.

32. List the main morphological changes that occurred in ancestors of modern man.

Ans. The following main morphological changes that occurred in the ancestors of man are :

(i) Narrowing and elevation of nose.

(ii) Formation of chin.

(iii) Flattening of face.

(iv) Reduction in body hair and brow ridges.

(v) Development of curves in vertebral column for erect posture.

(vi) Increase in brain size and intelligence.

(vii) Enlargement and rounding of cranium.

(viii) Bipedal locomotion.

33. List the evidences for common ancestry of great apes and man.

Ans. The evidences for common ancestry of great apes and man are as follows :

(i) Chromosomal evidences.

(ii) Evidence from blood proteins.

(iii) Evidences from blood groups.

(iv) Evidence from Haemoglobin.

34. (a) List any three drawbacks of Darwinism.

(b) State Hardy Weinberg's principle.

(c) Differentiate between Directional natural selection and Disruptive natural selection.

Ans. (a) (i) Lack of evidence or fossil records to support Darwinian evolution.

(ii) Sometimes survival of the fittest would not be possible because variations selected by nature may be harmful to the life of organism.

* **Frequently asked previous years Board Exam Question.**

Short Answers

(iii) The failure of developmental biology to explain why vertebrate embryos diverge from the beginning of development.

(b) According to Hardy-weinberg law, in a large randomly breeding population, the allelic frequencies of sexually reproducing organisms will remain constant.

$$p^2 + 2pq + q^2 = 1$$

Or

$$(p + q)^2 = 1$$

(c)

Directional natural selection	Disruptive natural selection
The directional selection describes the change that occurs when a population shows a particular trend through time.	Discruptive natural selection describes the changes in population genetics in which extreme values for a trait are favoured over intermediate values.

35. Write short notes on :

(i) Vestigial organs
(ii) Atavism
(iii) Recapitulation theory
(iv) *Archaeopteryx*
(v) Biogeography
(vi) Missing link
(vii) *Homo erectus*

Ans. (i) Vestigial Organs : The organs which were well developed and functional in their ancestors but have become rudimentary and non-functional in the present day animals (because of their no use to the present day animals) are called vestigial organs. For *e.g.*, the vestiges of pelvic girdle in python and whales indicate that the ancestor of snakes and whales were four-footed. As the four-footed ancestors gradually lost their feet and arms, pelvic girdles became rudimentary.

In the body of man, following organs have become functionless and reduced in size :

(a) Nictitating membrane, (b) Ear muscle, (c) Abdominal hairs, (d) Nipples in males, (d) Vermiform appendix, (e) Coccyx (tail bone).

It proves that evolution had taken place in the past. Vermiform appendix in man is homologous with the caecum in rabbit like animals, which is responsible for the digestion of cellulose by bacterial action. But since man uses soft and boiled food, appendix is not of any use in man; hence, it has become reduced.

(ii) Atavism : The sudden appearance of a character which is absent or vestigial in the individuals of the same species but was functional in remote ancestors, is called atavism or reversion.

Certain examples of atavism are the presence of hairs on body, large canines, additional mammae. Sometimes a child may be born with a tail, which is later removed by surgery.

(iii) Recapitulation theory : This theory was proposed by Von Baer (1828). But it was renamed as Biogenetic Law by a German biologist Ernst Haeckel (1866). It states that ontogeny repeats phylogeny *i.e.*, every organism during its development (Ontogeny) passes through its ancestral history (phylogeny) *e.g.* Presence of fish characters like gills, gill slits, tail with a tail fin, lateral line, sense organs, *etc.*, in the tadpole larva of frog.

(iv) *Archaeopteryx* : *Archaeopteryx* is the fossil bird. Its fossils were obtained in late Jurassic period in 1861 in Germany. It has lizard–like body with a long tail. It is of the size of a crow or pigeon. It shares both reptilian and avian features. Its reptilian features are toothed beak, long slender tail, scales on the body and limbs and hands with three-clawed digits.

Avian characters are forelimbs modified into wings, tail with rectrices and body with conyour feathers, V-shaped furcula and jaws produced into a beak.

On this basis *Archaeopteryx* is regarded as a connecting link between reptiles and birds. It shows that birds have evolved from reptiles.

(v) Biogeography : The study of patterns of distribution of animals and plants in different parts of the earth is called biogeography. It is believed that millions of years ago all the six continents were in the form of a single land mass Pangaea. Due to geological changes and movement of crystal plates below the surface of earth huge land masses broke off and drifted apart from one another. These land masses were separated by seas which acted as barriers for free movement of organisms among the continents. Over the centuries, due to differences in environmental conditions in each continent, different populations of plants and animals evolved in each biogeographical region. According to geologist, the whole world can be divided into six major "bio geographical realms" which are :

(a) Australian region, (b) Ethiopian region, (c) Nearctic region, (d) Neotropical region, (e) Oriental region, (f) Palaearetic region.

(vi) Missing link : They are the extinct organisms, which had the characters of two different groups of organisms. *e.g., Archaeoptery* (Link between reptiles and birds).

(vii) *Homo erectus* : *Homo erectus* is the fore runner of the modern humans. The Australopithecus has succeeded by large brained forms and was described as Java man and Peking man.

(a) Java man (*Homo erectus*)
(1) Cranial capacity about 900 c.c.
(2) More than five feet tall with a skeleton like the present day man.
(3) Forehead low, standing face was prognathous.
(4) Jaw massive with huge teeth.
(5) Chin absent, bony eyebrow ridges present.
(6) They have learnt to lit a fire and use tools.

(b) Peking man (*Homo erectus pekinensis*) :
(1) Very similar to Java man.
(2) Cranial cavity larger 850-1200 c.c.

Chapter 8. Human Health and Diseases

1. Mention the affected parts of the body due to infection of the following pathogens
(i) Mumps virus
(ii) Tetanus causing bacteria
(iii) *Mycobacterium leprae*

Ans. (i) Parotid salivary gland
(ii) Jaw and neck muscles
(iii) Lack of sensation (numbness) in fingers and toes

2. Name the disease caused by deficiency of (i) Iron, (ii) Nicotinic acid, and (iii) Vitamin D.

Ans. (i) Iron—Anaemia
(ii) Nicotinic Acid—Pellagra
(iii) Vitamin D—Rickets

3. List the common forms of Arthritis.

Ans. (i) Osteoarthritis
(ii) Rheumatoid arthritis
(iii) Gout

4. What are Koch's Postulates ? What are its limitations ?

Ans. Koch's Postulates are :
(i) The organism must be found regularly in animals that have the disease.
(ii) It must be isolated and grown in a pure culture on an artificial medium.
(iii) When a healthy animal is inoculated with this organism it must cause the disease and show the characteristic symptoms.
(iv) The same organism must be recoverable from the inoculated animals.

Limitation : It is not applicable to viral diseases because they cannot be cultured on an artificial medium.

5. Name the causative agents, symptoms and cure for the following diseases :
(i) Malaria
(ii) Filaria

Ans. (i) Malaria : Causative agent—*Plasmodium* (sp. Malarial parasite)

Symptoms : The attack of malaria is preceeded by yawning, tiredness, headache and muscular pain. During the fever, the patient feels chilly and shivers, and has acute headache, nausea and high temperature. After few hours, the body perspires freely and the temperature becomes normal, till the next attack take place after 48 hours.

Cure : Quinine, the oldest drug for malaria. Other anti-malaria drugs are Paludrine, Primaquin, Chloroquinine, Camoquin and Comoprima.

(ii) Filaria : Causative agent—*Wuchereria bancrofti*.

Symptoms : The living adult causes mechanical irritation and deposits metabolites. The dead worms block the lymph vessels. As a result periodic attacks of fever occur and the tissue, surrounding the lymph nodes and other organs of reticuloendothelial system such as liver, spleen, scrotum, vulva, legs and groin become enlarged producing tumour-like solidity. This condition is known as elephantiasis.

Cure : Once the swelling appears, there is no other treatment except surgical operation to reconstruct the affected part, if possible.

6. Name the causative agents, mode of transmission, symptoms, prevention and cure for Typhoid.

Ans. Causative agent : *Salmonella typhi.*

Mode of transmission : Direct and oral. Bacteria are spread through intestinal discharges of typhoid carriers.

Symptoms : Headache, haemorrhage, ulceration of intestine, rose coloured rash on the skin mainly on the chest.

Prevention and Cure :

(i) Vaccination.

(ii) Proper care for sanitary as personal hygiene.

(iii) Use of boiled water, protection of eatables from flies.

(iv) Administration of drugs.

7. Write short notes on Thalassaemia.

Ans. Thalassaemia : Thalassaemia is the name of a group of genetic, inherited disorders of the blood. More specifically, it is a disorder of the haemoglobin molecule inside the red blood cells. It is an inherited genetic disease *i.e.,* a disease that is passed from parents to children through the genes (genetic). It is not infectious and cannot be passed on from one individual to the other by personal or any other contact, nor through blood, transfusion, food or air. The two main types of thalassaemia are (beta) and (alpha). Both affect the production of normal haemoglobin, a key constituent of human red blood cells. Other 'abnormal' types of adult haemoglobin also known as Structural Haemoglobin. Variants have been identified and these include, mainly, Haemoglobin S (HbS), Haemoglobin E (HbE), HbC, HbD, and Lepore. In this patient needs blood transfusion every 20-22 days.

8. Persons suffering from G-6-PD deficiency are resistant to Malaria. Explain.

Ans. G-6PD is an enzyme glucose-6-phosphate dehydrogenase which decomposes H_2O_2 formed during metabolism. The deficiency of this enzyme is an inborn error of metabolism controlled by specific gene. When persons suffering from malaria are given Primaquine it causes haemolysis because H_2O_2 produced during metabolism is not decomposed due to the deficiency of G-6PD enzymes. This H_2O_2 haemolysis the blood. The haemoglobin of haemolysed blood collects on cell membrane and RBC's are deformed. In such distorted RBC's, malarial parasite fails to survive and multiply. It means persons suffering from G-6PD deficiency are resistant to malaria. Such persons are, therefore, favoured by natural selection in area infected with malarial parasite. For this reason G-6PD deficiency is found in 50% Negroes living in areas with malarial infection.

9. Write short notes on :

(i) Haemophilia (ii) Amoebiasis

Ans. (i) Haemophilia : It is a sex-linked inherited disease. It is a disease in which blood lacks the capacity to coagulate. It is found in males and rarely in females. This disease is inherited by a man from his mother. This disease is lethal in homozygous recessive condition.

(ii) Amoebiasis : It is also called amoebic dysentery. It is caused by a protozoan known as *Entamoeba histolytica.* In this case, the source of infection are cysts from the faecal matter of the infected person. The cysts get into food and reach the stomach. Its incubation period is five days in severe infection and several months in chronic cases.

Symptoms : Mild diarrhoea alternating with constipation. In severe cases dysentery with mucus and blood in stools. Abscesses may be formed in lungs and liver.

Prevention : Proper disposal of faecal matter, protection of sources of drinking water and field sanitation.

10. What are the physiological effect of sickle cell anaemia ?

Ans. Sickle cell anaemia is due to a recessive gene Hb^S. Its normal gene Hb^A is dominant. over Hb^S and is responsible for synthesis of haemoglobin. In sickle celled anaemia due to gene Hb^S the haemoglobin produced is defective and the oxygen carrying capacity of haemoglobin is lost. Because of oxygen debt the RBCs become sickle shaped.

11. What are the body's two lines of defence in a non-specific defence mechanism ?

Ans. (i) **First Line of Defence :** It is a physical barrier such as skin, oil, sweat, tears, network of fine hair in nostrils and ciliated epithelium of respiratory tract.

(ii) **Second Line of Defence :** Various chemicals are associated with this defence mechanism. It is also termed inflammatory response. It may be localized or systemic in which body temperature rises with the release of pathogens and number of WBC increases.

12. How do oil and sweat help in defence mechanism ?

Ans. Oil and sweat are secreted by sebaceous glands and sweat glands respectively. They are acidic in nature (pH 3 to 5). They inhibit the growth of micro-organisms on the surface of the skin. Sweat also contains an enzyme lysozyme which destroys the cell walls of many bacteria, thus help in protection.

13. What is immune system ? Mention its two types.

Ans. The organs and the processes of the body that provide resistance to infections and toxins constitute immune system. This system comprises two components :

(i) The Humoral Immune System : It is made up of a group of diverse proteins called antibodies that circulate in the body fluids (blood plasma and lymph). This system defends against viruses and bacteria that enter blood and lymph. It forms the body's third line of defence.

(ii) The Cell Mediated System : It is carried out by highly specialised cells and works against pathogens, including fungi and protista, that have entered host cells. The system also reacts against transplants. It is thought to be important in protecting the body from its own cells if they become dangerous.

14. Briefly describe the cells of immune system.

Ans. The most important cells of the immune system are the WBC's called Lymphocytes. These are critical for immune reactions and are two types— B-cells and T-cells. Both develop from stem cells located in the liver, in the foetus and in bone marrow cells of adults.

B-cells : The cells that continue in the bone marrow for differentiation are termed as B-cells. The young lymphocytes subsequently migrate to lymphoid tissues such as spleen, lymph nodes and tonsils where they undergo final maturation.

T-cells : Those cells that migrate to the thymus and differentiate under its influence are known as T-cells.

15. How do B-cells respond to antigens ?

Ans. When the antigen specific B-cells are triggered by the antigen it gives rise to a clone of plasma cells. These cells produce antibodies which counter the invading pathogen.

16. What is the limitation in organ transplantation ?

Ans. The severe limitation in organ transplantation is rejection by the body as it recognises a 'non-self' organ and its immune response is triggered.

17. What are the symptoms of allergic reactions ?

Ans. The symptoms of allergies :

(i) A person may suffer from high fever which affects the mucous membrane of eyes, nose and upper respiratory tracts.

(ii) Mucous membrane of the lower part of the respiratory tract which leads to cough and asthma.

(iii) Reddening of skin, appearance of blisters on the skin.

(iv) Accumulation of tissue fluid below skin.

(v) Watering of eyes and inability to breath.

18. What are stem cells ?

Ans. Stems cells are mother cells that have the potential to become any type of cell in the body. One of the main characteristics of stem cells is their ability to self-renew or multiply while maintaining the potential to develop into other types of cells like blood, heart, bones, skin, muscles, brain etc. There are different sources of stem cells but all types of stem cells have the same capacity to develop into multiple types of cells.

They are the basis for treating diseases such as Parkinson's disease, diabetes, cerebral palsy, heart disease in future. Stem cells may also be used for screening new drugs and toxins and understanding birth defects without subjecting human volunteers to the toxins and drugs.

19. Explain :

(i) Benign tumour
(ii) Cancerous tumour
(iii) Metastasis

Ans. (i) Benign tumour : These are non-malignant tumours. Such tumours grow slowly but later become quite large. They remain restricted to the place of origin and do not spread to other areas of the body. Mostly tumours are of this type and do not give rise to cancer.

(ii) Cancerous Tumour : (Malignant tumour). It begins as a small tumour growth at first, grows slowly and more rapidly later on. The tumour ultimately spreads off from the original site and migrates through blood to the other sites and they divide and redivide to form a secondary tumour.

(iii) Metastasis : The stage when a secondary tumour is formed and accumulates by repeated division is called *metastasis*. This stage is fatal and causes death sooner or later.

20. Write short notes on :

(i) B-cells
(ii) T-cells
(iii) AIDS

Ans. (i) B-cells : When antigens invade a tissue fluid, B-cells are stimulated to produce antibodies. The body has thousands of antigen-specific B-cells. The membrane of each type would have been sensitized by previous contact with the antigen. If this does not happen, the B-cells die quickly. However, there is a constant supply of new B-cells. Once an antigen specific B-cells is triggered by the antigen, it gives rise to a clone of plasma cells by quick multiplication. Most members of this clone produce antibodies at the rate of about 2000 molecules per second.

(ii) T-cells : T-cells respond to antigens by producing a clone of T-cells. Each T-cell recognises a specific antigen. Therefore, the body contains separate T-cells for every antigen that the body encounters.

(a) **The killer T-cells :** They directly attack and destroy antigens. They do so by moving to the site of invation and producing chemicals that attract phagocytes and stimulate them to feed more aggressively of antigens. They also produce substances which attract other T-cells.

(b) **Helper T-cells :** They act to stimulate antibody production by B-cells.

(c) **Suppressor T-cells :** They suppress the total immune reaction keeping it or to prevent it from attacking the body's own cells.

(d) **Memory T-cells :** They have the ability to produce more effect or cells.

(iii) AIDS : Acquired Immuno Deficiency Syndrome was first recognised in USA in 1981.

Its causative organism is Human Immunodeficiency Virus (HIV).

Symptoms : Headache, dizziness, unexplained weight loss, purple patches on the skin and mucous membrane inside the mouth, unexplained bleeding and prolonged diarrhoea. A person may suffer from any one of these symptoms or all of these symptoms.

Mode of Transmission :

(a) Sexual intercourse
(b) Blood transfusion
(c) Shared needles of drug users
(d) Mother to child during pregnancy or mother foetal transmission

Preventive Measures : There must be clean sex habits, by not sharing items that could become contaminated by blood, *e.g.*, razor, tooth brush or any skin piercing instrument. Blood used for transfusion should not be contaminated. Used needles must be avoided.

21. Write short notes on (i) Drug resistant, (ii) Sickle-cell anaemia.

Ans. (i) Drug resistance : Due to continuous use of drugs over any population shows no effect of against such drug, this is called drug resistance. *e.g.*, Original populations of mosquitoes had unique ability of resistance to DDT. But after continuous spray of DDT the new generation of mosquitoes becomes resistant to DDT.

(ii) Sickle-cell anaemia : This disease is caused by genetic disorder. RBCs become sickle-shaped and resulting in severe anaemia. Haemoglobin synthesised is defective and is controlled by its recessive homozygous genes.

22. List four reasons for drug addiction.*

Ans. Following could be the reason for drug addiction:

(i) **Stress :** To escape financial, career, family or social pressure, person abuse the drugs. Depression frustration and other emotional turbulence contribute to the reason for drug addiction.

(ii) **Curiosity :** Drugs advertisements in news-papers, literature and on radio make a person curious to have personal experience of the taste of a particular drug.

(iii) **Peer pressure :** Constant description by friends about the "good feeling" creates a temptation.

(iv) **To overcome frustrations and depressions :** The desire to get solace or relief from personal problems initiates the use of drugs.

23. List four effects of alcoholism on human health.*

Ans. Following could be the effect of Alcoholism :

(i) Excessive alcohol consumption leads to organ failure or prevent them from properly balancing blood sugar levels.

(ii) A person suffering from alcohol abuse commonly, becomes dull, anti-social, depressed, tired, aggressive, etc. leading to low self esteem.

(iii) Dilated pupils and mood swings, loss of sleep (insomnia), lowers immunity.

(iv) Continous alcohol usage ultimately leads to failure of vital organs like liver and kidneys.

Chapter 9. Strategies for Enhancement in Food Production

1. Which is apiculture ? How is it important in our lives ?

Ans. Apiculture is the rearing, caring and management of honey bee for obtaining honey bee and wax.

Importance :

(i) It produces bee wax used in industry for making cosmetics, polishes.

(ii) It provides honey which have high nutritive food.

2. What is meant by term breed ? What are the objectives of animal breeding ?

Ans. Breed refers to the group of animal having same ancestral characters, general appearence, size, etc.

Objectives of animal breeding

(i) To increase the quality of yield

(ii) To improve the described qualities of the produce.

3. Enumerate any four essential features of good and effective poultry farm management practices.*

Ans. Following poultry farm management practices are effective :

(i) **Farm Management :** Poultry farming involves both egg layers and broilers, for high yield.

(ii) **Shelter :** Poultry birds are needed to be maintained in a strictly hygienic manner with proper housing. A good and clean environment with proper ventilation should be provided.

(iii) **Food :** Broilers are raised for meat. They are fed with protein and fat rich food. Vitamin supplementary helps them for good growth rate. Egg layers are fed with diet rich in vitamins A and K which help in more egg production.

(iv) **Disease Management :** Poultry farming should have a disease-free condition. Regular and proper cleaning, sanitation, spraying of pesticides and disinfectants should be done. Vaccinations help to prevent the chance of occurrence and spreading of infections.

4. List the two steps that are essential for carrying out artificial hybridization in crop plants and why ?

Ans. (i) Selection of parents–only those plants should be selected which have desired traits.

(ii) Crossing over–Pollen grains from selected male plants is collected and transferred to the female plant after which is bagged.

5. IARI has released sexual varieties of crop plants that are biofortified. Give three examples of such crops and their biofortification.

Ans. (i) Bitter gourd enriched in vitamin C.

(ii) Carrots enriched in Vitamin A.

(iii) Spinach enriched in iron and calcium.

6. What is inbreeding ? What is the danger of inbreeding ?

Ans. The mating of more closely related individual with in the same breed for 4-6 generation called inbreeding. It may lead to inbreeding depression i.e. the loss of fertility, vigour and the productivity of hybrid.

* **Frequently asked previous years Board Exam Question.**

7. (i) Why are the plants raised through micro propagation termed as somaclones ?
(ii) Mention two advantages of this technique.

Ans. (i) The plants are genetically identical, hence the name somaclones.
(ii) (a) Large number of plants are produced in short duration.
(b) Virus free plants are produced.

8. A certain plant, when introduced into a new environment, neither produces seeds nor responds to vegetative propagation. How can more plants be produced of its kind from this plant ? State the method employed.

Ans. More plants of the same kind can be produced by micropropagation or tissue culture.

Tissue culture : Culture of an excised part in an aseptic culture medium to form a callus, in which differentiation of root and shoot buds is induced by providing suitable combination of auxins and cytokinins then the plantlets are transferred to the soil to develop into mature plants.

9. Give the advantages of propagation by tissue culture technique.

Ans. Advantages of propagation by tissue culture technique :
(i) Tissue culture provides rapid multiplication under favourable conditions. A small plant bearing five to six leaves is produced within a few weeks which under normal conditions may need several months.
(ii) By tissue culture techniques, large number of offsprings or plantlets can be produced every year.
(iii) The progeny of these plants can be produced in millions, which multiply with difficulty by conventional methods.
(iv) Cloning can be done throughout the year in a very small space under controlled conditions.
(v) Offsprings can be obtained of sterile plants or of rare hybrids of extra ordinary characters *e.g.*, Oil palm.
(vi) Production of disease free plants. In the virus infected plants only the meristimatic cells of the apical dome are devoid of virus. Therefore, by using shoot apical meristem culture, virus free plants can be raised.

10. What is a single cell protein? How is it significant for human welfare?*

Ans. Single cell protein (SCP) refers to dried microbial cells or protein extracted from pure microbial cell culture. It refers to the whole cells of bacteria, yeast, filamentous fungi or algae. SCP is not pure protein. It also contains carbohydrates, lipids, nucleic acids, mineral salts and vitamins. The composition depends upon the organism and the substrate on which it grows.

Significance for human welfare :
(i) They can be used as a protein supplement for both humans or animals. Microorganisms like algae, fungi, yeast and bacteria have very high protein content in their biomass.
(ii) It is cost effective and easy to digest.
(iii) Reduces pressure on agriculture.
(iv) The fat content is low and there is no pathogenicity and toxicity present in SCP.

11. Explain the term biofortification. How is this technique useful for the production of golden rice?*

Ans. Biofortification is the enrichment of crops with nutrients of higher levels of vitamins, minerals, proteins and fat content. Golden rice is an example of GM crop, the grains of golden rice are yellow as it contains β-carotene which is precursor of vitamin A and it is produced to control night blindness and to overcome vitamin A deficiency.

Chapter 10. Microbes in Human Welfare

1. Study the table given below. Do not copy the table, but write the answers in the correct order.

Scientific Name	Commercial Product	Use
(a)	Streptokinase	(b).................
Monascus purpureus	(c).................	(d).................
(e)...........	Lactic acid	(f).................

* Frequently asked previous years Board Exam Question.

Ans. (a) *Streptococcus.*
(b) to dissolve blood clots.
(c) Statins.
(d) Control body's Cholestrol level.
(e) Lactic acid bacteria (LAB).
(f) Food fermentation or curd production.

2. How was Penicillin discovered ?

Ans. Penicillin was an accidental discovery. Sir Alexander Fleming observed that, in unwanted culture plates of *Staphylococcus* a mould *Penicillium* was growing. This mould inhibited the growth of *Staphylococcus* around it. Later the antibiotic penicillin was isolated from this fungus.

3. What is sewage ? In which way can sewage be harmful to us ?

Ans. Sewage is the municipal waste water containing large quantities of human excreta and other organic waste. Sewage could be harmful to us as it contains many pathogens microbes and produce foul smell. It is the cause of many water borne disease.

4. How does methanogens help in producing biogas ?

Ans. Methanogens produce biogas by anaerobic decomposition of cellulose material *i.e.*, excreta of cattle.

5. Explain the significant role of the genus Nucleopolyhedrovirus in an ecological sensitive area.

Ans. Nucleopolyhedrovirus is very useful for species specific narrow spectrum insecticidal application (IPM). They have been seen to show no negative impact on plant, mammals, birds fish or even on-target insects.

6. What are methanogens ? How do they generate biogas ?

Ans. Anaerobic methane producing bacteria, Methanogens generate biogas when they act on cellulose rich biowaste.

7. Explain the role of Leguminous plant in increasing the soil fertility and crop yield.

Ans. Leguminous plant possess root nodules where nitrogen is fixed by symbiotic nitrogen fixing bacteria *Rhizobium* which fertilises the soil.

Chapter 11. Biotechnology : Principles and Processes

1. What are the types of artificial cloning ?

Ans. There are three different types of artificial cloning : gene cloning, reproductive cloning and therapeutic cloning.

Gene cloning produces copies of genes or segments of DNA. Reproductive cloning produces copies of whole animals. Therapeutic cloning produces embryonic stem cells for experiments aimed at creating tissues to replace injured or diseased tissues.

2. Name the two types of 'biological tools' used in recombinant DNA technology.

Ans. The two types of 'biological tools' used in recombinant DNA technology are :

(i) **Enzymes :** These include restriction enzymes (Endonucleases), which acts as molecular scissors; Ligases, which are responsible for joining DNA fragments together and Reverse Transcriptase, which is used for converting mRNA to cDNA.

(ii) **Vectors :** They serve as a medium or vehicle to carry recombinant DNA into the host cell. The most commonly used vectors are Plasmids, Cosmids and Bacteriophages.

3. What is the origin of replication in a plasmid ?

Ans. An origin of replication is a sequence of DNA at which replication is initiated on a chromosome.

4. What is the role of cosmid in recombinant DNA Technology ?

Ans. Cosmids act as vectors to insert DNA sequences (genes) into the genome of a bacteria, much like a plasmid. Cosmids in recombinant DNA technology, utilize phage (bacterial viruses) to insert DNA sequences of genes with up to 44,000 base pairs while normal plasmids are only able to efficiently carry 10,000 base pairs.

5. What is the function of a vector in genetic engineering ?

Ans. "Vector" is an agent that can carry a DNA fragment into a host cell. It is used for expressing certain gene in the DNA fragment.

6. How are plasmids used in Genetic Engineering ?

Ans. Plasmids are used in genetic engineering to amplify, or produce many copies of, certain genes. In molecular cloning, a plasmid is a type of vector that can transport foreign genetic material from one cell to another cell, where the genes can be further expressed and replicated. Plasmids are useful in cloning short segments of DNA and can be used to replicate proteins, such as for insulin. Additionally, plasmids are being investigated as a way to transfer genes into human cells as part of gene therapy.

7. What are the different methods used for genetic transformation in animals ?

Ans. The three principal methods used for the creation of transgenic animals are DNA microinjection, embryonic stem cell-mediated gene transfer and retrovirus-mediated gene transfer.

(i) DNA microinjection : This method involves the direct microinjection of a chosen gene from another member of the same species or from a different species, into the pronucleus of a fertilised ovum. The introduced DNA may lead to the over or under-expression of certain genes or to the expression of genes entirely new to the animal species.

(ii) Embryonic stem cell-mediated gene transfer : This method involves prior insertion of the desired DNA sequence by homologous recombination into an in vitro culture of embryonic stem (ES) cells. Stem cells are undifferentiated cells that have the potential to differentiate into any type of cell (somatic and germ cells) and therefore to give rise to a complete organism. ES cell-mediated gene transfer is the method of choice for gene inactivation, the so-called knock-out method.

(iii) Retrovirus-mediated gene transfer : To increase the probability of expression, gene transfer is mediated by means of a carrier or vector, generally a virus or a plasmid. Retroviruses are commonly used as vectors to transfer genetic material into the cell, taking advantage of their ability to infect host cells in this way. Offspring derived from this method are chimeric, *i.e.*, not all cells carry the retrovirus.

8. Write in brief about the physical delivery methods of gene transfer in plants :

(i) PEG
(ii) Micro-injection
(iii) Macro-injection
(iv) Liposome mediated gene transfer
(v) Electroporation
(vi) Microprojectiles

Ans. (i) PEG (polyethylene glycol) stimulated DNA uptake involves the use of 15-25% PEG, which stimulates uptake of DNA by endocytosis without any gross damage to protoplasts. The transformed protoplasts can be selected on a selection medium.

(ii) Micro-injection involves injection of DNA, using micropipettes with 0.5-10 µm diameter tip. In this technique, the recipient cells are immobilized on a solid support. (cover slip or slide, etc.) or artificially bound or held by a pipette under suction.

(iii) Macro-injection of DNA, employing needles with diameters greater than cell diameter, has been tried in rye (*Secale cereale*). DNA was injected into the stem below the immature floral meristem, so as to reach the sporogenous tissue, leading to the production of transgenic rye plants.

(iv) Liposome mediated gene transfer involves encasement of DNA in lipid bags, which can be made to fuse with protoplasts with the help of Peg.

(v) Electroporation or electromobilisation involve application of electric field to increase the permeability of cell membrane, to take up the DNA (electro transfer).

(vi) Microprojectiles are carried by a 'macroprojectile' (or bullet), and are accelerated into target cells, where they penetrate the cell wall, leaving DNA to be incorporated.

9. What is Human Genome Project ? What are the advantages of Human Genome Project ?

Ans. The Human Genome Project (HGP) is an international research program that aims to spell out the complete genetic inheritance of human beings and selected experimental animals. The HGP's goal is to decode the complete DNA inheritance, or genome, of human beings by 2003, and make them accessible for further biological study. The Department of Energy's Human Genome Program and the National Institutes of Health's National Human Genome Research Institute (NHGRI) together sponsored the U.S. Human Genome Project.

Advantages of human genome project are :

(i) to develop new and improved medicines.
(ii) to predict and prevent diseases.
(iii) to diagnosis accurately.
(iv) to improve forensic science.

10. What are the goals of Human Genome Projects ?

Ans. (i) A primary goal of the Human Genome Project is to make a series of descriptive diagrams, maps of each human chromosome at increasingly finer resolutions.

(ii) After mapping is completed, the next step is to determine the base sequence of each of the ordered DNA fragments.

(iii) The ultimate goal of genome research is to find all the genes in the DNA sequence and to develop tools for using this information in the study of human biology and medicine.

(iv) Create physical map of the 24 human chromosomes (22 autosomes, X and Y).

(v) Identify the entire set of genes and map them all to their chromosomes.

(vi) Determine the nucleotide sequence of the estimated 3 billion base pairs.

(vii) Analyze genetic variation among humans.

(viii) Map and sequence the genomes of model organisms.

11. Write a short note on Electrophoresis.*

Ans. When electric field is applied across two electrodes that are totally submerged in a colloidal solution, the particles (colloidal) tend to move towards the electrode depending on their electric charge. This movement of particles under the effect of electric field is known as electrophoresis. When extracted DNA is subjected to electro-phoresis following points are observed :

(i) DNA fragments are separated by agarose gel electrophoresis.

(ii) DNA fragments being negatively charged move towards positive electrode.

(iii) DNA Fragments are separated according to size/ charge ratio.

(iv) Separated fragments, when stained with ethidium bromide and exposed to UV rays, they form orange coloured bands.

(vi) The bands formed are extracted by elution or southern blotting.

Chapter 12. Biotechnology and its Applications

1. Give any three applications of genetic engineering.

Ans. Applications of Genetic engineering :

(i) Production of protein products like Insulin, vaccines etc.

(ii) Creating pest resistant and drought resistant crop varieties and genetically engineered crops. E.g. – Bt Brinjal, Bt Cotton etc.

(iii) Improving shelf life of plants and nutrient incorporation in plants and their products. E.g. Golden rice with vitamin A and Flavr Savr tomato with longer shelf life.

2. Give some examples of recombinant DNA technology applications.

Ans. Some examples of recombinant DNA technology applications are :

(i) **DNA fingerprinting** or DNA typing used for the identification of individuals. It is used in forensic sciences, immigration cases, study of population and ecological genetics, in the case of disputed parentage and to confirm the cell line identity.

(ii) **Gene Therapy :** It involves the treatment of genetic disease by replacing the defective gene by a normal one. Used in the treatment of several diseases such as SCID (Severe combined immunodeficiency), Cystic fibrosis, Sickle cell anaemia.

(iii) **Medical diagnosis** of genetic disease.

(iv) Production of **Pharmaceutical products** such as Insulin, human growth hormone, Hepatitis B vaccine, Erythropoietin.

3. List any four beneficial effect of GM plants.

Ans. (i) Increased tolerance against abiotic stresses (cold, drought, salt, heat).

(ii) Reduced post harvest losses.

* **Frequently asked previous years Board Exam Question.**

(iii) Reduced reliance on chemical pesticides.

(iv) Enhanced nutritional value of foods.

4. What are the uses of transgenic plants ?

Ans. Uses of transgenic Plants :

(i) Transgenic plants show higher crop production because they are resistant to herbicides, insects and viruses.

(ii) These plants are suitable for food processing. Delayed ripening and Bruise resistant tomato plants are produced by plant transgenics.

(iii) It produced male sterile lines of *Brassica napus* (Rape seed). Such plants can be directly used as female parents in hybridization experiments. It reduces the cost for emasculation for hybrid seed production.

(iv) These are used as bioreactors for large-scale production of commercially valuable products, specialized medicines, chemicals and antibodies. Such usage is known as Molecular farming.

(v) These plants are used for identification of regulatory sequences for many genes.

5. Write short notes on :

(i) Transgenic plants with herbicide resistance.

(ii) Transgenic plants with insect resistance.

Ans. (i) Transgenic plants with herbicide resistance : Due to increasing concern about contamination of environment due to herbicides, new herbicides are being developed that are safer and biodegradable. The herbicides affect processes like photosynthesis or biosynthesis of essential amino acids. Transgenic plants resistant to these herbicides have been produced, utilizing one of the following two approaches :

(a) Either the target protein is over–expressed (*e.g.,* EPSPS) or it should become insensitive to the herbicide (*e.g.,* ALS).

(b) A pathway is introduced, which will detoxify the herbicide.

(ii) Transgenic plants with insect resistance : Genes for insect resistance from three different sources are being transferred for developing insect resistance :

(a) *Bt2* gene encoding Bt toxin, derived from *Bacillus thuringiensis;*

(b) *Cowpea trypsin inhibitor gene* (CpTi) from cowpea (*Vigna unguiculata*) and

(c) genes for other insecticidal *secondary metabolites* derived from a number of legumes.

Such transgenic plants are unaffected by the insects, thus reducing the use of insecticides, which pollutes the environment.

6. Giving an example in each case, explain how genetic engineering strategies are used in the creation of the following transgenic organism :

(i) An insect resistant plant.

(ii) A virus free plant.

(iii) An oil eating bacteria.

Ans. (i) An insect resistant plant : *Bacillus thuringiensis* (Bt), is a gram-positive soil bacterium, with a genome size of 2.4 to 5.7 million base pairs. *Bacillus thuringiensis* forms parasporal crystals (Cry proteins) during the stationary phase of its growth cycle. These crystals are specifically toxic to certain orders and species of insects, like *Lepidoptera, Diptera,* and *Coleoptera.*

First, scientists identify a strain of Bt that kills the targeted insect. Then they isolate the gene that produces the lethal protein. That gene is removed from the Bt bacterium and a gene conferring resistance to a chemical (usually antibiotic or herbicide) is attached that will prove useful in a later step.

The Bt gene with the resistance gene attached is inserted into plant cells. Any plant cell that has the Bt gene must also have the resistance gene that was attached to it. Researchers grow the plant cells in the presence of the antibiotic or herbicide and select the plant cells that are unaffected by it. These genetically transformed plant cells are then grown into whole plants by a process called tissue culture. The modified plants produce the same lethal Bt protein produced by Bt bacteria because the plants now have the same gene.

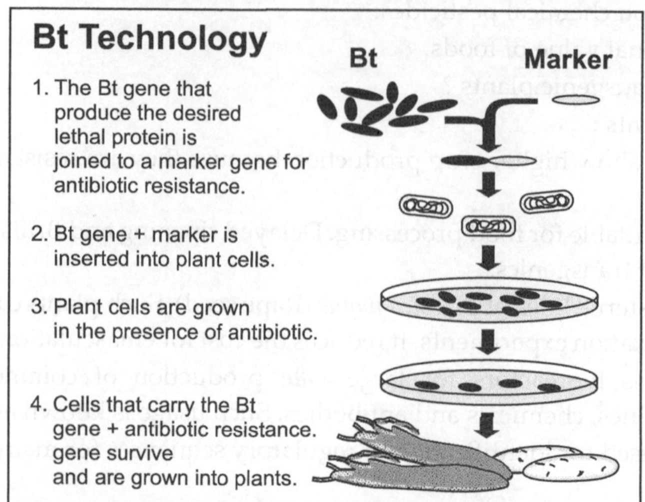

Fig. : Steps used in genetic engineering for creation of an insect resistant plant

(ii) A virus free plant : In certain crops, and particularly in ornamentals, it has become possible to produce virus free plants through tissue culture at the commercial level. This can be done by regenerating plants from cultured tissues derived from : (i) virus free plants, (ii) meristems which are generally free of infection, (iii) meristems treated with heat shocks (34-36°C) to inactivate the virus, (iv) callus, which is usually virus (CMV) free like meristems.

Example : Mottle virus has been eliminated from strawberry; potato virus X from cassava, and cauliflower mosaic virus from cauliflower.

(iii) A oil eating bacteria : Transgenic oil eating bacteria are produced by treating the bacteria (*e.g.*, Two of the major families in which these microbes are found are *Pseudomonas* and *Bacillus*) with chemicals and radiation, thus inducing a mutation.

7. What is a bioreactor? Explain important features of a Stirred tank bioreactor.*

Ans. A bioreactor is a vessel or a fermenter like an apparatus that is used to grow organisms such as bacteria or fungi under controlled environmental conditions e.g. pH, temperature, dissolved oxygen, etc. It can be used for the production of compounds of pharmaceutical importance, antibodies or vaccines.

A stirred tank reactor is cylindrical has a curved base that assists in the mixing of the reactor contents. The stirrer facilitates even mixing and oxygen availability throughout the bioreactor. Alternatively, air can be bubbled through the reactor. The function of the agitation system is to provide good mixing and thus increase mass transfer rates through the bulk liquid. The agitation system consists of the agitator and the baffles that are used to break the liquid flow to increase turbulence and mixing efficiency.

The components of a bioreactor are:

(i) An agitator system
(ii) An oxygen delivery system
(iii) A foam control system
(iv) Temperature control system
(v) pH control system
(vi) Sampling ports to withdraw small volumes of culture periodically.

8. Explain the process of RNA interference.

Ans. RNA interference takes place in all eukaryotic organisms as a method of cellular defence. It involves silencing of a specific mRNA due to complementary dsRNA molecules that bind to and prevent translation of the mRNA.

9. Explain giving an example how recombinant DNA technology can be used for the formation of the following :

(i) A vaccine.
(ii) A plant with delayed fruit ripening.

Ans. (i) A vaccine :

(a) Hepatitis B vaccine, makes use of just a small portion of a pathogen. For a virus, the vaccine can contain just a piece of the protein coat that surrounds the virus's DNA

* **Frequently asked previous years Board Exam Question.**

(or RNA). Even small portion of a virus is sometimes enough to stimulate an immune response in the body.

(b) A segment of the virus's DNA is responsible for the production of the virus's protein coat. A segment of DNA is added to the DNA of a yeast cell (which is in the yeast culture).

(c) The yeast cell, as it grows, will "read" the viral DNA incorporated in its own DNA and produce the protein that makes up the protein coat of hepatitis B.

(d) To finish making the vaccine, the proteins are needed to be separated from the yeast cells. A purifier is used to isolate the hepatitis B antigen produced by the yeast cells.

(e) The isolated hepatitis B protein, produced by the yeast cells, contains none of the viral DNA that makes hepatitis B harmful. Therefore, there is no possibility of it causing the disease.

(ii) A plant with delayed fruit ripening : Scientists have added to the tomato a new gene from the soil bacteria *Pseudomonas chlororaphis*. The gene encodes an enzyme called ACC deaminase, which causes the breakdown of one of the precursors of ethylene synthesis (ACC). The reduction in levels of this precursor causes reduced ethylene production and delayed ripening.

10. Suggest any two possible treatment that can be given to a patient exhibiting adenosine deaminase deficiency.

Ans. (i) Enzyme replacement therapy (in which functional ADA is injected)

(ii) Bone marrow transplantation.

11. The Gene therapy is the ultimate goal of the human genome project and is a boon to mankind. Justify this statement.

Ans. Gene therapy began with the Human Genome Project. Gene therapy is the introduction of genes into existing cells to prevent or cure a wide range of diseases. Gene therapy began with the Human Genome Project. The Human Genome Project has found gene locations for many diseases, like Huntington's disease, cystic fibrosis, ADA deficiency etc. Disease-causing gene is then corrected by various methods. One of the first tested methods was to obtain the faulty gene from the patient, introduce the corrected gene into cells, and finally inject the altered cells back into a blood vessel or altering the genes of an organ, Also, Human Genome database on the world wide web help to examine the gene by providing the related information. This data might help in identifying the usage of such gene in therapeutic applications.

12. Explain gene therapy, with reference to treatment of SCID.*

Ans. Gene therapy is an experimental technique in which genes are introduced into individual cells or tissues, to cure some hereditary diseases in which a defective allele of a gene is replaced with normal one. The technique of gene therapy was initiated in 1972.

Severe Combined Immune Deficiency (SCID) is a potentially fatal primary immuno deficiency in which there is combined absence of T-lymphocyte and B-lymphocyte function. There are at least 13 different genetic defects that can cause SCID. These defects lead to extreme susceptibility to very serious infections. This condition is generally considered to be the most serious of the primary immune deficiencies. Fortunately, effective treatments, such as stem cell transplantation, exist that can cure the disorder. The future holds the promise of gene therapy for several more types of SCID.

Another type of SCID is caused by mutations in a gene that encodes an enzyme called adenosine deaminase (ADA). ADA is essential for the metabolic function of a variety of body cells but especially T-cells. The absence of this enzyme leads to an accumulation of toxic metabolic by-products within lymphocytes that cause the cells to die. For patients with SCID due to ADA deficiency, replacement therapy with a modified form of the enzyme (from a cow, called PEG-ADA) has been used with some success.

13. What is the function of GEAC ?*

Ans. GEAC (Genetic Engineering Approval Committee) make decision regarding the validity of GM research. It also checks the safety of introduction GM organism for public services and for their large scale use.

14. What are the risks associated with genetically engineered plant, animals and humans ?

Ans. Some of the risks of genetically engineered plants are :

(i) The crops may not be safe to eat in the long term. It may cause toxicity.

(ii) GM crops may cross breed with wild plants and release their genes into the wild with unpredictable results.

(iii) There would be less variation in the plant's genes, and if there are any faulty genes in the plants, disease among the plants would be more common.

---

* **Frequently asked previous years Board Exam Question.**

Some risks of genetically engineered animals are :

(i) Animal products could be unsafe.

(ii) The well being of the animal could be affected *e.g.,* if the animal was modified to produce more meat it would be fatter and wouldn't be able to move around.

(iii) Something could go wrong and harm the animals.

Some of the risks of genetically engineered humans are :

(i) Something could go wrong and harm the person

(ii) Some people may realize than they can decide how they want their unborn baby to look and abuse the power of genetic engineering by creating 'designer babies'.

15. What are ethical issues associated with DNA fingerprinting ?

Ans. As soon as DNA identification tests have been introduced as a new powerful tool in criminalistics and in paternity testing, this new technology has immediately aroused a mixture of ethical concerns, suspicion and interest among scientists and non-scientists. The major concerns about the so called 'DNA fingerprints' were related first to the possible constitution of data based by the police agencies for the purpose of identifying and investigating individuals as potential criminal suspects, and secondly to the risk of a widespread use without safeguards for private investigation as establishing paternity. In this context, and in order to preserve civil liberties and the respect of individual privacy identification should be strictly limited to judicial use and performed by accredited laboratories.

Chapter 13. Organisms and Populations

1. What are the main causes of population explosion ?

Ans. The main causes of population explosion are as follows :

(i) Decline in death rate due to better medical facilities.

(ii) Better availability of food due to increased food production with agricultural development.

(iii) Lack of awareness and education about the need of controlling population.

(iv) Due to better storage and transport facilities for food, the famine-like condition do not occur.

2. Give three adaptations in organisms by which they avoid predation.*

Ans. Prey animals must constantly be on the lookout for their predators. To survive, they must escape from or defend themselves against those animals trying to eat them. Thus, they developed various adaptations to avoid predation.

1. **Camouflage:** Camouflage is an adaptation in which a species evolves to resemble its background. It is the technique used by many animals of resembling and blending in with their surroundings. It is a kind of colouring, body shape, and/or behaviour which animals use to confuse, distract, startle, or hide. For example, many species of frogs and insects camouflage in their surroundings and escape their predators.
2. **Highly developed senses:** Prey animals often rely on escape as their form of protection. Therefore, they must have a well-developed sense of sight, hearing, and smell so they can detect the presence of a predator in enough time to escape.
3. **Warning Signals:** Warning signals are used to scare away predators and to warn other animals. Some animals are poisonous, taste bad, or have a painful bite or sting. Many of these animals are brightly coloured with bold patterns. These colours serve as a warning signal to predators. For example, a skunk is associated with its bad smell, when threatened.
4. **Defensive Weapons:** Prey animals have adaptations to fight or avoid their predators. These weapons could be physical like thorns, wings, hooves, teeth, claws, antlers, etc or chemicals, like snakes have venom.
5. **Behavioural Adaptations:** Some animals trick their predators into thinking they are hurt or dangerous. Others will congregate into large groups or have highly adapted forms of hiding from their attackers.

3. Write short notes on :

(i) Growth curve

(ii) Age pyramid

(iii) Age distribution

* **Frequently asked previous years Board Exam Question.**

Ans. (i) Growth Curve : When the time is plotted on X-axis and the number of organisms (population on the Y-axis, the growth of organisms (population) follows a characteristic **S-shaped (sigmoid)** curve in the presence of environmental resistance. This growth curve shows the three regions :

(a) The lag phase of little growth.

(b) Exponential phase of rapid growth.

(c) Stationary phase of zero growth.

(1) Lag phase : It is the phase of slow growth when the organisms of a population adapt to the new environment.

(2) Exponential or Logarithmic phase : The population density in this phase increases rapidly due to ample amount of space and food being available.

(3) Decline phase : After reaching the maximum limit the population stays stable for sometime and then declines rapidly due to decrease in availability of food and space.

J-shaped growth curve : It is shown by the organisms in a natural predator free environment having plentiful food supply. It comprises of Lag phase, and Exponential phase. *e.g.*, Annual plants.

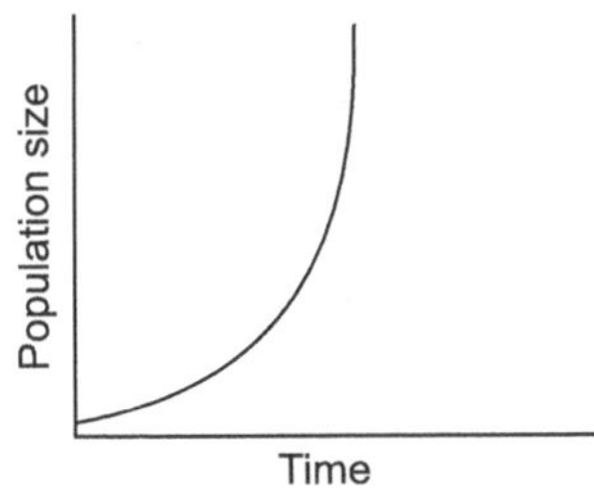

J shape graph

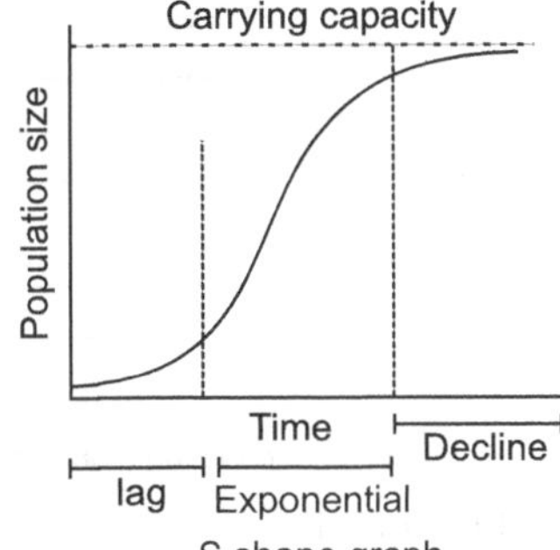

S shape graph

(ii) Age pyramid : An age pyramid is a graphic representation of proportion of various age groups of a population. The pre-reproductive age groups is at the base, reproductive age group in the middle and the post reproductive at the top.

There are three types of age pyramids—triangle shaped, bell-shaped and urn-shaped.

(a) **Triangle shaped age pyramid :** It is a graphic representation of a population which has a very high proportion of pre-reproductive individuals, moderate number of reproductive individuals and the postreproductive individuals are very less. It indicates the young and growing population.

(b) **Bell-shaped age pyramid :** This pyramid displays a bell-like appearance with pre-reproductive individuals little more in number than reproductive individuals. The post reproductive individuals are lesser. It represents stable population.

(c) **Urn-shaped age pyramid :** This pyramid exhibits smaller base represented by proportionately small number of pre-reproductive individuals a larger number of reproductive individuals, a larger number of reproductive individuals and a small-sized tip formed of post-reproductive individuals. The pyramid of population shows a negative growth in population.

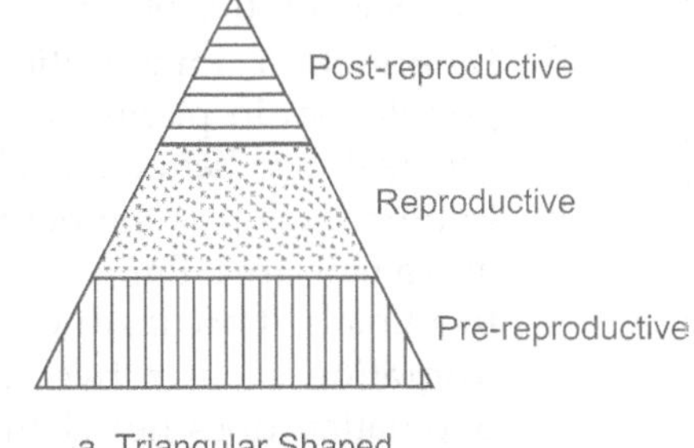

a. Triangular Shaped

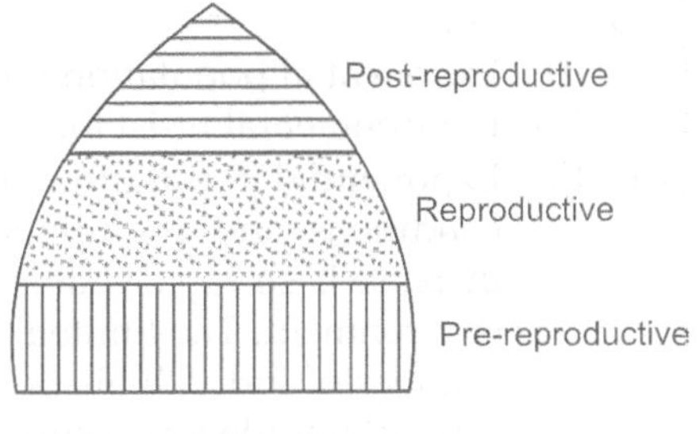

b. Bell Shaped

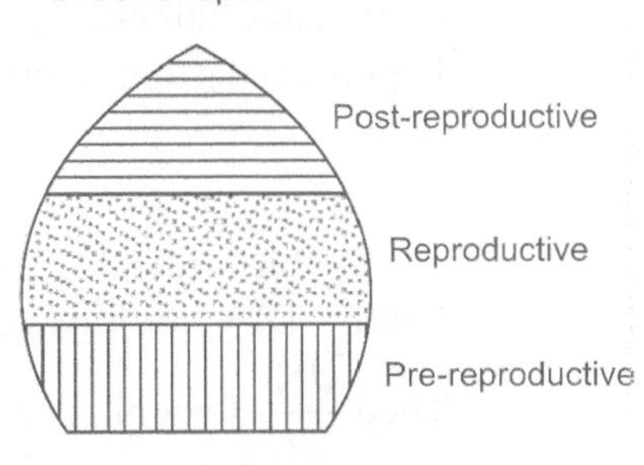

c. Urn Shaped

(iii) Age distribution : Also called as age structure. There are three ecological ages pre-reproductive, reproductive and post-reproductive.

(a) **Pre-reproductive individuals** are young individuals which will enter the reproductive age after sometime and contribute to increase in population. A high proportion of this age group in a population results in expanding population.

(b) **Reproductive individuals** are the ones which are actually adding new numbers to the population. In a stable population, there number is almost equal to the pre-reproductive individuals.

(c) **Post-reproductive individuals** are older individuals which no longer take part in reproduction. Their small number indicate a growing population, while their large number indicates a stable or declining population.

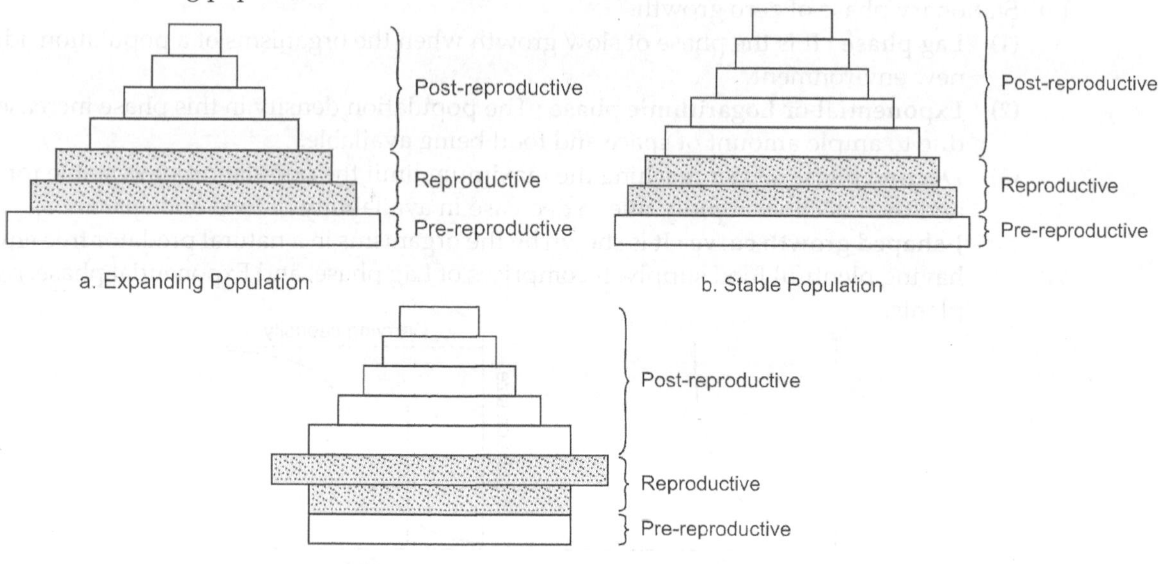

4. Write about Population Dispersal.

Ans. Population dispersal : It is the movement of individual or their propagation units into and out of an area. This movement changes the population size. Population dispersal is helpful in wider distribution of organisms colonization of new areas, protection from predators, over crowding and better prospects of food and shelter. It is of three types :

(i) **Immigration :** Immigration is permanent inward movement of individuals from outside into existing population. In plants, it occurs due to chance setting of dispersed seeds. In animals, immigration occurs due to better availability of food and shelter and less resistance from the local population.

(ii) **Emigration :** It is the permanent outward movement of individuals from a population for settling into a new area. Emigration decease the local population. It is absent in plants as they are not motile. In animals, it occurs due to better availability of food, shelter and to escape any natural calamity.

(iii) **Migration :** It is a two way movement of the whole population from and into an area. Size of population does not change. Migration is meant to avoid unfavourable conditions. *E.g.,* Siberian cranes migrate to warmer climate areas during extreme winter in Siberia.

5. Explain :

(i) Exponential population growth

(ii) Environmental resistance

Ans. (i) Exponential population growth : It is the period of rapid growth of a population due to favourable conditions and no competition. During this phase population density increases rapidly. The increase is so rapid that soon the population size grows much beyond the carrying capacity of the environment. The limited availability of food and space is responsible for the sudden decline in the number of individuals of the population. This is the crash phase. J-shaped or exponential growth form of population occurs in those cases where either the food chains are short or there is excessive human interference *e.g.,* of J-shaped growth form is the insect populations.

Exponential growth form can be represented mathematically by the following exponential equation :

$$\frac{dN}{dt} = (B - D) \times N$$

B = Birth rate, D = Death rate

Let (B – D) = r or intrinsic rate of growth.

Then $\frac{dN}{dt} = r \times N$

where d is the rate of change, N = number of existing individuals.

t = time and r = intrinsic rate of natural increase $\frac{dN}{dt}$ represents the rate of change in population size.

(ii) **Environmental Resistance :** The sum of total inhibitory environmental factors, both biotic and abiotic, which regulate population size and do not permit unlimited growth of population is called environmental resistance. Because of environmental resistance, the populations are unable to reach full biotic potential.

6. Enlist the measures that the Government of India has undertaken to spread the message of the need of population control.

Ans. (i) Mass media such as radio, T. V., newspaper, magazines, posters and hoardings.

(ii) National Programme of Family Planning has created awareness among the people and it has made various contraceptive devices available either free or at a very low cost.

Chapter 14. Ecosystem

1. How are productivity, gross productivity, net productivity and secondary productivity interrelated ?

Ans. Productivity is the rate of biomass production.

Gross productivity is the rate of production of biomass by green plants per unit time per unit area. It is the total amount of productivity.

$$NPP = GPP - R$$

Net primary productivity is the difference between the gross primary productivity and the loss due to respiration.

Secondary productivity is the rate of formatiojn of organic matter by the consumers.

2. Define decomposition. Explain main steps involved in the process of decomposition.*

Ans. Decomposition is the process that involves the breakdown of complex organic matter or biomass from the body of dead plants and animals with the help of decomposers into inorganic raw materials such as carbon dioxide, water and other nutrients.

The various steps involved in decomposition are as follows:

(i) **Fragmentation:** It is the first step in the process of decomposition. It involves the breakdown of detritus into smaller pieces by the action of detritivores such as earthworms.

(ii) **Leaching:** The fragmented particles may contain a lot of water-soluble nutrients which are inorganic in nature. These nutrients get dissolved in the water and seep into the soil and get precipitated in the process of leaching.

(iii) **Catabolism:** Once the complex material is broken down into smaller particles and the inorganic nutrients are removed, it is time to convert the detritus into simpler inorganic compounds. This process is carried out by various fungal and bacterial enzymes by the process of catabolism.

(iv) **Humification:** The next step is humification which leads to the formation of a dark-coloured colloidal substance called humus, which acts as a reservoir of nutrients for plants.

(v) **Mineralisation:** The humus is further degraded by the action of microbes, which finally leads to the release of inorganic nutrients into the soil. This process of releasing inorganic nutrients from the humus is known as mineralisation.

3. What is meant by saying that the energy flow in an ecosystem is unidirectional ?

Ans. The sun is an ultimate source of all energy. The energy passes from producer to primary consumer, secondary consumer, tertiary consumer and ultimately to the quaternary consumer which is in one direction. *i.e.*, energy cannot pass back in a reverse direction.

4. Justify the importance of decomposers in an ecosystem.

Ans. Decomposers are hetrotrophic organisms, mainly fungi, and bacteria which break down complex organic matter into inorganic substance like carbon dioxide, water and nutrients. They meet their energy and nutrient requirement by degrading dead organic matter or detritus. These are also called saprophytes. Decomposers secrete digestive enzymes that breakout dead and waste materials into simple inorganic materials, which are subsequently absorbed by them.

5. It is possible that a species may occupy more than one trophic level in the same ecosystem at the same time ? Explain with help of one example.

* **Frequently asked previous years Board Exam Question.**

Ans. Yes, for example sparrow is an omnivore, when it eats seeds, fruit or any other plant product, it occupies the primary trophic level. Whereas, when it eats worms and any other insect it occupies the secondary trophic level. Thus it occupies more than one trophic level in the same ecosystem.

6. List the parameters used for constructing ecological pyramids. Describe any one instance where the pyramids may look inverted.

Ans. The three parameters used are—

(i) Number of individual in a trophic level.
(ii) Biomass of individual in a trophic level.
(iii) Rate of flow of energy in a trophic level.

The pyramids of number may look inverted in a tree ecosystem where the number of consumer are numerous depending on a single producer.

7. Explain with the help of two examples, how the pyramid of number and the pyramid of biomass can look inverted.

Ans. The pyramid of biomass in sea is generally inverted because the biomass of fishes far exceeds that of phytoplankton and the number of big fishes eating the small fishes is also greater than the small ones. Also in pyramid of number, the number of insects feeding on a big tree is far greater than the tree. Now the number of small birds depending on the insect and the number of larger birds eating the smaller ones also increases in the order.

8. Write three causes and three effects of cultural eutrophication.*

Ans. Various causes for eutrophication include:

(i) Eutrophication is predominantly caused by the use of phosphorus and nitrate-rich fertilizers.
(ii) Direct sewage discharge and industrial wastes result in high amounts of chemical nutrients, thereby stimulating the dense growth of algae.
(iii) Natural events such as floods and the natural flow of rivers and streams can also wash excess nutrients off the land into the water systems, thus causing excessive growth of algal blooms.

Adverse effects of eutrophication are:

(i) The abundance of particulate substances such as phytoplankton, zooplankton, bacteria, fungi and debris on which the turbidity and colouration of the water depend.
(ii) Threatens the survival of fish and other aquatic life forms.
(iii) Extensive deterioration of water quality and a decline in the availability of clean drinking water.
(iv) Depletion of dissolved oxygen in the water body.

9. Define BOD. What is its significance in an aquatic ecosystem ?*

Ans. Biochemical oxygen demand (BOD) refers to the amount of dissolved oxygen needed by aerobic biological organisms required to break down organic material present in a given water sample at certain temperature over a specific period of time. It is commonly expressed in milligrams of oxygen consumed per litre of sample during 5 days of incubation at 20 °C and is often used as a "surrogate of the degree of organic pollution of water.

BOD is an important water quality parameter because it provides an index to assess the effect of discharged wastewater on the environment. The higher the BOD value, the greater the amount of organic matter or "food" available for oxygen consuming bacteria.

Chapter 15. Biodiversity and its Conservation

1. What are three different methods for calculation of diversity ?

Ans. Several methods for calculating diversity :

(i) **Alpha diversity** refers to diversity within a particular area, community or ecosystem and it is typically measured as the number of species within that area.
(ii) **Beta diversity** is the species diversity between areas and involves comparing the number of species that are unique to each area.
(iii) **Gamma diversity** is a measure of the overall diversity across a region.

2. What do you mean by Genetic Erosion ?

Ans. Genetic erosion refers to the process in which a plant or animal species faces a gradual or drastic diminishing or complete loss of its unique gene pool. A gene pool is a complete set of unique alleles that

* **Frequently asked previous years Board Exam Question.**

occur in the genetic matter of all members of a particular species. In genetic erosion, there is a loss of genetic diversity. This means there is a loss of particular individual genes or a loss of a certain gene combinants or gene complexes.

The disappearance of certain species can have an unfavourable effect on other species that might have depended on them in some manner for their survival and ultimately on the environment as a whole.

3. What are the main reasons behind the genetic erosion of plant species ?

Ans. The main reasons behind the genetic erosion of plant species :

(i) Loss of habitat is a cause of genetic erosion here.

(ii) Overgrazing of an area can lead to loss of plant species and also the spoiling of an environment by land clearing or chemical dumping.

(iii) Replacing local varieties of plants with those that would not occur there naturally, like plants from another area or genetically modified (GM) plants, can also lead to genetic erosion.

(iv) Modern agriculture is also responsible to a large extent for loss of genetic diversity. Farmers tend to grow a limited number of commercial crop varieties or GM crops, enforcing a uniformity in farming, and so there has been a noticeable reduction in the many crop varieties that were seen with traditional farming.

4. Which steps should be taken to slow down or stop the process of genetic erosion ?

Ans. Steps that should be taken to slow down or stop the process of genetic erosion :

(i) Preserve natural habitats.

(ii) Create wildlife sanctuaries and wildlife corridors that allow animals to migrate from one area to another.

(iii) Set up breeding programs in zoos.

(iv) In the case of critically endangered species, set up tissue and sperm banks.

(v) Release new members of a species in their natural habitat.

(vi) Be careful about introducing non-native species in local areas.

(vii) Set up seed banks.

(viii) Cultivate wide varieties of non-GM crops.

(ix) Guard against genetic pollution.

5. Discuss the role of wild plants in modern agriculture.

Ans. Role of wild plants in modern agriculture :

(i) Maintenance of gene bank for breeding programme in agriculture.

(ii) To get high yield from crop plants is not with the simple application of pesticides and fertilizers and proper irrigation but the crop plants must have the genetic potentiality to respond to improved inputs. Thus by examining wild relatives of crop plants, useful genes can be selected and introduced to breeding programme *e.g.,* Resistance gene has been successfully introduced to modern rice.

(iii) An average life of a crop variety is 5–15 years only. New varieties are being produced to meet the changing conditions such as pathogens, pests, climate and soil etc.

6. Write short notes on national parks, sanctuary and biosphere reserves.

Ans. A national park is an area which is strictly reserved for the betterment of the wildlife and where activities such as forestry, grazing or cultivation are not permitted. In India, there are 99 national parks.

A sanctuary is an area where protection is given especially to the fauna and operations like harvesting of timber and collection of minor forest products are permitted as long as they do not interfere with the well being of animals. In India, there are about 551 sanctuaries.

Biosphere reserves : They are large tracts of protected land with multiple use preserving the genetic diversity of representative ecosystem by protecting wildlife, traditional lifestyle of tribals and varied plant and animal genetic resources. They have been set up under (MAB-Man and Biosphere) programme of UNESCO in 1975. India's first biosphere reserve is Niligiri Biosphere Reserve set up in 1986.

7. Write a note on 'Biosphere reserves'.

Ans. Biosphere Reserves : They are multipurpose protected areas which are meant for preserving genetic diversity in representative ecosystem by protecting wild populations, traditional life style of tribals and domesticated plant and animal genetic resources. There are some 243 biosphere reserves in 65 countries of the world. In India 14 potential sites were identified in 1979 by Core Advisory Group but 9 biosphere reserves have been set up by now. Each biosphere reserve has a :

(i) **Core Zone :** No human activity is allowed.

(ii) **Buffer Zone :** Limited human activity is permitted.
(iii) **Manipulation Zone :** Human activity is allowed but ecology is not permitted to be disturbed.
(iv) **Restoration Zone :** Degraded area for restoration to near natural form.

8. Explain the Rivet Popper hypothesis.*

Ans. Ecologist Paul Ehrlich gave rivet popper hypo-thesis to help in understanding the contribution of species richness on the ecosystem. He compared each species with rivet in the body of an airplane. This hypothesis explains that ecosystem to be an airplane and the species to be the rivets joining all parts together. If every passenger travelling in the airplane starts taking rivets home (causing a species to become extinct), initially it may not affect flight safety (proper functioning of ecosystem), but over a period of time the plane becomes weak and dangerous (species become endangered and then extinct). In an ecosystem, the role of each species is important.

Chapter 16. Environmental Issues

1. How have human activities caused deforestation ? Explain.

Ans. The fertile top soil is removed by human activities like over cultivation, unrestricted grazing, deforestation and poor irrigation practices. This leads to formation of arid patches of land. These large barren patches when left for a long time form deserts.

2. A person has inhaled air having particulate pollutants of size less than 2.5 micrometres in diameter. What could be the impact of their pollutant on his respiratory system ?

Ans. Particulate pollutants inhaled deep into the lungs cause irritation, inflammation, damage to the lungs and premature death.

3. State any four measures to control noise pollution.*

Ans. Measures to control noise pollution :

(i) Using silencers to control noise from automobiles, ducts, exhausts etc.
(ii) Minimum use of loudspeakers and amplifiers especially near silence zones.
(iii) Proper lubrication will reduce noise from the machine.
(iv) Plantation of coniferous trees on highways.

4. Explain giving reasons why thermal power plant are not considered eco-friendly ?

Ans. The thermal power plant release particulate and gaseous pollutants in the environment. Inhalation of these pollutants can cause breathing or respiratory symptoms, irritation, inflammation damage to lungs and even premature death.

5. List any four effects of global warming.*

Ans. The various consequences of Global warming are :

(i) Warmer climate : Global warming on an average, has resulted in increase by approxmately 0.6% in 20th century.
(ii) Extinction of a large number of plant and animal species.
(iii) Decrease in Food production.
(iv) Rise in mean Sea-level.

6. Is it true that if the dissolved oxygen level drops to zero the water will become septic ? Give example which could lower the dissolved oxygen content of an aquatic body.

Ans. Yes, the water will become septic if the dissolved oxygen drops to zero. Organic pollution (biodegradable) is an example.

7. The crane had DDT level as 5 ppm in its body. What would happen to the population of such birds ? Explain giving reason.

Ans. The high concentration of DDT (5 ppm) in cranes disturb their calcium metabolism which causes thinning of egg shells and their premature breaking, thereby causing decline in the bird population.

Short Answers

* Frequently asked previous years Board Exam Question.

Scientific Reasoning

Set 8

Chapter 1. Reproduction in Organisms

1. Why is reproduction essential for organisms ?

Ans. Reproduction is essential :

(i) For multiplication and maintaining the identity of species.

(ii) To introduce variation among the individuals of a species.

(iii) To maintain and to inherit the genetic constitution or genetic make up.

2. Which is better mode of reproduction : sexual or asexual ? Why ?

Ans. Sexual mode of reproduction is better because it is biparental reproduction and introduces variation among offsprings (in a population), due to crossing over and recombination during gamete formation by meiosis.

3. Why is the offspring formed by asexual reproduction referred to as clone ?

Ans. In asexual reproduction the offspring is morphologically and genetically identical to the parent and to each other. Hence it is called clone.

4. Offspring formed due to sexual reproduction have better chances of survival . Why ? Is this statement always true.

Ans. Offspring formed due to sexual reproduction have better chances of survival because :

(i) The offspring retains its hybrid vigour which may adapt better with the changing environment.

(ii) Genetic variation is introduced among the offsprings, which increase the range of biological tolerance.

No, it is not always necessary that the offspring produced due to sexual reproduction has better chances of survival.

5. Higher organism have resorted to sexual reproduction inspite of its complexity. Why ?

Ans. Higher organism have resorted to sexual reproduction to :

(i) get over the unfavourable condition.

(ii) restore high gene pool in a population.

(iii) restore vigour and vitality of the race.

6. Explain why meiosis and gametogenesis are always interlinked ?

Ans. Gametogenesis (formation of male and female gametes) is associated with reduction in chromosome number thus the gamete formed contains half chromosomes set of the parental cell. So, gametogenesis is interlinked with meioses because in meiosis, reduction of chromosome number from diploid set (2n) to haploid set (n) takes place.

7. How is the chromosome number maintained in sexually reproducing organisms?*

Ans. In sexually reproducing organisms, the gametes undergo meiosis and therefore, each gamete contains only half set of chromosomes. During gametogenesis the chromosome number is reduced to half by meiosis and during fertilisation, the diploid number is restored in the zygote.

8. Why do vegetative propagules in sugarcane and ginger appear from the nodes ?

Ans. This is because the nodes have meristematics cells.

9. Why is coconut plant referred to as monoecious ?

Ans. A coconut plant has both male and female flowers present on the same individual thus it is known as monocieus.

10. Why is banana considered a good example of parthenocarpy ?

Ans. Because formation of fruit occur without fertilisation *i.e.,* there is no formation of seeds.

---

* **Frequently asked previous years Board Exam Question.**

11. Why large number of male gametes are produced as compared to female gametes ?

Ans. Several male gametes fail to reach the female gamete. So in order to compensate this loss of male gametes during transport, the number of male gametes produced is several thousand times more than the number of female gamete produced.

12. How do potato reproduce by asexual means ?

Ans. Potato reproduce by vegetative propagation of potato tubers. Vegetative propagation in potato is through eye (buds) present in tuber of potato. These eyes (buds) give rise to aerial shoots.

13. How does *Bryophyllum* plant propagate ?

Ans. In *Bryophyllum*, the leaves bear adventitious buds in the notches of the margins of leaves. When these leaves fall on the soil, the bud grows into a new plant.

14. A moss plant produces a large number of antherozoids but relatively only a few egg cell. Why?

Ans. Antherozoids are motile male gametophyte which have to swim on the water surface to fertilised the immotile female gametophyte *i.e.*, egg. Since during its transfer, many antherezoids get destroyed, a large number of them are produced.

15. Why are mosses and liverworts unable to complete their sexual mode of reproduction in dry condition ? Give reasons.

Ans. For sexual reproduction to take place in mosses and liverworts, the motile male gametophyte antherezoids, have to swim on the water surface to fertilise the immotile female gametophyte egg. In dry conditions, the antherozoid do not reach the egg and hence fertilisation cannot occur.

16. A liverwort plant is unable to complete its life-cycle in a dry environment. State two reasons.

Ans. (i) They need water as medium of gamete transfer for fertilisation.

(ii) For germination of spores.

17. Why do algae and fungi shift to sexual mode of reproduction just before the onset of adverse conditions ?

Ans. Algae and fungi shift to sexual mode of reproduction for survival during unfavourable conditions because fusion of gametes helps to pool their resources for survival. The zygote develops a thick wall that is resistant to dessiccation and damage which undergoes a period of rest before germination.

18. Why amoeba is called immortal ?

Ans. In amoeba, the parent divides completely into daughter cells and nothing is left. So amoeba is called immortal as it suffers no natural death.

19. Though earthworm is hermaphrodite, but it shows cross fertilisation. Why ?

Ans. Because, in earthworm, testis mature first than ovaries (protandrous), so the sperms from an individual are unable to fertilise the ovum from the same individual.

20. Why do birds have meroblastic cleavage ?

Ans. Birds lay megalecithal eggs *i.e.*, eggs which contain large amount of yolk. This yolk provides nourishment to the embryo. Such megalecithal eggs undergo meroblastic cleavage, where the line of segmentation do not completely pass through the egg and remain confined to a part. The yolk provides resistance to the cleavage.

21. Give reason for the following : Some organisms like honey-bees are called parthenogenetic animals.

Ans. They (drones/males) develop from unfertilised eggs.

22. Banana produces fruit but is propagated only by vegetative means. Why is it so ?

Ans. Banana does not produce seed, so is propagated only by vegetative means.

23. Why dogs and cats have oestrous cycles but human beings have menstrual cycle, though all are mammals ?

Ans. Dogs and cats are seasonal breeders. They have heat period of 2-4 weeks during which ovulation takes place. Human on the other hand, have this cycle monthly.

24. Why are offsprings of oviparous animals at a greater risk as compared to offspring of viviparous animals ?

Ans. In viviparous animals, the young ones develops inside the body of the female organism. As a result of this, the young ones get better protection and nourishment for proper development. In case of oviparous animals, they lay eggs and the young ones develop inside the calcareous shell, outside the body of the female so, the young ones are not effectively protected and nourished and are vulnerable to predators. So, they are at a greater risk as compared to the offsprings of viviparous animals.

Chapter 2. Sexual Reproduction in Flowering Plants

1. What is the receptacle of a flower and what are its functions ?

Ans. The stalk of the flower is called pedicel. Its tip enlarges to form the receptacle or thalamus. All floral appendages are attached to the receptacle.

2. What is a pollenkitt ? What is its function ?

Ans. It is a yellowish, viscous and sticky substance covering the exine of insect pollinated pollen grains. It helps in the pollination by insects.

3. How many meiotic divisions are required to produce 100 seeds of maize ?

Ans. To produce 100 pollen grains 25 meiotic divisions are necessary and for producing 100 embryo sacs, 100 meiotic divisions are required. So for the production of 100 seeds of maize, 125 meiotic divisions are required.

4. Why is an embryo sac diploid in agamospermy ?

Ans. Embryo sac is diploid in agamospermy because it is formed without reduction, division and fertilisation.

5. Why does the zygote undergo a period of rest after fertilisation ?

Ans. After fertilisation, the zygote undergoes a period of rest so that sufficient growth and accumulation of nutrients takes place in the endosperm. This ensures adequate supply of nutrients to the growing embryo.

6. Pollen grains of wind pollinated flowers are produced in large quantities. Give reason.*

Ans. Pollen grains of wind pollinated flowers are produced in large quantities because lot of them get wasted during the process of pollination. Hence, to increase the chance of pollination they are produced in large quantities.

7. What happens if a flower bud is emasculated and auxin applied on the stigma ?

Ans. It induces the conversion of unfertilised ovary into seedless, parthenocarpic fruits.

8. If the haploid number of chromosome in angiospermic plant is 8, then what will be ploidy in embryo, antipodals, seed coats and endosperm respectively ?

Ans. It will be 16 in embryos, 8 in antipodals, 16 in seed coats, and 24 in endosperms.

9. What is seed dormancy ? What are its types ?

Ans. Many seeds fail to germinate even when they are provided with favourable conditions. This phase when seed remains inactive is called dormancy. Dormancy is of various types :

(i) **Seed dormancy :** Occurs due to unfavourable environmental conditions. It is also called quiescence

(ii) **Innate Dormancy :** When dormancy is due to conditions like hard seed coat, immature embryo, growth inhibitors or requirement of chilling treatment.

10. Give reason for the following :

(a) Seeds are preferred as a source of staple food to vegetative parts of plants.

(b) Young fruits are green in colour.

(c) Quiescent seeds live for many years.

(d) Coconut trees mostly occur at the sea coasts.

(e) Propagation through grafting is not successful in monocots.

(f) Endosperm in angiosperms becomes triploid.

Ans. (a) Grains contain very small amount of water and remain dormant over a prolonged period. They are the source of protein, starch and oil reserves.

(b) The green colour of young fruits help them to remain hidden and protected from hostile animals. This also helps to protect immature seeds.

(c) The quiescent seeds can live for many years because they have low water content and hence no metabolic activity. They regain their activity and germinate when conditions are favourable.

(d) Coconut trees mostly occur at sea coasts because seed dispersal in them takes place through ocean currents.

(e) Grafting is not successful in monocots because of the absence of cambium and due to scattered vascular bundles. Grafting is successful mainly due to the activity and fusion of cambium which are absent in monocots.

(f) Endosperm in angiosperms become triploid because it is formed by the fusion of three haploid nucleic. Two of the nuclei are the polar nuclei of the female gametophyte and one nucleus is of the male gamete.

---

* **Frequently asked previous years Board Exam Question.**

Chapter 3. Human Reproduction

1. Why does the middle piece of sperm contain the mitochondria ?

Ans. The middle piece of human sperm contains mitochondria in order to provide energy for the movement of sperms. The absence of mitochondria makes the sperm immobile which in turn cause no fertilisation in the fallopian tube.

2. Suppose the acrosome of a mammalian sperm does not function normally. How would it effect fertilisation ?

Ans. In the absence of acrosome, spermatozoa will fail to enter the ovum hence no fertilisation takes place.

3. What will happen if the fallopian tubes are partially blocked and ovulated eggs are prevented from reaching the uterus ?

Ans. Fertilization may take place but the zygote may develop in the tube instead of uterus.

4. Why is an ovum immobile ?

Ans. An ovum is immobile because of presence of large amount of cytoplasm filled with reserve materials.

5. Why are polar bodies formed during oogenesis ?

Ans. Formation of polar bodies helps in the formation of ovum with large amount of cytoplasm and haploid nucleus.

6. Give reasons for the following :
   (a) Sterility is produced if testes fails to descend in scrotum.
   (b) Spermatids possess a haploid chromosome number.
   (c) The human ejaculate contains large number of sperms though one sperm is sufficient to fertilise the ovum.
   (d) There is no menstrual cycle in pregnancy.
   (e) Human males are reproductively active throughout their life span.

Ans. (a) Testes are suspended in scrotum located outside the main body cavity behind the penis. The scrotum maintains the temperature of testes lower by 2° C to 3° C than the body temperature, which is favourable for sperm production. If testes do not descend into the scrotum, then the sperms in the testes become infertile and immature at higher temperature of the body and cause sterility.

(b) The spermatids are formed from the spermatogonia of the germinal epithelium of seminiferous tubules in the testes. The spermatogonia give rise to spermatocytes which ultimately undergo meiosis to produce haploid spermatids. Spermatids possess a haploid chromosome number as they are formed due to meiotic process.

(c) The large number of sperms increase the probability of fertilisation because sufficient number of sperms die before they reach the vicinity of ovum. Fertilisation is purely a matter of chance and so large number of sperms are released in human ejaculation.

(d) There is no menstrual cycle during pregnancy due to the secretion of progesterone hormone by the corpus luteum. The progesterone hormone inhibits the production of the gonadotropic hormones from the pituitary and prevents the uterine lining from sloughing and supports pregnancy.

(e) Human males are reproductively active throughout their life because they produce sperms throughout their life. The human female's ovaries begin to produce one mature egg every month at puberty and continue to do so until menopause when they shrink and stop releasing egg. So the females do not remain reproductively active throughout their life.

7. Give reason for the following :
   (a) Human egg undergoes holoblastic cleavage.
   (b) Eggs of birds have large quantities of yolk.
   (c) Human egg contains small amount of yolk.
   (d) Size of morula remain almost the same as that of the ovum.

Ans. (a) Human egg undergoes holoblastic cleavage because the entire zygote divides completely with each cleavage.

(b) Egg of birds contain large quantity of yolk and are referred as megalecithal. The large quantity of yolk provides nourishment to the growing embryo, inside the egg

(c) A human egg contain small amount of yolk because the developing embryo depends upon the mothers body for its nutrition and not on the yolk. The human beings are viviparous and so their embryos derive their nutrition from the mother's body during the entire course of development and not from the reserved yolk of the egg.

(d) Size of morula remains same as that of ovum because the zona pellucida of egg remains intact throughout the cleavage division and hence there is no increase in the protoplasmic mass of the blastomeres.

8. What is the purpose of implantation of embryo ?

Ans. The purpose of implantation of embryo is to attach to the uterine wall to derive nourishment and undergo development process. Implantation leads to pregnancy.

Chapter 4. Reproductive Health

1. Indiscrimination diagnostic practices using X-rays etc., should be avoided. Give reasons.

Ans. These practices act as carcinogens which convert normal cells to neoplastic cells by harmful mutation or chromosomal aberration.

2. Is sex education necessary in schools. Why ?

Ans. Yes, sex education is necessary in schools because :

(i) It will provide proper information about reproduction organs, adolescence, safe and hygienic sexual practices and sexually transmitted diseases (STDs).

(ii) It will provide right information to avoid myths and misconceptions about sex-related queries.

3. Is the use of contraceptive justified. Give reasons.

Ans. Yes, use of contraceptive is justified because it helps to control the rapid growth of human population. It will also help in preventing unwanted pregnancies and STD's contraceptives.

4. Why do intensely lactating mothers not generally conceive ?

Ans. Due to suppression of gonadotropins, ovulation and menstrual cycle do not take place during lactation period.

5. Why is tubectomy considered a contraceptive method ?

Ans. Tubectomy involves cutting a piece of the fallopian tube and tying its ends. This way, the sperms are not able to reach the egg and it act as contraceptive method.

6. Why is CuT considered a good contraceptives device to space children ?

Ans. CuT is an ideal contraceptive device for human female because :

(i) Cu ions released supress sperm motility and fertilizing capacity of sperms.

(ii) It makes uterus unsuitable for implantation and cervix hostile to the sperm.

7. Why is hormone releasing IUD considered a good contraceptive to space children ?

Ans. Hormone releasing IUD's are considered a good contraceptive because :

(i) They make the uterus unsuitable for implantation.

(ii) They increase the phagocytosis of sperm within the uterus and the Cu ions released suppress sperm motility and the fertilizing capacity of sperms.

8. Why do some women use "Saheli" pills ?

OR

Why is Saheli considered to be an improved form of oral contraceptive for human females ?

Ans. "Saheli" contains a non-steroidal preparation and is a once a week pills, with high contraceptives value with very less side-effects. Therefore, it is considered an improved form of contraceptive pills.

9. Why is medical termination of pregnancy (MTP) carried out ?

Ans. MTP is carried out to get rid of unwanted pregnancies. It is also essential when the foetus is suffering from an incurable disease or when continuation of the pregnancy could be harmful or even fatal to the mother or foetus.

10. Why is the term test tube baby a misnomer ?

Ans. The term test tube is a misnomer because the baby is not developed in the test tube, only fertilisation is carried out in the laboratory conditions (*in vitro*). The fertilised egg (zygote) or early embryo is then transferred into the fallopian tube or uterus of the mother where it develops and a normal baby is born.

Chapter 5. Principles of Inheritance and Variation

1. Why are traits acquired during the life-time of an individual not inherited ?

Ans. Any changes in somatic cells (or non-reproductive tissues) cannot be passed to DNA of a germ cells Hence, these traits are not inheritable.

2. Why was Mendel successful in proposing the laws of heredity while his predecessors were unsuccessful ?

OR

What were the reasons behind the Mendel's success in genetic experiments ?

Ans. Mendel's success in the genetic experiments can be attributed to the following :

(i) Mendel concentrated on the results of one trait at a time.

(ii) When the behaviour of one trait was established, only then he considered two characters together.

(iii) He conducted a large number of crosses (2000-3000) to eliminate the chance factor and to obtain a valid and accurate explanation.

(iv) Most of all, he actually counted the number of offsprings of each category and maintained accurate records for each generation in each experiment.

(v) *Pisum sativum* is with seven homologous pairs of chromosomes. The unit factors for the seven characters identified are located on different homologous pairs in this plant.

3. How can epistasis decrease the number of phenotypes observed in a population ?

Ans. In epistasis, one gene affects the expression of another. The gene interaction may hide the expression of some phenotypes from a population.

4. How does crossing over unlink genes ?

Ans. Crossing over separates alleles that occurred together on the same chromatid, so that alleles that were previously linked are no longer transmitted together.

5. Crossing over is observed in the allosomal genes in an experiment. How is it possible ?

Ans. Pairing and crossing over is possible for the genes present on homologous regions of X and Y chromosomes (XY linked genes) like autosomes.

6. Why are sex linked recessive characters more in male human being ?

Ans. Sex linked recessive characters mostly occur in males because these need only one copy of the mutant allele to express the phenotype and are mostly due to gene mutation in X-chromosome since males have only 1 X-chromosome, these characters are mostly seen in male human beings.

7. If two different but linked genes are located very far apart on a chromosome, how may the inheritance pattern create the appearance of independent assortment ?

Ans. Since the genes are very far apart on the chromosome, they have a high probability of being separated by crossing over. Thus, there is a probability of their independent assortment.

8. How do patterns of inheritance differ for unlinked versus linked pairs of genes ?

Ans. When pairs of genes are linked, they are carried on the same chromosome and are inherited together. Crossing over complicates the inheritance of linked genes; sometimes allele combinations differ from either parent. The inheritance pattern of non-linked genes is more predictable since it is not affected by crossing over. The inheritance of non-linked genes can be visualized by using a Punnett Square.

9. Why is *Drosophila* a convenient animal for the study of linked genes ?

Ans. The fruit fly *drosophila* is suitable for the study of Genetics because it presents many distinct traits and have only four chromosomes (one sex chromosome and three autosomes.)

10. Diploid forms of *Drosophila* are never intersex. Why ?

Ans. In diploid *Drosophila*, the karyotype is AAXX or AAXY. In AAXX condition, the ratio is 1·0, so the sex of individual is fertile female. In AAXY condition, the ratio is 0·5, so the sex of individual is male.

11. Why are sex linked dominant characters more in female human being ?

Ans. Females are more likely to be affected by sex-linked dominant characters as the females have 2 X-chromosomes and the mutual allele is mostly on the X-chromosome. So the females have double chance to inherit the mutated allele.

12. Do you agree with the statement that all alleles on the Y-chromosome are dominant ? Why or why not ?

Ans. No. because the Y-chromosome is not homogonously paired, so all alleles on the Y-chromosome are expressed. However, recessive alleles are still "recessive" even if no dominant allele can mask them.

A recessive allele encodes a non-functional protein. If an allele on the Y-chromosome encodes a non-functional protein, then the allele is recessive.

13. Why do males and females express recessive X-linked alleles differently ?

Ans. Each female has a pair of X-chromosomes, whereas a male has only one X-chromosome. Any trait a male has on its X-chromosome will be expressed. Recessive alleles on a X-chromosome of a female may be masked by dominant alleles on its homologous X-chromosome.

14. How are X-linked genes inherited differently in male and female humans ?

Ans. A female inherits two X-chromosomes, a male inherits his single X-chromosome from his mother. A male expresses every allele (dominant or recessive) on his X-chromosome because he lacks a second allele that could mask the expression of recessive alleles.

Chapter 6. Molecular Basis of Inheritance

1. Why Griffith has choose *Streptococcus/Diplococcus* for his experiment ?

Ans. Griffith chose *Streptococcus* (*Diplococcus*) for his experiment on transformation because the bacteria exists in two distinct strains :

(i) **Virulent or S-strain,** which possesses a covering of mucilage (capsule) forming a smooth or regular colony and the ability to cause disease of pneumonia.

(ii) **Non-virulent or R-strain** which lacks mucilage sheath (capsule) and forms rough or irregular colonies. This strain is incapable to cause the disease, pneumonia.

2. Why the second process of protein synthesis is termed as translation ?

Ans. The process is referred as translation as it involves the changing of coded genetic message of mRNA into a polypeptide chain.

3. Why is the genetic code said to be "degenerate" ?

Ans. There are more codon combinations than there are amino acids. The genetic code is degenerate because more than one codon sequence can code for the same amino acid.

4. How DNA of virulent viruses reproduces ?

Ans. The DNA in these type of viruses reproduces through the lytic cycle. When these viruses reproduce, they break, open, or lyse, their host cells, resulting in the destruction of the host. The whole cycle can be complete in 20-30 minutes depending on a variety of factors such as temperature.

5. Why gene regulation is essential for viruses, prokaryotes and eukaryotes ?

Ans. Gene regulation is essential for viruses, prokaryotes and eukaryotes as it increases the versatility and adaptability of an organism by allowing the cell to express protein when needed.

6. Why allolactose is known as inducer ?

Ans. Allolactose is called an inducer because in its presence it turns on, or induces the expression of the lac genes.

Chapter 7. Evolution

1. Why were some plants and animals much more likely to be preserved as fossils than others ?

Ans. The organisms with hard body structures had more of a chance to be fossilized than the soft bodied animals or herbaceous plants which decayed readily.

2. Two closely related species lived in the same area and yet no hybrids were produced, why ?

Ans. It is because of the mechanism of "Reproductive Isolation".

3. Give reason :

(a) After the development of chlorophyll, the atmosphere on earth was still reducing for some years.

(b) Life does not originate under present day conditions.

(c) Existence of life is possible under the Antarctic sheet of ice.

Ans. (a) The first autotrophs resembled bacteria. They were anaerobic and utilised hydrogen from sources other than water. Therefore no oxygen was evolved and atmosphere remained reducing in nature.

(b) The abiotic origin of life today is not possible on today's earth because it needs abundant energy in the form of heat, cosmic rays and lightning, together with anaerobic reducing atmosphere. However, the present atmosphere on the earth is aerobic and oxidising type which would prevent abiotic origin of life.

(c) Water expands on freezing and floats as a sheet of ice over the surface and protects the underlying water as well as living organisms from freezing. Hence, aquatic life exists even below the antarctic sheet of ice.

4. Why is the wing of a bat said to be homologous to the wing of a bird and analogous to the wing of an insect ?

Ans. As bat's wing and bird's wing have similar structure and origin *i.e.,* they have bony structures hence, they are called homologous organs. Bat's wing is analogous to the wing of an insect, because the wing of an insect does not have a bony structure, it has membranous wing. So bat's wing and insect's wing are analogous, because they have similar functions but do not have the same origin and structure.

5. What are the causes of evolution ? Enlist main causes of evolution.

Ans. The process which disturbs the Hardy Weinberg equilibrium gives evolutionary trend to a population. These are called evolutionary agents.

The main evolutionary agents are :

(i) Mutations
(ii) Migration
(iii) Genetic drift
(iv) Non-random mating
(v) Natural selection.

Chapter 8. Human Health and Diseases

1. How communicable diseases are caused ?

Ans. They are caused by micro-organisms present in the atmosphere, food, water, etc.

2. How are diseases transmitted from the reservoir of infection to a healthy person ?

Ans. (i) **By direct transmission** which includes contact with an infected person, droplet infection, contact with soil and bite of an animal.

(ii) **By indirect transmission** vectors, vehicle borne agents, air borne, formite borne and unclean hands.

3. How does the cow help the calf by licking it ?

Ans. Saliva of the cow contains lysozyme which kills harmful bacteria.

4. If the first line of defence is unable to prevent the infection, how does the body respond to the infection ?

Ans. When the first line of defence fails, then the second line of defence *i.e.,* the inflammatory response, takes over to prevent infection.

5. Why fever is sometimes caused with inflammation ?

Ans. In systemic inflammatory response, the number of white blood cells increases and fever is caused by the toxins released by pathogens or by compounds called pyrogens. These are released by WBCs to regulate body temperature. Fever if moderate stimulates the phagocytes and hence, inhibits growth of micro-organism.

6. Why does the body often reject transplanted kidney or heart ?

Ans. The body recognises 'non-self' organ and its immune system starts functioning against it.

7. Why are some childhood diseases not caused second time ?

Ans. On first encounter with certain antigen, our immune system produce antibodies against them. However, simultaneously, memory cells of T-cells and B-cells are stored in the immune system, ready to fight off the disease causing antigen, on second encounter.

8. How the proto-oncogenes are activated into oncogenes ?

Ans. Proto-oncogenes can be activated by viral infection, continuous irritation, temperature, ultraviolet rays, X-rays, etc.

9. What will happen if :

(a) An Rh -ve woman bears a Rh^+ foetus for the second time.

(b) Phenylalanine in the body fails to get converted into tyrosine.

Ans. (a) Rh^- mother will be already having anti-Rh factor (antibodies) which would have developed when she bears the first Rh positive child. Thus, at the time of second Rh^+ pregnancy, the mother's anti Rh^- antibodies will kill RBC's of the embryo. Thus, the child will be acutely anaemic and will die.

(b) If phenylalanine is not converted to tyrosine, it will result in severe disturbances in body metabolism because tyrosine acts as a substrate for a large number of metabolic products needed for nervous system. Hence, it will lead to mental retardation and other damage to brain.

Chapter 9. Strategies for Enhancement in Food Production

1. Why are beehives kept in crop field during flowering period ? Name any two crops field where it is practiced.

Ans. Beehives are kept in crop field during flowering period to increase pollination efficiency and to increase honey yield. This is practiced in the field of Sunflower, *Brasica*, Apple, Pear etc.

2. In animal husbandry if two closely related animals are mated for a few generations it results in loss of fertility and vigour. Why is this so ?

Ans. The phenomenon being referred to is called inbreeding depression and results in loss of fertility and vigour. This happens because the recessive alleles tend to get together and get expressed harmful effect in the progeny.

3. Why is selection process after hybridization very crucial in breeding programmes ?

Ans. (i) This step yields plants are superior to both of parents.

(ii) These plants are self pollinated to raise homozygous on pure lines for breeding purpose.

4. Why it is necessary to emasculate a bisexual flower in a plant breeding programme ? Mention the condition under which emasculation is not necessary.

Ans. Emasculation is necessary to ensure that only the desired pollen grains are used for pollination and the stigma is protected from contamination (from unwanted self pollen). The anthers are removed followed by bagging so the plant now behaves as a female plant. The pollen grains from the anthers of the desired male plant can be dusted on the stigma of flower of the female plant to obtained desired results. Emasculation is not required if the plant produces unisexual flowers.

5. Give a reason : Bagging is essential in artificial hybridisation.*

Ans. In order to prevent contamination from unwanted or foreign or self-pollen grains.

6. Why is the south Indian Sugarcane preferred by agriculturalists ?

Ans. South Indian Sugarcane has thicker stem and higher sugar context.

7. Why is mutation breeding necessary for producing disease resistance varieties ?

Ans. Because there is limited availability of disease resistance genes in the crop plant and their wild relative.

8. Why are apical and axillary meristems used for tissue culture ?

Ans. Because these are free from virus and show rapid cell division.

Chapter 10. Microbes in Human Welfare

1. Why is distillation required for producing certain alcoholic drinks ?

Ans. Distillation increases the alcohol content in alcoholic drinks.

2. Why are drinks like whisky and rum more intoxicating than wine ?

Ans. This is because wine is produced without distillation and whisky and rum are produced by distillation of the fermented broth, thus increasing their alcoholic content.

3. Why are some molecules called bioactive molecules ? Give two examples of such molecules.

Ans. This is because microbes like bacteria or fungi are used in their production.

E.g., Citric acid – Acetic acid
Butyric acid – Lactic acid
Ethanol – Lipase
Streptokinase – Cyclosporine A

4. Why are biogas plant more suitable and advantageous in rural areas ?

Ans. (i) Raw material for biogas plant is mainly cow dung which is available in plenty in rural areas where cattle are used.

(ii) The biogas is used for lighting and cooking in these areas as distribution is only in short distances.

(iii) The spent slurry from the biogas plant is used as a fertilizer for agriculture hence are more suitable in rural areas.

*** Frequently asked previous years Board Exam Question.**

5. Why do farmers prefer biofertilizers to chemical fertilizer these days ? Explain.

Ans. Chemical fertilizers have significantly increased the soil pollution, reducing its quality. Therefore, farmers use biofertilizers which enrich the quality of the soil.

6. Why is *Rhizobium* categorized as a 'Symbiotic bacterium' ? How does it act as a biofertilizers ?

Ans. *Rhizobium* is present in the root nodules of leguminous plants. There is a symbiotic relationship in which the bacterium obtained food and the shelter from the plant and the plant gets fixed nitrogen in return These bacteria fix atmospheric nitrogen into organic forms which is used by the plant as nutrients.

7. Why are antibiotics always sold in combination with *lactobacillus* ?

Ans. Antibiotics may kill even the useful bacteria present in the digestive tract. LAB is given to protect useful microbes in the digestive tract and enhance the production of Vitamin B_{12}.

8. Cyanobacteria increase the productivity in paddy fields.*

Ans. Cyanobacteria is a nitrogen fixing bacteria and thus help in the management of the nutrient in the soil and maintain the soil structure, increasing the productivity of paddy crops.

Chapter 11. Biotechnology : Principles and Processes

1. Do clones ever occur naturally ?

Ans. Yes. In nature, some plants and single-celled organisms, such as bacteria, produce genetically identical offspring through a process called asexual reproduction. In asexual reproduction, a new individual is generated from a copy of a single cell from the parent organism.

Natural clones, also known as identical twins, occur in humans and other mammals. These twins are produced when a fertilised egg splits, creating two or more embryos that carry almost identical DNA. Identical twins have nearly the same genetic makeup as each other, but they are genetically different from parent.

2. Why is it important that the plasmids used for genetic engineering contain antibiotic resistance genes ?

Ans. The plasmid used for genetic engineering should contain antibiotic resistance gene so that :

(i) The transformed bacteria can produce antibiotics.

(ii) The plasmid can replicate in the bacteria.

(iii) The bacteria which have taken up the plasmid can be identified.

(iv) The bacteria will grow fast.

(v) The bacteria can produce insulin.

3. Why does DNA move towards the anode in gel electrophoresis ?

Ans. The DNA molecule contains phosphate groups which give it a negative charge. The DNA fragments being negatively charged move towards the positively charged anode under the influence of an electric field in gel electrophoresis.

Chapter 12. Biotechnology and its Applications

1. Why do scientists want to sequence the human genome ?

Ans. DNA sequence contains the instructions for everything a cell does, from the moment of conception until death. Changes in DNA spellings can increase the chances of developing an illness, protect it from getting sick, or predict the way body will handle medicines. Once scientists can read the DNA instruction book, they will be able to better understand and treat diseases.

2. Why are yeast used extensively for functional expression of eukaryotic genes ?

Ans. Yeast are simplest unicellular eukaryotic organism and like bacteria they are genetically well characterized easy to grow and manipulate. They can be readily cultured in small culture vessels as well as in large-scales bioreactors.

* **Frequently asked previous years Board Exam Question.**

3. Why do the toxic insecticidal proteins secreted by *Bacillus thuringiensis* kill the insect and not the bacteria itself ?

Ans. The Bt toxin protein exist as inactive protoxin but once an insect ingest the inactive toxin, it is converted into an active form of toxin due to the alkaline pH of the gut which solublise the crystals. Therefore it does not kill the bacteria.

4. Why is the gene encoding for 'Cry' protein inserted into a crop plant ?

Ans. 'Cry' protein producing gene is transferred to the plant to provide resistance against insect larvae.

5. Why transgenic plants are used in factories or bioreactor ?

Ans. Transgenic plants are also proposed to be used as factories or bioreactors for manufacturing special chemicals or pharmaceuticals.

6. Why are transgenic animal so called ?

Ans. Transgenic animal are so called because these animals have had their DNA manipulated.

7. Why is the functional insulin thus produced considered better than the once used earlier by diabetic patients ?

Ans. The insulin prepared by rDNA technology does not produce sensitive allergic reaction and immunological reaction where as those used earlier produced allergic reaction and others complications to the foreign protein as earlier they were extracted from pancreas of slaughtered cattle or pigs.

8. Why do children cured by enzyme-replacement therapy for adenosine deminase deficiency need periodic treatment ?

OR

Why is the introduction of genetically engineered lymphocytes into an ADA deficiency patient not a permanent cure ? Suggest a possible permanent cure.

Ans. Introduction of genetically engineered lymphocytes into ADA deficiency patients is not a permanent cure because these cells are not immortal and the patient requires periodic infusion of such genetically engineered lymphocytes. A possible permanent cure can be isolating the gene producing adenosine deaminase (ADA) from bone marrow cells and introducing it into cells at early embryonic stages.

9. Biopiracy should be prevented. State why and how ?

Ans. Biopiracy is unauthorized exploitation of bio-resources of developing or under-developed countries. Hence it should be prevented. It can be prevented by developing laws to obtain proper authorization and by paying compensatory benefits.

Chapter 13. Organisms and Populations

1. What wiln an aquatic ecosystem, the biomass of the fish (top consumer) is greater than the producers (phytoplankton).ll happen if the factors of environmental resistance do not produce zero growth ?

Ans. The population experiences exponential growth.

2. What does the lag phase of sigmoid growth curve indicate and why is it so ?

Ans. In the initial stages as population settles down and adjust itself, the growth curve shows a very little increase *i.e.,* because of slow growth.

Chapter 14. Ecosystem

1. Why is an earthworm called a detritivores ?

Ans. This is because earthworm break down detritus into smaller particles.

2. Why is earthworm considered a farmer's friend ?

Ans. Earthworm help in breakdown of complex organic matter as loosening of the soil. This help in the proper growth of the crops. Therefore, they are considered farmer's friend.

3. Why are green plants not found beyond a certain depth in the ocean ?

Ans. Beyond a certain depth in the ocean, sunlight is not able to penetrate. Due to which green plants cannot perform photosynthesis and thus do not survive.

4. Why is the length of a food chain in an ecosystem generally limited to 3-4 trophic levels ? Explain with an example.

Ans. The amount of energy flow decrease with successive trophic levels as only 10% of energy is transferred from one trophic level to the next successive level. The energy is lost in the form of respiration and other vital activities to maintain life. If more trophic levels are present, the residual energy will be limited and decreased to such an extent that it cannot further support any trophic level by the flow of energy. So the food chain is generally limited to 3-4 trophic levels only.

For e.g.

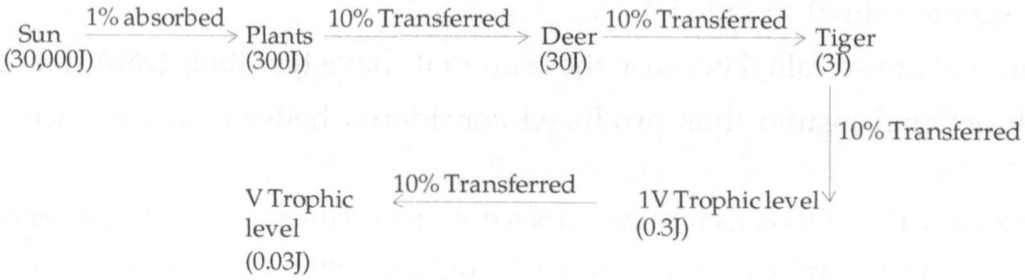

Sun (30,000J) —1% absorbed→ Plants (300J) —10% Transferred→ Deer (30J) —10% Transferred→ Tiger (3J)

Tiger (3J) —10% Transferred→ 1V Trophic level (0.3J) —10% Transferred→ V Trophic level (0.03J)

5. Why is the rate of assimilation of energy at the herbivore level called secondary productivity ?

Ans. It is because the biomass available to the consumer for the consumption is a resultant of the primary productivity from plants.

6. Why is food web formed in nature ?

Ans. Many organism occupy position in different food chains and several food chains become inter connected to form a food web.

7. Why the pyramid of energy is always upright ? Explain.

Ans. Pyramid of energy is never inverted when energy flows from a particular trophic level to the next trophic level, some energy is always lost as heat at each step. Each bar in the energy pyramid indicated the amount of energy present at each trophic level in a given time.

8. Why is the pyramid of biomass inverted in a pond ecosystem ?

Ans. The pyramid of biomass is inverted in a pond ecosystem because the biomass of fish (top consumer) is larger than the producers (phytoplanktons).

9. Why are the pyramids referred to as 'upright' or 'inverted' ?

Ans. Pyramid of energy is always upright as only 10% energy is transferred from one trophic level to the next. Inverted pyramid shows less number/ biomass of producer when compared to primary consumers.

10. Why is secondary succession faster than the primary succession ?

Ans. Secondary succession is faster than the primary succession because of presence of (i) soil for growth, which also result in quick attainment of climax. (ii) Water, the environment and the seeds or other propagules which are easily available.

11. Equilibrium of a forest ecosystem can be disturbed by uncontrolled hunting of big predators. Give reason.*

Ans. Hunting creates an imbalance in the natural elements of the environment. If a predator, like a wolf or mountain lion is hunted down, their prey, deer will increase in number. They inturn will feed on vegetation. In the long term, their uncontrollable number will lead to shortage of vegetation, which will affect other herbivores. This will increase interspecific and intraspecific competition and ultimately disturb the food chains.

12. Climax stage is achieved quickly in secondary succession as compared to primary succession. Why ?*

Ans. The rate of succession is much faster in secondary succession as the substratum (soil) is already present as compared to primary succession where the process starts from a bare area (rock).

13. The shape of the pyramid of biomass in an aquatic ecosystem is inverted.*

Ans. In an aquatic ecosystem, the biomass of the fish (top consumer) is greater than the producers (phytoplankton).

*** Frequently asked previous years Board Exam Question.**

Chapter 15. Biodiversity and its Conservation

1. How do ecologists estimate the total number of species present in the world ?

Ans. The diversity of living organisms present on the Earth is very vast. According to an estimate by researchers, it is about seven millions. The total number of species present in the world is calculated by statistical comparison between a species richness of a well studied group of insects of temperate and tropical regions. Then, these ratios are extrapolated with other groups of plants and animals to calculate the total species richness present on the Earth.

2. Give three hypothesis for explaining why tropics show greatest levels of species richness.

Ans. There are three different hypothesis proposed by scientists for explaining species richness in the tropics :

(i) Tropical latitudes receive more solar energy than temperate regions, which leads to high productivity and high species diversity.

(ii) Tropical regions have less seasonal variations and have a more or less constant environment. This promotes the niche specialization and thus, high species richness.

(iii) Temperate regions were subjected to glaciations during the ice age, while tropical regions remained undisturbed which led to an increase in the species diversity in this region.

3. What are the main reasons behind the genetic erosion of animal species ?

Ans. The main reasons behind the genetic erosion of animal species are :

(i) Members of the endangered species cannot meet and breed as result of habitat loss, habitat fragmentation or geographical distance.

(ii) The individuals do breed, but because they are restricted to a certain area and cannot travel to meet different members of the species, there is low genetic diversity and inbreeding occurs. Inbreeding leads to physical defects that weaken the entire species.

4. Why would it be unethical to destroy wild species ?

Ans. It is an established fact that rich diversity of todays organisms are the outcome of natural evolution stretching about 3·5 billion years. A species once lost cannot be retrieved. Therefore, it would be unethical on the part of man to be responsible for destruction of a species. Man has an evolutionary responsibility to conserve biological diversity for our generations to come.

5. If extinction is a natural process, why should we make an effort to save endangered species ?

Ans. We should make an effort to save endangered species because we can no longer attribute the accelerating extinction of plants and animals to natural causes.

Chapter 16. Environmental Issues

1. Why CNG considered as better fuel than diesel for automobiles ?

Ans. CNG is considered a better fuel because it burns most efficiently, is cheaper and cannot be siphoned off by thieves and adulterated like petrol or diesel.

2. Why are the use of unleaded petrol recommended for motor vehicles equipped with catalytic converters ?

OR

Why are owners of motor vehicles equipped with catalytic converters advised to use unleaded petrol ?

OR

Why is it desirable to use unleaded petrol in vehicles fitted with catalytic converters ?

Ans. Lead in petrol inactivates the catalysts (platinum, palladium and rhodium) which convert harmful pollutants (CO, unburnt hydrocarbons, nitric oxide) to lesser harmful pollutants (CO_2, H_2O, N_2).

3. Why are lichens regarded as pollution indicators ?

Ans. Lichens are regarded as pollution indicators because they do not grow in area that are polluted. So their presence indicates no pollution in that area and their absence indicates that the area is polluted.

4. Explain giving reason why thermal power plant are not considered eco-friendly.

Ans. Thermal power plant release particulate and gaseous pollutants in the environment. Inhalation of the pollutants can cause breathing or respiratory symptoms irritation. Inflammation cause damage to lung and even premature death.

5. Why are the colourful polystyrene and plastic packagings used for protecting the food, considered ar environmental menance ?

Ans. The colourful polystyrene and plastic packaging are non-biodegradable and non-ecofriendly.

6. In spite of being non-polluting why are there great apprehensions in using nuclear energy for generating electricity ?

Ans. There are great apprehensions in using nuclear energy because of accidental leakages and finding ways for safe disposal to radioactive waste.

7. Why is *Eichhornia crassipes* nick named as "Terror of Bengal" ?

Ans. *Eichhornia crassipes* is an aquatic weed that grows abundantly and very fast in Eutrophic water bodies and imbalances water ecosystem. It causes oxygen depletion leading to death of aquatic life (Eutrophication).

8. Why is the concentration of toxin found to be more in the organism occupying the highest trophic level in the food chain in a polluted water body ? Explain with the help of a suitable examples.

Ans. The concentration of toxic materials like heavy metals and pesticides increases at each trophic level of food chain and is more in organism of highest trophic level due to the accumulation at each trophic level. For examples, when DDT was used to control mosquitoes in a lake of USA. 800 times more DDT was found in the phytoplankton's than in the water of lake zooplanktons had about 13 times more DDT than phytoplankton's. It was also observed that the fishes population had 9-40 times more DDT than zooplankton's and fish eating birds has 25 times more DDT than fishes.

9. Why ozone hole forms over Antartica ?

Ans. In Antarctica, winter months receive no sunlight and thus the temperature is extremely low (–85 °C). It facilitates the formation of ice clouds which provide the catalytic surface on which the chlorine atom react with ozone and degrade it. This happens with the return of sun to Antarctica in springs (September and October). Therefore an ozone hole appears in Antarctica in spring.

10. Why do ozone hole form in spring and not in winter ?

Ans. Sunlight is necessary for ozone degradation and it is available only during spring.

Differentiate Between

Set 9

Chapter 1. Reproduction in Organisms

1. Differentiate between :
   (i) Oviparous animals and Viviparous animals.
   (ii) External fertilisation and Internal fertilisation.

Ans. (i)

S. No.	Oviparous Animals	Viviparous Animals
1	Oviparous animals lay fertilised eggs.	Viviparous animal give birth to young ones.
2	The fertilised eggs have calcareous shell and young ones hatch out after a period of incubation.	The fertilised egg (zygote) has no shell and develops into a young one inside the body of the female organism.
3	Chances of survival of young one is less as the female lays egg in the environment.	Chances of survival of young one is more because of proper embryonic care and protection inside the mother's body.
	For example – reptiles, birds, etc.	For example – majority of mammals including human.

(ii)

S. No.	External Fertilisation	Internal Fertilisation
1	When the fusion of gametes occur outside the body of an organism. It is called external fertilisation.	When the fusion of gamete occurs inside the body of an organism it is called internal fertilisation.
2	Large number of both male and female gametes are released into the surrounding in order to enhance chances of syngamy.	The number of sperms produced are very large and number of eggs produced are less.
3	The offsprings are extremely vulnerable to predators, threatening for their survival. Examples –Bony fishes, Frog and majority of algae.	The offsprings are well protected comparatively. Examples – Birds, mammal

2. Give three significant differences between asexual reproduction and sexual reproduction.*

Ans.

Asexual Reproduction	Sexual Reproduction
Single parent is involved.	Two parents are involved.
There is no fusion of gametes.	Fusion of male and female gametes occurs.
Somatic cells act as reproductive units	Sex cells act as reproductive units.
Genetic makeup of offspring is exactly similar to the parent	Offspring are genetically different from parents
Meiotic divisions may or may not present.	Meiotic divisions are present
Rate of production of offspring is rapid	Rate of production of offspring is slow.

* Frequently asked previous years Board Exam Question.

3. Differentiate between a zoospore and a zygote.

Ans.

S. No.	Zoospore	Zygote
1	These are endogenously, asexually produced unicellular, naked and motile spore having one or two flagella.	Zygote is a diploid cell formed by fusion of male and female gametes.
2	It may be haploid or diploid.	It is always diploid.
3	Zoospore takes part in dispersal.	Zygote does not have significant role in dispersal. It develops to form organism.

4. How does the progeny formed from asexual reproduction differ from those formed by sexual reproduction ?

Ans.

S. No.	Progeny from asexual reproduction	Progeny from sexual reproduction
1	The progenies have similar make up.	The progenies have different genetic make up.
2	The progenies are not only identical to one another but are exact copies of their parents *i.e.* clone of the parent.	The progeny are different from each other and also dissimilar to the parent.
3	Variation is absent.	Variation is a common phenomenon.
4	Variation may occur due to mutation.	Variation occur due to mutation, crossing over and recombination.
5	Progeny is less adaptive to change in environment.	Progeny is more adaptable to change in environment.

5. Differentiate between gametogenesis and embryogenesis.

Ans.

S. No.	Gametogenesis	Embryogenesis
1	It is the formation of gametes from meiocytes (gametes mother cell).	It is the formation of embryo from zygote cell.
2	This is a pre-fertilisation event.	This is the post-fertilisation event.
3	The cell division during gametogenesis is meiotic in diploid organism.	The cell divison during embryogenesis is mitotic in diploid organism.
4	It occurs inside reproductive organs.	It occurs outside or inside the female body.
5	It produces haploid gamete.	It give rise to diploid embryo.

6. Differentiate between menstrual and oestrus cycle.

Ans.

S. No.	Menstrual Cycle	Oestrus Cycle
1	It occurs in female members of primates of class Mammalia.	It occurs in females of non-primates of class Mammalia.
2	Blood loss occurs during menstruation, due to sloughing off of broken endometrium.	No blood loss. The sloughed lining of reproductive tract is reabsorbed.
3	There is no irresistible sexual urge during menstrual cycle.	The female develops a strong irresistible sexual urge during oestrus.
4	The interval between two successive cycles is about four weeks.	The interval between two successive cycles varies between a few months to a year.
5	The cycle consist of three phases : menstrual phase, proliferative phase and secretory phase.	The cycle consist of short period of oestrus (heat) followed by non-oestrus period.

Chapter 2. Sexual Reproduction in Flowering Plants

1. Mention one significant difference between each of the following :
- (i) Quiescent seed and dormant seed
- (ii) Mesogamy and porogamy
- (iii) Isogamy and Anisogamy
- (iv) Pollination and fertilisation
- (v) Aggregate fruits and composite fruits

Ans. (i)

Quiescent Seed	Dormant Seed
This type of seed can live for many years but germinates when soaked in water under suitable temperature and presence of oxygen.	This is an inactive seed which is unable to germinate even under favourable conditions.

(ii)

Mesogamy	Porogamy
The entry of pollen tube into embryo sac (ovule) through integuments and funiculus is called mesogamy. *E.g. Cucurbita, Populus, Pistacia.*	The entry of pollen tube into embryo sac (ovule) through micropyle is called porogamy. It is the most common way of entry of pollen tube into the embryo sac in angiosperms. *E.g.* Lily.

(iii)

Isogamy	Anisogamy
During fertilisation if the fusing gametes are identical in size, structure and appearance, the phenomenon is called isogamy.	When the fusing gametes are different in one or more aspects, the phenomenon is called anisogamy.

(iv)

Pollination	Fertilisation
The process of transfer of pollen grains from the anther to the stigma is called pollination.	The process of fusion of the male gamete with the female gamete is termed as fertilisation.

(v)

Aggregate Fruits	Composite Fruits
An aggregate fruit is a fruit that develops from the merger of several ovaries that were separate in a single flower. E.g. Strawberry.	Composite fruits are the fruits that develop from an inflorescence *e.g.* Mulberry

2. What is the difference between the endosperm of gymnosperms and angiosperms ?

Ans. The endosperm of gymnosperms is a simple haploid tissue or gametophyte whereas in angiosperms endosperm consists of triploid cells.

3. Differentiate between the following :
- (i) Self and Cross pollination.
- (ii) Anemophilous and Entomophilous Flower.
- (iii) Autogamy and Allogamy.
- (iv) Gamopetalous and Polypetalous.
- (v) Apomixis and Amphimixis.
- (vi) Mound layering and Air layering.
- (vii) Apocarpous ovary and Syncarpous ovary.
- (viii) True Fruit and False fruit.
- (ix) Stock and Scion.

Ans. (i)

S. No.	Self Pollination	Cross Pollination
1	It is the transfer of pollen grains from anther to the stigma of the same flower or to the another flower of same plant.	It is transference of pollen grains from anther to the stigma of another flower of different plant.
2	It occurs always through touch or wind.	It always occurs through an external agent.
3	Both anthers and stigma mature simultaneously.	The anthers and stigma mature at different time.
4	It can occur in closed and open flowers.	It occurs only when the flowers are open.
5	Race is almost constant *i.e.* homozygous.	The race is changing *i.e.* heterozygous.
6	It gives rise to pure lines.	It gives rise to offsprings having variation.
7	It preserves the parental characters.	It does not preserve the parental characters.
8	Yield of plant falls with time.	Yield of plant does not fall.
9	Origin of new species is not possible.	It produces new races and varieties.

(ii)

S. No.	Anemophilous Flower (Wind)	Entomophilous Flower (Insects)
1	The flowers are small.	The flowers are either large or if small they are grouped to form a mass.
2	Flowers are neither brightly coloured nor emit any odour.	Flowers are brightly coloured and emit odour.
3	The flowers are devoid of any nectar and edible pollen.	The flowers usually possess nectar or edible pollen.
4	Pollen grains are produced in large number and are dry, non-sticky and light weight.	Pollen grains are rough, sticky and often with spinous outgrowths.

(iii)

S. No.	Autogamy	Allogamy
1	Pollination by the pollen grains of the same flower is termed as autogamy.	Pollination by the pollen grains of another flower of a different plant is termed allogamy.
2	It does not provide opportunity for genetic recombination and genetic variability.	It provides opportunity for genetic recombination and genetic variability.
3	Autogamous plants breed true to their type e.g. Pea, Wheat, Rice.	Allogamous plants show hybrid vigour e.g., Maize, Cannabis, Mulberry.
4	It does not require any pollinating agent.	It requires pollinating agents like wind, insects, birds, etc.

(iv)

Gamopetalous	Polypetalous
Flowers in which petals are partially or completely fused are termed as gamopetalous flowers e.g., Tobacco, *Petunia*.	Flowers in which petals are free called polypetalous flowers e.g., *Hibiscus rosa sinensis*.

(v)

Apomixis	Amphimixis
Any asexual method of reproduction, not involving normal production of embryo by fertilisation is known as apomixis. E.g. : Vegetative propagation through bulbs, tubers, bulbils, adventitious buds.	Sexual reproduction by fusion of gametes and formation of normal embryo is known as amphimixis. E.g. : Rice, Wheat.

(vi)

S. No.	Mound Layering	Air Layering
1	A branch is bent downwards and buried into the soil in this method of propagation.	In this process, branch is girdled and covered with a mixture of moist soil and mosses and is wrapped in polythene covering.
2	It is used for multiplying plants with soft and flexible branches, e.g. Jasmine, Strawberry.	It is used for multiplying plants with woody stem e.g. Citrus, Indian Rubber plant.

(vii)

Apocarpous Ovary	Syncarpous Ovary
It is an ovary made up of separate carpels e.g. custard apple (*Annona squamosa*).	It is an ovary having a number of partially or wholly fused carpels e.g. Tomato (*Lycopersicum esculentum*).

(viii)

S. No.	True Fruit (Eucarp)	False Fruit (Pseudocarp)
1	It develops from a fertilised ovary.	It develops from the floral parts other than ovary.
2	Other floral parts are not developed into the fruit. e.g. Maize, Mango.	Other floral parts like fused perianth (as in Mulberry), thalamus (as in apple) or peduncle (as in cashewnut) developed into the fruit e.g. Apple, Mulberry.

(ix)

S. No.	Stock	Scion
1	A rooted stem into which a scion or a bud is grafted is called stock.	A piece of young stem or bud which is inserted into a root stock is called a scion.
2	Stock is selected for its adaptability, resistance to diseases and efficient absorbing system, e.g. *Rosa indica*.	Scion is selected for good quality of fruits, flowers or latex contents and resistant to disease of aerial parts e.g. *Rosa hybrida*.

4. Illustrate the different modes of entry of pollen tube into ovule.

Ans. After pollination the intine of pollen grain forms pollen tube through germ pore. The pollen tube with two male gametes and tube nucleus runs through the style and finally enters into the ovule. There may be three possible modes of entry of pollen tube in to the ovule.

Porogamy : When the pollen tube enters through micropylar end of the ovule for fertilisation, it is called porogamy e.g., in most angiospermic plants.

Chalazogamy : In some plants, pollen tube enters the embryo sac through the chalaza of the ovule. It is called chalazogamy E.g. *Casuarina, Betula, Juglans regia,* etc.

Mesogamy : When the pollen tube enters the ovule by piercing the nucellus then it is called mesogamy. E.g. *Cucurbita, Pistacia,* etc.

5. Give four chemical differences between a dicot leaf and monocot leaf.

Ans.

S. No.	Monocot leaf	Dicot leaf
1	Shows isobilateral symmetry.	Shows Dorsiventral symmetry.
2	Stomata are distributed equally on both the surfaces. (Amphistomatic)	Stomata are present on the lower surface of leaf (Hypostomatic).
3	Bulliform cells are present on upper epidermis.	Bulliform cells are usually absent.
4	Mesophyll is made of only spongy parenchyma with very small inter-cellular spaces.	Both palisade and spongy parenchyma are present with large intercellular spaces.

Chapter 3. Human Reproduction

1. Mention one significant difference between each of the following :

(i) Microgametes and Macrogametes
(ii) Binary fission and Multiple fission
(iii) Syngamy and Conjugation
(iv) Hologamy and Isogamy
(v) Prostate gland and Cowper's gland
(vi) Spermatocytes and Oocytes
(vii) Primary and Secondary sex organs
(viii) Egg and Zygote
(ix) Spermatogenesis and Spermiogenesis
(x) Oogenesis and Ovulation
(xi) Menarche and Menopause.
(xii) Sertoli cells and Spermatogonial cells.

Ans. (i)

S. No.	Microgametes	Macrogametes
1	Generally male gametes are referred as microgamete, since they are smaller in size.	Female gamete is usually larger in size than the male gametes; it is thus referred to as macrogamete.
2	They are produced in large number.	It is usually produced in single number.

(ii)

S. No.	Binary Fission	Multiple Fission
1	It involves the division of parent into two equal sized daughter individuals.	It involves the simultaneous division of parent into many small daughter individuals.
2	It is common in protozoans and also occurs in flatworms.	It occurs in protozoans such as *Plasmodium*.

(iii)

Syngamy	Conjugation
In syngamy, the fusion of two gametes takes place completely and permanently. It is most common type of sexual reproduction in plants and animals. e.g. Plants	Conjugation is temporary union of two individuals of same species to exchange their pronuclei. e.g. *Paramecium*.

(iv)

Hologamy	Isogamy
Entire mature organism start to act as gamete and fusion of individuals occurs, this is known as hologamy. e.g. Lower algae	Isogamy is a kind of reproduction which involves fusion of morphological identical gametes e.g. Primitive aquatic organism.

(v)

S. No.	Prostrate Gland	Cowper's Gland
1	Prostrate gland is a collection of 30-40 tubuloalveolar glands, lying at the base of the bladder and surrounds the prominent part of urethra.	Cowper's glands or bulbourethral gland is a pair of exocrine glands in the male reproductive system, and lie at the base of penis.
2	It provides nutrition to sperms and neutralises the acidity of urine.	They secrete a mucus-like substance that lubricates the penis for its easy movement.

(vi)

S. No.	Spermatocytes	Oocytes
1	These are formed when the spermatogonia in the seminiferous tubules of testis divides mitotically.	These are formed in the graafian follicle of the ovary.
2	Each spermatogonium on mitotic division forms the two primary spermatocytes.	Each maturing graafian follicle contains diploid primary oocyte at its centre.
3	Each primary spermatocyte undergoes meiosis-I and forms the two haploid secondary spermatocytes.	Each primary oocyte on meiotic division-I forms two haploid (n) cells. The large one is known as secondary oocyte and smaller as polar body.
4	The secondary spermatocytes undergo meiosis-II and form the haploid spermatids.	The secondary oocyte (N) undergoes meiosis-II and forms one ovum and polar body.
5	Each primary spermatocyte forms four haploid spermatids as a result of spermatogenesis.	Each primary oocyte gives rise to one ovum and three polar bodies.

(vii)

S. No.	Primary Sex Organs	Secondary Sex Organs
1	These are the organs which produce gametes and sex hormones.	These are the organs which do not produce gametes and sex hormones but are essential for performing vital functions in reproduction.
2	Testes are the primary sex organs in males while ovaries are the primary sex organs in females.	These include prostate gland, seminal vesicle, vas deferens, epididymis penis, cowper's glands in males while fallopian tubes, uterus, vagina and mammary glands in females.

(viii)

S. No.	Egg	Zygote
1	Egg is the female gamete produced by the process of gametogenesis called 'oogenesis'.	Zygote is the result of fertilisation of egg (female gamete) with sperm (male gamete).
2	It has haploid number of chromosomes.	It has diploid number of chromosomes.

(ix)

S. No.	Spermatogenesis	Spermiogenesis
1	It is a process of formation of spermatozoa from germinal cells.	It is the process of differentiation of spermatozoan from a spermatid.
2	It involves conversion of a diploid structure into haploid structures.	It changes a haploid structure into another haploid structure.
3	There is growth and division during spermatogenesis.	There is reconstruction during spermiogenesis. Division and growth are absent.
4	A spermatogonium forms four spermatozoa.	Here a spermatid forms a single sperma tozoan.

(x)

Oogenesis	Ovulation
Oogenesis is the process where ova are formed and developed in the germinal epithelium of the ovary.	Ovulation is the process where ovum is matured and released from the ovary in the abdominal cavity. It occurs midway between two menstrual periods and last for about 6 hours.

(xi)

Menarche	Menopause
It is the onset of menstrual cycle in a young females at about the age of 13 years.	It is the permanent stoppage of menstrual cycle in female at the age of 45 to 50 years.

(xii)

Sertoli Cells	Spermatogonial Cells
These are non gametic cells in the seminiferous tubules which provide nourishment to the developing sperms.	These are germinal cells which lie along the periphery of the seminiferous tubules and closely packed. These cells ultimately give rise to the sperms.

3. Differentiate between :
(i) Spermatogenesis and oogenesis.
(ii) Proliferative phase and secretory phase of menstrual cycle.

Ans. (i)

S. No.	Spermatogenesis	Oogenesis
1	It occurs in the male gonads *i.e.,* testes.	It occurs in the female gonads *i.e.,* ovaries.
2	Sperms are produced in large number.	Ova are produced in small number.
3	The sperms are small, active and motile.	The ovum is large sized, non-motile and receptive in nature.
4	Each primary spermatocyte gives rise to 4 spermatozoans.	Each primary oocyte gives rise to one ovum and three polar bodies.
5	All the four spermatozoans are functional.	Only ovum is functional and polar bodies degenerate.

(ii)

S. No.	Proliferative Phase	Secretory Phase
1	It is the period between the end of menses and ovulation (from 5th to 14th day of menstrual cycle).	It is the period between ovulation and onsent of next menses (from 16th to 28th day of menstrual cycle).
2	Follicular development takes place in the ovary under the influence of FSH secreted by anterior pituitary, wherein one ovarian follicle grows into a Graafian follicle..	Formation of corpus luteum from the cells of ruptured Graafian follicle takes place under the influence of LH secreted by anterior pituitary.
3	Regeneration and proliferation of endometrium takes place under the influence of oestrogen.	Thickening of endometrium and secretion of juices takes place under the influence of progesterone.
4	Contraction of uterine muscles increases.	Contraction of uterine muscles decreases.
5	The epithelium of fallopian tube gets thickened and its cilia and their movement increase.	The epithelium of fallopian tube becomes normal and its cilia and their movement decrease.

4. Mention one significant difference between each of the following :
(i) Gametogenesis and embryogenesis
(ii) Blastulation and gastrulation
(iii) External fertilisation and internal fertilisation
(iv) Fertilizin and antifertilizin
(v) Morula and blastula
(vi) Zona pellucida and corona radiata
(vii) Troph ectoderm and ectoderm

(viii) Animal pole and vegetal pole
(ix) Blastocoel and archenteron.

Ans. (i)

Gametogenesis	Embryogenesis
The production of male and female gametes is termed as gametogenesis.	The process of development of fertilised egg into full fledged individual is termed as embryogenesis.

(ii)

Blastulation	Gastrulation
The morula undergoes repeated mitotic divisions and form blastula. A fluid filled cavity blastocoel is formed.	Movements and rearrangement of embryonic cells take place. The formation of three germinal layers takes place.

(iii)

External Fertilisation	Internal Fertilisation
Sperms and ova by male and female respectively are discharged outside simultaneously. Fertilisation takes place outside the body of the animal e.g. frogs, fishes.	Sperms are deposited inside the body of the female. Fertilisation takes place inside the body of the animal (female) e.g. Mammals.

(iv)

Fertiliszin	Antifertilizin
It is a proteinaceous chemical which occurs on the surface layer of egg. Its molecular weight is 30,000D.	It is a proteinaceous substance found on the surface of sperm. Its molecular weight is 10,000D.

(v)

Morula	Blastula
It is a solid ball of 16-32 blastomeres. It consists of a group of centrally located cells, called inner cell mass, which is covered with a layer of clear cells called trophoblast.	When morula enters uterus, the glands of uterina wall secrete a glycogen-rich fluid which enters the morula. This fluid is collected between trophoblast and inner cell mass forming a cavity called blastocoel. With appearance of blastocoel, morula becomes blastula.

(vi)

Zona Pellucida	Corona Radiata
The ovum is surrounded by a non-cellular, transparent layer called zona pellucida.	Outside the zona pellucida, there is a thick layer of follicular cells, which is referred as corona radiata.

(vii)

Trophectoderm	Ectoderm
It differentiates during blastulation. It is the outermost layer of cells of blastodermic vesicle. It does not form embryo proper. It forms protective and trophic membrane.	It differentiates during gastrulation. It is outermost layer of cells of the gastrula. It forms embryo properly. It gives rise to nervous system, epidermis and sense organs.

(viii)

Animal Pole	Vegetal Pole
The area of the blastocyst where embryonal knob is present is referred as Animal pole.	The region diagonally opposite to animal pole is the vegetal pole.

(ix)

Blastocoel	Archenteron
Blastocoel is the fluid filled cavity of blastula stage.	Archenteron is the cavity of gastrula and also called primitive gut.

5. How does cleavage differ from ordinary mitosis ?

Ans.

S. No.	Cleavage	Ordinary Mitosis
1	Cleavage is a special type of mitosis that occurs during early segmentation of zygote.	Normal mitosis occurs during division of somatic cells of the body.
2	It is rapid with short interphase.	It is comparatively slow with a long interphase.
3	Newly formed or cleaved cells or blastomeres do not grow in size.	Daughter cells formed after normal mitosis grow in size.
4	There is a progressive decrease in size.	Cell size is maintained. It become equal to that of parent cell.

Chapter 4. Reproductive Health

1. Differentiate between Vasectomy and Tubecstomy.

Ans. In males, a small part of the deferens is removed and tied up through a small incision in the scrotum. This is called vasectomy.

In females, a small part of the fallopian tube is removed and tied up through a small incision in the abdomen or vagina. This is called tubectomy.

2. Differentiate between ZIFT and IUT.

Ans. The zygote or embryo with upto 8 blastomeres is transferred into the fallopian tube this is called Zygote Intra Fallopian Transfer (ZIFT).

Embryo with more than 8 blastomeres are transferred into the uterus, this is called Intra Uterine Transfer (IUT).

Chapter 5. Principles of Inheritance and Variation

1. Differentiate between the following :
   (i) Dominant trait and Recessive trait
   (ii) Homozygous and Heterozygous
   (iii) Monohybrid Cross and Dihybrid Cross
   (iv) Test cross and back cross.

Ans. (i)

Dominant Trait	Recessive Trait
The trait expressed irrespective of the presence or absence of the recessive factor, is called the dominant trait. Example : Tall plant, round seeds, yellow pods.	The trait that is expressed only in the absence of dominant factor, is called the recessive trait. Example : Dwarf plant, wrinkled seeds, green pods.

(ii)

S. No.	Homozygous	Heterozygous
1	It contains two similar alleles for a particular trait.	It contains two different alleles for a particular trait.

2	Genotype for homozygous possess either dominant or recessive, but never both the alleles. For example, RR or rr.	Genotype for heterozygous possess both dominant and recessive alleles. For example, Rr.
3	It produces only one type of gamete.	It produces two different kinds of gametes.

(iii)

Monohybrid cross	Dihybrid cross
Monohybrid cross involves cross between parents, which differs in only one pair of contrasting characters. For example, the cross between tall and dwarf pea plant is a monohybrid cross.	Dihybrid cross involves cross between parents, which differs in two pairs of contrasting characters. For example, the cross between pea plants having yellow wrinkled seeds with those having green round seeds is a dihybrid cross.

(iv)

S. No.	Test cross	Back cross
1	Test cross is the cross between progeny in F_1 generation with its homozygous recessive parent.	Backcross is the cross between progeny in F_1 gen eration with any of its parent.
2	Test cross is done to know the genotype of progeny of F_1 generation.	Bac kcross is done to know the genotype of parents along with the progeny.
3	All test cross are back cross.	All back cross are not test cross.

2. Differentiate between first filial generation (F_1) and second filial generation (F_2).

Ans. The progeny obtained as the result of crossing the plants with contrasting characters is called the first filial generation (F_1). Whereas the progeny of F_1 plants intrabreed to self fertilisation represents the second filial generation (F_2).

3. Mention one significant difference between each of the following :

(i) Co-dominance and incomplete dominance
(ii) Dominance and epistasis
(iii) Epistasis and complementary genes
(iv) Pleiotropy and epistasis

Ans. (i)

Co-dominance	Incomplete dominance
In co-dominance, both allelic genes of a genetic trait are equally expressive, thus both the characters appear side by side in F_1 hybrids.	In incomplete dominance, dominance is absent and the hybrid individuals resemble neither of the parent but are intermediate between the two parents.

(ii)

Dominance	Epistasis
Out of a pair of allelomorphic characters, the one which appears in F_1 generation, is known as dominant character and this phenomenon is known as dominance.	In epistasis, non-allelic genes are involved wherein one gene masks or inhibits the expression of another non-allelic gene. The gene that suppresses the other gene, is called epistatic gene.

(iii)

Epistasis	Complementary genes
Epistasis genes are those non-allelic pair of genes, where a gene masks or inhibits the expression of another non-allelic gene. The gene that masks the effect of another is said to be epistatic to the other and the gene whose expression is prevented is said to be hypostatic.	Complementary genes are those in which two separate pair of genes interact to produce the phenotype in such a way that dominant is gene not expressed unless the other one is present. The effect of one dominant must be complemented by the other dominant for expression.

(iv)

Pleiotropy	Epistasis
The term pleiotropy is the ability of a single gene to influence more than one trait.	The term epistasis is the interaction between two or more non-allelic genes.

4. How are Mendel's principles different from the concept of blending inheritance ?

Ans. According to Mendel's principles, if an individual contains two different alleles, then the individual's gametes could contain either of these two alleles (but not both). Blending inheritance proposes that offspring are the result of blended genetic material from the parent and the genetic factors are not discrete units. Once blended, the combined genetic material could not be separated from each other in future generations.

5. Differentiate between :

(i) Autosomes and Sex chromosomes

(ii) Linkage and crossing over*

(iii) Single crossing over and double crossing over

(iv) Sex determination and sexual differentiation

(v) Autosomes and allosomes

(vi) Heterogametic and homogametic

Ans. (i)

S. No.	Autosomes	Sex Chromosome
1	Autosomes are the chromosomes that determine all the characterstics other than the sex determination.	Sex Chromosomes are the chromosomes that determine the sex of the orgsanisms.
2	In autosomal pairs, both chromosomes are of same length.	In sex chromosome, Y-chromosome is shorter than X-chromosome.
3	The position of centromere is same.	The position of centromere may not be same.
4	In human genome, there are 22 pairs of autosome.	In human genome, there is only one pair present (XX or XY).

(ii)

S. No.	Linkage	Crossing Over
1	Linkage ensures to keep the genes in a chromosome to inherit together.	Crossing over facilitates the separation of genes present in a chromosome and segregate into different gametes.
2	The strength of linkage between two genes is inversely proportional to the distance between them in the chromosome.	The chance of crossing over between two genes is directly proportional to the distance between them in the chromosome.
3	The strength of linkage decreases with increase in distance between the genes.	The chance of crossing increases with increase in distance between the genes.
4	Linkage ensures the maintenance of parental trait in the offspring.	The crossing over causes alternations in the parental traits in the offspring.
5	Linkage reduces the chance of creation of variability with sexual reproduction.	Crossing over increases the chance of creation of variability with sexual reproduction.

(iii)

S. No.	Single crossing over	Double crossing over
1	There is only one chiasma where chromatids of corresponding homologous chromosomes exchange characters.	The crossing over occurs at two points on the non-sister chromatids.
2	The non-sister chromatids break and rejoin only at one point along their entire length.	The non-sister chromatids break and rejoin at two or more points.

* Frequently asked previous years Board Exam Question.

Differentiate Between

(iv)

S. No.	Sex Determination	Sexual Differentiation
1	Process of determining the sex.	Process of development of the differences between males and females.
2	It is operated by chromosomes.	It is operated by hormones.

(v)

S. No.	Autosomes	Allosomes
1	Body (somatic) chromosomes.	Sex chromosomes.
2	More in number.	Less in number.
3	They determine the body characters.	They determine sex of an individual.

(vi)

S. No.	Heterogametic	Homogametic
1	It is the condition in which there are two different types of gametes formed.	It is the condition in which similar type of gametes are formed.
2	In human male, XY chromosomes are present. It produces two types of sperms : (a) Sperm with X-chromosome (b) Sperm with Y-chromosome	In human female XX chromosomes are present. The female produces only one type of egg *i.e.*, the eggs with X- chromosomes.

6. Sex determination is based on particular chromosomes in both birds and humans. State two points of difference between their mechanisms of sex determination.

Ans.

S. No.	Sex determination in humans	Sex determination in birds
1	It is determined by XX and XY chromosomes.	It is determined by WZ and ZZ chromosomes.
2	The males are heterogametic (XY) and females are homogametic (XX).	The males are homogametic (ZZ) and females are heterogametic (WZ).

Chapter 6. Molecular Basis of Inheritance

1. Differentiate between the following :
   (i) Chromosomal DNA and Plasmid DNA
   (ii) RNA and DNA
   (iii) Endonuclease and Exonuclease
   (iv) Prokaryotic and Eukaryotic replication
   (v) Transcription and Translation
   (vi) Transition and transversion*

Ans. (i)

S. No.	Chromosomal DNA (Single Copy)	Plasmid DNA (Multiple Copy)
1	Chromosomal DNA is chromosomal DNA where the genetic material is present.	Plasmid DNA is extra-chromosomal DNA in bacteria and some yeasts, i.e., Plasmid.
2	It is primary DNA in all living organisms.	It is secondary DNA.
3	It is linear in eukaryotes whereas circular in prokaryotes.	It is circular.
4	As it encodes genetic information, it is much larger than plasmid.	It is smaller.
5	Chromosomal DNA is organized with proteins called histones.	Plasmid DNA is not with histones.
6	It contain sessential genes which codes for functional and structural proteins.	It contains non-essential genes.
7	It can be transferred only through cell division within the same soecies.	It can be transferred through horizontal way of gene transfer between same or different species.

* **Frequently asked previous years Board Exam Question.**

(ii)

S. No.	RNA	DNA
1	RNA is a single stranded molecule.	DNA is a double stranded molecule.
2	It contains ribose sugar.	It contains deoxyribose sugar.
3	The pyrimidines in RNA are adenine and uracil.	The pyrimidines in DNA are adenine and thymine.
4	RNA cannot replicate itself.	DNA molecules have the ability to replicate.
5	It is a component of the ribosomes.	It is a component of the chromosomes.
6	RNA expand as Ribonucleic acid.	DNA expand as Deoxyribonucleic acid.

(iii)

S. No.	Endonuclease	Exonuclease
1	Endonuclease refers to an enzyme that cleaves the polynucleotide chain separating nucleotides other than the two end ones.	Exonuclease refers to an enzyme that cleaves. polynucleotide chain from the end of the chain by removing the nucleotides one by one.
2	They cleave nucleic acids at the middle of the nucleic acids.	They cleave nucleic acids at the ends.
3	They result in oligonucleotides.	They result in single nucleotides.
4	They may form either blunt ends or sticky ends.	They form sticky ends.

(iv)

S. No.	Prokaryotic Replication	Eukaryotic Replication
1	There is only one origin of replication (ori).	There are several origin of replication (ori).
2	There are three types of DNA polymerases involved.	There are five types of DNA polymerases involved.
3	Synthesis or extension of Okazaki fragments on RNA primer is carried out by polymerase III.	Extension of RNA primer is carried out by polymerase δ.
4	Removal of RNA primers is carried out by polymerase I.	Removal of RNA primers is carried out by polymerase δ.

(v)

S. No.	Transcription	Translation
1	Transcription is the first step of gene expression process.	Translation is the second step of gene expression process.
2	Synthesises RNA copies of the genetic instructions written in the genome.	Synthesises proteins from RNA which are copied from genes.
3	Genes in the genome act as template.	mRNA act as template.
4	It occurs in the nucleus.	It occurs in the cytoplasm.
5	RNA polymerase act as the enzymes.	Ribosomes act as enzymes.
6	They produces several functional forms of RNA.	It produce protein.
7	It is localised into prokaryotes cytoplasm and eukaryotes nucleus.	It is localised into prokaryotes cytoplasm and eukaryote ribosomes on the endoplasmic reticulum.

(ii)

Transition	Transversion
In transition, a purine base is substituted with a purine base or a pyrimidine base is substituted with a pyrimidine base. (C ↔ T or A ↔ G)	In transversion, a purine base is substituted with a pyrimidine or a pyrimidine with a purine.

Differentiate Between

2. State two differences between leading strand and lagging strand.

Ans.

S. No.	Leading Strand	Lagging Strand
1	It is one of two strands of DNA found at the replication fork, being replicated continuously.	The other strand found at the replication fork, replicating discontinuously from the 5′ to the 3′ direction.
2	It grows continuously.	It grows discontinuously by forming Okazaki fragments.
3	It opens up in the 3′ to 5′ direction.	It opens up in the 5′ to 3′ direction.
4	It grows in the 5′ to 3′ direction.	It grows in the 3′ to 5′ direction.
5	It requires a single primer for the synthesis.	It requires a new primer to start each Okazaki fragment.
6	It starts to grow at the beginning of replication.	It starts to replicate a bit later.
7	It grows towards the replication fork.	The Okazaki fragments grow away from the replication fork.
8	It is formed at high speed.	It is formed slowly
9	It does not require DNA ligase.	It requires DNA ligase to ligate Okazaki fragments together.

3. Differentiate :

(i) Cistron and Muton
(ii) Conjugation and transduction
(iii) Deoxyvirus and ribovirus
(iv) Exons and introns
(v) Generalised and specialised transduction
(vi) One gene one enzyme hypothesis and one gene one polypeptide hypothesis
(vii) Virulent and avirulent phages
(viii) Operators and Repressors

Ans. (i)

Cistron	Muton
Cistron refers to the continuous segment of DNA which specifies one polypeptide chain. It is the region where mutants show a cis-trans position effect.	A muton is the smallest length of DNA capable of giving rise to new form by mutation.

(ii)

Conjugation	Transduction
Exchange of genetic material by direct cell-to-cell contact in bacteria is known as conjugation.	When transfer of genetic material from one cell to another takes place by a bacteriophage, then it is known as transduction.

(iii)

Deoxyviruses	Riboviruses
The viruses which contain DNA are called Deoxyviruses.	Virus having RNA are known as Ribovirus.

(iv)

Exons	Introns
The region of a split gene that becomes part of mRNA and codes for the different amino acids is known as exons.	The region of gene that do not form a part of mRNA and are removed during RNA processing during mRNA synthesis are known as introns.

(v)

Generalised Transduction	Specialised Transduction
In generalised transduction, as the phage begins lytic cycle, viral enzymes hydrolyse the bacterial chromosome into many small pieces of DNA. Any part of the bacterial chromosome may be incorporated into the phage head during phage assembly.	In specialized transduction, certain temperate phage strains transfer only a few restricted genes of the bacterial chromosome.

(vi)

One gene-one enzyme hypothesis	One gene-one polypeptide hypothesis
One gene-one enzyme hypothesis states that one gene controls the production of one enzyme.	One gene-one polypeptide hypothesis states that one gene controls the production of only one polypeptide chain of an enzyme molecule.

(vii)

Virulent (lytic) Phage	Avirulent (temperate) Phage
The viruses which always multiply by using the host machinery when they enter the host cell and then causes the lysis of cells are called virulent (lytic) phages. Example : T_4 Phages.	The viruses whose nucleic acids gets incorporated in bacterial DNA and replicate along with it to constitute lysogenic cycle are said to be avirulent (temperate) phages. Example : Lambda Phages.

(viii)

Operators	Repressors
Operators are regions of DNA (15 nucleotides long) that control RNA access to promoter.	Repressor are regulatory proteins that bind to operator and turn genes off (negative control mechanism).

Chapter 7. Evolution

1. Mention one significant difference between each of the following :
   (i) Abiogenesis and biogenesis
   (ii) Chemical evolution and biological evolution
   (iii) Homologous organs and analogous organs
   (iv) Missing links and connecting links
   (v) Artificial selection and natural selection
   (vi) Intraspecific struggle and interspecific struggle
   (vii) Allopatric speciation and sympatric speciation.

Ans. (i)

Abiogenesis	Biogenesis
The origin of life from non–living matter is termed Abiogenesis.	The origin of life from pre–existing life is termed biogenesis.

(ii)

Chemical Evolution	Biological Evolution
Chemical evolution of life involves formation of complex macromolecules and nucleoproteins which are supposed to be building blocks of living system.	Biological evolution involves formation of coacervates, formation of primary organisms autotrophs and eukaryotes.

(iii)

Homologous Organs	Analogous Organs
The organs which have the same structure and origin but look different and perform different functions are called homologous organs. Example : wings of bird, forelimbs of horse.	The organs which have the same function and are superficially alike but are quite different in there fundamental structure and embryonic origin are called analogous organs. Example : wings of insect and bird.

Differentiate Between

(iv)

Missing Links	Connecting Links
They are extinct organisms. They had the characters of two different groups of organisms. Example : *Seymouria, Archaeopteryx*.	They are living organisms and they have the characters of two different groups. Example : *Protopterus, Peripatus*.

(v)

S. No.	Artificial Selection	Natural Selection
1	Through artificial selection, man perpetuate the selected characteristics, which leads to the formation of new species in few years.	Through natural selection, most favourable variation is selected by nature.
2	It involves selective breeding of new populations.	It takes a long time and leads to great diversity in nature.

(vi)

Intraspecific Struggle	Interspecific Struggle
It is the struggle between the individuals of same species for food, shelter, breeding places which are similar for each. Most human wars are examples of intraspecific struggle.	It is the struggle between members of different species normally for food and shelter. For Example : a fox hunts out for a rabbit, while the fox is preyed by a tiger.

(vii)

Allopatric Speciation	Sympatric Speciation
The splitting of a species into two or more geographically separated species.	When a few individuals of a species within the same geographical area suddenly become reproductively isolated and form new species.

2. Differentiate between a fossil and a living fossil.

Ans. A fossil is a impression or dead remains of ancestral rock or prehistoric plants and animals while living fossils are living organisms at the verge of extinction such as Cycads, Ginkgo.

3. What is the main difference between Darwin's theory of natural selection and De Vries mutation theory ?

Ans. The mutation theory differs from Darwin's theory in nature of the variations. Continuous variations or fluctuations are the raw material for the formation of new species in Darwin's theory of natural selection; but in De Vries theory, discontinous variation or mutations are the raw materials.

4. State three structural features by which Neanderthal man different from *Homo sapiens*.

Ans.

S. No.	Neanderthal Man	*Homo sapiens* (Modern Man)
1	The skull bones were thick with prognathous forehead and prominant eyebrow ridges.	The skull bones were thin with orthognathous forehead and low eyebrow ridges.
2	The lower jaw was strong with strong jaw muscles with no chin.	The lower jaw was not as strong as Neanderthal man.
3	The cranial capacity was about 1450 cc (1350 – 1700 cc).	The cranial capacity was 1350 cc (1350 – 1500 cc).

5. Compare between apes and man.

Ans.

S. No.	Apes	Man
1	Forehead is flat and skull too flat.	Forehead is large and protruded; skull dome shaped.
2	Lower jaws are heavy, teeth large, chin round.	Lower jaws small, teeth also small, chin flat.

3	Big toes in the feet are opposable.	Big toes in the feet are not opposable.
4	Cranium (brain box) small, cranial capacity varies species to species.	Cranium large with cranial capacity of about 1450 cc.
5	The foramen magnum is directed backward.	The foramen magnum is directed downward.

6. Differentiate between Apes and Man with respect to the following characteristics :

(i) Posture

(ii) Brow ridges

(iii) Cranial capacity

Ans. (i) **Posture :** Man has upright stance and bipedal movement whereas apes are actually quadrupeds on the ground and brachiators (arm swingers) in trees with semi erect posture.

(ii) **Brow Ridges :** In man brow ridges are inconspicuous whereas in apes they are prominent.

(iii) **Cranial Capacity :** In man the cranial capacity is of 1450 cm^3 whereas in apes it is 650 cm^3.

Chapter 8. Human Health and Diseases

1. Mention one significant difference between each of the following :

(i) Antigen and Antibody.

(ii) B-cells and T-cells.

(iii) Active immunity and Passive immunity.

(iv) Infection and Infestation.

(v) Antibodies and Interferons.

(vi) SCID and AIDS.

(vii) Innate immunity and Acquired immunity.

(viii) Communicable and Non–communicable diseases

(ix) Osteoarthritis and Rheumatoid arthritis

(x) Klinefelter's syndrome and Turner's syndrome

Ans. (i)

Antigen	Antibody
Antigen is a foreign material that triggers antibody formation. They are proteins or polysaccharides.	These are the proteins produced by the living organisms in response to invasion by disease causing micro-organisms. They are plasma proteins.

(ii)

B-Cells	T-Cells
B-cells that form humoral or antibody mediated immune system are processed in bone marrow. They release antibodies that attack antigens.	T-cells form cell-mediated immune system that are matured in the thymus gland and do not release but hold on to the antibodies. The whole body directly attacks the antigens.

(iii)

Active Immunity	Passive Immunity
When a person's own cells produce antibodies, in response to infection or vaccination. It is harmless and it is for a longer duration.	When the antibodies produced in other organisms are injected into a person to counteract antigens. The foreign antibodies may be killed by the antibodies produced by the body. It is for a shorter duration.

(iv)

Infection	Infestation
The introduction or transmission of a disease producing agent or organism to a disease free organism.	Presence of a large number of disease causing organism of the same type in the body of the host.

(v)

Antibodies	Interferons
They are slow acting and long lasting and act outside the cells.	They are quick acting and temporary and act inside the cells.
They act against bacteria and viral infections.	They act against viruses only.

(vi)

SCID (Severe Combined Immuno Deficiency)	AIDS (Acquired Immuno Deficiency Syndrome)
In this disorder both B-cells and T-cells are absent and hence, this congenital disease makes child highly susceptible to even minor infections.	This syndrome is caused by human immuno deficiency virus (HIV) that breaks down the body's immune system, as the name suggests this is an acquired disease.

(vii)

Innate Immunity	Acquired Immunity
Innate immunity is inborn immunity mechanism.	Acquired immunity is that which person develops after birth.

(viii)

Communicable Diseases	Non-communicable Diseases
These diseases are caused by invasion of pathogen in body and are transmitted from infected person to healthy person through agencies.	These diseases are caused by deficiency of nutrient or degeneration of tissue or some other reason but do not spread from infected to healthy person.

(ix)

Osteoarthritis	Rheumatoid Arthritis
In osteoarthritis the joints become partially or completely ossified and lose their flexibility and hence, movement becomes painful. This disease attacks in ageing people.	Rheumatoid arthritis is a painful inflammation of synovial membranes of joints. It also involves erosion of joints and starts at age of 20 to 40 years.

(x)

Klinefelter's Syndrome	Turner's Syndrome
Persons with this syndrome are found to contain 47 chromosomes and their genotype as XXY. Males are underdeveloped and may have female features.	Persons with this syndrome have 45 chromosomes with genotype XO. The females are underdeveloped with short stature and congenital malformations.

2. Give one significant difference between each of the following pairs :
   (i) Humoral immunity and cell mediated immunity.*
   (ii) Benign tumour and malignant tumour.*
   (iii) Primary lymphoid organs and secondary lymphoid organs.*

* Frequently asked previous years Board Exam Question.

Ans. (i)

S. No.	Humoral Immunity	Cell mediated Immunity
1	Mainly involves B-lymphocytes which produce the antibodies.	Consists of T-lymphocytes.
2	Generation and maturation occurs in bone marrow.	Originates in bone marrow and complete development occurs in thymus.
3	Defends body against extracellular microbes and their toxins.	Defends body against intracellular microbes, viruses, parasites (*leishmania*), bacteria and tumour cells.
4	Onset is rapid.	Onset is delayed.

(ii)

S. No.	Benign tumour	Malignant tumour
1	It is non-cancerous.	It is cancerous.
2	It occurs by expansive growth.	It occurs by invasive growth.
3	Mitotic activity of cells is not high.	Mitotic activity of cells is frequently high.
4	There are no or few symptoms.	Symptoms are associated with pain and disability.
5	No metastasis.	Frequent metastasis.
	Example : Papiloma, Adenoma.	**Example :** Liposarcoma, Adenocarcinoma.

(iii)

S. No.	Primary Lymphoid Organs	Secondary Lymphoid Organs
1	They are organs of the immune system where lymphocytes are formed and mature.	They are organs of the immune system which maintain mature naive lymphocytes and initiate an adaptive immune response.
2	It allow lymphoid stem cells to proliferate, differentiate, and mature	It allow lymphoid cells to become functional
3	It contain either T cells or B cells	It contain both T cells and B cells
4	It have no contact with antigens	It have contact with antigens
5	It undergo atrophy with age	It increase size with age
6	It includes : (a) Thymus (b) Bone Marrow.	It includes : (a) Spleen (b) Lymph nodes (c) Appendix (d) Peyer's Patches (e) Tonsils.

Chapter 9. Strategies for Enhancement in Food Production

1. Differentiate between (i) Inbreeding and heterosis (ii) Inbreeding and outbreeding.

Ans. (i)

Inbreeding	Heterosis
The cross between two individuals of a species that are related by descent is called inbreeding.	The cross between two unrelated individuals or species or genus may developed an important hybrid called Heterosis.

(ii)

S. No.	Inbreeding	Outbreeding
1	It is breeding between same breed of animals.	It is breeding between different breeds of animals
2	Advantage : helps in accumulation of superior gene.	Advantage : help overcoming inbreeding depression.

2. Differentiate between outbreeding and outcrossing.

Ans. Outbreeding is breeding of unrelated animals (having no ancestors for 4–6 generation) belonging to same breed or different breed or different species.

Outcrossing is breeding within animals of same breed having no common ancestors for 4-6 generation on either side of pedigree.

3 Differentiate between an inbreed line and a hybrid variety of crop.

Ans.

Inbreed Line	Hybrid Variety
The continuous inbreeding in a cross-pollinated crops develops a homozygous line called inbreed line.	A hybrid variety is produced by crossing two different species where progeny obtained is used for raising the next seasonal crop.

Chapter 10. Microbes in Human Welfare

1. Differentiate between broad spectrum antibiotics and narrow spectrum antibiotics.

Ans. A broad spectrum antibiotic is one which can inhibit the growth of both gram positive and gram negative bacteria. For example : Tetracycline.

While a narrow spectrum antibiotics are those which can inhibit the growth of certain bacteria Example : Glycopeptides and bacitracin.

2. Differentiate between manure and biofertilizers.

Ans. Manure– They are semi-decayed organic remains of various plants.

Biofertilizers are the micro-organism (bacteria, fungi) which enrich the nutrient (nitrogen, phosphorus etc) quality of soil.

Chapter 11. Biotechnology : Principles and Processes

1. Differentiate between rDNA and cDNA.

Ans.

rDNA	cDNA
It is the DNA which is formed by joining together the DNA from two different organism.	It is the DNA which is obtained from RNA template catalysed by reverse transcriptase enzyme.

Chapter 12. Biotechnology and its Applications

1. Differentiate between Gene therapy and Gene cloning.

Ans.

Gene Therapy	Gene Cloning
The process of replacing defective gene responsible for hereditary disease by the normal gene is called gene therapy.	The technique to produce identical copies of a particular segment of DNA or a gene.

2. Differentiate between Insulin and Proinsulin.

Ans.

S. No.	Insulin	Proinsulin
1	It is made up of two short polypeptide chains. A and B linked by disulphide bridges.	Along with the two polypeptide chains in insulin. It contains an extra stretch called C peptide.
2	It is functional.	It is non-functional.

3. Differentiate between PCR and Gene Cloning.

Ans.

S. No.	PCR	Gene Cloning
1	DNA is required as a biological material.	Restriction enzyme, ligase, vectors, bacterial cells, DNA are the biological materials.
2	It is a process of making number of copies of a particular gene segment.	Gene cloning is forming multiple identical copies of any template DNA.
3	Automation is present.	Automation is absent.
4	It is less expensive.	It is highly expensive.
5	Less error probability.	More error probability.
6	Users skill is not required.	Users skill is required.

4. Distinguish between Cry and *cry*.

Ans. Cry stands for crystal of toxic protein whereas *cry* is the gene responsible for synthesis of toxin.

Chapter 13. Organisms and Populations

1. Mention one significant difference between each of the following :

(i) Natality rate and Mortality rate

(ii) Immigration and Emigration

Ans. (i)

Natality Rate	Mortality Rate
It is the average number of new individuals produced by a population in a unit of time (per year).	It is the average number of individual that die and are reduced from the population in a unit of time (per year).

(ii)

Immigration	Emigration
The permanent movement of individuals into the population is known as immigration. It results in an increase in the population.	The permanent movement of individuals out of the population is known as emigration. It results in a decrease in the population.

2. State difference between population size and population density.

Ans.

S. No.	Population Size	Population Density
1	It is the total number of individuals of a species found in a geographical area.	It is the number of individuals of a population found per unit area at a given time.
2	It is nearly uniform at a given time.	It varies from area to area.

3. Distinguish between J-Shaped curve and S-Shaped curve.

Ans.

S. No.	J-Shaped Curve	S-Shaped Curve
1	It is divided into three phases *i.e.,* Lag phase, Exponential phase and Senescence. Last phase stops abruptly due to mass starvation and mortality.	It has three phases *i.e.,* carry phase of no growth, middle phase of rapid growth and stationary phase of zero growth.

2	When the food supply ceases then mass starvation takes place leading to death.	When the food supply ceases, then growth becomes stationary *i.e.,* newly produced cells are approximately equal to the number of dying cells.
3	It is applicable to few animals like reindeers and microscopic organisms.	It is applicable to most of the plants, animals and microscopic organisms like yeast cells.

4. How is diapause different from hibernation ?

Ans.

S. No.	Hibernation	Diapause
1	Under unfavourable conditions, the animals that fail to migrate avoid the stress by escaping in time and showing winter sleep is called hibernation.	Under unfavourable conditions, many species is lakes and pond are known to enter a stage of suspended development called diapause.
2	It occurs usually in winters.	It occurs both in summer and winter.
3	Example–Bear goes into hibernation during winter.	Example, zooplanktons undergo diapause in lakes and pond under unfavourable conditions.

5. How does mutualism differ from commensalism ? Give one example of each.

Ans.

S. No.	Mutualism	Commensalism
1	It is an interspecific interaction in which both the species (individuals) are mutually benefitted.	It is an interspecific interaction in which one species is benefited and other is neither harmed non-benefitted.
2	The two individuals may be physically or physiologically associated. Example-*Rhizobium* and legume plants.	The two individuals may be physically associated. Example- Sucker fish and sharks.

6. Differentiate between parasitism and competition, giving one example of each.

Ans.

Parasitism	Competition
It is the interaction in which one species is benefited and other in harmed.	It is the interactions in which both species are harmed.
For example - Tapeworm in humans, ticks on dogs.	For example – Herbivores and plants in an area.

7. Distinguish between the following—

(i) Hibernation and aestivation

(ii) Ectotherms and endotherms

Ans. (i)

S. No.	Hibernation	Aestivation
1	It is the condition of passing the winter in a resting or dormant condition.	It is the state of inactivity during hot and dry summer.
2	Animal rest in a warm place.	Animal rest in a cool and shady place.
3	It lasts usually for the whole winter season.	It is generally lasts for hot dry day time because nights are often cooler.
4	It is also called winter sleep.	It is also called summer sleep.

(ii)

S. No.	Ectotherms	Endotherms
1	They are also called cold-blooded animals	They are also called warm-blooded animals.
2	They are unable to regulate their body temperature which changes with temperature of environment.	They can regulate their body temperature.
3	They exhibit both hibernation and aestivation.	Their activities are uncommon.
4	They are less active animals.	They are more active animals.

8. Differentiate between the following interspecific interactions in a population :

(i) Mutualism and competition. (ii) Commensalism and amensalism.

Ans. (i)

S. No.	Mutualism	Competition
1	This interaction benefit both the interacting species.	In this interaction, both the interacting species suffer negatively.
2	The two individuals may be physically or physiologically associated.	There is no physical association between the competitions.
3	E.g.- Lichens represent mutualism between fungus and algae where fungus absorbs nutrition and provide protection and algae prepared food.	E.g.– In some American lakes, visiting flamingoes and resident species compete for their common food.

(ii) Commensalism is the interaction between two species where one species is benefitted and the other is neither harmed nor benefitted e.g. an orchid growing as an epiphyte on a mango tree benefits by getting shelter and nutrition but the mango tree is not affected.

Amensalism is the interaction between two different species in which one species is harmed and the other is neither benefitted nor harmed e.g. *Penicillium* produces a toxin killing other micro-organism but is not affected itself.

9. Differentiate between conformers and regulators.

Conformers	Regulators
Aquatic animals and plants in which the osmotic concentration/body temperature of body fluids changes according to the ambient conditions/ environment of water/ environment are called conformers.	Some organism are able to maintain homeastasis by physical means which ensures constant body temperature, constant osmotic concentration etc. Such organism are known as regulators.

Chapter 14. Ecosystem

1. Differentiate between :

(i) Food chain and Food web

(ii) Grazing food chain and Detritus food chain

(iii) Gaseous cycle and Sedimentary cycle.

Ans. (i)

S. No.	Food Chain	Food Web
1	Transfer of energy from producer to top consumers through a series of organism is called food chain.	A number of food chain interconnected with each other forming a web-like pattern is called food web.
2	One organism holds only one position.	One organism can hold more than one position.
3	The flow of energy can be easily calculated.	The flow of energy is very difficult to calculate.
4	It is always a straight and proceed in a progressive straight line.	Instead of straight line it is a series of branching lines.
5	Competition is limited to member of same trophic level.	Competition is amongst members of same and different trophic levels.

(ii)

S. No.	Grazing Food Chain (GFC)	Detritus Food Chain (DFC)
1	It start with green plants called producers as first trophic level.	It begins with dead organic matter and decomposers called saprophytes as first trophic level.
2	A much lesser fraction of energy flows through this types of food chain.	A much larger fraction of energy flow through this type of food chain.
3	Energy of food chain comes from the sun.	Energy for the food chain comes from organic remains or detritus.

(iii)

S. No.	Gaseous cycle	Sedimentary cycle
1	The reservoir is the atmosphere.	The reservoir is the earth crust.
2	The nutrient occurs as gas or vapour.	The nutrients are non-gaseous.
3	It is comparatively quick or fast.	It is quite slow.
	For example - carbon cycle, nitrogen cycle.	For example - phosphorus cycle, sulphur cycle.

2. What properties distinguish a pioneer community from a climax community ?

Ans.

S. No.	Pioneer Community	Climax community
1	The species which invade a base area or land to initiate succession is called pioneer community.	The last or final stage in a succession constitute the climax community.
2	The pioneer species have high reproduction rate.	The climax species have low reproductive rate.
3	The pioneer species have short life span.	The climax species have long life span.
4	They are replaceable.	The climax species are stable and not replaced.

3. Differentiate between phosphorus cycle and carbon cycle.

Ans.

S. No.	Phosphorus cycle	Carbon cycle
1	Phosphorus returns to the soil.	Carbon is released into the atmosphere.
2	The atmospheric unit of phosphorus through rainfall in much smaller.	The atmospheric input of carbon from rainfall is greater.
3	There is no gaseous exchange of phosphorus between organism and environment.	Carbon gas is exchanged between organism and atmosphere during respiration.

4. State the difference between the first trophic levels of detritus food chain and grazing food chain.

Ans.

S. No.	First trophic level of detritus food chain	First trophic level of grazing food chain
1	Decomposers are the first trophic level	Producers are the first trophic level.
2	They breakdown the complex organic matter into simpler form by secreting enzymes.	They prepare complex organic molecular from simple inorganic material with the help of sunlight.

5. Differentiate between a detritivore and a decomposer giving an example of each.

Ans.

Detritivores	Decomposers
They are organisms which feed on detritus and break them into matter particles e.g. earth worm.	They are organisms which by secreting enzymes breakdown complex organic matter into inorganic substances e.g. some bacteria and fungi.

6. Differentiate between two types of pyramids of biomass with the help of one example of each.

Ans.

Upright pyramid of biomass	Inverted pyramid of biomass
The biomass of producer is more than that of consumer. For example–Forest ecosystem.	The biomass of the producer (phyto- planktons) is less than that of consumer (fish) For example–Aquatic ecosystem.

7. Give three differences between seral stages and climax community during succession.

Ans.

S. No.	Seral stages	Climax community
1	It is the sequential replacement and establishment of species in the process of succession.	It is the community which gets established at the terminal stage of succession.
2	The species composition changes from time to time.	The species are stable and species composition do not change.
3	Simple food chain and food web are found.	Complex food chains and food web are found.

8 Distinguish between :

(i) Production and Decomposition

(ii) Litter and Detritus

(iii) Upright and Inverted pyramid

(iv) Primary and Secondary productivity

Ans. (i)

S. No.	Production	Decomposition
1	It is the process of formation of fresh biomass from inorganic matter by producer (plant) using sunlight	It is the process of degradation of waste biomass into its constituent by decomposers.
2	It traps energy.	It release energy.
3	It needs sunlight.	It does not require sunlight.
4	It is an anabolic process.	It is a catabolic process.

(ii)

S. No.	Litter	Detritus
1	It is made of dried fallen plant matter.	It is freshly deposited organic matter *i.e.,* remains of plant and animals.
2	It is found above the ground.	It is found both above and below the ground.

(iii)

S. No.	Upright Pyramid	Inverted Pyramid
1	The base bar comprises of producer in large number.	The base bar comprises producer in smallest number.
2	The number of consumer decreases and become least in top consumer level.	The numbers of consumer increases and become highest in top consumer level.
3	Pyramid of energy is always upright.	Pyramid of number and biomass may be inverted.

(iv)

S. No.	Primary Productivity	Secondary Productivity
1	It is the rate at which organic matter is built up by producers.	It is the rate at which organic matter is built up by consumers.
2	It is the result of synthesis of fresh organic matter from inorganic materials.	It is the result of synthesis of organic matter from plant organic matter.

Chapter 15. Biodiversity and its Conservation

1. How do zoological parks differ from National park concerned with species conservation?

Ans. Zoological park – *ex-situ* conservation and National park-*in-situ* conservation.

2. Differentiate between :

(i) *In situ* conservation and *Ex situ* conservation.

(ii) Keystone species and endangered species.

(iii) Genetic diversity and species diversity.

Ans. (i)

S. No.	*In-situ* Conservation	*Ex-situ* Conservation
1	It is conservation and protection of biodiversity in its natural habitat.	It is conservation of selected threatened plant and animal species in place outside their natural habitat.
2	Population is conserved in the surroundings where they have developed their distinctive features. For Example : National park, biosphere reserve, wildlife, sanctuaries, etc.	Population is conserved under simulated conditions that closely resemble their natural habitats. For Example : Botanical gardens, zoological parks, wildlife safari, gene banks, etc.

(ii)

S. No.	Keystone species	Endangered species
1	It is a species of lower abundance that plays vital role in sustaining the community.	It is a species of lower abundance that may or may not have a vital role in sustaining the community.
2	The species is not threatened e.g. *Ficus*.	The species is at high risk of extinction in near future e.g. Red panda.

(iii)

S. No.	Genetic diversity	Species diversity
1	It represent the variety of genetic information present in an organism.	It is the variety of species and their relative abundance found within a region.
2	It is trait of an organism.	It is trait of biotic community.

Chapter 16. Environmental Issues

1. Differentiate between BOD and COD.

Ans.

S. No.	Biological Oxygen Demand (BOD)	Chemical Oxygen Demand (COD)
1	It is the amount of oxygen required for microbial breakdown of organic matter.	It is the amount of oxygen required for oxidation of both biodegradable and non-biodegradable organic matter in water.
2	Its value is less than COD	Its value is more than BOD.

2. Differentiate between 'bad' ozone and 'good' ozone.

Ans.

S. No.	Bad Ozone	Good Ozone
1	The ozone found in the lower atmosphere is called bad ozone.	The ozone found in the stratosphere is called good ozone.
2	It harms both plants and animals.	It absorb UV radiations and protect the living organism from its harmful effects.

Long Answers

Set 10

Chapter 1. Reproduction in Organisms

1. Describe, in brief, the various methods of artificial vegetative propagation.

Ans. The various methods of artificial vegetative propagation are :

(i) **Cutting :** The division of any plant organ used for propagation is called as cutting. A cutting of an optimal length bearing nodes and suitable diameter is generally taken from the stem and is grown in the soil to regenerate the missing parts. Roots strike out from the base of the stem but in some plants, roots do not come out readily and have to be treated with hormones.

Examples : Sugarcane, grapes, cocoa, rose, *Bougainvillea* and carnations.

Layering

(ii) **Layering :** In this technique, the roots are induced on a stem before it is detached from the parent plant. In mound layering, the lower branch of a plant is bent inside the moist soil. After few days, the roots are produced from the covered portion of the stem. Then, this rooted branch is cut out from the main plant as a new plant.

In air layering the stem is shaved off at an upward angle and is covered with *Sphagnum* moss or cotton as moistening medium for rooting. The wrapping is made by polythene film. After few days, the roots are induced from the injured portion. The branch is removed and planted in a pot.

Examples : Jasmine and strawberry.

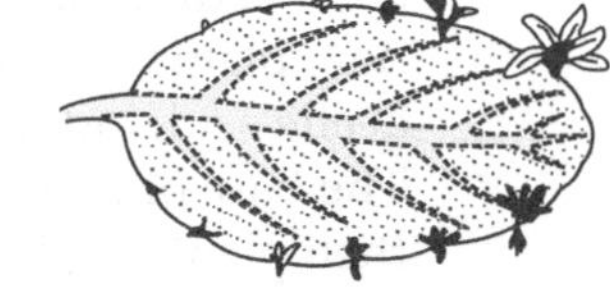

Bryophyllum
Propagation by leaves

(iii) **Propagation by leaves :** In certain leaves of plants, the plantlets develop in the notches of intact leaves. These plantlets develop into independent plant when they are detached and grown into moist ground.

Example : *Bryophyllum* and *Begonia*.

(iv) **Propagation by roots :** The intact roots of many plantlets develop shoots and produce new plants. In sweet potato, adventitious buds are formed that develop into new plants when placed in a moist soil.

Example : Sweet potato, *dahlia, asparagus, tapioca*.

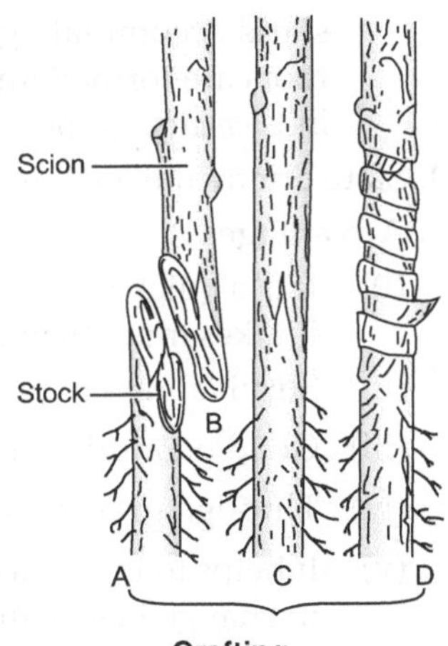

Grafting

(v) **Grafting :** This technique is largely used by horticulturist to improve the breeds of many fruit crops. A scion is derived from a plant of desired character and is grafted on the rooted stock with a cord. After few days, the cambium and vascular bundles of stock and scion fuse together and induce the same growth rate. In bud grafting, a single bud with adjacent bark is taken from a selected plant as a scion and is inserted in the cut made on the stock under the bark. After a few days, the bud gets fixed in the stock and give rise to new branch.

Examples : Mango apple, pear, citrus and guava.

2. Name and describe the method of artificial propagation employed by garden farmers for the following :

(i) Sugarcane
(ii) Litchi
(iii) Apple
(iv) Jasmine
(v) Lemon
(vi) China rose

Ans. (i) **Sugarcane :** It is mostly propagated by stem cuttings. 10-15 cm long stem cuttings are taken. Care is taken to put the morphologically lower side of the cutting in the soil. Soon adventitious roots grow at the lower end. Usually soft wood cutting (stem cuttings taken from current year growth and which are often green) form roots quickly. Applications of growth hormones like indole acetic acid help to promote rooting on cuttings.

(ii) **Litchi :** The process of air layering or gootee is used in propagation of this plant. It is done because the stem branches of litchi cannot be bent to the ground for common mound layering. In this process the stem is girdled *i.e.*, a ring of bark tissues is removed from it. It is then covered with moist moss or cotton and wrapped in polythene to prevent drying. This region is kept moist continuously. Auxins may be applied to the girdled region to promote root formation. Adventitious roots are formed above the girdled region. The stem is now cut off from below the girdle and planted.

(iii) **Apple and (iv) Jasmine :** For both these plants the method of mound layering is practiced. In it a lower branch of the plant is bent down to the ground and covered with moist soil in such a way that its growing tip remains above the soil. This branch is now called a layer. In a few days adventitious roots grow from the covered part. The rooted branch is then cut off from the parent plant and grown separately.

(v) **Lemon :** In Lemon, the artificial vegetative propagation is by cutting. In this process, the plants are generally propagated by placing their stem cuttings in moist soil. The cuttings soon strike roots and grow into plants.

(vi) **China rose** : In China rose, the artificial vegetative propagation is by cutting. In this process, the plants are generally propagated by placing their stem cuttings in moist soil. The cuttings soon strike roots and grow into plants.

3. Describe with suitable example the following artificial methods of vegetative propagation :

(a) Stem cutting (b) Layering

Ans. (a) **Stem cutting :** Plants like rose, grape, *Chrysanthemum, Bougainvillea,* etc., are grown from stem cuttings. In this process pieces of stem with one or several buds are taken. The branch is cut across below a node and the lower leaves are removed. These pieces are planted in moist, porous soil. Adventitious roots arise from the part embedded in the soil while the aerial buds develop into shoot. In sugarcane, stem cuttings are placed horizontally under the surface of the soil. Buds on these pieces develop into new plants having their own adventitious roots and new aerial shoots. Many chemicals like Indole acetic acid, indole butyric acid and Naphthalene acetic acid stimulate root initiation in stem cuttings. The base of cuttings is dipped in very dilute solution of these growth promoting substances and are then planted. This produces roots more rapidly and in great number.

(b) **Layering :** In this process a branch of the plant in the lower region is bent down. A ring of bark is removed and this region is covered by soil and kept in place by some pressure. The upper part of the branch is kept free. Due to ringing, translocation of food downward below the girdle stops. Food starts accumulating just above the girdle and this induces root formation in that region. Once the roots are formed, the branch is cut from the parent plant and is grown separately. This done mostly in *Clematis*, grapevine, jasmine, etc.

4. State briefly the advantages and disadvantages of vegetative propagation.

Ans. Advantages :

(i) It is a quick method of multiplication *e.g.*, potato crop is raised in only three months from tubers but it takes more than one year to raise it from seeds.

(ii) Survival rate of the progeny is much higher than of those formed by seeds.

(iii) Good quantities of the plant can be preserved indefinitely.

(iv) Through grafting, desirable qualities of two varieties can be combined.

(v) It helps to introduce plants in new areas where seed germination fails to produce natural plants due to change in soil and environmental conditions.

(vi) It helps in rapid spread of a plant over an area.

(vii) Quick fruitening of plants can occur by vegetative propagation in plants which otherwise takes years to reach maturity.

(viii) Common infections can be removed from plants through pruning, micropropagation, etc.

(ix) Vegetative propagation gives rise to clones. In case of plants propagated through seeds, variations occur which may not be good.

Disadvantages :

(i) There are no variations, so adaptability to changed environment gradually decreases.

(ii) Good qualities can not be included or bad characters cannot be eliminated in plants multiplying through vegetative propagation.

(iii) The plants may show degeneration due to absence of sexual stimulus and variations.

(iv) As these plants are not dispersed, vegetative propagation causes overcrowding and this could damage the plants.

(v) Vegetative propagules can get easily decayed and are more prone to various virus, fungal and bacterial infections.

Chapter 2. Sexual Reproduction in Flowering Plants

1. Explain the process of development of ovule and the female gametophyte.

Ans. A gynoecium is the female reproductive part of the flower. It consists of ovary, style and stigma. The ovules are formed in the ovary. The stalk of the ovule is called funiculus and is attached to the ovary at placenta. The ovule is surrounded by one or two protective layers called integuments, leaving a small opening at one end termed as a micropyle.

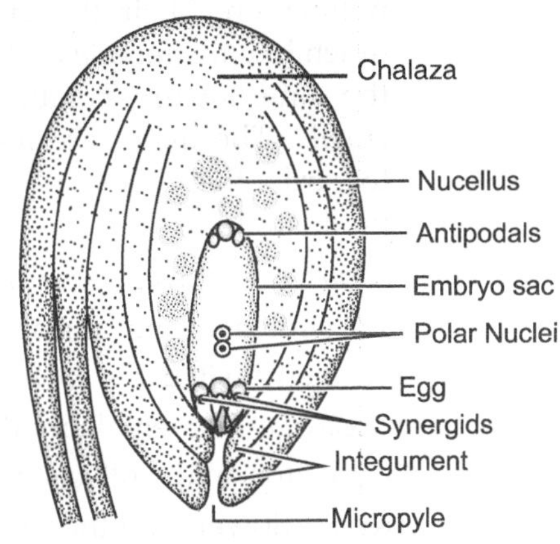

Structure of an Ovule

The ovule is composed of a multilayered cellular tissue called the nucellus. A hypodermal cell of the nucellus enlarges to form an archesporia cell which transformed into a megaspore mother cell. Each diploid megaspore mother cell undergoes meiosis to form linear tetrad of megaspores. Three megaspores degenerate and remaining functional one enlarges at the chalazal end.

The functional megaspore undergoes three subsequent mitotic divisions and forms the eight nuclei. The eight nuclei arrange themselves into three groups. Three nuclei migrate at the micropylar end and forms the egg apparatus, consisting of two synergids and a round egg cell. The other three nuclei arrange themselves at the chalazal end and form the antipodal cells. The remaining two nuclei come together as polar nuclei in the centre of the embryo sac. All the cells of the embryo sac are haploid.

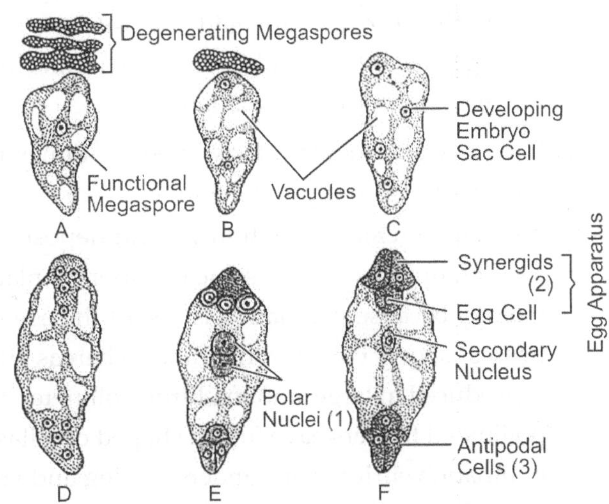

Development of female gametophyte

2. What is pollination ? Give the adaptation of flowers pollinated by insects with a suitable example.

Ans. Pollination is the transfer of pollen grains from the anther of flower to the stigma of the same flower or of another flower of same or allied species.

Adaptations of Insect Pollinated Flowers :

(i) These flowers are large and brightly coloured, if the flowers are small they are grouped into large conspicuous inflorescence to attract insects. *e.g.*, Sunflower and other flowers of compositae.

(ii) If the flower is small and inconspicuous the bracts may become large and colourful which help to attract insects *e.g. Bougainvillea.*

(iii) The flowers produce scent or nectar for the insects, some flowers like *Cestrum* (queen of the night) emit a strong odour at night and these are pollinated by moths which are active during night.

(iv) The nectaries are present at the base of the petals. So that insects coming in search of nectar has to put their proboscis and head into the centre of the flower *e.g., Petunia.* In this process their bodies get dusted with pollen grains. When the same insect visits another flower the dusted portion of its body comes in contact with the stigma of that flower, facilitating pollination.

(v) The pollen grains of such flowers are sticky or have spinous outgrowths on the exine, their stigmas are also sticky.

(vi) Some flowers like *Anagallis, Hyacinthus* and *Verbascum* secrete juices which cause the pollen to stick to the insect bodies.

(vii) In *Salvia,* the corolla is bilipped. The lower lip acts as a platform on which the bees alight. The filament of anther forms a sort of lever which swings the anther. When a bee enters the flower in search of nectar, the lever pushes the fertile lobes of the anther hitting it on the back of the insect and pollen grains are dusted on it.

(viii) Several species of the orchid genus *Ophrys* require a particular moth for pollination. Their flowers look very much like the female moths and have the same texture, colour pattern and smell. The males are thus attracted towards these flowers and help in pollination.

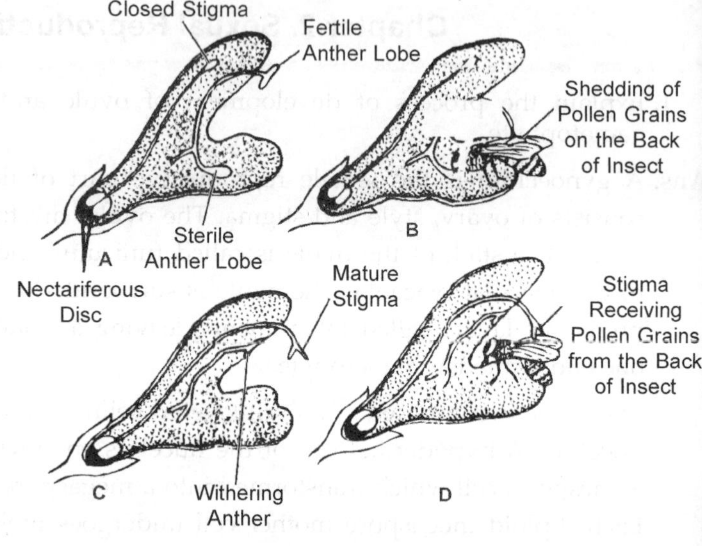

Insect pollination in Salvia

(ix) In *Yucca,* the flowers are fully expanded and emit perfume during night. They are then visited by the moth-*Pronuba yuccasella.* The female moth has a long ovipositor and collects much pollen from a flower. She then visits another flower pierces the wall of the ovary with the ovipositor and deposits the eggs below the ovules. Then the moth climbs the top and pushes the pollen into the stigma-opening bringing about pollination.

3. In what ways does the study of pollination enrich our understanding of biology and enable us to apply it for increasing crop productivity ?

Ans. Pollination refers to the transfer and deposition of pollen on the stigmatic surface of the same flower or a different flower of the same or different plant. The study of pollination enriches our understanding of biology because pollination is one of the essential functions of the flower. The insect-pollinated flowers possess showy colour or fragrance whereas wind pollinated flowers are odourless and in them the pollen is produced in large amount. Birds pollinated flowers have a nectar, composed of sugars. Some of the bird pollinated flowers have funnel shaped corollas and produce copious amount of nectar. Thus, the study of pollination enriches our understanding and knowledge about biology.

The study of pollination is very essential to increase the productivity of crop plants. During cross-pollination, genetic recombination and variation occurs which results in producing the new and improved offsprings. Cross-pollination results in producing the offspring that is better adapted in terms of healthy and vigorous seeds. Mustard, safflower, sunflower and some of the pomaceous fruit crops give higher yields if the bees are available for pollination.

4. Give a brief account of various adaptations which promote cross pollination.

Ans. Adaptations promoting cross pollination :

(i) **Unisexuality :** The flowers are unisexual *i.e.,* having one sex only. These unisexual flowers are present on different plants thus promoting the cross-pollination.

(ii) **Dichogamy :** In bisexual flowers anthers and stigma mature at different times. Dichogamy is of two types : **Protandry** and **Protogyny.** In protandry the anthers mature first but the stigma of the same flower is not receptive at that time (**Sunflower, Cotton** and ***Salvia***). In protogyny the stigma of the carpel matures earlier than the anthers of the same flower *e.g.*, *Ranunculus* and *Mirabilis jalapa* (Four o'clock plant).

(iii) **Suppression of one sex :** In certain bisexual flowers one sex organ, either stamen or carpel is completely suppressed and becomes sterile. Anther does not form pollen grains and carpel does not produce ovule. Thus, facilitating cross-pollination.

(iv) **Prepotency :** In many plants the pollen grains from a flower, when present on the stigma of another genetically different flower, germinate more quickly as compared to the pollen grains of the same flower. Thus promoting cross pollination *e.g.*, Grape, Pear and Apple.

(v) **Self-sterility or self-incompatibility :** Pollen grains of a flower are incapable of effecting fertilisation even if they are placed on the stigma of the same flower due to mutual inhibition e.g., Potato, Tobacco and *Petunia axillaris*.

(vi) **Heterostyly :** The occurrence of two or more types of flowers having different length of styles and stamens. In Primrose, there are two types of flowers while in *Oxalis* three types, based on the length of styles and stamens. An insect visiting a flower will get dusted with pollen grains on a particular area of its body depending upon the height of anther. Now this insect will pollinate only that flower in which stigma lies at the level of anther of the first flower. This process involves pollen and stigma from two or more flowers having the similar height.

(vii) **Herkogamy :** In some flowers there may be some physical barriers between the anther and the stigma so that pollination between them becomes difficult or even impossible *e.g.*, in *Calotropis* the pollens are present in a sac like structure called pollinium and the pollination is entirely at the mercy of insects.

5. Discuss the events that take place during fertilisation in an angiosperm.

OR

Describe the events that take place between pollination and fertilisation in plants.

Ans. After pollination stigma secretes sugar *i.e.*, when a pollen grain reaches stigma, it absorbs and swells. It results in the rupture of the exine at certain thin or weak points called germ pore. The intine grows out in the form of a tube called the pollen tube. The vegetative and generative nuclei migrate to the distal end of the pollen tube.

The pollen tube gradually elongates and pierces its way through the style into the cavity of the ovary. From here it grows and moves towards the micropyle of the ovule. When the pollen tube penetrates the ovule at the chalazal end such fertilisation is said to be chalazogamous. When the pollen tube enters through the micropyle, it is known as porogamous. By this time the generative nucleus divides into two equal, non-motile gametes and the vegetative nucleus disintegrates.

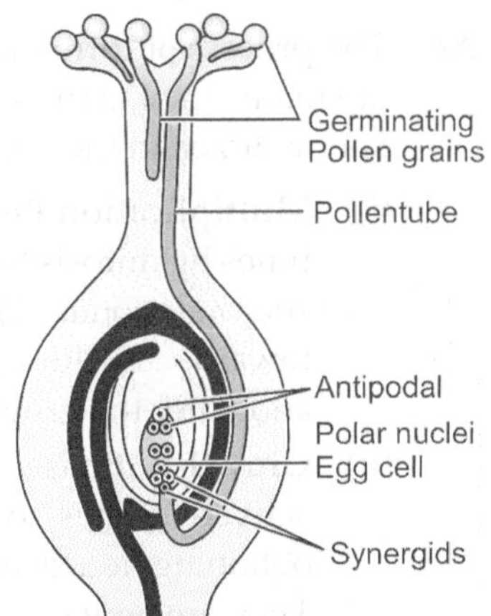

Longitudinal section of ovary showing growth of pollen tube towards embryo sac

The fully developed embryo sac at this stage consists of three antipodal cells, two synergids, one egg cell and one secondary nucleus. The pollen tube pierces through the nucellus and finally penetrates the wall of the embryo sac. The tip of the pollen tube disintegrates and the two male gametes are set free in the embryo sac. Of the two male gametes one fuses with the egg cell and the other with the secondary nucleus on the polar nuclei. The fertilised egg now becomes zygote or oospore and the secondary nucleus or fusion forms primary endosperm nucleus, which is the result of triple fusion. The triple fusion along with the fertilisation of the egg is known as double fertilisation. The zygote or oospore develop into the embryo and the endosperm nucleus gives rise to the endosperm.

6. Give a brief account of the dispersal of fruits and seeds.

Ans. Dispersal : The movement of fruits and seeds, away from the parent plant, with the help of many different agencies such as air, water, animals etc. is called dispersal of fruit and seeds.

(i) **Dispersal By Wind :** *(Anemochary)* Such fruits and seeds have following adaptations :

(a) They are light in weight and have small size *e.g.,* seeds of orchids, grasses.

(b) They have wings *e.g., Acer, Shorea, Terminalia, Pinus Moringa, Jacranda,* etc.

(c) They develop some parachute mechanism, as the development of hair-like pappus in compositae.

(d) They have special mechanism of dehiscence e.g., in *Argemone* or Poppy.

(e) They have hairs *e.g. Calotropis,* Cotton.

(f) They have persistent hairy styles *e.g., Clematis.*

(g) They develop balloon like appendages *e.g.,* Balloon vine, *Physalis.*

(ii) **Dispersal by Water** *(Hydrochory)* : Such fruits develop floating devices in the form of spongy or outer coats. They are protected from rotting due to presence of waxy coating. *e.g., Nymphea* (Water lily), Lotus, Water Chestnut, *Cocos nucifera,* etc.

(iii) **Dispersal by Animal** *(Zoochory)* : They have following peculiarities.

(a) They are edible *e.g. Ficus benghalensis, F. religiosa, Achras sapota* (Cheeku), etc.

(b) They are sticky *e.g., Boerhaavia, Cleome,* etc.

(c) They have hook like bristles or spines *e.g., Xanthium, Achyranthes, Tribulus* etc.

(iv) **Explosive Mechanism (Autochory) :** Example of such fruits and seeds are camel's foot climber (*Bauhinia vahlii*), Pansy (*Viola*), *Ruellian, Geranium,* Squirting cucumber (*Ecballium etalerium*), etc.

Chapter 3. Human Reproduction

1. Explain the process of spermatogenesis in humans.*

Ans. The process of formation of sperms is called spermatogenesis. It occurs in the male gonads-testes. Testes are made up of many seminiferous tubules lined by germinal epithelium. Cells of this layer divide to form spermatozoa in the following four steps:

(i) **Multiplication Phase:** At sexual maturity, the undifferentiated primordial germ cells divide several times by mitosis to produce a large number of spermatogonia. These spermatogonia move towards the lumen of seminiferous tubules and enter the growth phase.

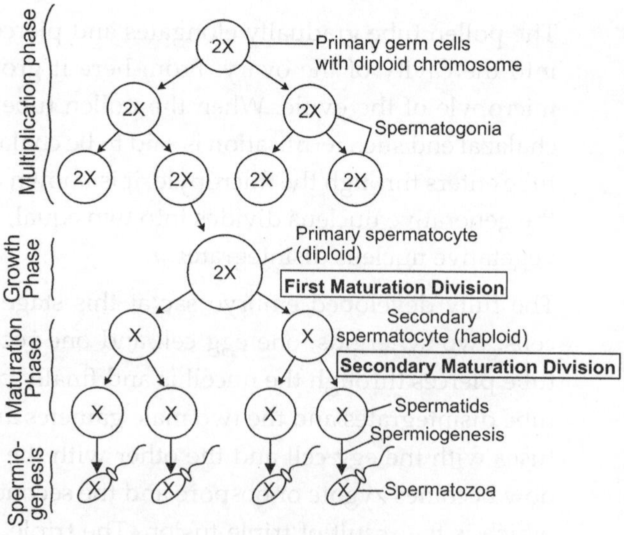

(ii) **Growth Phase:** Each spermatogonium actively grows to a larger spermatocyte by obtaining nourishment from the nursing cells. They are now called primary spermatocytes.

(iii) **Maturation Phase:** Each primary spermatocyte undergoes two successive divisions, called maturation divisions. The first maturation division is reductional or meiotic. Hence, the primary spermatocyte divides into two haploid daughter cells called secondary spermatocytes. Both secondary spermatocytes now undergo second maturation division which is an ordinary mitotic division to form, four haploid spermatids, by each primary spermatocyte.

(iv) **Spermiogenesis:** It is the process of transformation of circular spermatid to a spermatozoan. The

* **Frequently asked previous years Board Exam Question.**

spermatozoa are later on known as sperms. In this process the nucleus of the spermatid becomes the head of the sperm, it shrinks with closely packed DNA and the golgi apparatus, containing proteolytic enzymes, becomes the acrosomal cap. Axial filaments are formed from distal centriole and from the mitochondria, mitochondrial spiral is formed. Also the loss of cytoplasm takes place and only a sheath remains around mitochondria.

2. What is oogenesis ? Describe the process in brief.

Ans. The phenomenon of formation of haploid ova (egg) from the diploid egg mother cells of ovary (female gonad) is called oogenesis. It occurs in three phases *i.e.,* Multiplication, Growth and Maturation.

Process of oogenesis includes the following steps :

(i) **Multiplication phase :** Some cells of the germinal epithelium of ovary function as primary germ cells and are called oogonia, which multiply by repeated mitotic division and become primary oocytes. At this stages cell division ceases.

(ii) **Growth phase :** This is growth phase of oogenesis is very prolonged one and the increase in size of primary oocyte is very considerable. It is a slow process and may take a long period. In human female, the oogonia are already formed at the time of birth and show growth only after attaining sexual maturity.

(iii) **Maturation phase :** The primary oocytes undergo meiosis upto the interphase or upto prophase I (diakinesis) during embryonic growth. It then resumes development at puberty. The 1° oocyte (2n) undergoes meiosis-I to produce two haploid cells (n) larger one is called 2° oocyte and the smaller one is 1st polar body. 2nd oocyte undergo the process of meiosis-II and proceeds upto metaphase-II only. Further development will start only after the arrival of spermatozoa (sperm) by breaking down the MPF (Metaphase promoting factor) and switching on APC (Anaphase promoting complex). Completion of meiosis-II result in the formation of ovum and 2nd polar body.

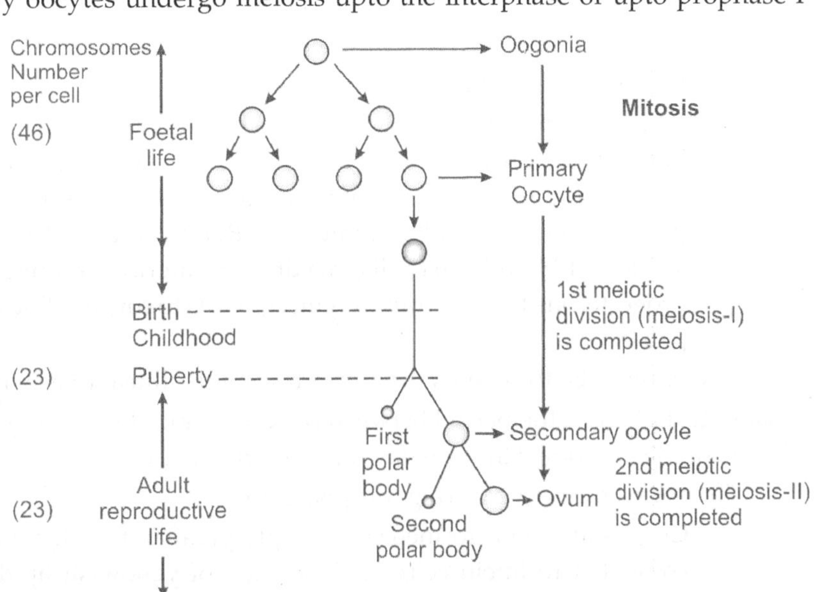

3. Discuss various phases of menstrual cycle in females.

Ans. The events occurring during the menstrual cycle can be divided into four phases :

(i) **Menstrual phase :** It lasts for 3-5 days, and extends from 1st and 4th day of the menstrual cycle. When ovum remains unfertilised, corpus luteum starts degeneration, decreasing the level of progesterone. The uterine tissue fail to be maintained. The unfertilised ovum along with ruptural uterine epithelium, 50-100 ml of blood and some mucus is discharged through the vaginal orifice and is called Menstruation.

(ii) **Follicular/Proliferative Phase :** It follows the Menstrual phase and lasts for about 10-12 days (from 5th to 14th days of the menstrual cycle). It involves following the changes —

(a) Under the stimulation of GnRH of hypothalamus, there is increased secretion of FSH from the anterior pituitary. The FHS stimulates the ovary and bring about the maturation of a primary follicle into a Graafian follicle. The follicular cells of Graafian follicle secrete estrogen.

(b) This estrogen stimulates the growth, maintenance and normal functioning of secondary sex organs.

(c) The estrogen helps in repairing the endometrial lining for 2 days after which it becomes thick (2-3 mm), more vascular, and glandular in the next 7-9 days. This is preparation for supporting the implantation of fertilised egg.

(d) The epithelium of fallopian tubes becomes thick and densely ciliated to conduct the released egg

(e) The end phase inhibits FSH secretion and stimulates the secretion of LH about 6-10 times.

(iii) **Ovulatory phase :** This phase involves the ovulation from the Graafian follicle of ovary. It occurs midway between two menstrual cycles (on 15th day of onset of menstruction). The released egg is received by fimbriae of the fallopian tube and reaches the ampulla site of fertilisation in about 24 hours. Ovulation is controlled by increased level of LH. The ovum is viable for 2 days. Two days after ovulation is called fertility phase.

The ovulatory phase witnesses the maximum concentration of both estrogen and progesterone. LH starts the change of empty Graafian follicle into corpus luteum and secretion of progesterone from corpus luteum is responsible for the maintenance of the endometrium.

(iv) **Luteal/Secretory/Post ovulatory phase :** It lasts for about 12-14 days (from 16th-28th days of menstrual cycle). Corpus luteum is formed from empty Graafian follicle which increases in size. The follicular cells are converted to Lutein cells by deposition of yellowish lipid inclusions, giving it a yellow colour. This phase is the Luteal phase.

Corpus luteum secretes progesterone by appressing the secretion of FSH and LH. It inhibits the maturation of follicle and ovulation.

It also inhibits uterine movements. The uterine endometrum proliferates (5-6 mm) and is ready for implantation. If fertilisation occurs, the uterine endometrium implants the zygote after 7 days by maintaining the high level of progesterone. If fertilisation does not occur, the high concentration of progesterone and Inhibin inhibits GnRH production from hypothalamus. In the absence of GnRH, levels of LH and FSH falls, making the uterine endometrium start detaching. Due to decreasing progesterone level, corpus luteum converts to corpus albicans and prepares to start the next menstrual phase.

4. Briefly describe the events that occur during the luteal phase of the menstrual cycle.

Ans. Luteal phase is the period between ovulation and the onset of the next menstrual flow. It lasts for about 12-14 days (from 16th–28th day) of menstrual cycle.

The events occurring during this phase are —

(i) Corpus luteum is formed from empty graafian follicle which increase in size. The follicular cells are converted to lutein cells by deposition of yellowish lipid inclusions, giving it a yellow colour, so called Luteal phase.

(ii) It secretes progesterone by suppressing the secretion of FSH, and LH inhibits the maturation of follicle and ovulation. It also inhibits uterine movements.

(iii) The uterine endometrium proliferates (5-6 mm) and is ready for implantation.

(iv) If fertilisation occurs, the uterine endometrium implants the zygote after 7 days by maintaining high levels of progesterone.

(v) If fertilisation does not occur, the high progesterone concentration and inhibin inhibits GnRH production from hypothalamus. In absence of GnRH, levels of FSH and LH falls, thus the endometrum starts detaching. Due to decreasing progesterone level, corpus luteum converts to corpus albicans and prepares to start the next menstrual phase.

5. Describe the physico-chemical events that take place during fertilisation in humans.*

Ans. Fertilisation is defined as a process of union of a haploid male gamete or sperm with a haploid female gamete or ovum to form a diploid cell, the zygote.

Process of Fertilisation : During copulation (coitus), male discharges the semen (coagulated form) into the vagina of the female, where the sperms reach the ampulla partly by the movement of their tails and partly by the action of uterus by peristalsis. The sperms present in the semen travel a long way from

* **Frequently asked previous years Board Exam Question.**

vagina through the uterus to the fallopian tube.

The sperm can survive in the female's reproductive tract for 1 to 3 days and it can fertilise the ovum in 12 to 24 hours following ovulation.

Events of Fertilisation :

(i) **Activation of sperm and ovum :** Sperm secrete the chemical hyaluronidase and surface protein called antifertilizin (composed of acidic amino acid). The ovum secretes a chemical named fertilizin which mixes with the water to form egg water which attracts the sperms of its own species. Thus, the liquefication of coagulated semen occurs in this way. The fertilizin of an egg interacts with the anti fertilizin of sperm of the same species. This attraction between fertilizin and anti-fertilizin makes the sperms stick to the egg surface. This process is also known as insemination.

(ii) **Capacitation of sperm :** Capacitation is the process of acquiring the capacity to fertilise the egg by the sperm. Freshly ejaculated sperms are unable to fertilise the ovum. Thus, they undergo a series of changes collectively known as capacitation. Capacitation occurs while sperm reside in the female reproductive tract for a period of time. It is associated with removal of adherent seminal plasma proteins, reorganisation of plasma membrane lipids and proteins. It also seems to involve an influx of extracellular calcium, increase in cyclic AMP and decrease in intracellular pH.

(iii) **Penetration of sperm :** The ovum is surrounded by three membranes such as corona radiata, zona pellucida and the vitelline membranes. At first the sperm passes through corona radiata to reach zona pellucida. There, it releases the enzyme hyaluronidase or sperm lysin from its acrosome. This enzyme dissolves zona pellucida as a result of which the sperm reaches the plasma membrane of the egg. The above changes on the sperm head are called ***acrosome reaction***. Acrosin digests zona pellucida and calcium ions are needed for this.

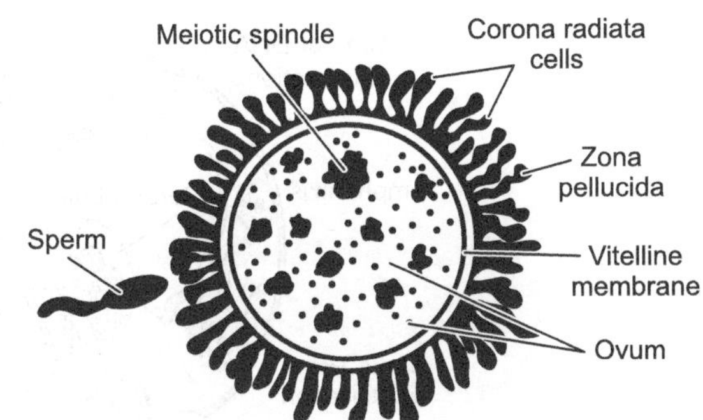

Fig. : Penetration of sperm

(iv) **Cortical Reaction :** Prior to fertilisation, the egg is in a quiescent state, arrested in metaphase of the second meiotic division. Upon binding of a sperm, the egg rapidly undergoes a number of metabolic and physical changes that collectively are called *egg activation*. Prominent effects include a rise in the intracellular concentration of calcium, completion of the second meiotic division and the so-called cortical reaction.

(v) **Zonal Reaction :** Once one sperm has entered the egg (ovum) the vitelline membrane thickens and is converted into fertilisation membrane. This membrane is rigid and never allows other sperms to pass through this membrane. The egg cytoplasm extends around the head of sperm to form a fertilisation cone that engults the sperm so that the head and the mid piece enter the ovum. The formation of fertilisation membrane around the egg blocks polyspermy and allows only single sperm to enter.

(vi) **Amphimixis :** A sperm consists of three parts: head, middle piece and tail. Shortly before or after entering the egg, the sperm loses its tail. After the sperm enters the egg, the membrane of head and middle piece dissolves, liberating nucleus, centrosome and mitochondria. The sperm nucleus enlarges to form the male pronuclear and the nucleus of the ovum becomes female pronuclear.

The male pronuclear moves inwards and then change its direction to meet the egg nucleus. The chromosomes (haploid set) of the sperm and the chromosomes (haploid set) of the egg or ovum are set free by the breakdown of their nuclear envelops.

Mixing up of the chromosomes of a sperm and an ovum resulting in a diploid zygote nucleus is known as amphimixis or karyogamy. The fertilised egg is now called the zygote.

6. Describe the stages of embryonic development :

(a) Cleavage (c) Implantation (e) Organogenesis.

(b) Blastulation (d) Gastrulation

Ans. (a) **Cleavage :** It is the process which transforms the fertilised ovum into a sphere of aggregated cells and this sphere is called morula. The fertilised ovum divides mitotically by holoblastic cleavage starting from animal pole to vegetal pole. This, divides the fertilised ovum into two daughter cells called blastomeres. Immediately, it undergoes second cleavage division at right angle to the first and four dumb–bell shape cells are formed. Then, subsequent cleavage divisions resulting in forming blastomeres which gradually become smaller and smaller and after completion of several cleavages, the blastomeres assume the shape of solid ball of cells as morula.

(b) **Blastulation :** During blastulation the arrangement of blastomeres occur and a central cavity is formed inside the morula. This embryonic stage with central cavity is called blastodermic vesicle. This newly developed central cavity of blastula is called blastocoel. The fluid from the uterine cavity filters into blastocoel and zona pellucida envelope disintegrates. At this stage, outer cell mass forms the outer wall of the blastocyst, called as trophoblast or trophoectoderm. It does not take part in the formation of embryonic body but forms the protective and trophic membranes of blastocoel which eventually develop into foetal portion of the placenta. The cluster of inner cell mass differentiates and aggregates at one pole of the distended blastodermic vesicle. This cluster of cells is called the embryoblast. It is primarily destined to form the body of the developing embryo.

(c) **Implantation :** It is the embedding of the blastocyst into endometrium of uterus. The blastocyst comes in contact with the endometrium in the region of embryonal knob. The surface cells of trophoblast secrete lytic enzymes which cause corrosion of endometrial lining. These give rise to finger-like projections called chorionic villi, which helps in fixation and absorption of nutrients. This is implantation, which begins on about 7th day of fertilisation and takes 3 days to complete.

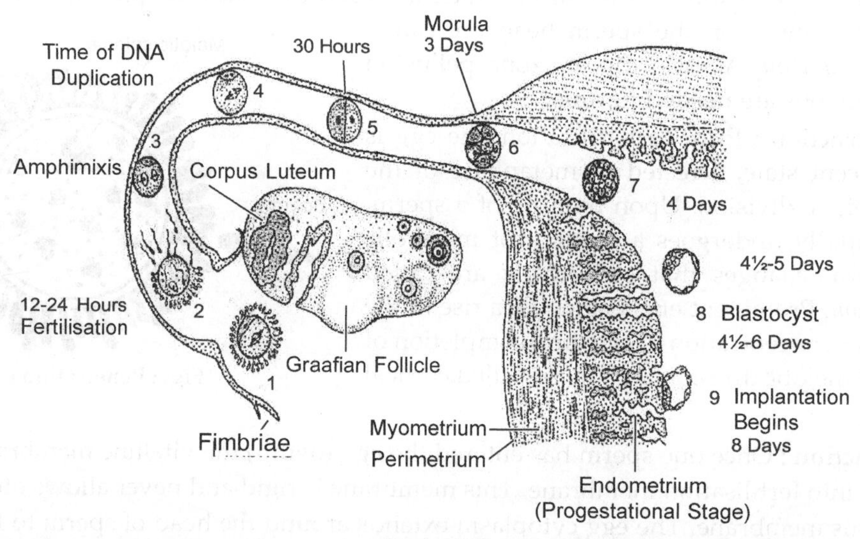

Implantation

(d) **Gastrulation :** It is a process in embryonic development which is characterized by movement of cells in small masses or sheets so as to form primary germinal layers. There are three primary germinal layers namely, endoderm, mesoderm and ectoderm. Since they lead to initiation of morphogenesis. The product of gastrulation is gastrula.

In the late blastocyst stage, the inner cell mass becomes differentiated into epiblast, hypoblast with an amniotic cavity filled with amniotic fluid. The inner cell mass cells get rearranged to form a flat embryonic germinal disc. Gastrulation is associated with the formation of primitive streak on the epiblast surface. After the formation of primitive streak, the epiblast cells more inward and displace the hypoblast to form endoderm. This converts the latter into primary yolk sac. The trophoblast of this region grows to form extra embryonic mesoderm. It splits up and produces extra embryonic coelom. Similarly embryonic endoderm forms a tube of its own which represents the gut tract of the embryo. The growing embryo/foetus develops four membranes called extra-embryonal/ foetal membranes. Those are chorion, amnion, allantois and yolk sac.

(e) **Organogenesis :** It is the formation of organs and tissues in the embryo. It begins in the ninth week of pregnancy and is completed within a month. With the beginning of organogenesis, the embryo is converted into a foetus. The first sign of organogenesis is notogenesite (differentiation of notochrod) followed by neuralation or appearance of rudiments of nervous system. After the first month of pregnancy, the heart is formed. Limbs and digits develop by the end of second month. By the end of 12 weeks (first trimester) most of the organ systems develop along with external genitals.

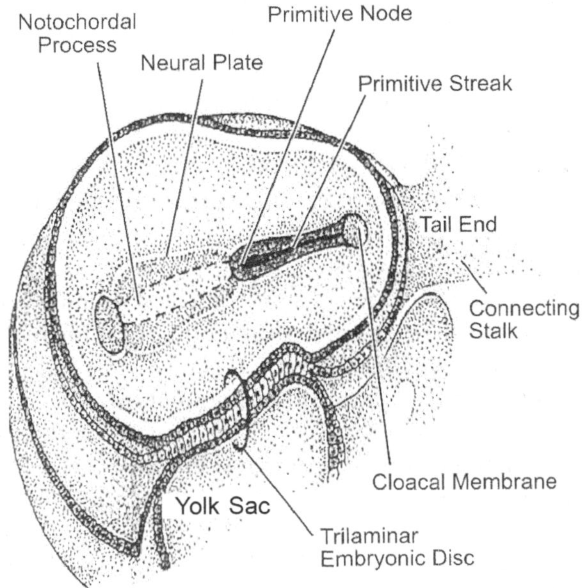

Organogenesis

7. What is parturition ? Explain its different stages.

Ans. Parturition is the process of giving birth to a baby at the end of gestation period. Parturition involves uterine and abdominal contractions, dilation of cervix and passage of foetus. These physical activities are together called labour. Labour is accompanied by localized sensation of discomfort or agony called labour pains.

Signals of parturition originate in the form of mild uterine contraction. The foetus signals that is ready (mature) by secreting certain hormones which diffuse through placenta into mother's blood and cause secretion of oxytocin which initiate uterine contractions. These are called foetal ejection reflexes. It is accompanied by rise in estrogen to progesterone ratio, increase in oxytocin receptors in uterine muscles, increase in level of oxytocin secretion by both mother and foetus. The labour period is divided into three stages.

(i) **Dilation stage :** It is the time from the onset of labour to the complete dilation of cervix. It may last upto 12 hours. Uterine contractions begin from top. They force the baby towards cervix. Contractions are accompanied by pain caused by compression of blood. Vessels and subsequent hypoxia of uterine muscles. These contractions are under positive feedback. Oxytocin induce uterine contractions which in turn caused increased oxytocin secretion. Relaxin hormone produced by the placenta increases flexibility of pubic symphysis and dilates uterine cavity which helps in pushing the baby down into the cervix. Dilation occurs in vagina too. This results in rupturing of amniotic membrane and flowing out of amniotic fluid.

(ii) **Explulsion stage :** It is the period from complete dilation cervix to delivery of the baby. In this stage, intensity of uterine and abdominal contraction further increases. The foetus passes out through cervix and vagina with head in forward direction. It causes intense labour pain and may take 20-60 minutes. The umbilical cord is tied and cut off close to navel.

(iii) **After-birth/placental stage :** Within 10-45 minutes of the delivery of the baby, the placenta separates from uterus and is expelled out due to series of strong uterine contractions.

Chapter 4. Reproductive Health

1. What do you think is the significance of reproductive health in a society ?

Ans. In a society, if the people are aware of birth control methods, sexually transmitted disease, the importance of breast feeding and the post-natal care of the mother and baby, then the society can have healthy children who are the future citizen of a nation.

2. Suggest some methods to assist infertile couples to have children.

Ans. The reason of infertility could be physical, congenital disease a drugs immunological or even psychological. Specialized health care unit called infertility clinic could help in diagnosis and corrective treatment of some of these disorders. Though special techniques called Assisted Reproductive Technologies (ART), an infertile couple could be assisted to have children. ART include (i) In vitro fertilisation, (ii) Gamete Intra Fallopian Transfer (GIFT), (iii) Intra Cytoplasmic Sperm Injection (ICSI) (iv) Artificial Insemination (AI)

3. What is the role of amniocentesis in detecting genetic defects ? How it is performed ?

Ans. Amniocentesis is a test done during pregnancy to get information about the fetus. In this test amniotic fluid is removed from the uterus for testing or treatment. Amniotic fluid is the fluid that surrounds and protects a baby during pregnancy. This fluid contains fetal cells and various chemicals produced by the baby. It helps in developing baby to move in the womb, which allows for proper bone growth. It is done to find genetic abnormalities in the foetus, prenatal diagnosis of genetic defects and sex of the foetus. Thus, this technique is misused in the present time.

Amniocentesis can be done for various reasons :

(i) To test certain conditions — such as Down's syndrome and spina bifida.
(ii) To determine whether the baby's lungs are mature enough for birth.
(iii) To evaluate a baby for infection or other illness.
(iv) Sometimes it is used to decrease the volume of amniotic fluid.

Procedure :

The first step in the procedure is to perform an ultrasound of the uterus. With ultrasound, the doctor can see the fetus, amniotic fluid, and the placenta.

After cleaning the abdomen, the doctor inserts a needle and uses the ultrasound picture to guide the needle safely into a pocket of amniotic fluid. The doctor removes a small amount of fluid about one ounce (30 mL).

Occasionally, the doctor will need to insert the needle more than once to get enough fluid. If there is more than one fetus, the doctor may need to take a sample of fluid from around each fetus.

The sample of amniotic fluid is then sent to a laboratory for testing.

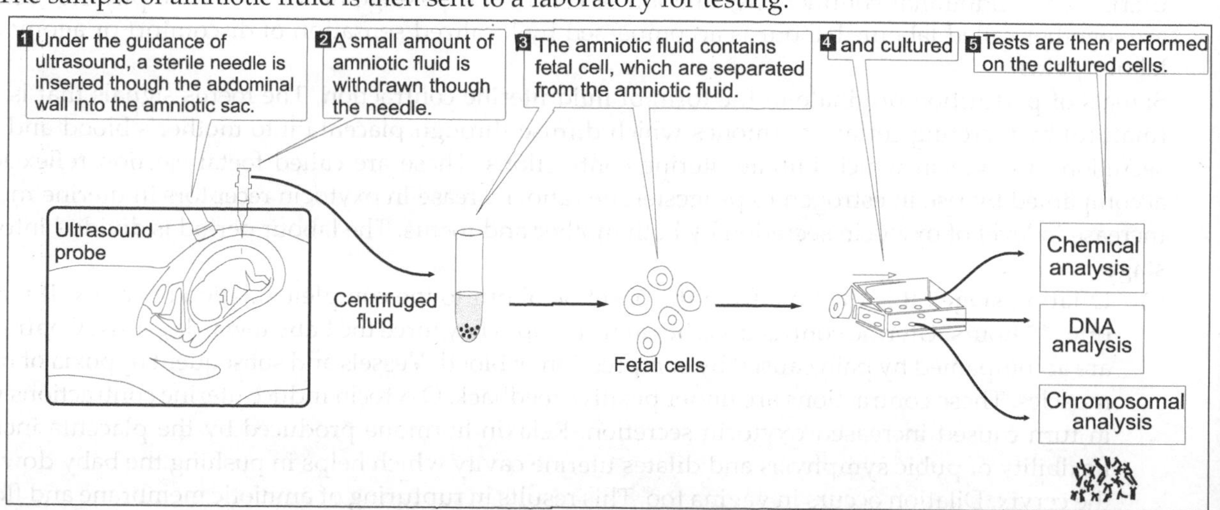

Amniocentesis is a procedure for obtaining fetal cells for genetic testing.

4. What is meant by ART ? Explain the different ARTs in brief.

Ans. ART means Assisted Reproductive Technologies; it is a group of special techniques that help in reproduction in infertile couples by retrieving oocytes from the ovary or sperms from testis, bringing about artificial/in vitro fertilisation and development of embryo.

The main ARTs are as follows :

(i) **IVF (In Vitro Fertilisation):** It involves the fertilisation of ovum with sperm outside the body of mother in a glass container. The embryo transfer (ET) is then done in the uterus of the mother through IUT (Intrauterine transfer) to start the development and finally leading to normal birth. Also called test tube baby technique.

(ii) **GIFT (Gamete Intrafallopian Transfer) :** It involves transfer of two of the retrieved unfertilised oocytes/ova and the sperms into fallopian tube through laparoscope. Fertilisation occurs in vivo, in fallopian tube. This technique is used in cases where husband have very low sperm count.

(iii) **ZIFT : (Zygote intrafallopian transfer)** : It involves the transfer of zygote or embryo at 8 blastomere stage in the fallopian tube through laparoscope. It is more successful than GIFT as fertilisation is already achieved.

(iv) **AIT (Artificial Insemination Technique) :** In case, husband's sperms are defective, artificial insemination donor is required. This semen from donor is artificially introduced into the vagina or into uterus (intra uterine insemination).

(v) **ICSI (Intra Cytoplasmic Sperm Injection) :** It case the sperm count is very low, a single healthy sperm is injected into an ovum. This is intra Cytoplasmic sperm injection.

(vi) **IUT (Intra Uterine Transfer) :** In case of females with blocked fallopian tubes, the eggs retrieved are allowed to get fertilised in vitro and the zygote in 32-celled stage is inserted into the uterus for implantation.

The aim of these techniques is to find a solution to infertility and have children.

5. Write a short note on MTP.

Ans. MTP or Medical Termination of Pregnancy is an intentional or voluntary termination of pregnancy before the foetus becomes viable. It is one of the most widely used methods of birth control in the world. MTP or induced abortion is safe during the first trimester (*i.e.*, up to 12 weeks) of pregnancy. It is riskier in second trimester as by then the foetus is intimately associated with the maternal tissue :

MTP is done to avoid birth of unwanted child conceived due to :

(i) Casual unprotected intercourse.

(ii) Failure of contraceptive during intercourse.

(iii) Pregnancy due to rape.

(iv) In conditions, where continuation of pregnancy may pose threat to the foetus, the mother or both.

MTP was legalised in India by an Act, The Medical Termination of Pregnancy Act, 1971, in order to prevent deaths by unsafe abortions, by untrained persons. This act was put to misuse by performing abortions of female/girl foetus. The sex of the foetus is known by amniocentesis or sonography conducted prior to aborting the foetus. To prevent this malpractice of female foeticide, government enacted a law, Prenatal Diagnostic Techniques (Regulation and Prevention of Misuse) Act, 1994.

Chapter 5. Principles of Inheritance and Variation

1. Which cross is useful in determining the genetic purity of gametes ? Show with an example.

Ans. Test cross is beneficial in determining the genetic purity of gametes. It helps to tell whether the F_1 individual is homozygous or heterozygous.

Let us consider that on crossing tall and dwarf plant, a tall plant is produced in F_1 generation, whose genetic composition is unknown.

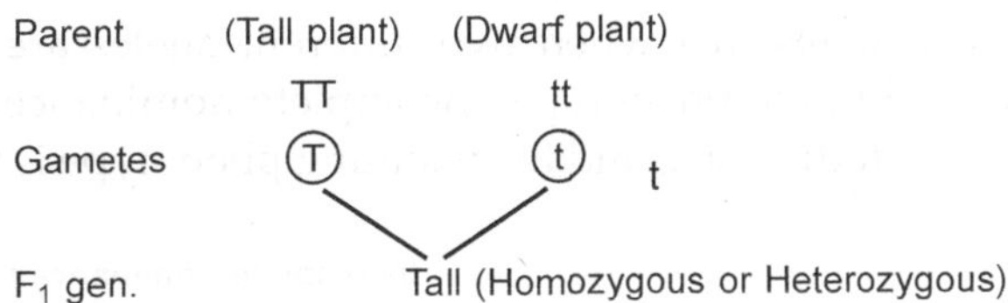

If F_1 individual is considered as heterozygous—Tt

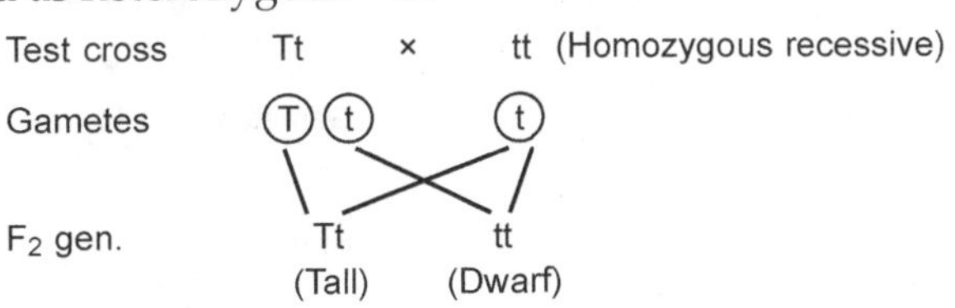

Ratio of tall and dwarf = 1 : 1

The appearance of both tall and dwarf plants in the ratio of 1 : 1 after conducting the test cross indicates that the F_1 offspring is a heterozygote/hybrid.

If F_1 individual is considered as homozygous—TT

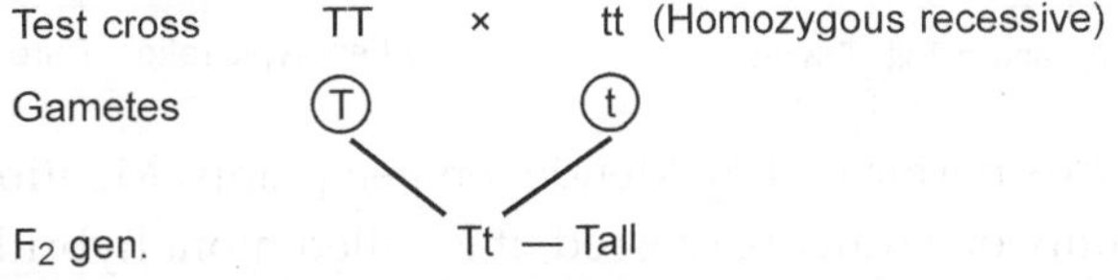

Appearance of only tall plants after conducting the test cross indicates that the F_1 hybrid is a homozygote.

2. What is monohybrid cross ?

Ans. Monohybrid cross is the cross involving only a pair of contrasting traits of a single character. It is conducted by crossing individuals from two parent having pair of contrasting character (pure lines). Each parent exhibits one of the two contrasting traits of a character.

For example : In the pea plant Tall (TT) and dwarf (tt) represent contrasting traits of the character height of the stem. These tall and dwarf are pure lines. The offspring produced on crossing the two pure line parents of contrasting traits represents the F_1, or first filial generation. The F_1 generation, on crossing a homozygous tall plant to a dwarf plant, consisted of hybrid tall plants (Tt).

Parents	→	Tall	×	Dwarf
Genotype	→	TT	×	tt
		↓		↓
Gametes	→	T, T	×	t, t

F_1 Plants

	T	T	
t	Tt Tall	Tt Tall	All are tall
T	Tt Tall	Tt Tall	

On self crossing the F_1 generation

F_1 gen. Tt × Tt

Gametes (T) (t) (T) (t)

F_2 gen.

	T	t
T	TT	Tt
t	Tt	tt

Phenotypic ratio = Tall : Dwarf
3 : 1

Genotypic ratio = TT : Tt : tt
= 1 : 2 : 1

3. What is the concept of dominance ? How does dominance differ from incomplete dominance ?

Ans. The concept of dominance states that when two different alleles are present in a genotype, only the dominant allele is expressed in the phenotype. Incomplete dominance occurs when different alleles are expressed in a heterozygous individual, and the resulting phenotype is intermediate to the phenotypes of the two homozygotes.

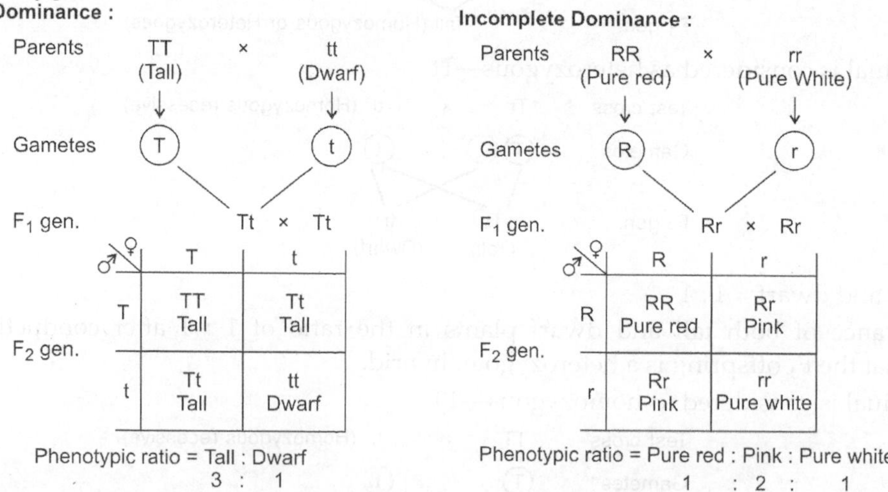

4. Explain monohybrid cross performed by Mendel on pea plants. Mention the law of segregation.

Ans. If parents differing in only one trait are crossed, it is called monohybrid cross.

Example : When a tall plant and dwarf plant were crossed, all the F_1 hybrids were tall.

The dwarf trait did not appear in the F_1 progeny. F_1 plants when self-pollinated, F_2 progeny is obtained.

In F_2 generation, dwarf trait appeared again. The ratio between tall and dwarf plants is 3:1 which is called as monohybrid ratio.

F_2 generation phenotypic ratio = 3 : 1 (3 tall, 1 dwarf)

F_2 generation genotypic ratio = 1 : 2 : 1 (1TT, 2Tt, 1tt)

Parents carrying similar genes (TT/tt) are homozygous plants. Parents carrying dissimilar genes (Tt) are heterozygous plants. Mendel observed that in F_1 progeny tallness is dominant over the dwarf character. The character which is expressed in F_1 generation is called dominant character. The character which is unexpressed is called recessive character. Pair of genes controlling contrasting characters (T/t) is called alleles.

5. In cucumbers, orange fruit colour (R) is dominant over cream fruit colour (r). A cucumber plant homozygous for orange fruits is crossed with a plant homozygous for cream fruits. The F_1 are intercrossed to produce the F_2. Give the genotypes and phenotypes of the parents, the F_1, and the F_2.

Ans. The cross of a homozygous cucumber plant that produces orange fruit (RR) with a homozygous cucumber plant that produces cream fruit (rr) will result in an F_1 generation which is heterozygous for the orange fruit phenotype.

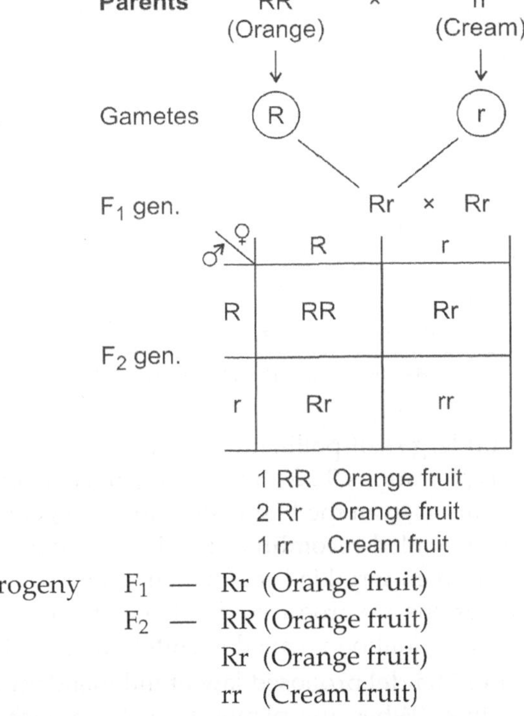

Progeny F_1 — Rr (Orange fruit)

F_2 — RR (Orange fruit)

Rr (Orange fruit)

rr (Cream fruit)

6. What is two factor test cross?

Ans. A two factor test cross is the test cross conducted to know the genotype of the F_1 progeny in a dihybrid cross.

Example : (Yellow round seeds) (Green wrinkled seeds)

Parents YYRR yyrr

F_1 YyRr

Two factor test cross :

Parents YyRr yyrr

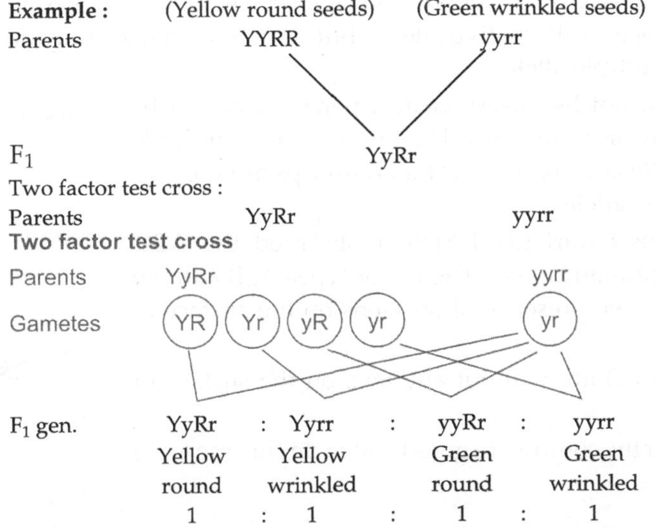

The ratio of 1 : 1 : 1 : 1 indicates that F_1 progeny is heterozygous.

7. Explain dihybrid cross performed by Mendel on pea plants. Mention the law of independent assortment

OR

How do Mendel's Experiments show that traits are inherited independently ?

Ans. A cross involving two parental plants differing in two characters is called dihybrid cross.

Example : When a homozygous pea plant having yellow, round seeds (YYRR) is crossed with anothe: homozygous pea plant having green, wrinkled seeds (yyrr), all the F_1 progeny were found to have yellow round seeds (YrRr).

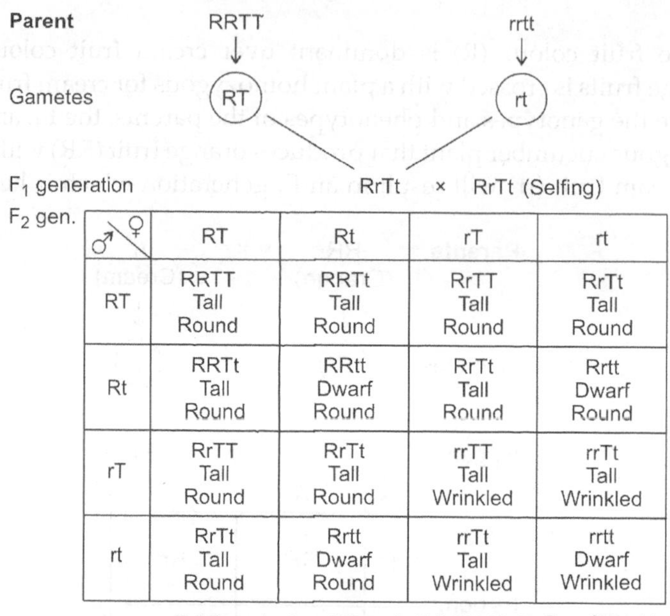

♂ \ ♀	RT	Rt	rT	rt
RT	RRTT Tall Round	RRTt Tall Round	RrTT Tall Round	RrTt Tall Round
Rt	RRTt Tall Round	RRtt Dwarf Round	RrTt Tall Round	Rrtt Dwarf Round
rT	RrTT Tall Round	RrTt Tall Round	rrTT Tall Wrinkled	rrTt Tall Wrinkled
rt	RrTt Tall Round	Rrtt Dwarf Round	rrTt Tall Wrinkled	rrtt Dwarf Wrinkled

Tall Round : Dwarf Round : Tall Wrinkled : Dwarf Wrinkled

9 : 3 : 3 : 1

The F_1 hybrids were allowed to undergo self-pollination, the F_2 progeny showed four kinds of phenotypes in a definite pattern, *i.e.*, phenotypic ratio is 9 : 3 : 3 : 1, and genotypic ratio is 1 : 2 : 2 : 4 : 1 : 2 : 1 : 2 : 1. When F_2 generation progeny is raised from the first generation, the genes combine independently. Such genotypes with new characters are called **recombinants**. The F_2 progeny showed not only the parental combinations (yellow round and green wrinkled) but also the new combinations (yellow wrinkled and green round). These new combinations are produced due to genetic recombination among the factors by independent assortment. This recombination is called **Mendelian recombination.**

Law of independent assortment : Mendel proposed law of independent assortment based on the results of the dihybrid cross. It states that "When the plants differ from each other in two or more pairs of contrasting characters, (or factors) then the inheritance of one pair of factors is independent to that of the other pair of factors".

8. What are multiple alleles ? Describe multiple alleles with the help of ABO blood groups ?

Ans. Generally a gene has two alternative alleles. But sometimes a gene may have more than two forms. These are referred to as multiple alleles.

Multiple alleles can not be observed in a person, but can be observed in a family or population. The inheritance of multiple alleles is called *multiple allelism*. ABO blood groups are the best example for multiple alleles.

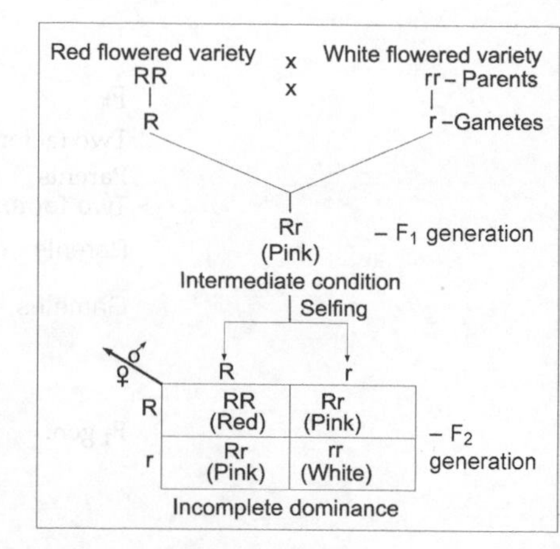

ABO Blood Groups : Karl Land Steiner observed the ABO blood grouping in human beings. The phenotypes A, B and AB are characterised by the presence of antigens on the surface of red blood cells.

(i) The phenotypes O are without antigens on the surface of RBCs.

(ii) These are under the control of a gene located on chromosome nine (9) .

(iii) The four phenotypes of blood groups are due to the antigens A and B.

(iv) Blood group A has antigen A on the RBC and anti B antibody in plasma.

(v) Blood group B has antigen B on RBC and anti A antibody in the plasma.

(vi) Blood group AB has both antigen A and anti gen B on RBCs but no antibodies in the plasma.

(vii) Blood group O is without any antigens on RBCs but with both anti A antibody and anti B antibody in plasma.

Character	Group A	Group B	Group AB	Group O
Antigen on RBC	Antigen A	Antigen B	Antigens A & B	Absent
Antibodies in plasma	Antibody b	Antibody a	Absent	Antibodies a & b

The genetic basis of ABO blood grouping is mainly dependent on the three Alleles I^A, I^B, I^O of the gene I. Isoagglutininis are the antibodies of an individual of a species causing agglutination reactions with antigen of another individual of the same species. The alleles I^A, I^B are mainly responsible for the production of isoagglutinogens.

Out of the three alleles I^A, I^B, I^O, the alleles I^A and I^B are dominant and I^O is recessive, when I^A, I^B are present, co-dominance occurs.

I^A and I^B alleles are dominant to the allele I^O but both are co-dominant to each other. 6 genotypes are possible with three alleles I^AI^A, I^AI^O, I^BI^B, I^BI^O, I^AI^B, I^OI^O

Phenotype	Genotype
A	I^AI^A, I^AI^O
B	I^BI^B, I^BI^O
AB	I^AI^B
O	I^OI^O

9. In cats, blood type A results from an allele (I^A) that is dominant over an allele (I^B) that produces blood type B. There is no O blood type. The blood types of male and female cats that were mated and the blood types of their kittens follow. Give the most likely genotypes for the parents of each litter.

Male parent	Female parent	Kittens
(a) blood type A	blood type B	4 kittens with blood type A, 3 with blood type B
(b) blood type B	blood type B	6 kittens with blood type B
(c) blood type B	blood type A	8 kittens with blood type A

Ans. (i) **Male with blood type A Female with blood type B :** Because the female parent has blood type B, she must have the genotype i^Bi^B. The male parent could be either I^AI^A or I^Ai^B. However, as some of the offspring are kittens with blood type B, the male parent must have contributed an i^B allele to these kittens. Therefore, the male must have the genotype of I^Ai^B.

(ii) **Male with blood type B Female with blood type B :** Because blood type B is caused by the recessive allele i^B, both parents must be homozygous for the recessive allele or i^Bi^B. Each contributes only the i^B allele to the offspring.

(iii) **Male with blood type B Female with blood type A :** Again, the male with type B blood must be i^Bi^B. A female with type A blood could have either the I^AI^A or I^Ai^B genotypes. Because all of her kittens have type A blood, this suggests that she is homozygous for the for I^A allele (I^AI^A) and contributes only the I^A allele to her offspring.

10. How is incomplete dominance is different from co-dominance ?

Ans. Incomplete dominance : This phenomenon occurs when the alleles received by parents are neither dominant nor recessive but blend together and produce a physical trait that is somewhere in between the two traits. For example, the cross between a plant with red flowers and a plant with white flowers may produce a plant with flowers that are neither white no red but are pink which is a colour made by blending red and white together.

Codominance : Codominance is a similar phenomenon where offspring receives neither dominant nor recessive genes. Both alleles get mixed up and show in the offspring. In the above example of red and white flowered plants, the offspring may show up white flowers with freckles of red spots which is codominance in the sense that both genes are showing up but neither is dominant. Example - Inheritence of AB Blood Group.

11. Explain polygenic traits with one example.

Ans. Polygenes are two or more different pairs of non-allelic genes, present on different loci, influencing a single phenotypic character and have cumulative effect. Single phenotypic character governed by more than one pair of genes is called polygenic character.

Example : Skin colour in man.

Skin colour is man depends on the presence of melanin pigment. The amount of melanin depends on three pairs of genes A, B, C. These are present on three different loci and each one is dominant. The effect of all dominant genes is cumulative.

NEGROS AABBCC × ALBIONS aabbcc

↓

MULATTO (INTERMEDIATE) AaBbCc

	ABC	ABc	AbC	Abc	aBC	aBc	abC	abc
ABC	AABBCC 6	AABbCC 5	AaBBCC 5	AABBCc 4	AABbCc 5	AaBbCC 4	AaBBCc 4	AaBbCc 3
ABc	AABbCC 5	AAbbCC 4	AaBbCC 4	AABbCc 3	AAbbCc 4	AabbCC 3	AaBbCc 3	AabbCc 4
AbC	AaBBCC 5	AaBbCC 4	aaBBCC 4	AaBBCc 3	AaBbCc 4	aaBbCC 3	aaBBCc 3	aaBbCc 4
Abc	AABBCc 4	AABbCc 3	AaBBCc 3	AABBcc 2	AABbcc 3	AaBbCc 2	AaBBcc 2	AaBbcc 1
aBC	AABbCc 4	AAbbCc 4	AaBbCc 4	AABbcc 3	AAbbcc 4	AabbCc 3	AaBbcc 3	Aabbcc 2
aBc	AaBbCC 4	AabbCC 3	aaBbCC 3	AaBbCc 2	AabbCc 3	aabbCC 2	aaBbCc 2	aabbCc 1
abC	AaBbCc 4	AaBbCc 3	aaBBCc 3	AaBBcc 2	AaBbcc 3	aaBbCc 2	aaBBcc 2	aaBbcc 1
abc	AaBbCc 3	AabbCc 2	aaBbCc 2	AaBbcc 1	Aabbcc 2	aabbCc 1	aaBbcc 1	aabbcc

Genotype		No. of offspring	Phenotype
6	Dominant Genes	– 1	Pure Black Or Negro
5	Dominant Genes	– 6	Very Dark Brown
4	Dominant Genes	– 15	Dark Brown
3	Dominant Genes	– 20	Mulatto (Intermediate)
2	Dominant Genes	– 15	Fair
1	Dominant Genes	– 6	Very Fair
	No Dominant Genes	– 1	Pure White (Albino)

12. Which similarities are noted by Sutton and Boveri between chromosomes and genes ?

Ans. Sutton and Boveri have noticed following similarities between the chromosomes and genes :

(i) Chromosomes occur in pairs like the alleles of the Mendelian factor.

(ii) Homologous chromosome separate during meiosis like the pair of similar or dissimilar alleles of a Mendelian factor separate at the time of gamete formation.

(iii) During the formation of gametes, different types of (non-homologous) chromosomes segregate independently.

(iv) The paired condition of both chromosome and Mendelian factors is restored during meiosis like Mendelian factors.

(v) Both Mendelian factors and chromosomes maintain their individuality from generation to generation.

13. Discuss various types of linkage.

Ans. T.H. Morgan and his co-workers have found two types of linkage.

(i) **Complete linkage** : Genes located on a chromosome are not altered and are inherited as such from generation to generation without any cross over is called complete linkage. In this type of linkage, genes are closely associated and tend to remain together. Example, male Drosophila and female silk worm (*Bombyx mori*).

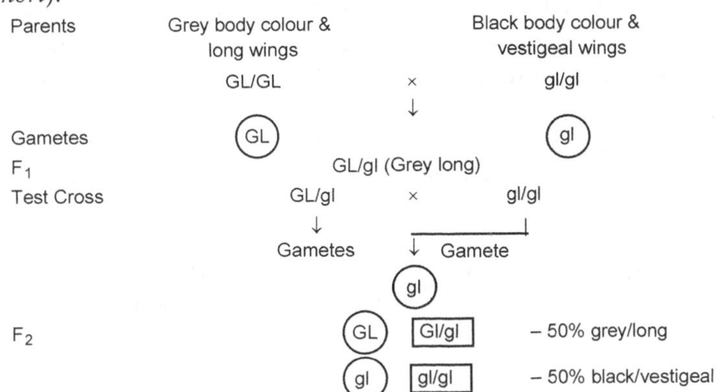

Experiment to show Complete Linkage

100% parental combinations indicated that the gene for grey body colour is completely linked with long wings. In this dihybrid, F_2 phenotypic ratio is 3 : 1 and test cross ratio is 1 : 1 (like a monohybrid).

(ii) **Incomplete linkage** : The linked genes do not always stay together because homologous non-sister chromatids may exchange segments of varying length with one another during meiosis. This is known as crossing over. The linked genes may have chances of separation by crossing over, are called incompletely linked genes and the phenomenon of their inheritance is called incomplete linkage. It produces both parental and recombinant types in variable ratio.

Bateson and Punnett studied *Lathyrus Odoratus* and defined coupling and repulsion of dominant and recessive gene. In cis arrangement or coupling condition the incomplete linkage ratio was 7:1:1:7 (14-parental 2-recombinants). In trans arrangement or repulsion case the ratio was 1: 7: 7: 1 (parental 14, 2-recombinants).

Example : In maize, incomplete linkage was observed by Hutchinson with respect to seed coat colour and seed shape. The results show that parental combination of alleles (CS/CS and cs/cs) appears in about 96% cases. The other two are new combinations (Cs/cs and cS/cs) appear in about 4% cases. Thus in about 4% cases, crossing over has occurred between linked genes. Another example was demonstrated by Morgan etal., while working with *Drosophila*.

Parents   Coloured & full   ×   Colourless & shrunken
CS/CS   cs/cs

Gametes   (CS)   ×   (cs)
↓

F_1 CS/cs
Coloured & full

Test Cross   Cs/cs   ×   cs/cs

Gametes   cs

Result :		
CS	CS/cs	– Coloured & full 48·2%
Cs	Cs/cs	– Coloured & shrunken 1·8%
cS	cS/cs	– Colourless & full 1·8%
cs	cs/cs	– Colourless & shrunken 48·2%

Experiment to show Incomplete Linkage

14. What are some characteristics that make *Drosophila melanogaster* a good model genetic organism ?

Ans. (i) These fruit flies have a relatively short generation time of about 10 days, and produce large numbers of progeny, with females producing 400 to 500 eggs in a 10-day period.

(ii) They are easily and inexpensively cultured in the laboratory, and being small, require little laboratory space.

(iii) They have a complex life cycle and morphology and are large enough that males and females are easily distinguished, and many morphological mutations may be observed with just a hand lens or dissecting microscope.

(iv) *Drosophila melanogaster* also has very large polytene chromosomes in the salivary glands, which have facilitated cytological studies of chromosomes.

15. Describe the experiment conducted by Morgan to explain sex linkage.

Ans. Sex linkage was discovered by T.H. Morgan in fruit fly, *Drosophila melanogaster*.

Morgan observed that white eye colour of *Drosophila* is sex linked character. Normally *Drosophila* shows red eyes. Red eye colour gene is dominant whereas the white eye colour gene is recessive. White eye gene is located on X-chromosome.

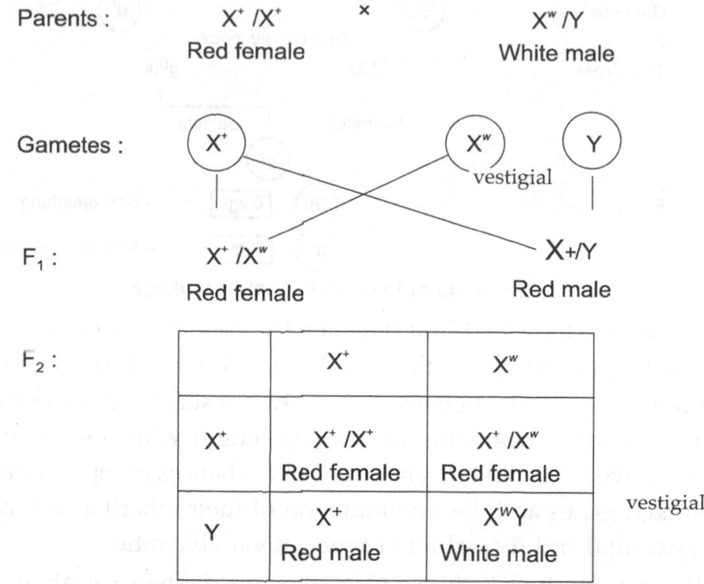

	X+	Xw
X+	X+ /X+ Red female	X+ /Xw Red female
Y	X+ Red male	XwY White male

vestigial

Inheritance of eye colour in Drosophila

When white eyed male is crossed with red eyed female, in the first generation, all the males and females are red-eyed. When first generation red-eye female was crossed to a red-eyed male, in the second generation all the females were red-eyed and 50% of males were white-eyed. This pattern of inheritance is called Criss cross inheritance in which a gene is transmitted from a male parent to a female child and to male grandchild.

In a reciprocal cross, in which a white-eyed female was crossed to red-eyed male, the results were different. The first generation males were white-eyed. Morgan observed that the difference was due to the inheritance of recessive gene located on the X-chromosome. Males always inherit the sex-linked recessive traits from the female parents.

F_1 Cross : $X^w X^w \times X^{w+} Y$

	X^w	X^w
X^{w+}	$X^{w+} X^w$	$X^{w+} X^w$
Y	X^w Y	X^w Y

Ratios :

Female – 100% wild eyes (red)

Male – 100% white eyes

Sex linked inheritance in Drosophila

16. Describe the chromosomal theory of sex determination.

Ans. In most of the animals a pair of chromosomes is responsible for the determination of sex. These two chromosomes are called sex chromosomes or allosomes. The chromosomes other than the sex chromosomes are called Autosomes. If the two sex chromosomes are similar the individual is described as homogametic, if the two sex chromosomes are different or it contains one sex chromosome, individual is described as heterogametic.

Heterogametic sex determination : It is proposed by Correns. It is based on number and nature of sex chromosomes.

(i) **Sex determination based on number of sex chromosomes :** In this method of sex determination chromosomal number is different in male and female.

(a) **XX-XO method of sex determination :** In the insects of *Hemiptera* (bugs) and *Orthoptera* (grasshoppers and cockroaches), females have two X-chromosomes and males have one X-chromosome. Unpaired X-chromosome determines the male sex. Karyotype of female is AAXX and male is AAXO. All the ova formed from females have AX chromosomes. So females are homogametic. But the sperms are of two types . Half of them have AX and other half have "A" only. Thus males are heterogametic. The sex of offspring depends on the fertilizing sperm.

Organism	Hetero-gametic sex	Gametes		Zygotes	
		Sperm	Eggs	Female	Male
Mammals and most insects	Male	X and Y	All X	XX	XY
Grasshoppers	Male	X and O	All X	XX	X
Birds, some reptiles and amphibians	Female	All Z	Z and W	ZW	ZZ
Fumea	Female	All Z	Z and O	ZO	ZZ

(b) **ZO – ZZ method of sex determination :** It is shown by *Fumea* (moth), where the female is heterogametic with one Z chromosome (ZO) and male is homogametic with two Z-chromosomes (ZZ). All the sperms are similar in their karyotype (AZ type only) but ova differ in their karyotype (AZ type and AO type). The sex of the offspring depends on the fertilizing ovum.

(ii) **Sex determination based on differences in sex chromosomes :** Both males and females have the same number of chromosomes.

(a) **XX-XY method of sex determination** : In human beings and insects like *Drosophila*, both females and males have the same number of chromosomes. Females are homogametic with XX chromosomes. They produce ova having X chromosome. All ova are similar. Males are heterogametic with X and Y-chromosomes. They produce two kinds of sperms, one type with X-chromosome and other with Y-chromosome. On fertilisation, the zygotes may have either the XX or XY. The zygote with XX becomes the female and the zygote with XY becomes the male.

The sex of the offspring depends on the fertilising sperm. In the case of human beings, a single gene (SRY) on the Y-chromosome determines maleness.

(b) **ZW-ZZ method of sex determination :** In birds, reptiles, some fishes, butterflies etc., females are heterogametic with ZW chromosomes and males are homogametic with ZZ-chromosomes. All sperms are similar with allosome Z. Ova are of two different kinds. One kind of ova is with allosome Z and other with allosome W. The sex of the offspring depends on the fertilising ovum.

17. Explain in brief about the characters of X-linked recessive traits.

Ans. Characters of X-linked recessive traits :

(i) Males are much more likely to be affected because they need only one copy of the mutant allele to express the phenotype.

(ii) Affected males get the disease from their mothers.

(iii) If a female is homozygous affected, all her sons are affected.

(iv) Daughters of affected male are either carriers or affected.

(v) Affected male pass his X-chromosome to his daughters.

(vi) X – linked recessive disorders are typically passed on from an affected father to 50% of his grandson through carrier daughters.

(vii) The X-linked recessive traits follow criss cross pattern of inheritance. Duchenne muscular dystrophy (DMD), haemophilia and colour blindness are some of the important sex linked recessive traits in the human beings.

18. Describe the sex determination in Drosophila with the help of Genetic balance theory.

Ans. Sex determination in Drosophila by Genetic balance theory : The sex determination method in Drosophila is XY method. The allosomes are X and Y chromosomes. Males and females have the same number of chromosomes *i.e.* 8 of these, six are autosomes and two are allosomes. Female is homogametic AAXX and male is heterogametic AAXY, where A is the single set of autosomes. Calvin Bridges in 1916 showed that Y chromosome has no role in the determination of sex in *Drosophila.* He proposed that both the X chromosomes and autosomes together play a critical role in the determination of sex in *Drosophila.* He studied the offspring resulting from the non-disjunction of X chromosomes during meiosis in females. Non-disjunction is the failure of the paired chromosomes to segregate or separate during the anaphase stage of the first or second meiotic divisions of gametogenesis. The result is the production of abnormal gametes. One of which contains one extra chromosome (AXX) and other contains one chromosome less (A). Fertilisation of such gametes with normal gametes produces aneuploid zygotes (2n+1 or 2n–1).

3A + XXX ♀ triploid X 2A + XY diploid ♂

Functional gametes	gametes: A + X	gametes: A + Y
2A + XX	3A + XXX triploid female	3A + XXY triploid intersex
A + X	2A + XX triploid female	2A + XY diploid male
2A + X	3A + XX triploid intersex	3A + XY supermale
A + XX	2A + XXX super female	2A + XXY diploid female

Results obtained from a cross of a triploid (3A + XXX) _ fly, and a diploid _ (2A + XY) fly in Drosophila

Bridges crossed a triploid female (AAAXXX) to a normal male and observed many combinations of autosomes and sex chromosomes in the offspring. Triploid female produces gametes like 'AAXX', 'AX', 'AAX', and 'AXX'. These gametes fertilise with normal sperms of the male producing different karyotypes. From these karyotypes, Bridges found that the XXY flies were normal females, and the XO flies were sterile males. The presence of Y chromosome in the XXY flies did not cause maleness, and its absence in the XO flies did not produce femaleness. He concluded that Y chromosome in Drosophila lacks male-determining factors, but contains gametic information essential to male fertility as the XO males were sterile. He realised that the factor in determining the sex in Drosophila is the ratio of X chromosomes to the number of haploid sets of autosomes. Normal females have a ratio equal to 1.0. If the ratio is >1.0, females are metafemales, which are infertile. Normal males have a ratio equal to 0.5. When the ratio decreases the males are metamales, which are infertile. If the ratio is between 0.5 and 1.0, the flies are intersexes. This mode of sex determination is explained as the genic balance theory of Bridges. This theory explains that genes for maleness are located on autosomes and for femaleness on X-chromosomes in *Drosophila.*

Balance of X-chromosome and autosomes in Drosophila

Sex		X Chromosomes	Sets of autosomes (A)	Sex index (X/A)
	Super female	3	2	1·5
	tetraploid	4	4	1·0
Normal female	triploid	3	3	1·0
	Diploid	2	2	1·0
	haploid	1	1	1·0
	Intersex	2	3	0·67
	Normal male	1	2	0·50
	Supermale	1	3	0·33

19. What is chromosomal mapping ? Explain the mapping technique.

Ans. The representation in figure of relative position of genes on the chromosomes is known as chromosome map and the process of identifying gene loci is called mapping.

Mapping technique : The assumption of consistent gene order along the chromosomes coupled with the fact that crossing occurs, makes it almost certain that gene loci can be determined in relation to each other. If genes are located in a consistent linear order along a chromosome, then the distance between any two genes and the amount of crossing over should be in direct proportion.

Homologous chromosomes cross over which leads to an exchange of chromosome segments. The percentage of crossing over therefore gives direct measure of the chromosomal crossing over. In turn, it directly measure the distance between the genes involved.

Suppose for examples genes A,B,C,D are linked. By carrying out-breeding phenotypes, the following cross over frequencies are determined.

Cross over between genes	A & B	B & C	C & D	A & C	B & D
Cross over frequency %	8	8·5	18·5	16·5	10

This information allows us to map the locus of each gene. Drawing a line to represent the chromosome, we can assign arbitary position to genes A and B, subject to the stipulation that the distance between them is 8 map units.

Chromosome

8

Genes B and C have 8·5 cross over frequency. Two possibilities clearly exist. The first locus possibility predicts a 16·5% crossover between A and C and second is 0·5% since the data given shows 16·5%, it follows that the first locus chosen for C is correct.

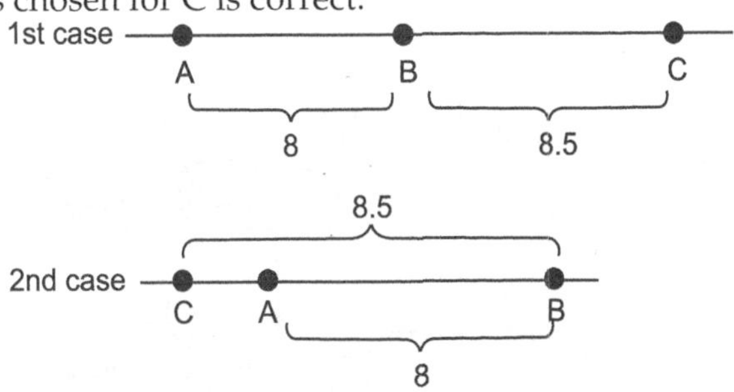

The locus for D must be 18·5% map units. Two loci satisfy this requirement, but only one which places gene D near gene A, is consistent with 10% crossing over frequency given for B and D. Thus, the final chromosome map for these genes are as follows :

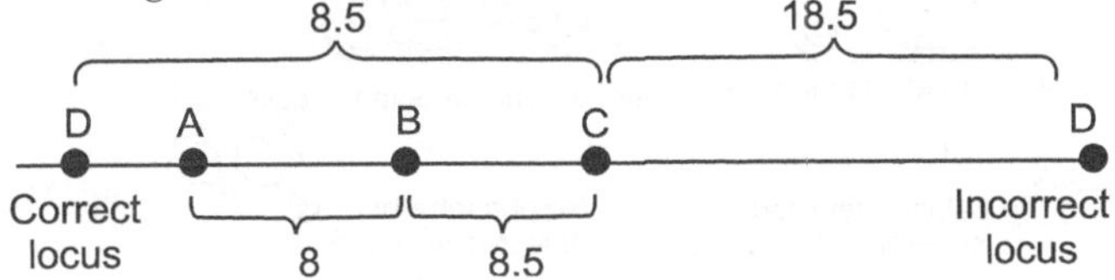

20. What is mutation ? What are its different types ?

Ans. Mutation includes sudden inheritable changes which alter the phenotype of an individual. It is of the following types :

(i) **Spontaneous and induced mutations :** Mutations that occur in nature referred as spontaneous whereas mutations brought about critically are induced mutations.

(ii) **Gene mutation :** It is the alteration in the sequence of nucleotides in nucleic acids or change in the sequence of triplet bases. It can occur by (a) Frame shift mutation, (b) Substitution, (c) Tautomerisation.

(iii) **Point mutation :** If gene mutation arise due to change in single base pair of DNA, it is called point mutation.

(iv) **Chromosomal mutations/aberrations :** Any change in the chromosomal morphology. It is of 4 types : (a) Deficiency or deletion, (b) Duplication, (c) Translocation, (d) Inversion.

(v) **Genomatic mutation :** It refers to numerical changes in chromosome number two types : (a) Aneuploidy, (b) Euploidy.

Chapter 6. Molecular Basis of Inheritance

1. Give an account of Hershey and Chase experiment.

OR

Give an account of Hershey-Chase experiment proving that DNA, not the protein coat of virus is the infecting agent.

Ans. Hershey and Chase conducted experiments on bacteriophage to prove that DNA is the genetic material.

(i) Some bacteriophage viruses were grown on a medium that contained radioactive phosphorus (^{32}P) and some in another medium with radioactive sulphur (^{35}S).

(ii) Viruses grown in the presence of radioactive phosphorus (^{32}P) contained radioactive DNA.

(iii) Similar viruses grown in presence of radioactive sulphur (^{35}S) contained radioactive protein.

(iv) Both the radioactive virus types were allowed to infect E. coil separately.

(v) Soon after infection, the bacterial cells were gently agitated in blender to remove viral coats from the bacteria.

(vi) The dulture was also centrifuged to separate the viral particle from the bacterial cell.

Observations and Conclusions (i) Only radioactive ^{32}P was found to be associated with the bacterial cell, whereas radioactive ^{35}S was only found in surrounding medium and not in the bacterial cell. (ii) This indicates that only DNA and not protein coat entered the bacterial cell. (iii) This proves that DNA is the genetic material which is passed from virus to bacteria and not protein.

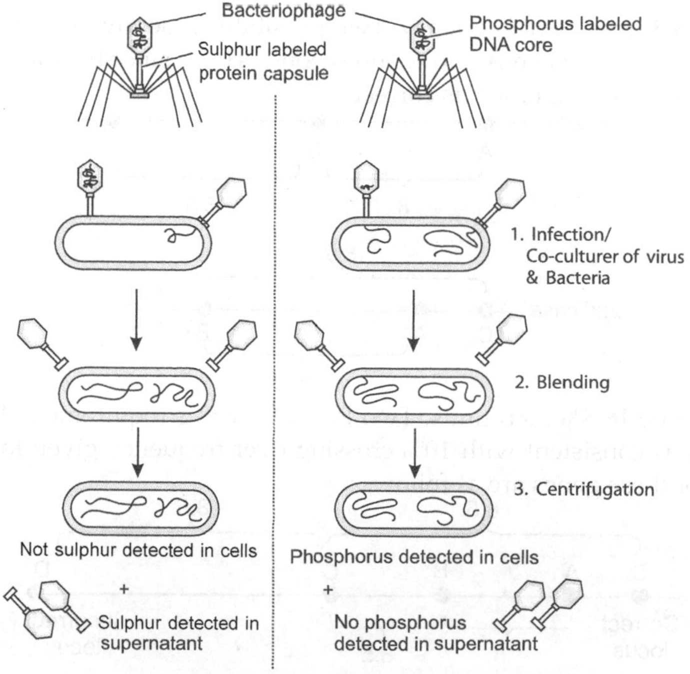

Fig : The Hershey-Chase Experiment

2. Which experiment was conducted by O. T. Avery, Macleod and McCarty ?

OR

Describe Avery, Mcleod and Mc Carty Experiment. State its significance ?*

Ans. When Griffith conducted experiments on transformation, protein was believed to be genetic material. Since transforming material in Griffith's transformation principle was the genetic material, three scientists, Oswal T. Avery, Coin Macleod and Maclyn McCarty (1944) set out to find the chemical nature of the transforming material. Their experiment was :

(i) They killed S-type bacterial strain with heat and separated its components. DNA proteins and carbohydrates. DNA components was subdivided into two-one part with hydrolysing enzyme DNAase and the other without the enzyme.

(ii) They added the four different components to the culture medium supporting R-type bacteria.

(iii) After an interval, the bacteria were analysed.

* **Frequently asked previous years Board Exam Question.**

(iv) There was no change in the three cultures having additions of S-type carbohydrates, S-type proteins and S-type DNA with DNAase.

(v) The fourth culture having S-type DNA without DNAase was found to have some bacteria of S-type. They must have been formed from R-type bacteria with the help of DNA of S-type.

Thus, this experiment of Avery, Macleod and McCarty gave experimental proof that DNA is the genetic material in most living organisms.

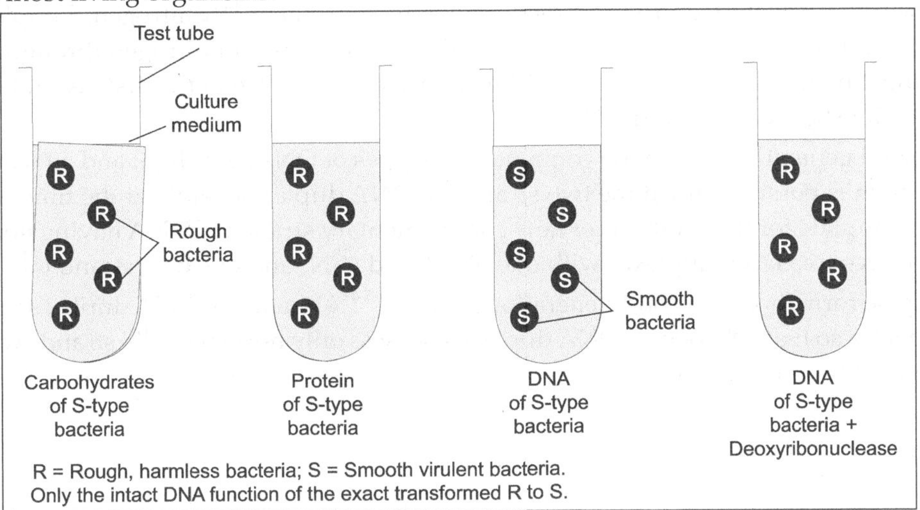

Experiment of Avery and others showing that DNA is the Transforming Agent

3. Write a note on semiconservative mode of replication.

Ans. In the replication process, while the original strands are unwound, new nucleotides are added to form pairs with the bases of the original strand. To begin with, DNA duplex is nicked by the enzyme DNA topoisomerase, allowing part of the molecule to untravel to form a replication fork. Next, the enzyme DNA helicase splits the two strands by breaking the hydrogen bonds. This exposes the bases. Because of the characteristic Y-shape of the replicating DNA, it is often referred to as a "replication fork."

DNA polymerase enzyme then moves along the exposed base sequences, and creates a new complementary strand. DNA polymerase reads the exposed code from the 3′ to the 5′ end and therefore assembles the new strand from the 5′ to the 3′. The DNA strand that is synthesized in the 5′ to 3′ direction is called the leading strand. The opposite strand is the lagging strand, and it is synthesized in the 3′ to 5′ direction. This brings us to the first rule of DNA replication : DNA synthesis only occurs in one direction, from the 5′ to the 3′ end.

Several molecules of DNA polymerase act simultaneously, each assembling a separate section of the new strand of DNA. Each DNA polymerase is lead by an RNA polymerase enzyme, which constructs an RNA primer to guide the action of the DNA polymerase. All Okazaki fragments are subsequently joined together by DNA ligase to form a long continuous DNA strand.

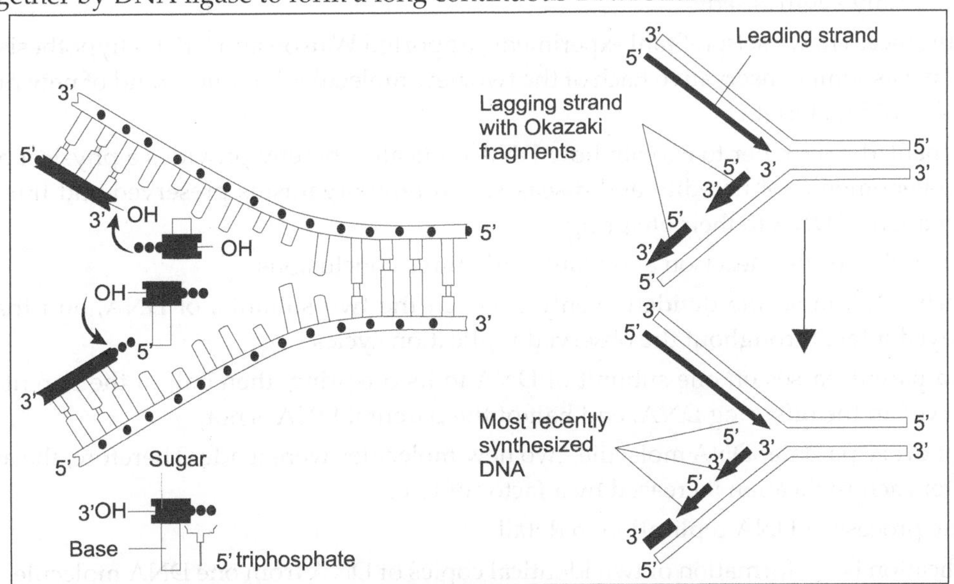

Replication of DNA

4. (i) Give an account of Meselson and Stahl's experiment.*

(ii) What is the significance of Meselson and Stahl's experiment?*

Ans. (i) Semiconservative replication of DNA was proved by the work of Mathew Meselson and Franklin Stahl (1958). They grew *Escherichia coli* for many generations in a medium having a heavy isotope of nitrogen ^{15}N, till the bacterial DNA became completely labelled with heavy isotope. The labelled bacteria were then shifted to a fresh medium having normal or ^{14}N nitrogen. Samples were taken for each generation and the DNA was tested for the heavy isotope of nitrogen through density gradient centrifugation using caesium chloride. They found that the DNA of the first generation was hybrid or intermediate between ^{15}N and ^{14}N.

The second generation of bacteria contained two types of DNA, 50% light and 50% intermediate. This observation is possible only if the two strands of DNA duplex separate at the time of replication and act as a template for the synthesis of new complementary strands of DNA having normal or ^{14}N. This will produce two DNA duplexes with one old strand (^{15}N) and one new strand (^{14}N).

During the formation of second-generation, ^{15}N and ^{14}N strands of DNA duplex separate to function as templates so that 50% of new DNA duplexes possess only normal or ^{14}N strands while another 50% have both ^{15}N and ^{14}N strands.

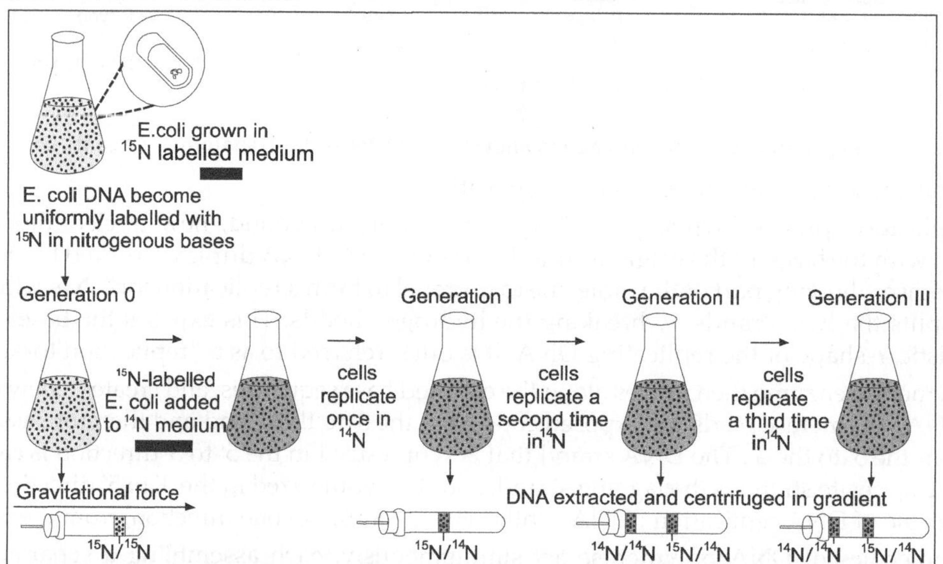

Experiment of Meselson and Stahl demonstrating semi-conservative replication

In this way at each replication, one strand of parent DNA is conserved in the daughter while the second is freshly synthesised.

(ii) **Significance:** The Meselson-Stahl experiment supported Watson and Crick's hypothesis that DNA replication was semi-conservative Each of the two new molecules has one strand of new material and one strand of old material.

This enabled researcher to explain how DNA replicates thereby provides a physical basis for the genetic phenomenon of heredity and diseases, and how organisms preserved and transferred genetic information of DNA to their offspring.

Based on the result, Meselson and Stahl made three conclusions:

1. Each DNA molecule divided evenly between the two subunits of DNA, and that the subunits stayed intact throughout the observed replication cycles
2. If a parent passes on one subunit of DNA to its offspring, then half of the parental DNA is conserved in the offspring DNA, and half of the parental DNA is not.
3. For every parental DNA molecule, two new molecules were made. Therefore, the amount of DNA after each replication increased by a factor of two.

5. Explain the process of DNA replication in detail.

Ans. DNA replication is the formation of two identical copies of DNA from one DNA molecule. It is a complex,

* **Frequently asked previous years Board Exam Question.**

multi-step process that requires a number of enzymes protein factors and ions. It always proceeds in 5′-3′ direction. The main enzyme which takes part in the formation of new strands of DNA by adding deoxyribose nucleotides is **DNA polymerase**. It is called DNA dependent DNA polymerase, because it employs a DNA template for combining of deoxyribonucleotides. Prokaryotes have three types of DNA polymerase. I, II and III. Eukaryotes have 5 types of DNA polymerases α, β, γ, δ, ε.

Mechanism of DNA replication :

(i) **Origin of Replication :** Replication begins at a particular region of DNA which has a particular nucleotide sequence called Autonomic replicating sequence or ARS. It is called Origin of replication, replication or Ori. Prokaryotes have a single origin of replication while eukaryotes have several of them.

(ii) **Activation of deoxyribonucleotides :** There are four types of deoxyribonucleotides.

$$\begin{matrix} dAMP \\ dGMP \\ dTMP + 2H_3PO_4 \\ dCMP \end{matrix} \xrightarrow[\text{energy}]{\text{Phosphorylase}} \begin{matrix} dATP \\ dGTP \\ dTTP \\ dCTP \end{matrix}$$

(iii) **Initiation of replication :** The origin of replication is recognized by origin recognition complex. The unwinding of DNA molecules start at a specific point—the Initiation point. These are identified by specific initiator proteins. Large DNA molecules (in eukaryotes) contain many initiation points.

(iv) **Unwinding of helix :** The enzyme helicase unwinds the DNA helix and unzips the double strands of DNA. This process is ATP dependent. It takes place by the breaking of it bonds. Unwinding of DNA molecule into two strands results in the formation of DNA replication bubbles which lator extend a Y-shaped structure called replication fork.

(v) **Formation of primer strand :** Replication fork exposes two different ends of the two DNA strands, 3′end and 5′end. At the free 3′end of one strand, a small RNA is synthesized with the help of DNA dependent RNA polymerase enzyme or primase. The synthesized RNA is called RNA primer and it functions as 5′ end of the new strand.

(vi) **Elongation of new strand :** (a) Once the primer strand is formed, DNA replication occurs in 5′ → 3′ direction, *i.e.,* during synthesis of new strand, deoxyribonucleotides (dATP, dGTP, dTTP and dCTP) are added only to the free 3′ OH end. Thus, nucleotide, at 3′ end is always most recently added.

$$\begin{matrix} dATP \\ dGTP \\ dCTP \\ dTTP \end{matrix} \xrightarrow{\text{Phosphorylase}} \begin{matrix} dAMP + 2Pi + energy \\ dGMP + 2Pi + energy \\ dCTP + 2Pi + energy \\ dTMP + 2Pi + energy \end{matrix}$$

The release of 2 phosphate molecules (pyrophosphate) and energy help in the formation of H-bonds.

(b) The activity of DNA polymerase III in prokaryotes and polymerase and in eukaryotes along with source of energy (ATP/GTP/TTP) and Mg^{2+} ions is required for the elongation process.

(c) As the DNA replication proceeds on the two parental strands, synthesis of daughter or new strand occurs continuously along the parent 3′ → 5′ strand. It is called leading daughter strand.

(d) Synthesis of another daughter strand along the other parental strand, takes place in the form short pieces because of the opposite arrangement of nucleotides. Where new DNA strand is formed in the form of small segments, a new RNA primer is formation everytime in the region of opening of chain.

(e) This strand is called lagging daughter strand and the short pieces of DNA are called Okazaki fragment named after its discoverer Okazaki (1968).

(f) The RNA primers are removed by polymerase I and enzymes in prokaryotes and eukaryotes respectively.

Since replication is continuous over one strand and discontinuous over the other, it is called semi-continuous/semi-discontinuous. Discontinuous pieces of the lagging strand are joined together by the enzyme DNA ligase (after removing the primers) to form continuous daughter strand.

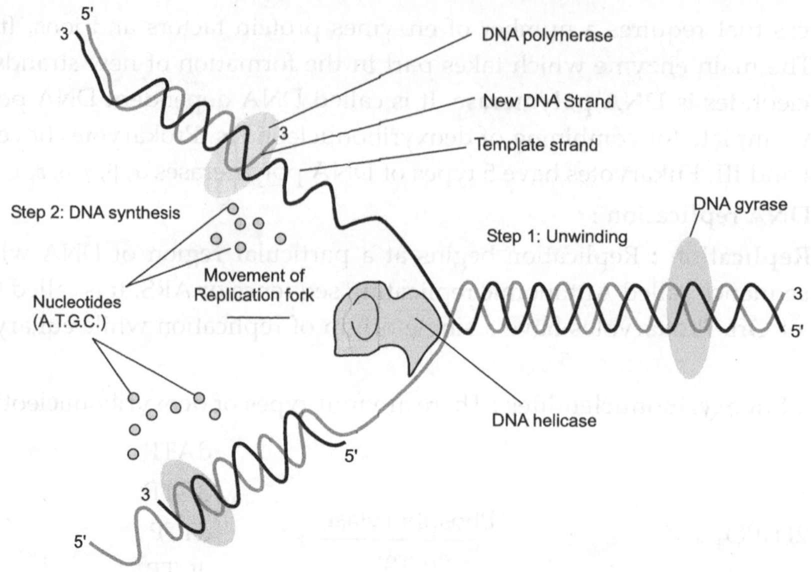

6. Give a brief account of the different steps involved in the transcription of DNA into mRNA.

Ans. Transcription is the process of transfer of genetic information from DNA to mRNA. It is done by formation of RNA over the template of DNA. It creates single stranded RNA which has a coded information similar to the sense or coding strand of DNA (with the exception of U in place of T).

The DNA strand which function as template for RNA synthesis is the template or Antisense strand. It has 3′ → 5′ polarity and transcription proceeds in 5′ → 3′. The segment of DNA that takes part in transcription is called Transcription unit. It has 3 components :

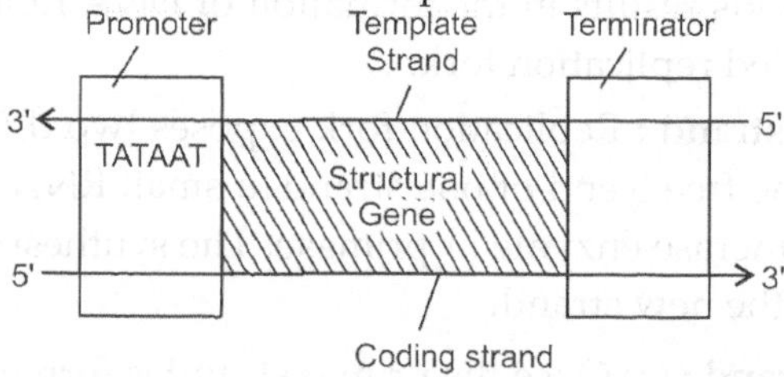

(i) **Promoter region :** It is the proximal area of transcription unit, which provides site for attachment of transcription factors (σ factor) and RNA polymerase. It is present upstream at 5′ end of coding strand. It has special structure; an AT rich area called **TATA box or Pribnow box.**

(ii) **Structural gene :** It is the area of template strand involved in transcription. It is monocistronic in eukaryotes and polycistronic in prokaryotes.

(iii) **Terminator region :** It is the distal end of transcription unit. It is present downstream at 3′ end of coding strand. It provides attachment of termination factor [rho (ρ) factor] for release of RNA polymerase. It has either palindromic sequences or poly A sequences.

The main enzyme taking part in transcription is the RNA polymerase enzyme. In prokaryotes, there is a single RNA polymerase enzyme. It is a 5-subunit complex. It has a sigma (σ) factor for recognizing the start signal of promoter region. Remaining part of RNA polymerase is called core enzyme, which continues the transcription Rho (ρ) factor terminates the process.

In eukaryotes, there are three types of RNA polymerases. RNA polymerase I (for synthesis of rRNA except 5S) RNA polymerase II [for synthesis of hnRNA (mRNA) and many snRNAs)].

RNA polymerase III (for synthesis of tRNAs, 5S rRNA and some snRNAs, scRNAs).

Mechanism of transcription :

(i) **Activation of ribonucleotides :** Four types of ribonucleotides take part in synthesis of RNA over DNA. These are AMP, GMP, UMP & CMP. These are found in nucleoplasm. They are converted to active triphosphates by energy, phosphate and phosphorylase enzyme.

(ii) **DNA template :** The antisense strand is used for transcription, which starts at promote region and ends at terminator region. The separation of template strand here does not require any specific enzymes and compounds, as in DNA replication. Only the RNA polymerase (core enzyme part) travels along the template strand, with the help of protein factor NuSA protein.

(iii) **Initiation :** RNA polymerase enzyme binds to the promoter region. The σ factor recognises the promoter site and initiates transcription, by using triphosphates.

Long Answers

(iv) **Elongation :** The activated ribonucleotides come to lie opposite to the template strand complementary and starts pairing. This releases phosphate radical and energy. The enzyme with the help of energy and Mg^{2+} ions, bonds the adjacent nucleotides and form the RNA chain. The RNA remains attached to the enzyme in a small region. As the enzyme moves along the DNA, the RNA chain elongates, till terminator region is reached.

(v) **Termination :** When enzyme reaches terminator region, the RNA gets detached along with RNA polymerase, by means of ρ (rho factor). This results in termination of transcription. ρ (rho) factor has ATPase activity. After RNA separation the sense and antisense strands of DNA restablish H-bonds.

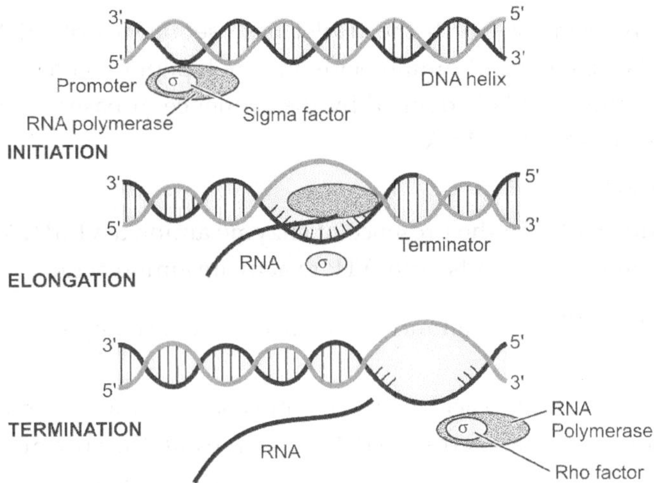

7. Describe post transcriptional processing of RNA in eukaryotes.*

Ans. Messenger RNA production requires synthesis of a pre-mRNA by RNA Pol II and processing of the nascent precursor by 5′ capping, splicing of introns, and 3′ cleavage or poly-adenylation to make mature mRNA.

(i) **Splicing :** In molecular biology, splicing refers to editing of the precursor messenger RNA (pre-mRNA) transcript into a mature messenger RNA (mRNA) wherein introns are removed and exons are joined together i.e., they are ligated. Splicing occurs within the nucleus after transcription.

(ii) **Capping :** While the pre-mRNA is synthesized, a 7-methylguanosine cap is added to the 5′ end of the growing transcript by a 5′-to-5′ phosphate linkage. In molecular biology, the five-prime cap (5′ cap) is a specially altered nucleotide on the 5′ end of some primary transcripts, this process is known as mRNA capping. This is a highly regulated and vital process for forming stable and mature messenger RNA able to undergo translation during protein synthesis. The cap protects the nascent mRNA from degradation. Moreover, initiation factors involved in protein synthesis recognize the cap to help initiate translation by ribosomes.

* Frequently asked previous years Board Exam Question.

(iii) **Tailing :** An enzyme called poly (A) polymerase (PAP) is a part of the protein complex that cleaves the pre-mRNA and it immediately adds a string of approximately 200 A nucleotides, called the poly (A) tail, to the 3′ end of the just-cleaved pre-mRNA. The poly (A) tail protects the mRNA from degradation, aids in the export of the mature mRNA to the cytoplasm, and is involved in binding proteins involved in initiating translation.

8. Give a brief account of the different steps involved in the translation of mRNA into a polypeptide in prokaryotes.

Ans. Translation is the mechanism by which the triplet base sequences of mRNA molecules (codons) are converted into a specific sequence of amino acids in a polypeptide chain. It occurs on ribosomes. The order and sequence of amino acids is defined by the sequence of bases in mRNA. The amino acids are transferred to the ribosomes by the tRNA.

Mechanism of translation :

(i) **Activation of amino acids :** In the presence of enzyme amino acyl tRNA synthetase from DHU loop of tRNA, specific amino acid binds with ATP to activate amino acid.

$$\text{Amino acid + ATP} \xrightarrow[+\,Mg^{2+}]{\text{Enzyme}} \text{amino acids-AMP-Enzyme complex + Pyrophosphate}$$

(ii) **Charging of tRNA :** The amino acid AMP-enzyme complex formed in first step reacts with specific tRNA. Thus, amino acid is transferred to tRNA. As a result, the enzyme & AMP are liberated.

$$\underset{\text{Complex}}{\text{Amino acid AMP Enzyme}} + \underset{\text{(uncharged)}}{\text{tRNA Q}} \rightarrow \underset{\text{(Charged tRNA)}}{\underset{\text{tRNA}}{\text{Amino acid}}} + \text{AMP} + \text{Enzyme}$$

(iii) **Formation of polypeptide chain :** It is completed in three steps :

(a) **Chain initiation :** It requires 3 initiation factors in prokaryotes (IF_3, IF_2, IF_1). The charged tRNA shifts to the ribosomes. The mRNA binds with smaller subunit of ribosomes (30S in prokaryotes). The binding is facilitated by IF_3 in prokaryotes. After this, tRNA attaches to this smaller subunit of mRNA complex. The initiation is brought about by AUG codon which codes for methionine. Non-formylated methionine is attached with tRNA in case of eukaryotes and formylated methionine (t-met) in prokaryotes. Factor responsible for attachment is IF_2 in prokaryotes.

$$\text{30S + mRNA complex +} \xrightarrow[GTP]{IF_2} \text{50S mRNA-tRNA (t-met), complex to this complex)}$$

Then larger subunit of ribosome attaches complex with the help of IF_1, factor in prokaryotes.

(b) **Chain elongation :** The larger subunit of ribosome has 2 sites for binding charged t-RNA.

Amino acyl or A-site (Acceptor site) for initiation and peptidyl site or P-site (Donor site) for peptide chain. A second charged t-RNA molecule alongwith its appropriate amino acid approaches the A-site close to the P-site. Its anticodon binds to the complementary codon of mRNA chain. A peptide bond is formed between the Ist and 2nd amino acids. This formation of peptide bond requires energy and is catalysed by peptidyl transferase. All the newly coming amino acyl t-RNA bind to A-site and thus elongation of chain continues. The elongation factor required for prokaryotes are EF-Tu and EF-To.

(c) **Translocation :** The ribosome shows movement along mRNA towards the 3′ end by a distance of one codon; thus shifting the dipeptdyl tRNA from A site to P site, releasing empty tRNA. This movement is called translocation. It is facilitated by **translocase** and is associated with EF-G in prokaryotes. Energy for transloation is given by GTP.

(d) **Chain termination :** The termination of polypeptide is signalled by one of the three stop codons in the mRNA (UAA, UAG, UGA). A GTP dependent factor known as **Release factor** is associated with termination codon. It is RF_1, RF_2 and RF_3 (prokaryotes).

At the time of termination, the terminal codon immediately follows the last amino acid codon and terminates the polypeptide chain. After termination, tRNA, mRNA are released and ribosome subunits gets dissociated.

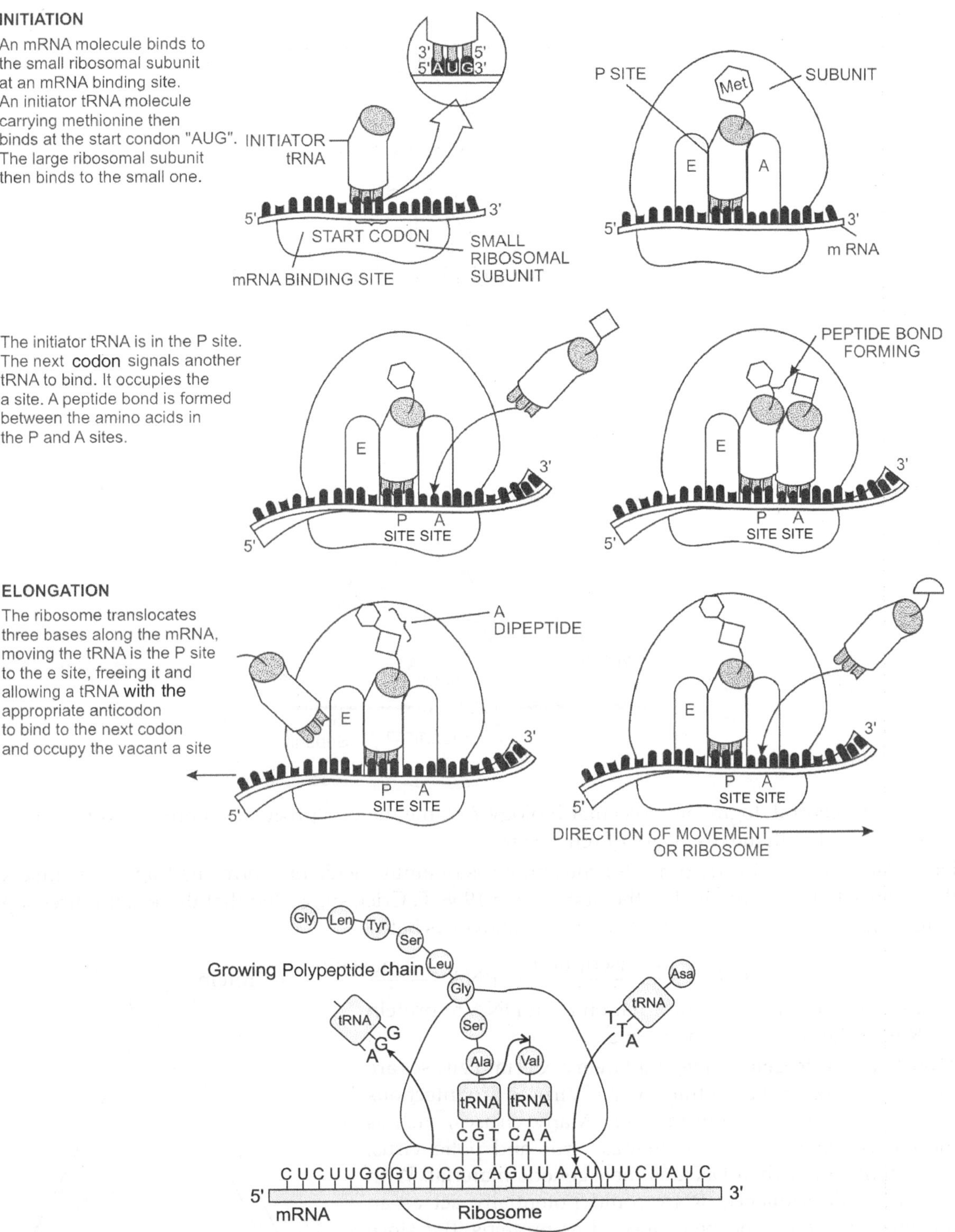

Termination : It is signalled by the stop codon in mRNA with released factor After termination, tRNA, mRNA are released and ribosome unit gets dissociate.

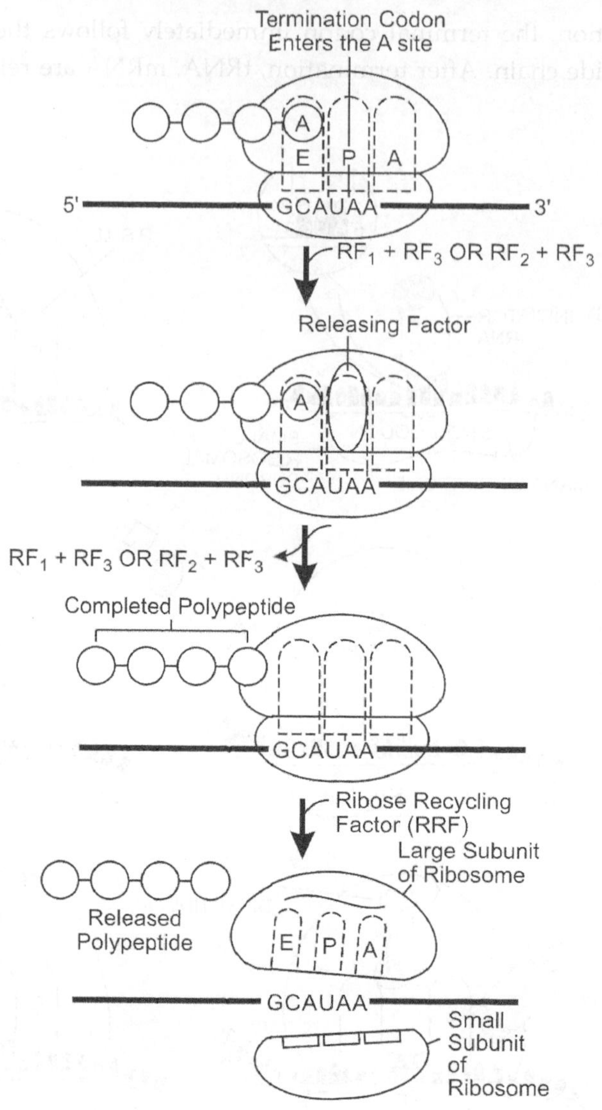

9. What is the central dogma in molecular biology ? Name the two processes represented in it. How has it been modified with the discovery of retrovirus ?

Ans. DNA itself cannot directly order for the synthesis of amino acids but forms its transcripts first which is then translated into protein. For the first time in 1958, F. Crick suggested that there is unidirectional flow of information from DNA to RNA to protein shown as below :

$$\text{DNA} \xrightarrow{\text{Transcription}} \text{RNA} \xrightarrow{\text{Translation}} \text{Protein}$$

This sequential transfer of information from DNA to protein via RNA is known as central dogma.

Modifications to Central Dogma Theory : Certain viruses are rare exceptions to the central dogma. Viruses are infectious particles that reproduce inside cells. Many viruses, such as the tobacco mosaic virus, influenza virus and polio virus, have RNA rather than DNA as their genetic material. With its nucleotide sequence, RNA could potentially act as an information carrier and be expressed as proteins. But since RNA is usually single-stranded, its replication is a problem. The viruses generally solve this problem by transcribing from RNA to RNA, making an RNA strand that is complementary to their genome. This "opposite" strand is then used to make multiple copies of the viral genome by transcription. The Human Immunodeficiency Virus (HIV) and certain rare tumor viruses also have DNA as their genome, but do not

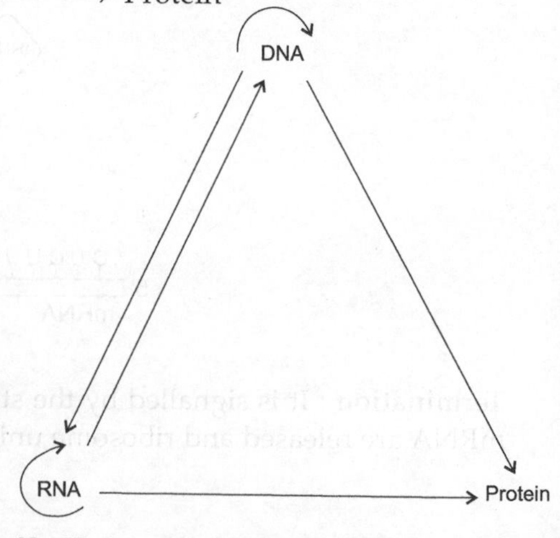

Modified central dogma *i.e.* RNA dependent DNA synthesis

Long Answers

replicate it as RNA-to-RNA. Instead, after infecting a host cell, they make a DNA copy of their genome and use it to make more RNA. This RNA is then used both as genomes for more copies of the virus and as mRNA to produce viral proteins. Synthesis of DNA from RNA is called reverse transcription and such viruses are called retroviruses.

10. Give an account of the Beadle and Tatum's one gene – one enzyme hypothesis.

Ans. The concept that genes have the information to produce enzymes which are proteinaceous in nature was proved by George Beadle and Edward Tatum's in the early 1940s with their work on the fungus *Neurospora crassa*.

This mould can normally grow in a simple minimal medium containing salts and sugar, making all other chemicals such as amino acids, purines, pyrimidines etc., through enzyme catalysed reaction. Beadle and Tatum's used X-rays to induce mutations in the mould and these mutants could not grow on a minimal medium. Each type of mutant required some extra nutrient and as such the nutritional mutants are called auxotrophs.

Beadle and Tatum's obtained different nutritional mutants requiring the amino acids ornithine, citrulline or arginine for growth. In the cell, ornithine is converted into citrulline which is in turn converted into arginine, and each step is catalysed by an enzyme.

$$\text{Precursor} \xrightarrow[\text{Enz. A}]{\text{Gene A}} \text{Ornithine} \xrightarrow[\text{Enz. B}]{\text{Gene B}} \text{Citrulline} \xrightarrow[\text{Enz. C}]{\text{Gene C}} \text{Arginine}$$

Beadle and Tatum's reasoned that the mutant requiring ornithine for growth is defective in the enzyme producing ornithine as a result of a gene mutation. Similarly, a defect in another gene may lead to a defect in the enzyme catalysing the conversion of ornithine to citrulline and therefore this mutant would require citrulline for growth.

Therefore, each gene probably has the information to produce one enzyme and thus, the one-gene-one enzyme hypothesis was given.

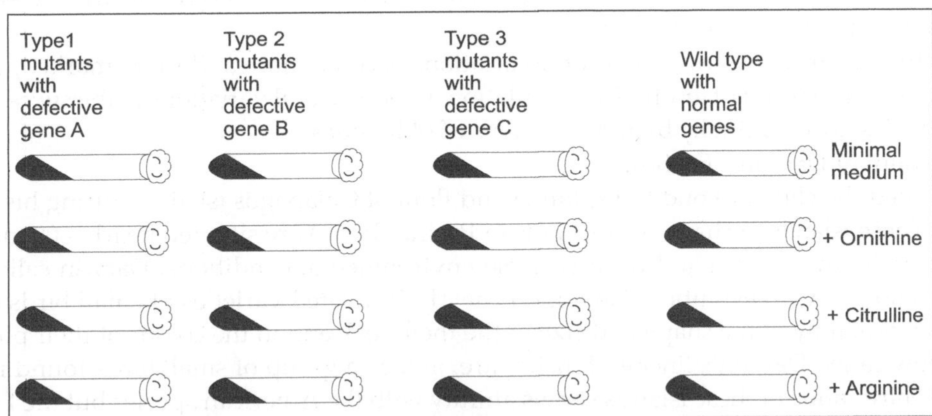

Beadle and Tatum's Experiment on *Neurospora crassa*

In the above figure, Type 1 mutants have defective gene A and lack enzyme A, hence need ornithine, citrulline or arginine for growth. Type 2 mutants have defective gene B and lack enzyme B, hence need cirtulline or arginine for growth. Type 3 mutants have defective gene C and lack enzyme C, hence need arginine for growth. Wild type grows in minimal medium and does not need any additional material for growth.

Chapter 7. Evolution

1. How does palaeontological evidence support the theory of organic evolution ? Explain with an example.

Ans. The study of fossils is called palaeontology and it reveals the existence of life in the past and illustrates the course of evolution of plants and animals. The sedimentary rocks with fossils can be compared with the leaves of a book with evolutionary history written in the language of fossils. The fossils provides direct evidence of evolution, *e.g.*, *Archaeopteryx lithographica* is a connecting link between reptiles and birds. Its fossil was obtained by Andreas Wagner in 1861 from limestone rocks of solenhofen in Bavaria. It possess teeth in jaws, fingers with claws, long tail with free caudal vertebrae and sternum without keel like reptiles and feathers, wings, beak and V-shaped furcula like birds. Hence, it is an strong evidence of reptilian origin of birds.

2. Explain the Lamarck theory of Evolution.

Ans. Lamarck theory of Evolution : Jean Baptist Lamarck, a French naturalist published his book 'Philosophic Zoologique' in 1809 giving the first modern theory of evolution. Lamarck theory is called the Theory of Inheritance of Acquired Characters, through the use and disuse of organs. It is also termed as Lamarckism. Lamarck believed that organisms possess a natural ability to adapt themselves advantageously to their environment, so that their organs and functions get modified. The inheritance of acquired characters caused variations in species. This result in evolution as the effects are cumulative over period of time. Plants growing in shades have large leaves while plants growing in dark fail to develop chlorophyll and have weak, long and thin stems.

3. Explain the Darwin's theory of natural selection.

Ans. Darwin's theory of natural selection : Charles Darwin, a naturalist proposed a theory to explain the process of evolution. The main points of Darwin's Theory are as follows :

(i) **Rapid multiplication and over crowding :** The animal and plants have a great power of reproduction. They produce large number of eggs or seeds respectively; thus more offspring can actually survive.

(ii) **Limited food and space :** The space and food supply are limited in the universe. Every organism requires a suitable space to live and proper food to eat.

(iii) **Struggle for existence :** Due to rapid multiplication and limited food and space, there will be tough interspecific and intraspecific competition. In other words, there will be seen struggle for existence among individuals.

(iv) **Variation :** Variations are helpful in struggle for existence and the organisms possessing useful variations have a better chance of living than the other one which do not possess them.

(v) **Survival of fittest :** The individual possessing useful variations survive whereas the others are eliminated. Nature selects the best suited and discards the others.

(vi) **Inheritance of useful variations :** The useful variations possessed by the fittest survivors are transmitted to the next generation.

(vii) **Formation of new species :** The useful variations become more and more marked, generation after generation. As a result of gradual accumulation of these useful variations in the successive generation new species arise gradually by modifications of older ones.

4. Give an account of Darwin's finches.

Ans. Darwin studied the climatic conditions, fauna and flora of Galapagos islands, during his voyage around the world. These islands (a chain of 14 islands in Pacific Ocean) are situated nearly 600 miles away of the West coast of South America and have variable environmental conditions. Darwin called these islands as a living laboratory of evolution. He noticed nearly 20 related varieties of small birds in these islands which differed mainly in the shape and size of the their beak and in the colour of their plummage. These birds are now called Darwin's finches. Finches are, in fact, a group of small birds found in these islands. The general bird fauna of these islands shows affinity with the American species but the Darwin's finches differ considerably from American finches. However, a related species of Darwin's finches was reported from South American mainland. Therefore, Darwin concluded that the American mainland species was the original one from which different forms migrated to the different islands of Galapagos and adapted to the different environmental conditions of these islands. These adapted forms are considered to be the new species. Thus, Darwin's finches provide a beautiful example of adaptive radiation supporting the concept of evolution.

5. What is variation and explain about various sources of variation ?

Ans. Variation refers to the characteristics in an individual different from the others of the same kind. Variation is the first fundamental factor for evolution. However, only heritable variation can take part in the evolution of a new species. Sources of evolution are :

Mutation : A mutation is a heritable change in genetic material. Different forms of the same gene resulting from mutations are known as alleles. Alleles are the forms of a gene located at the site or locus of the chromosome and are the basis of heritable variation. Over the course of evolution, many mutations have occurred giving rise to considerable genetic allelic variation within species, and the process of meiosis ensures random distribution of particular parental alleles to offspring.

Recombination : It is the process by which a molecule of DNA is broken and joined to one another. It is one of the main sources of genetic diversity in sexual organisms. During recombination, the crossing over of genetic material between corresponding homologous chromosomes produces daughter cells with a different allelic composition from that of the parent cells.

Migration : Migration is an important source of new genetic variation; when individuals arrive from distant parts they will often have different genotypes from the local population; they thus provide new genetic variation.

Both recombination and migration work with existing allelic variation; they put existing variation into new genetic, or geographic, combinations. Recombination between identical chromosomes produces the same identical chromosomes over and over again.

Genetic Drift : Genetic drift is changes in allele frequencies of a gene pool due to chance or random events. This can occur in large or small populations. Genetic drift causes gene pools of two isolated populations to become dissimilar as some alleles are lost and other are fixed. Genetic drift acts faster and has more drastic results in smaller populations. This effect is particularly important in rare and endangered species. Genetic drift can contribute to speciation. For example, a small isolated population may diverge from the larger population through genetic drift.

Gene flow : Gene flow moves alleles among populations through interbreeding as well as by migration of breeding individuals. Gene flow increases variation within a population by introducing new alleles produced in another population. Continued gene flow tends to decrease the diversity among populations, causing gene pools to become similar. Reduction or restriction of gene flow between populations is essential for the development of new species.

6. Explain Hugo de Vries mutation theory.

Ans. According to his theory, new species arise suddenly showing abrupt deviations in characters from the normal forms. Thus, evolution is not a slow and gradual process but a sudden discontinuous and jerky process. The mutation theory differs from Darwin's theory in the nature of the variations, whereas continuous variations are the raw materials for the formation of new species in Darwin's theory of natural selection; but in De Vries theory, discontinuous variation or mutations are the raw materials.

7. Differentiate between aneuploidy and euploidy, including the different conditions of variations.

Ans. Aneuploidy and euploidy are different types of genomatic mutations.

(i) **Aneuploidy :** The abnormal conditions where one or more chromosomes of a normal set of chromosomes are missing or present in more than their usual number of copies. This type of variation involves one or few chromosomes but not the entire set.

(a) **Monosomy :** The loss of a single chromosome; individuals are called monosomics and their chromosomal composition is 2N-1.

(b) **Nullisomy :** The loss of both pairs of homologous chromosomes; individuals are called nullisomics and their chromosomal composition is 2N-2.

(c) **Trisomy:** the gain of an extra copy of a chromosome; individuals are called trisomics and their chromosomal composition is 2N+1. For example aneuploidy in humans is syndrome in which the individual has one extra copy of the chromosome 21 or trisomy-21.

(d) **Tetrasomy :** The gain of an extra pair of homologous chromosomes; individuals are called tetrasomics and their chromosomal composition is 2N+2.

(ii) **Euploidy :** Variation that involves entire sets of chromosomes are known as euploidy. Some major euploids are as follow:

(a) **Monoploidy and haploidy :** Ploidy refers to the number of sets of chromosomes in a biological cell. A distinction should be made between monoploidy and haploidy. Monoploids have a single basic set of chromosomes *e.g.,* 2n = x = 7 in barley or 2n = x = 10 in corn. Haploids, on the other hand represent individuals having half the somatic chromosome number found in normal individual. These latter kind of haploids obtained from polyploids are often called polyhaploids in order to distinguish them from monoploids. Haploids can be artificially produced by any one of the following methods : (i) X-rays treatment, (ii) delayed pollination, (iii) temperature shocks, (iv) colchicine treatment, (v) distant (interspecific or intergeneric) hybridisation, (vi) anther or pollen culture.

(b) **Polyploidy :** Polyploidy is caused by the addition of one or more complete chromosome sets to the normal diploid complement. They are of two types :

(1) **Autopolyploid :** Those polyploids that has an additional set of chromosomes that are identical to parental species. An autotriploid would have the chromosomal composition of AAA and an autotetraploid would be AAAA. Both of these triploids are in comparison to the diploid with the chromosomal composition of AA.

(2) **Allopolyploids :** Polyploidy may also result from doubling of chromosome number in a F hybrid which is derived from two distinctly different species. This will bring two differen sets of chromosomes in F_1 hybrid. The number of chromosomes in each of these two sets may differ. Let A represent a set of chromosomes (genome) in species X, and let B represen another genome in a species Y. The F_1 will then have one A genome and another B genome The doubling of chromosomes in this F_1 hybrid (AB) will give rise to a tetraploid with two A and two B genomes. Such a polyploid is called an allopolyploid or amphidiploid.

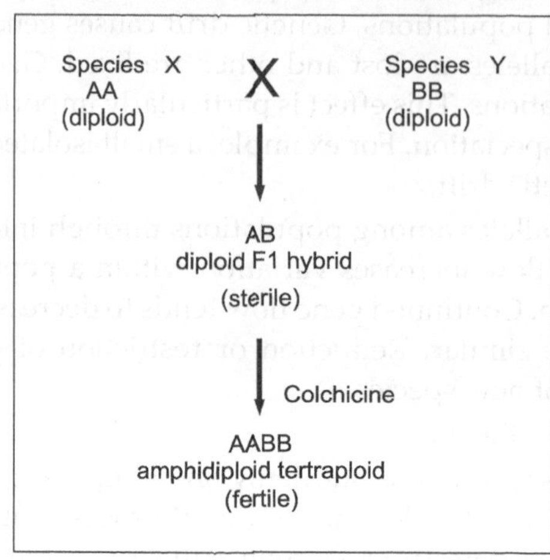

Derivation of a tetraploid amphidiploid from two diploid species.

8. (i) Describe the Oparin Haldane Theory of origin of life.*

(ii) The tadpole larva of amphibians resembles fishes. How does this observation support evolution?*

Ans. (i) Oparin-Haldane theory is also called a chemical theory or naturalistic theory. They believed that the origin of life was a gradual process of changes thought to have occurred about 4000-3600 million years ago. According to Oparin Haldane theory of evolution, life originated on early earth through physicochemical processes of atoms combining to form molecules, molecules react to produce inorganic and organic compounds. Organic compounds interacted to produce different forms of macromolecules which organised to form the first living system or cells. Thus, according to the theory 'life' originated on earth spontaneously from non-living matter. First inorganic compounds and then organic compounds were formed in accordance with ever-changing environmental conditions. This is referred to as chemical evolution which cannot occur under present environmental conditions upon the earth. The conditions suitable for the origin of life existed only upon primitive earth.

1. Free atoms	(Hydrogen, Carbon, Nitrogen)
↓	
2. Inorganic molecules	(H_2, H_2O, CH_4, NH_3, CO_2, HCN)
↓	
3. Simple organic molecules	(Sugars, Fatty acids, Glycerol, Amino Acids, Nitrogen bases *i.e.*, Purines, Pyrimidines)
↓	
4. Complex organic molecules	(Polysaccharides, Proteins, Lipids, Nucleotides, Nucleic acids)
↓	
5. Coacervates	(Cluster of complex organic molecules bounded by fatty acid membrane, division by budding)
↓	
6. Eobionts	(Nucleic acid controlled coacervates)
↓	
7. Prokaryotes	(Without organised nucleus)
↓	
8. Eukaryotes	(With well organised nucleus)

(ii) Haeckel formulated his theory as "Ontogeny recapitulates phylogeny", also known as the recapitulation theory. Ontogeny is the growth (size change) and development (structure change) of

* **Frequently asked previous years Board Exam Question.**

an individual organism; phylogeny is the evolutionary history of a species. Haeckel claimed that the development of advanced species passes through stages represented by adult organisms of more primitive species.

The resemblance of tadpole larva of amphibians with fishes shows that the amphibians might have evolved from fishes. For example, cold blooded animals, respiration by gills early in life, laying eggs in water, etc.

However, with due course of their evolution, amphibians divert from fishes with more adaptation towards land habitat, such as presence of lungs, 3-chamberd heart etc. Thus, amphibians are evolved from fishes.

Chapter 8. Human Health and Diseases

1. Mention the factors influencing infections caused by diseases.

Ans. Factors influencing infections caused by diseases are as follows :

(i) **The virulence of the organism :** Virulence is pathogen's ability to cause damage to the host. It is refer as the degree of damage caused by microbe. Similarly, pathogenicity of an organism is determined by its virulence factor. This is why some strains of influenza virus are easily spread and have serious consequences, while others are spread less readily and have a mild effect.

(ii) **The number of invading organisms :** A person whose body is invaded by a large number of organisms is of course more likely to be infected than one invaded by only a few. All of us come in contact with the germs of tuberculosis, yet if we live with a person having an active case of the disease we are more likely to get the infection because a greater number of bacteria will enter our bodies through this close association.

(iii) **The resistance of the body to the disease :** We all know that the chance of throwing off a threatened infection varies with the body condition, when we are properly nourished, have had sufficient sleep and exercise, and are not overtired by work or exposure to extremes of weather or nervous exhaustion, we are in a position to counteract microbe invaders successfully. The cold virus is prevalent during the cooler months of the year and we are quite likely to be harbouring it in our nasal passages during this time, but our body resistance is able to keep it under control. But if, for whatever reason, we lower our resistance, the virus multiplies and we catch cold. Apart from strict avoidance of organisms that cause infection this is the only factor influencing infection over which we have any control.

(iv) **Immunity :** Whether a person exposed to a disease takes the infection depends on the degree of his immunity. If he has recently had the disease or has been vaccinated or inoculated against it, he will have a high degree of immunity. Heredity also plays an important part. Some people are resistant to certain diseases simply because they inherit a high degree of natural immunity. Age also is very important in this regard. Adults are highly immune to childrens' diseases, yet may suffer greatly if they contact them; young people between the age of 17 and 25 are most susceptible to tuberculosis; elderly people seem susceptible to pneumonia.

2. How does the human body protect itself from infection ?

Ans. Pathogen produces disease by invasion followed by toxin production. Human body protects itself from infection of pathogen by the following way :

Skin forms the first line of defence of the body against pathogens. It's tough outer horny layer, formed of dead cells, prevents entry of pathogens. Mucus membrane lining the respiratory tract, urogenital tract, digestive tract and throat also prevents entry of pathogens.

Bacteria entering the respiratory tract through nostrils are filtered out by fine hairs along with the dust particles. Remaining ones are trapped by mucus which is expelled out due to movement of cilia.

The skin and mucus membrane secretes some chemicals which inhibit the growth of pathogens *e.g.* :

(i) Oil and sweat secreted by the sebaceous and sweat glands are acidic. This checks the growth of micro-organisms.

(ii) Some friendly bacteria living on the skin release acids which check the growth of micro-organisms.

(iii) Lysozyme present in sweat, saliva and tears destroys many micro-organisms.

(iv) HCl, secreted by the gastric glands, kills many micro-organisms.

(v) Bile of liver prevents the growth of bacteria in intestine.

(vi) Lysozyme of nasal secretions inactivates certain bacteria and viruses.

(vii) Certain bacteria living in the human female vagina produces lactic acid from glycogen. This kill the foreign bacteria.

The body also defends against infection by increasing the number of certain types of whith blood cells (neutrophils and monocytes), which engulf and destroy invading microorganisms. The increase can occur within several hours, largely because white blood cells are released from the bone marrow, where they are made. The number of neutrophils increases first. If an infection persists, the number of monocytes increases. The blood carries white blood cells to sites of infection.

When an infection develops, the immune system also responds by producing several substances and agents that are designed to attack the specific invading microorganisms

- Killer T cells (a type of white blood cell) that can recognize and kill the invading microorganism
- Antibodies that target the specific invading microorganism

Antibodies attach to and immobilize microorganisms. They kill them outright or help neutrophils target and kill them.

3. Give an account of Helminths diseases.

Ans. Helminths causes numerous diseases such as Taeniasis, Ascariasis, Filariasis.

Taeniasis : Causative agent *Taenia solium.* It is the parasite of human intenstine.

(i) **Mode of infection :** Eating pork of an infected pig.

(ii) **Signs and symptoms :** (a) abdominal pain, (b) Weakness, (c) Loss of weight, (d) diminished appetite, (e) chronic indigestion.

(iii) **Prevention and control :** (a) Well cooked pork eaten, (b) Disposal of infected sewage.

(iv) **Treatment :** Administering drugs to kill worms.

Ascariasis : Causative agent *Ascaris lumbricoides.* It is an infection of the intestinal tract.

(i) **Mode of infections :** Contaminated food and drinks.

(ii) **Signs and symptoms :** (a) diarrhoea, (b) vomiting, (c) slight fever.

(iii) **Prevention and control :** (a) Checking contaminated food and drinks (b) From anti-ascaris medicines.

(iv) **Treatment :** Drugs like piperazine, mebendazole, etc.

Filariasis : Causative agent of this disease is *Wuchereria bancrofti.*

(i) **Mode of infection :** Biting of infected mosquito.

(ii) **Signs and symptoms :** (a) Inflammation of lymph glands, (b) Enormous swelling of affected part such as arms, legs, etc.

(iii) **Prevention and control :** (a) Vector control, (b) Chemotherapy.

(iv) **Treatment :** A drug DEC (diethyl carbamazine).

4. Write an account on blood transfusion in man.

Ans. In cases of serious loss of blood due to accident, disease, haemorrhage or in case of severe anaemia, shock and depression, infusion of blood into the patient or blood transfusion may be the only means to save life. It helps to remove the threat and restore the normal vital activities of the body. A person who donates blood is called the donor and who receives it is called the recipient. Before blood transfusion it is made sure that the bloods of these two individuals is compatible and does not agglutinate on mixing. For this the two blood samples are diluted with saliva and mixed. If the corpuscles clump together than the donor's blood is unsuitable. The reaction between the antigen on the donor's red blood cells and the antibody in the recipients plasma is considered. The donor's RBC's must not carry the antigen for which the recipient has a corresponding antibody. Person having blood group O are universal donors as their blood could be transfused to any blood group without any effect. This is because there are no antigens in blood of O group. Persons belonging to AB blood group are called universal recipients as they can receive blood from any other blood group person. This is because these individuals do not have antibodies in their plasma. Blood group A can receive blood only from group O and A and not form AB or B group because its plasma contains antibody b which cannot co-exist with B or AB. Group B can only receive blood from group B or O as it contains antibody 'a' which cannot co-exist with blood group A or AB which have antigen 'A' on the RBCs.

Blood group Blood tranfusions

Blood group	Antigen on RBC	Antibody in plasma	Donation	Receiving
A	Antigen A	Anti-A	A, AB	A, O
B	Antigen B	Anti-B	B, AB	B, O
AB	Antigen AB	Nil	AB	A, B, AB, O
O	Nil	Anti-A, Anti-B	A, B, AB, O	O

Rh factor is also taken into consideration before blood transfusion. Blood of Rh positive person can be given to Rh negative person as there is normally no antibody for Rh factor. But if Rh negative blood is transfuse into Rh positive blood, his body will produce anti Rh antibodies which would destroy the Rh positive RBCs.

5. Give a brief account of Rh Blood group incompatibility.

Ans. Rh factor is an antigen-protein present on the surface of human RBC's. It is found in the RBC's of about 85 per cent people and they are called Rh positive. The remaining 15 per cent who lack the Rh factor are called Rh negative. Rh factor is a dominant character controlled by a pair of gene allele (R and r). Hence, Rh positive people are RR or Rr. The presence or absence of Rh factor is inherited and is important in blood transfusions and pregnancy.

(i) **Incompatibility in blood transfusion :** In blood transfusion, Rh –ve blood can be given safely to a Rh +ve individual. When Rh +ve blood is transfused into Rh –ve individual the recipient develops anti Rh antibodies in his blood. Usually no complication develops after one transfusion but if more Rh +ve blood is transfused, the antibodies formed will destroy the RBCs of the +ve blood. To avoid this, Rh factor is determined before such transfusions.

(ii) **Incompatibility during pregnancy :** There is no abnormality when both the mother and father are Rh +ve or Rh –ve or mother is Rh +ve and father is Rh –ve. Complications arise when the Rh –ve mother bears an Rh +ve child. In such a case the mother may pick up some of the baby's RBC's and antibodies will develop in her blood. There is usually no problem with the first pregnancy but in subsequent pregnancies, accumulated antibodies from the maternal blood may enter the blood stream of the unborn child and cause agglutination of the foetal blood cells. This abnormality is called erythroblastosis foetalis. It may cause death of the child before birth or neonatal death due to massive destruction of RBC's and production of excess bilirubin which damages the child's brain.

Chapter 9. Strategies for Enhancement in Food Production

1. What is "Biofortification"? Write its importance. Mention the contribution of Indian Agricultural Research Institute towards it with the help of an example.

Ans. Biofortification – It is the method for developing crops with higher levels of vitamins, minerals, protein and healthier fat to improve public health. The objective of biofortification is to improve:

(i) Protein content and quality
(ii) Oil content and quality
(iii) Vitamin content
(iv) Micronutrient and mineral content

The Indian Agricultural Research Institute has released several vegetable crops that are rich in vitamin and minerals. For example vitamin A enriched bitter gourd.

2. (i) What is protoplast ?
(ii) Name the two enzyme used in producing protoplasts.
(iii) What is explants ?
(iv) Mention the usefulness of somatic hybridisation.

Ans. (i) The plant cell without cell wall is called protoplast.
(ii) Pectinase and cellulose.
(iii) The part of the plant from which a new plant is generated by tissue culture is called explants.
(iv) **Uses :**
(a) Somaclonal variation can be created by this method.
(b) Healthy plant can be recovered from diseased plant.

3. Describe the tissue culture technique in plants.*

Ans. Plant tissue culture is an advanced technique of plant breeding that developed around the 1950s. Plant tissue culture makes use of parts of a plant to generate multiple copies of the plant in a very short duration. The technique is based on the property of totipotency of plant cell which means that any cell from any part of the plant can be used to generate a whole new plant.

Types of Plant tissue culture :

(i) Seed Culture (ii) Embryo Culture (iii) Callus Culture
(iv) Organ Culture (v) Protoplast Culture (vi) Anther Culture

An example of use of plant tissue culture in the lab :

Step 1 : A new gene of interest is introduced into tissue of a plant. Thus altering its genotype.

Step 2 : The transformed plant cells are then isolated and placed into sterile nutrient agar.

Step 3 : Each cell grows into a cluster of cells called a callus.

Step 4 : Calluses are separated and a plant hormone called cytokinin is added which stimulates growth of small leaf shoots.

Step 5 : Shoots are separated and the addition of another plant hormone called auxin stimulates the growth of roots.

Step 6 : New plants are transferred to soil and grown to observe how the alteration of its genotype has affected its phenotype under extreme environmental conditions such as high salt, increased CO_2 and drought.

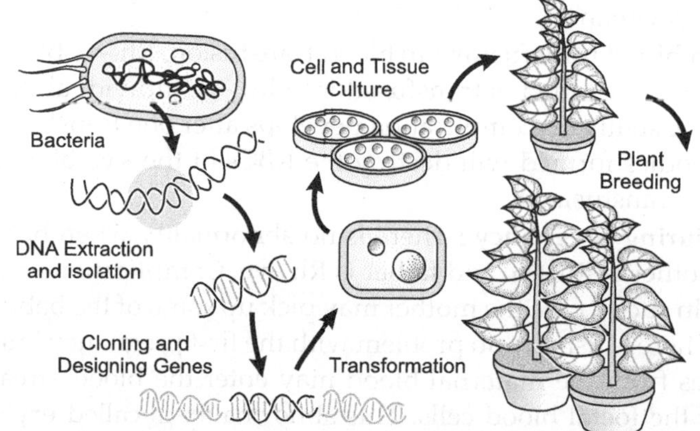

Tissue culture is used to develop thousands of genetically identical plants from one single parent plant known as soma clones, and this process is known as micro-propagation. The method offers an advantage over other methods as it can be used to develop disease free plants from disease-rode plants by using their meristems (apical and axillary) as explants.

Chapter 10. Microbes in Human Welfare

1. Explain the process of sewage water treatment before it can be discharged into natural water bodies. Why is this treatment essentials ?

Ans. Sewage water treatment is carried out in two stages :

(i) **Primary treatment :** This involves physical removal of particles from the sewage through filtration and sedimentation. Initially, floating debris is removed by sequential filtration. Then the grit (soil and small pebbles) are removed by sedimentation. The settled solid form the primary sludge and the supernant forms the effluent.

(ii) **Secondary treatment or biological treatment :** The primary effluent is passed in aeration tanks and constantly agitated mechanically. Air is pumped into it. This allow vigorous growth of useful aerobic microbes which consumes organic matter in the effluent. This significantly reduces the BOD (Biochemical Oxygen Demand) of the effluent.

2. What are the harmful effect of chemical pesticides ?

Ans. The harmful effects of chemical pesticides are :

(i) They pollute the soil and water.
(ii) They kill many useful organism in the soil.
(iii) They enter the food chain and cause disease at various trophic levels.
(iv) They affect the salinity of the soil.

* **Frequently asked previous years Board Exam Question.**

Chapter 11. Biotechnology : Principles and Processes

1. What are the main steps involved in DNA Fingerprinting ?

Ans. Steps involved in DNA Fingerprinting :

(i) **Isolating the desired DNA** : It can be done either chemically, by using a detergent to wash the extra material from the DNA, or mechanically, by applying a large amount of pressure in order to "squeeze out" the DNA.

(ii) **Cutting the DNA into several pieces of different size :** This is done using one or more restriction endonuclease enzyme.

(iii) **Sorting the DNA pieces by size :** The process by which the size separation is done is called gel electrophoresis. The DNA is poured into a gel, such as agarose, and an electrical charge is applied to the gel, with the positive charge at the bottom and the negative charge at the top. Because DNA has a slightly negative charge, the pieces of DNA will be attracted towards the bottom of the gel. The smaller pieces, however, will be able to move more quickly and thus, further towards the bottom than the larger pieces. The different-sized pieces of DNA will therefore, be separated by size, with the smaller pieces towards the bottom and the larger pieces towards the top.

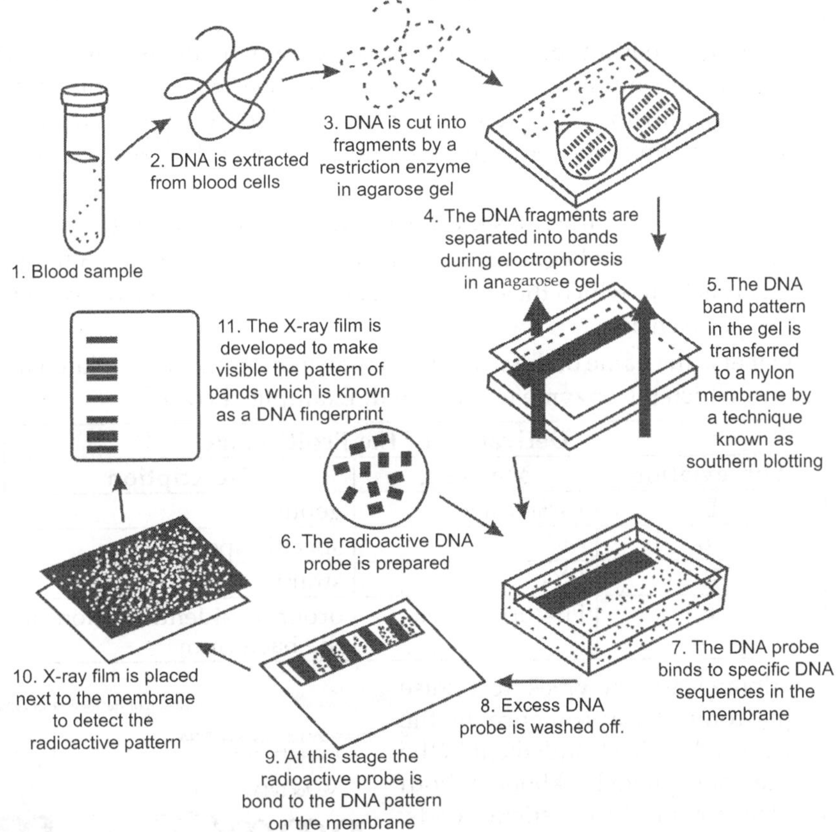

The DNA fingerprinting process

(iv) **Denaturing** the DNA fragments, so that all of the DNA is rendered single-stranded. This can be done either by heating or chemically treating the DNA in the gel using alkali.

(v) **Blotting the DNA** : The gel with the size-fractionated DNA is applied to a sheet of nitrocellulose paper or nylon membrane, and then baked to permanently attach the DNA to the sheet. The Southern Blot is now ready to be analyzed.

In order to analyze a Southern Blot, a radioactive genetic probe is used in a *hybridisation reaction* with the DNA. If an X-ray is taken of the Southern Blot after a radioactive probe has been allowed to bind with the denatured DNA on the paper, only the areas where the radioactive probe binds will show themselves on the film (autoradiography). This allow researchers to identify, in a particular person's DNA, the occurrence and frequency of the particular genetic pattern contained in the probe.

2. What are the applications of DNA Fingerprinting ?

Ans. Application of DNA Fingerprinting :

(i) **Diagnosis of Inherited Disorders :** DNA fingerprinting can be used in medical situations to diagnose inherited disorders. For example, DNA fingerprinting is used to diagnose inherited disorders such as cystic fibrosis, hemophilia, sickle cell anaemia and Huntington's disease in prenatal and newborn babies. Early detection of disorders allows for families to prepare themselves both mentally and physically as well as allows them to seek treatment early, to minimize and/or cure the disease.

(ii) **Forensics :** In addition to this, DNA fingerprinting can also be used in forensics to identify criminals. When a crime is committed, DNA can be extracted from hair, blood, or semen left at the scene of the crime. By performing DNA fingerprinting, the DNA fingerprint of the criminal can be obtained and compared to that of the suspects. This serves as biological evidence, to draw conclusions as to whether one of the suspects is indeed the criminal.

(iii) **Paternity Tests :** DNA fingerprinting can also be used in paternity tests, to identify the biological father of a child. DNA samples are collected from possible fathers and the mother and their DNA fingerprints are compared against the DNA fingerprint of the child. Each band on the DNA fingerprint of the child must match a band in at least one of the parents. Based on this, the biological father of the child can be determined.

3. (i) What are restriction endonucleases? Give the rules of their nomenclature.*

(ii) Explain the mechanism of action of restriction endonucleases that makes them suitable for genetic engineering.*

Ans. (i) Restriction Endonuclease is a type of restriction enzyme that cuts within the DNA at a specific site. This enzyme acts as an important tool in genetic engineering for cutting DNA sequences from a specific site. They play an important role in gene manipulation. The enzymes recognise a specific six-base pair sequence known as the recognition sequence or palindromic sequence and cut the sequence at a specific site.

Smith and Nathans proposed the rules for the Nomenclature of restriction endonucleases :

1. The enzyme names should begin with a three-letter acronym in which the first letter was the first letter of the genus from which the enzyme was isolated and the next two letters were the first two letters of the species name.
2. Fourth alphabet is the name of the strain from which the enzyme is isolated and the roman number is the order in which the enzyme was isolated. Example : EcoRI.

Derivation of the EcoRI name		
Abbreviation	**Meaning**	**Description**
E	*Escherichia*	genus
co	*coli*	specific species
R	RY13	strain
I	First identified	order of identifica-tion in the bacterium

(ii) Restriction endonuclease enzymes recognise a specific palindromic base sequence in the DNA and produce a double-stranded cut in the DNA at the sugar-phosphate backbone of both strands. This digestion produces "sticky" ends. For *e.g.*,

G|AATTC
CTTAA|G

Recognition sequences in DNA differ for each restriction enzyme, producing differences in the length, sequence and strand orientation (5′ end or 3′ end) of a sticky-end or overhang.

When cut by the same restriction enzyme, the resultant DNA fragments have the same kind of 'sticky-ends' and these can be joined together using DNA ligases. The foreign DNA and vector DNA are cut by same restriction enzymes.

4. (i) Explain what are the desirable characteristics of an ideal cloning vector used in rDNA technology.**

(ii) Describe two vectorless methods of gene transfer used in rDNA technology.*

* **Frequently asked previous years Board Exam Question.**

** **Frequently asked previous years Board Exam Question.**

Ans. (i) A cloning vector is a small piece of DNA, taken from a virus, a plasmid, or the cell of a higher organism, that can be stably maintained in an organism and into which a foreign DNA fragment can be inserted for cloning purposes. The vector therefore contains features that allows the convenient insertion or removal of a DNA fragment to or from vector.

An ideal cloning vector must have following characteristics :

(a) It must have an Origin of Replication (ori). It is the sequence from which replication starts and the foreign gene linked to it will clone itself and it also controls the copy number of the linked DNA.

(b) It must be small in size less than 10 kB.

(c) It must be self-replicating inside host cell.

(d) It must possess restriction site for Restriction Endonuclease enzymes.

(e) It must have the cloning sites for the ligation of the alien DNA which is carried out at the recognition site.

(f) Introduction of donor DNA fragment must not interfere with replication property of the vector.

(g) It must possess some marker genes for later identification of recombinant cell. These will help in identifying and eliminating non-transformants and selectively permitting the growth of transformants.

(ii) The term direct or vector less transfer of DNA is used when the foreign DNA is directly introduced into the genome.

(a) **Electroporation :** Electroporation involves the use of high field strength, electrical impulses to reversibly permeabilise the cell membranes for the uptake of DNA. This technique can be used for the delivery of DNA into intact plant cells and protoplasts. Electroporation has been successfully used for the production of transgenic plants of many cereals *e.g.*, rice, wheat, maize.

(b) **Particle Bombardment (Biolistics) :** Particle (or micro projectile) bombardment is the most effective method for gene transfer. This method is versatile due to the fact that it can be successfully used for the DNA transfer in mammalian cells and microorganisms. In this the gold or tungsten beads coated with DNA are bombarded with high velocity on the host tissue. The foreign gene gets incorporated in host DNA.

(c) Other methods include microinjection where a micropipette is used to inject the DNA into the nucleus. Transformation of bacteria can also be done by making the bacteria more competent to take up the foreign gene by placing them in cold calcium chloride or by heat stock. Liposome mediated transfer can also be used.

5. Give an account of the Blue-White Method of selection of recombinants.*

Ans. Blue-white method of selection is a powerful method of screening for the presence of recombinant plasmid. This method is based upon the insertional inactivation of the lac Z gene present on the vector (*e.g.*, PUC19). This gene expresses the enzyme β-galactosidase whose activity can cleave a colourless substrate called X-gal into blue coloured product. Isopropyl b-D-1-thiogalacto-pyranoside (IPTG) is used along with X-gal for blue-white screening. IPTG is a non-metabolizable analog of galactose that induces the expression of lac Z gene. It should be noted that IPTG is not a substrate for β-galactosidase but only an inducer. If the lac Z gene is inactivated due to the presence of the insert (express active enzyme), then the enzymes is

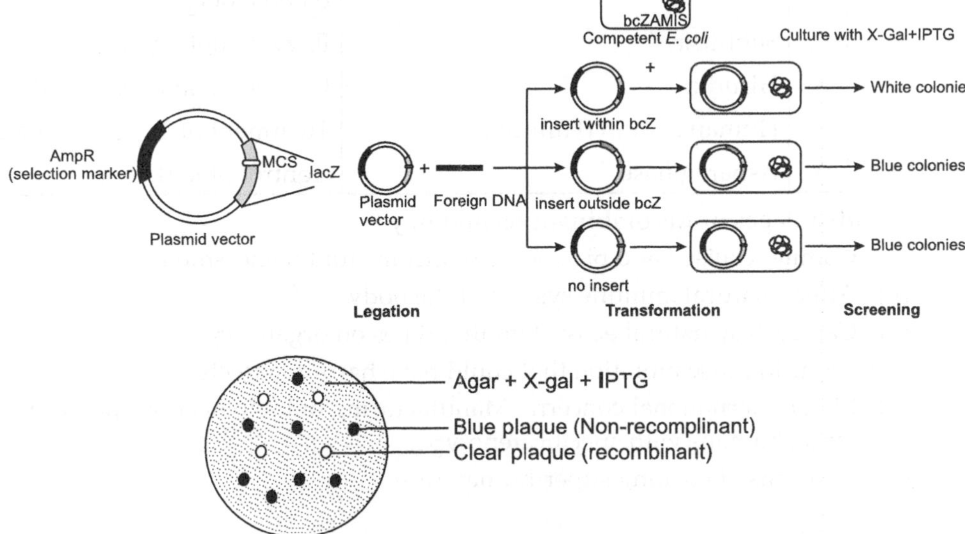

* Frequently asked previous years Board Exam Question.

not expressed. Hence, if after a transformation experiment the *E. coli* host cells are plated on an ampicillin and X-gal containing solid media plate then colonies which appears blue are those which have transformed cells (antibiotic resistant) but not have the insert (express active enzyme). Colonies which appear white are both ampicillin resistant and have the insert rDNA and thus are cells to be used for future experiment.

6. What does PCR stand for ? Describe the different steps of PCR.*

Ans. PCR (Polymerase Chain Reaction) method for synthesis of new DNA strands was developed by **Kary Mullis** in the 1980s. The technique of PCR is ability of DNA polymerase enzyme to synthesize new strand of DNA complementary to the template strand. DNA polymerase can add a nucleotide only onto a pre-existing 3′-OH group, so it needs a primer to which it can add the first nucleotide. At the end of the PCR reaction, the specific sequence will be accumulated in billions of copies (amplicons).

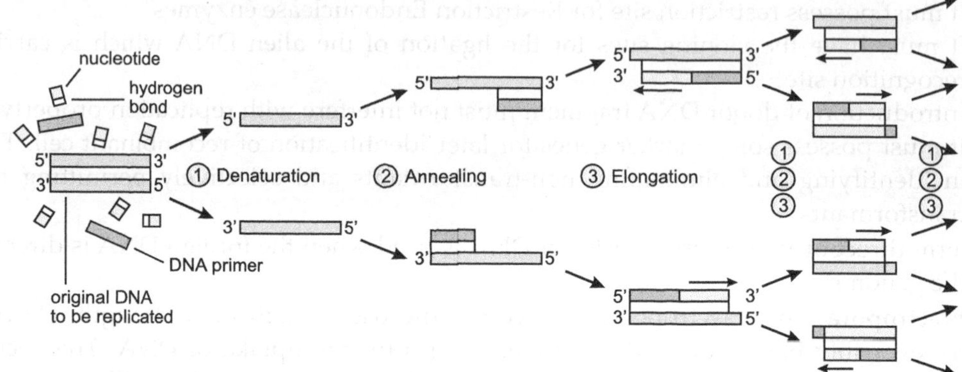

Process : There are three main steps in the process Denaturation, Annealing and Extending.

(i) **Denaturation :** The double stranded template of DNA is heated at a temperature of 94-95°C to separate them into two different strands.

(ii) **Annealing :** During this step the temperature is cooled to 50-54°C and the primers are annealed to single stranded DNA.

(iii) **Extending :** This step is final step and the temperature is increased to 72°C and the DNA is elongated with the help of heat stable enzyme Taq DNA polymerase.

The process is repeated 20-40 times to get the desired amount of DNA.

7. What is a recombinant protein ? What are the disadvantages of recombinant DNA technology ?

Ans. The protein that is produced by using recombinant DNA technology is called as Recombinant protein. It is a manipulated protein which is generated in a laboratory by modifying the gene sequences so as to produce greater amount of protein which are commercially useful.

Some of the examples of recombinant proteins and the uses are listed below :

	Recombinant Proteins	Uses
1.	Tissue plasminogen activator (TPA)	Treatment of myocardial infarction by lysing the blood clots
2.	Subtilisin	Enzyme detergent
3.	Hirudin	Used as an anticoagulant
4.	Human Insulin (Humulin)	Treatment of Type 1 Diabetes
5.	Asparaginase	Control of leukemia

Disadvantages of Recombinant technology :

(i) Commercialised and big source of income for businessmen.

(ii) Affects natural immune system of the body.

(iii) Can destroy natural ecosystem that relies on organic cycle.

(iv) Prone to cause mutation that could have harmful effects.

(v) Major international concern : Manufacturing of biological weapons such as botulism and anthrax to target humans with specific genotype.

(vi) Concerns of creating super-human race.

* **Frequently asked previous years Board Exam Question.**

Chapter 12. Biotechnology and its Applications

1. You have to identify a useful gene in a bacteria. Make a flow chart of the steps that you would follow to transfer this gene to a plant.

Ans. After identifying a useful gene in bacteria. Following steps should be undertaken :

Isolation of useful gene using restriction endonuclease

↓

Transferring gene to a suitable vector to create a recombinant DNA molecule to the target cells

↓

Screening of cell for transformation

↓

Selection of transformed cells

↓

Regeneration of plant from the transformed cell to get transgenic plants.

2. How does the process of RNA interference help to control the nematode from infecting roots of the tobacco plants ? Explain.

Ans. Using Agrobacterium vectors, nematode specific genes are introduced into host plants. The introduction of DNA produces both sense and antisense RNA in host cells. These two RNAs being complementary form a double stranded RNA (dsRNA) that initiates the RNAs and silence the specific mRNA of the nematode. As a result of this, the parasite can not survive in the transgenic host expressing specific interferring RNA.

3. Write in brief about biopatent and for what is it granted ?

Ans. It is the government license or legal rights giving someone the sole right to make, use or sell an invention for a set period. In other words, it is a right granted by the government to prevent others from commercial use of invention of someone.

A patent is granted for :

(i) an invention including product,
(ii) an improvement in the earlier invention,
(iii) the invention of the process generating product and
(iv) a new concept or design.

In a older days patents were granted for various industrial inventions from a particular company *e.g.* patent medicines. However, now-a-days, patents or legal rights are granted for biological entities and their products. Such patents are called biopatents. *e.g.*, neem and its products, haldi and its products etc. Developed and industrialised countries like U.S.A., Japan, and European Union countries grant the biopatents. These are granted for the following :

(i) Genetically modified plants and animals.
(ii) Various strains of micro-organisms.
(iii) Cell lines.
(iv) DNA sequences.
(v) The proteins enclosed by DNA sequences.
(vi) Biotechnological products.
(vii) Processes of production.
(viii) Applications of a product.

Some of the biotechnological patents have broad coverage. e.g., patent for transgenic plants of family cruciferae (mustard family). Such broad biopatents are unfair because they would enable financially powerful companies to have their monopoly and thereby to have control over the whole biotechnological processes. In such cases these companies would have control over the direction of whole agricultural research. This would cause serious threat to the food security of the world.

4. What is biopiracy ? Enumerate various examples of biopiracy.

Ans. When strong and developed countries or powerful organisations and multinational companies exploit biological resources of other nations without proper authorisations from the nations concerned, it is called biopiracy.

Following are the examples of biopiracy :

(i) The multinational companies of developed nations collect and patent the genetic resources themselves. *e.g.*, a patent granted in U.S.A. covers entire 'Basmati' rice germplasm indigenous to our country.

(ii) The bio-resources are analysed for identification of valuable biomolecules. A biomolecule is an active constituent of a living organism. *e.g.*, medical properties of 'Haldi' are mentioned in Ayurveda and in India this knowledge is used traditionally generation after generation. Therefore, no one cared to publish research paper to take patent on the active constituent of Haldi. In U.S.A. one company gave the chemical analysis of Haldi and a patent was awarded. Fortunately we won the battle against both the patents (of Basmati and Haldi) in the court of law.

(iii) Useful genes are isolated from the bio-sources and patented and thereafter, used to generate commercial products.

(iv) Sometimes even traditional knowledge itself, of other countries is patented *e.g.*, native plant of west Africa *Pentadiplandra brazzeana* is a source of protein called brazzein which is about two thousand times sweeter than sugar and has low calorie value. Local people have known and used the super-sweet berries of this plant for centuries. However, the protein brazzein was patented in U.S.A. The gene responsible for the production of said protein was isolated, sequenced and patented. This gene was transferred into the maize plant and it was expressed in maize grain. The protein can be extracted from the maize kernels and can replace sugar for its sweetness. This can give serious threat to sugar exporting countries.

Bioresources of the third world countries have always been commercially exploited by the developed and powerful nations like U.S.A. without giving them adequate compensation. Progress in biotechnological techniques has enhanced this exploitation a lot. It is a time for the developing countries to come forward and insist on making law against this biopiracy.

5. Explain the various steps involved in Recombinant DNA technology.*

Ans. Recombinant DNA technology involves several important steps that include:

(i) **Isolation of Gene of Interest:** The first step in rDNA technology is to isolate the desired DNA in its pure form *i.e.* free from other macromolecules. Since DNA exists within the cell membrane along with other macromolecules such as RNA, polysaccharides, proteins, and lipids, it must be separated and purified which involves enzymes such as lysozymes, cellulase, chitinase, ribonuclease, proteases etc.

(ii) **Restriction Enzyme Digestion:** Once DNA is purified; restriction enzymes are used to isolate the gene of interest. Restriction enzymes act as molecular scissors that cut DNA at specific locations. This step involves the incubation of the purified DNA with the selected restriction enzyme, at conditions optimal for that specific enzyme. The vector DNA is also processed using the same procedure.

(iii) **Amplification Using PCR:** Polymerase Chain Reaction or PCR is a method of making multiple copies of a DNA sequence using the enzyme – DNA polymerase in vitro. It helps to amplify a single copy or a few copies of DNA into thousands to millions of copies. The cut fragments of DNA can be amplified using PCR and then ligated with the cut vector.

(iv) **Ligation of DNA Molecules:** The purified DNA and the vector of interest are cut with the same restriction enzyme and ligated using the enzyme 'DNA ligase'. The resulting DNA molecule is a hybrid of two DNA molecules – the interest molecule and the vector. Hence, this new hybrid DNA molecule is also called a recombinant DNA molecule.

(v) **Insertion of Recombinant DNA into Host:** In this step, the recombinant DNA is introduced into recipient host cells mostly, a bacterial cell. This process is called transformation. Hosts are treated to make them competent to accept new DNA. The processes used may be thermal shock, Ca^{2+} ion treatment, electroporation etc.

(vi) **Isolation of Recombinant Cells:** The transformation process generates a mixed population of transformed and non-transformed cells. The selection process involves filtering the transformed host cells only. Gel electrophoresis is often used to separate pieces of DNA which in turn makes it possible to determine which plasmids took up the gene of interest. Several other methods involve growing the cells in antibiotic (e.g. tetracycline) agar plates.

* **Frequently asked previous years Board Exam Question.**

6. Explain the steps involved in the production of human insulin by Recombinant DNA technology.*

Ans. The first major product of genetic engineering was human insulin. It can be produced using recombinant DNA technology by the following steps:

(i) The gene responsible for producing human insulin protein is isolated.

(ii) Then, a circular piece of DNA called a plasmid is removed from a bacterial cell. Special enzymes called restriction enzymes are used to cut the plasmid ring.

(iii) With the open plasmid ring, the gene for insulin is inserted into the plasmid ring and the ring is closed.

(iv) The bacterial DNA now contains the human insulin gene and is inserted into a bacteria.

(v) Many plasmids with the insulin gene are inserted into many bacterial cells. When the bacterial cells reproduce by dividing, the human insulin gene is also reproduced in the newly created cells.

(vi) Human insulin protein molecules produced by bacteria are gathered and purified. By culturing the genetically engineered bacteria, limitless supplies of insulin may be produced.

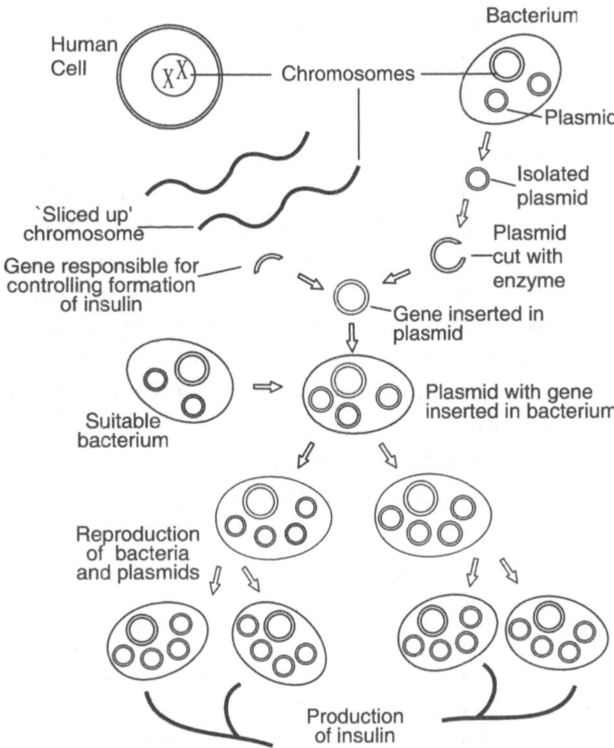

Long Answers

Chapter 13. Organisms and Populations

1. What is population growth ? State various methods to discourage population growth.

Ans. Population growth is the total number of increase in the organisms occupying a certain area. It includes natality or birth-rate and mortality or death-rate. The inherent ability of a population to increase the maximum possible rate of reproduction for a given species under optimum conditions is called potential natality. Under natural conditions, Potential Natality is rarely attained because of many adverse factors. Thus actual birth rate under existing conditions is called Realized Natality.

Thus, population growth is determined by the number of organisms added to the population (through birth or immigration) minus the number of organisms lost through death or emigration. When the number of individuals added to the population is equal to number of individuals lost, the population shows a zero population growth. When the factors of environmental resistance do not produce zero growth, the population shows exponential growth.

The population growth (R) is calculated by the following formula :

$$R = \frac{\text{No. of births + No. of immigrants – No. of deaths – No. of emigrants}}{\text{No. of individuals in population/Time}}$$

The maximum growth rate is called the intrinsic rate of growth.

Methods to discourage population growth : Population growth can be discouraged by the following family planning measures. These are :

(i) **Natural family planning**—By recognising the rhythm of their menstrual cycle, women can calculate the time when the egg is produced by the ovary and can be fertilised. During this unsafe time the couple need to abstain from intercourse.

(ii) **Artificial family planning**—This includes use of certain barriers like condoms and diaphragm, spermicidal creams and foams and intra uterine devices which inhibit implantation, use of pills and injections which inhibit formation and thus release of eggs from the ovaries.

(iii) **Sterilisation**—It includes vasectomy in males and tubectomy in females. In it the vas deferens and oviducts are cut and tied to prevent the movement of sperms and ovum so that no fertilisation

* **Frequently asked previous years Board Exam Question.**

occurs. It is also done by castration *i.e.*, surgical removal of testis in male and surgical removal of ovaries in female.

(iv) **Medical termination of pregnancy**—It can be done if the pregnancy is likely to result in birth of congenitally malformed child or the pregnancy is likely to harm the mother and also in case of contraceptive failure.

2. Write in detail about Natural Family Planning methods.

Ans. Natural family planning (NFP) is the method that uses the body's natural physiological changes and symptoms to identify the fertile and infertile phases of the menstrual cycle. Such methods are also known as fertility-based awareness methods. Natural family planning methods are generally the preferred contraceptive method for women who do not wish to use artificial methods of contraception for reasons of religion, or who, due to rumours and myths, fear other methods. However. Natural family planning methods are unreliable in preventing unwanted pregnancy. It takes time to practise and use then properly, which adds to their unreliability. Additionally, natural family planning methods do not protect against sexually transmitted infections (STIs), including the human immunodeficiency virus (HIV). Some of the methods are as follows :

A. Periodic Abstinence

(i) **Calendar method :** The calendar method of birth control focuses on tracking a woman's natural menstrual cycle to predict ovulation, according to Planned parenthood. It is a natural method of both preventing pregnancy as well as planning for conception.

(ii) **Temperature method :** To use the temperature method, one should take basal body temperature (BBT) at the same time everyday upon waking, and before getting out of bed, eating or drinking. This can be done vaginally, rectally or orally (the least accurate) with a special basal or ovulation thermometer that can give an accurate measurement to within one-tenth of a degree (0·1° Celsius). On charting daily temperature, one notice that following ovulation, your BBT will rise by 0·2-0·4 degrees Celsius. Once you have recorded 3 consecutive temperature rises. You can safely have intercourse. This means that you can only have sex in the phase of your menstrual cycle after ovulation (*i.e.* for about 12 out of 28 days).

(iii) **Mucus (Billings) method :** To use the mucus (Billings) method, you must learn to distinguish between your fertile and infertile days on the basis of the type of cervical mucus being produced by your body. During the first part of your cycle (immediately after your period—the infertile phase) you will have a feeling of vulval dryness. This is followed by a period of vulval wetness, where your body produces clear, stretchy, slippery mucus, which corresponds to ovulation (fertile phase). A post-ovulatory phase then occurs when progesterone is produced, which causes the vulval area to have thicker, cloudier mucus, before returning to dryness once again. One should not have sexual intercourse during this fertile phase, and until 3 consecutive dry days have passed.

B. **Lactational amenorrhoea method (LAM) :** The lactational amenorrhea method (LAM) is the use of breasteeding as a contraceptive method. *Lactational* means breastfeeding and *amenorrhea* means not having monthly bleeding. In this case, there is a delay in ovulation caused by the action of prolactin hormone from the effect of lactation or breastfeeding. An infant's suckling of the nipple sends neural signals to the mother's hypothalamus (part of the brain), which influences the anterior pituitary gland to secrete prolaction to stimulaste the breast for milk production. This, in turn, inhibits the secretion of follicle stimulating hormone (FSH) and luteinizing hormone (LH), and as a result ovulation does not occur. While women are exclusively breastfeeding, prolactin continues to be secreted and pregnancy is unlikely. When prolactin levels decrease, the woman's monthly bleeding may return, and if she continues to have unprotected sexual intercourse she may get pregnant.

C. **Coitus interrupts (withdrawal or pulling out) method :** Coitus interrupts or withdrawal in a traditional family planning method in which the man withdraws or pulls out his penis from his partner's vagina and ejaculates outside, keeping his semen away from her genitalia. Coitus interrupts prevents fertilisation by stopping contact between spermatozoa in the sperm and the ovum or egg.

3. Write about few artificial methods used for birth control.

Ans. Artificial methods for Birth control :

(i) **Barrier methods of Birth control :** Barrier methods of birth control block sperm from entering the uterus. Using a spermicide with a barrier method gives you the best possible barrier method protection.

(a) The spermicide kills most of the sperm that enter the vagina.

(b) The barrier method then blocks any remaining sperm from passing through the cervix to fertilise an egg.

Barrier methods include the diaphragm, cervical cap, cervical shield, male condom, and female condom and spermicidal foam, sponges, and film. Unlike other methods of birth control, barrier methods are used only when you have sexual intercourse.

(ii) **Intrauterine Contraceptive Devices (IUCDs) :** The intrauterine contraceptive device (IUCD) is a small plastic T-shaped device that is inserted into the uterus. An IUCD's contraceptive action begins as soon as the device is placed in the uterus and stops as soon as it is removed. IUCDs have an effectiveness rate of close to 100%. They are also a reversible form of contraception. Once the device is removed, a woman regains her fertility.

(iii) **Pills and Injection method :** Certain birth control methods contain man-made forms of hormones that are normally made in a woman's ovaries. These hormones are called estrogen and progestin.

(a) Both estrogen and progestin prevent a woman's ovaries from releasing an egg during her menstrual cycle (called ovulation). They do this by changing the levels of the natural hormones the body makes.

(b) Progestins also help prevent sperm from entering the uterus by making the mucus around a woman's cervix thick and sticky.

Birth control pills are one way of receiving these hormones. However, you must remember to take the pills everyday.

Other methods to prevent pregnancy use the same hormones, but these hormones are released slowly over time. Besides, a single progestogen injection is an effective contraceptive for a period of one to three months.

(iv) **Sterilisation :** There are sterilisation procedures for both men and women. These are permanent and for the most part irreversible (or intended to be).

Male sterilisation : Male sterilisation is achieved through a surgical procedure, called vasectomy, that is intended to be a permanent method of birth control (there is no guarantee that it can be reversed).

Female sterilisation : Permanent female sterilisation can be achieved through two procedures :

(a) **Tubal ligation :** It involves cutting the fallopian tubes and then tying and cauterizing the cut ends, so the egg does not get out of the tube into the uterus.

(b) **Hysteroscopic sterilisation (Essure procedure) :** A hysteroscope, a lighted telescopic instrument, is inserted into the uterus through the cervix. The hysteroscope is used to insert spring-like coils called micro-inserts in the opening of the fallopian tubes which blocks the tubes completely.

(v) **Medical termination of pregnancy :** There are several different surgical techniques for early termination of pregnancy (abortion in the first three months). Several drugs can also be prescribed alone or in combination to terminate early pregnancy. This is called medical abortion, and uses the hormones prostaglandins and/or mifepristone (an antiprogesterone often called RU486), and/or methotrexate. The review of trails found that medical methods for abortion in early pregnancy can be safe and effective, with the most evidence of effectiveness for a combination of mifepristone and misoprostal (a prostaglandin).

Chapter 14. Ecosystem

1. Describe the advantages for keeping the ecosystem healthy.

Ans. By keeping the ecosystem healthy we can take advantage of ecosystem services which are the products of ecosystem.

Following are economic and environmental benefits that we obtain from the ecosystem. They

(i) Purify air and water
(ii) Mitigate drought and floods
(iii) Cycle nutrient
(iv) Gennate fertile soil
(v) Pollinate crop
(vi) Provide wildlife habitat

(vii) Provide stable food chain
(viii) Maintain biodiversity
(ix) Provide economically useful forest producer
(x) Provide sustainable biological legacy to future generations.

2. Justify the importance of decomposer in an ecosystem.

Ans. Decomposer which are heterotrophic organisms mainly fungi and bacteria, breakdown complex organic matter into inorganic substance like carbon dioxide, water and nutrient. They meet their energy and nutrient requirement by degrading dead organic matter or detritus. These are also known as saprotrophs. Decomposers secrete digestive enzymes that breakdown dead and waste material into simple, inorganic material which are subsequently absorbed by them.

3. Construct an ideal pyramid of energy when 2,00,000 Joules of sunlight is available. Label all its trophic level.

Ans. After a 1% energy transfer from sunlight to the primary producer, and then, only 10% energy at the higher trophic gets successfully transferred to the next trophic. So, producers get 20,000 J, of the energy in sunlight.

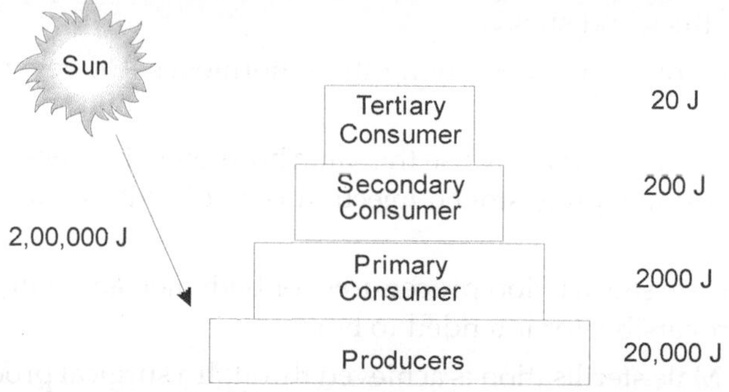

Chapter 15. Biodiversity and its Conservation

1. Why Biodiversity is important ?

Ans. Biodiversity is important for several reasons. It provides economic benefits, protects human health and safety, and offers recreational or aesthetic enjoyment. A healthy biodiversity provides a number of natural services for everyone :

(i) Ecosystem services, such as
(a) Protection of water resources
(b) Soils formation and protection
(c) Nutrient storage and recycling
(d) Pollution breakdown and absorption
(e) Contribution to climate stability
(f) Maintenance of ecosystems
(g) Recovery from unpredictable events

(ii) Biological resources, such as
(a) Food
(b) Medicinal resources and pharmaceutical drugs
(c) Wood products
(d) Ornamental plants
(e) Breeding stocks, population reservoirs
(f) Future resources
(g) Diversity in genes, species and ecosystems
(h) Social benefits
(i) Research, education and monitoring
(j) Recreation and tourism
(k) Cultural values

2. What is genetic conservation ? Explain in brief about the ex-situ and in-situ conservation.

Ans. Genetic conservation is the conservation of gene pool of a population with the objective to conserve and restore biodiversity.

Conservation of genetic resources is crucial to sustain food and ecological security. It can be achieved by the following ways : (i) To maintain them in the wild places like forests and nature preserves; (ii) To conserve them in protected areas, like national parks, nature reserves and botanical gardens; (iii) To feed them into the agricultural and horticultural trade so that they are cultivated by people at large; and (iv) To preserve them in the form of seeds or some other suitable material.

For conservation there are two different approaches that are used, in situ or ex situ gene conservation.

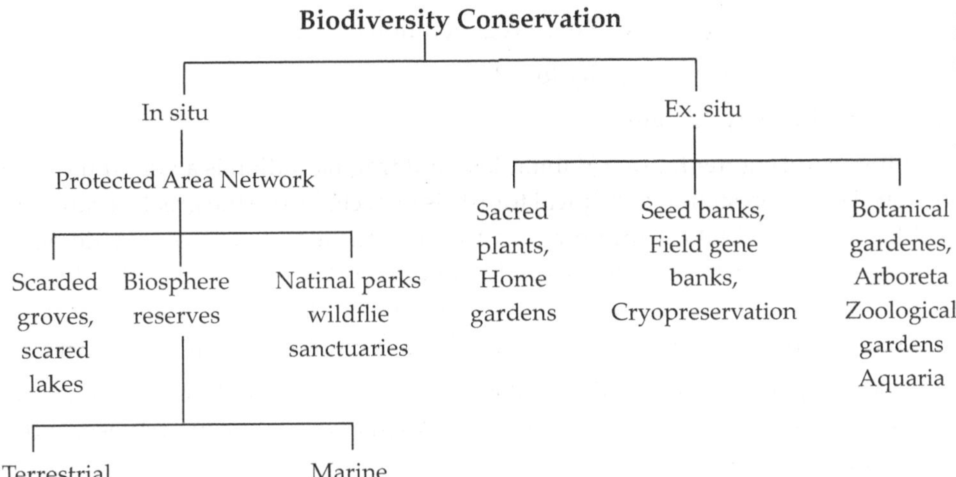

The *in-situ* and *ex-situ* approaches of conserving biodiversity in India.

***In-situ* conservation methods :** *In-situ* conservation, the conservation of species in their natural habitats. It is considered the most appropriate way of conserving biodiversity.

(i) In situ conservation is carried out through Biosphere reserves. National parks and wild sanctuaries and other protected area by Ministry of Environment.

(ii) The national Bureau of plants, animals and fish genetic resources collect, conserve and store germplasms of plants and animals in seed gene banks or field gene banks.

(iii) Botanical Gardens and Zoological Parks have large collection of plant and animal species.

Advantages of *in-situ* conservation :

(i) It aids in maintaining gene banks.

(ii) It is vital for maintenance of species and check its extinction.

(iii) It helps in preservation of genetic variability.

(iv) It helps in preserving the quality of life for millions of forest dwellers.

(v) In situ conservation helps to supply materials to restore plants.

Limitations of *in-situ* conservation : The forests are being destroyed to make dams, reservoirs, and established industries. By doing this we are losing the wild plants which are so important for in situ conservation.

***Ex-situ* Conservation Methods :**

Ex-situ conservation is the preservation of components of biological diversity outside their natural habitats. This involves conservation of genetic resources, as well as wild and cultivated species, and draws on a diverse body of techniques and facilities. Some of these include :

(i) Gene banks, *e.g.*, seed banks, sperm and ova banks, field banks.

(ii) *In vitro* plant tissue and microbial culture collections.

(iii) Captive breeding of animals and artificial propagation of plants, with possible re- introduction into the wild.

(iv) Collecting living organisms for zoos, aquaria, and botanical gardens for research and public awareness.

Advantages of *ex-situ* conservation :

(i) Efficient and reproducible.
(ii) Feasible for medium and long-term secure storage.
(iii) Wide diversity of each target taxonomy conserved.
(iv) Easy access for characterization, evaluation, and utilization.
(v) Little maintenance.

Disadvantages :

(i) Problem in storing seed of recalcitrant species.
(ii) Freezes evolutionary development.
(iii) Genetic diversity may be lost with each regeneration cycle.

3. What are the main causes of biodiversity loss ?

Ans. Main causes of biodiversity loss are :

(i) **Habitat loss and fragmentation :** Habitat loss and fragmentation is a very important cause of species extinction. The deforestation of tropical forests is expected to be the greatest cause of mass extinctions caused by human activity. Cutting down forests and hedgerows to provide agricultural land has increased. The loss of tropical forests, which holds 50% of the world's biodiversity proving to be the most significant threat to species in the modern age. This leads to fragmentation of habitats and species which require large habitats - such as bears cannot survive if the area is too small.

(ii) **Invasive alien species :** The introduction of exotic species that replace local and native species is cited as the second largest cause of biodiversity loss. Alien invasive species replace, and often result in the extinction of native species.

(iii) **Over-harvesting :** Over-harvesting and over-hunting for food, fashion and medical purposes has caused significant reduction in species and endangers many more. Whales and African elephants are two examples of species at most risk, with commercial hunting and poaching, being their main threats.

(iv) **Pollution :** Industrial, agricultural and waste-based pollutants can have catastrophic effects on many species. Those species which are more tolerant of pollution will survive; those requiring pristine environments (water, air, food) will not. Thus, pollution can act as a selective agent.

(v) **Climate change :** Observed recent changes in climate, especially warmer regional temperatures, have already had significant impacts on biodiversity and ecosystems, including causing changes in species distributions, population sizes, the timing of reproduction or migration events.

4. What are the different categories of plants and animals which need to be conserved ?

Ans. Different categories as laid down by World Conservation Union (WCU) of plants and animals which need to be conserved are :

(i) **Threatened species :** The species of plants and animals that are abundant in some parts but severely declining in other parts. Examples—Horned Rhino, Grizzly bear, species of whale population.

(ii) **Extinct species :** The species that have completely died out and there isn't a single individual left of this species. *E.g.*, Dinosaurs, Dodo bird.

(iii) **Extinct in wild :** A species is said to be extinct in the wild when exhaustive surveys fail to record even one alive individual in the expected habitat.

(iv) **Critically endangered :** A species or taxon facing an extremely high risk of extinction in immediate future. *E.g.*, Tiger, Sarus crane.

(v) **Endangered species :** A species that is facing high risk of extinction throughout all or significant portion of its range. *E.g.*, bald eagles, pandas, Asiatic lion, Great Indian Bustard.

(vi) **Vulnerable species :** These are species whose population had declined to levels from where its is likely to move into the endangered category in the near future if causative factors are not removed. *E.g.*, Deer and black buck.

(vii) **Rare or low risk species :** These include species which are uncommon or scarce. They have small population localised in specific geographical areas or habitats. Due to their small size, they are not vulnerable or endangered in present, but carry the risk of entering into these categories *E.g.*, Elephant.

Chapter 16. Environmental Issues

1. Study the graph given below. Explain how oxygen concentration is affected in the river when sewage is discharged into it.

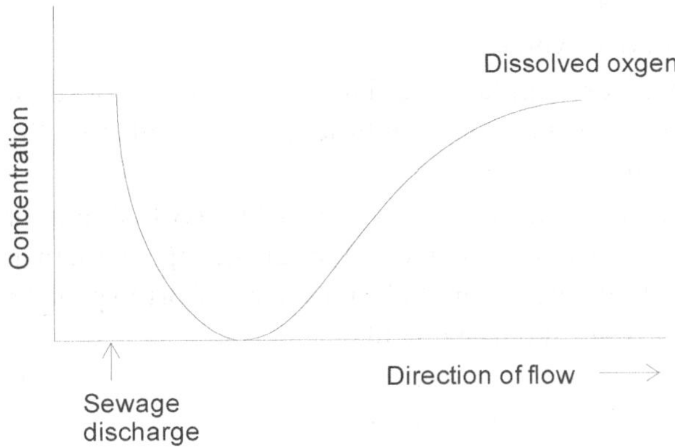

Ans. When sewage is discharged into the river, the oxygen concentration declines sharply because a large amount of oxygen is consumed by aerobic micro-organism in river to decompose the organic matter in river. When the amount of organic matter reduces, the amount of dissolved oxygen again increases.

2. Study the given aquatic food chain and answer the following question.

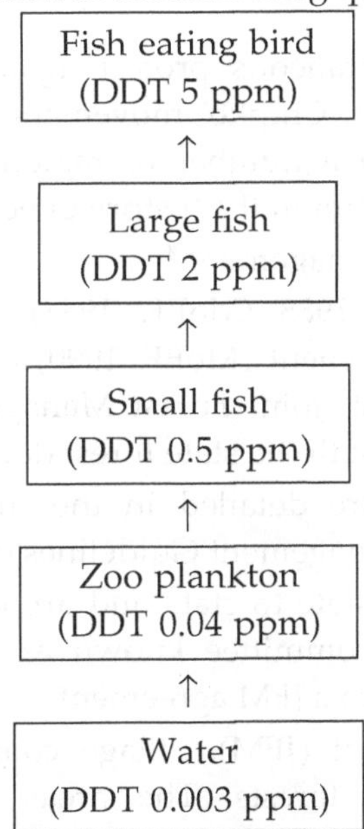

(i) Give the reason why there is a continuous increase in the DDT content in different trophic levels of the chain.

(ii) Name the phenomenon responsible for the increase in DDT content.

Ans. (i) DDT enters the food chain via water. Since they are not degradable these get accumulated progressively at each trophic level. As they reach higher trophic levels the concentration of DDT also increases.

(ii) Biological magnification

3. (i) State the cause of depletion of ozone layer.

(ii) Specify any two ill effects that it can be cause in human body.

Ans. (i) UV radiation acts upon CFC's (Chlorofluoro Carbon) releasing Cl atom in the stratosphere. These Cl atoms combine with O_3 and degrade it.

$$ClF_2C - Cl \xrightarrow{h\nu} ClF_2C^{\bullet} + Cl^{\bullet}$$

$$Cl^{\bullet} + O_3 \longrightarrow ClO^{\bullet} + O_2$$

$$ClO^{\bullet} + O_3 \longrightarrow Cl^{\bullet} + 2O_2$$

(ii) Ill effect of ozone depletions :
   (a) UV-B damages DNA and protection of living organism causing mutation.
   (b) UV rays skin aging, skin cell damage and skin cancer.
   (c) UV-B is absorbed by human eye and at high dose it causes inflammation of Cornea. This is called snow-blindness cataract.

4. Write a short note on Chipko Movement.*

Ans. (i) The Chipko movement or Chipko Andolan was a forest conservation movement. The name of the movement comes from the word 'embrace', as the villagers hugged the trees, and prevented the contractors from felling them.

(ii) During the movement that dates back to 1730 AD, 363 Bishnois of Khejarli village of Rajasthan, sacrificed their lives to save Khejri trees. A large group of them from 84 villages led by a lady called Amrita Devi Bishnoi laid down their lives in an effort to protect the trees. This incident became forebear of 20th centuary chipko movement.

(iii) All over the world it created a precedent for starting of non-violent protest in India. Sunderlal Bahuguna Started Chipko movement/Andolan in 1980.

(iv) This inspired to slow down the rapid deforestation, expose vested interests, increase ecological awareness, and demonstrate the viability of people power.

(v) Above all, it stirred up the existing civil society in India, which began to address the issues of tribal and marginalized people. The success achieved by this protest led to similar protests in other parts of the country.

(vi) From their origins as a spontaneous protest against logging abuses in Uttar Pradesh in the Himalayas, supporters of the Chipko movement, mainly village women, have successfully banned the felling of trees in a number of regions and influenced natural resource policy in India. This became a turning point in the history of eco-development.

5. Write a short note on Joint Forest Management.*

Ans. (i) The Indian Forest Policy of 1988 (MoEF, 1988) and the subsequent government resolution on participatory forest management (MoEF, 1990) emphasize the need for people's participation in natural forest management. Joint Forest Management (JFM) officially refers to partnerships in forest movement involving both the state forest departments and local communities.

(ii) The policies and objectives are detailed in the Indian comprehensive National Forest Policy of 1988 and the Joint Forest Management Guidelines of 1990 of the Government of India.

(iii) Although schemes vary from state to state and are known by different names in different Indian languages, usually a village committee known as the Forest Protection Committee (FPC) and the Forest Department enter into a JFM agreement.

(iv) Under joint forest management (JFM), village communities are entrusted with the protection and management of nearby forests. The areas concerned are usually degraded or even deforested areas.

(v) Villagers agree to assist in the safeguarding of forest resources through protection from fire, grazing, and illegal harvesting in exchange for which they receive non-timber forest products and a share of the revenue from the sale of timber products.

(vi) The communities are required to organize forest protection committees, village forest committees, village forest conservation and development societies, etc. Each of these bodies has an executive committee that manages its day-to-day affairs.

❐

* **Frequently asked previous years Board Exam Question.**

Identify the Following

Set 11

Chapter 1. Reproduction in Organisms

1.

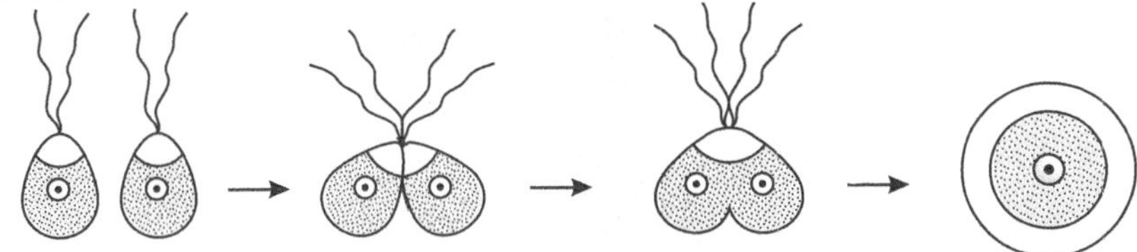

(i) State the type of gametes shown in the diagram.
(ii) Identify the process taking place and the resultant structure.
(iii) Name an organism that reproduces in this manner.

Ans. (i) Isogametes
(ii) Fertilisation is taking place and zygote is the resultant structure.
(iii) *Cladophora/Chlamydomonas*

Chapter 2. Sexual Reproduction in Flowering Plants

1. Study the following figure representing the structure of a mature pollen grain and answer the questions that follow :

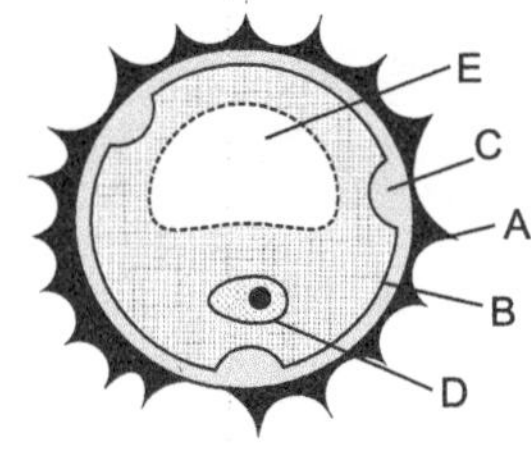

(i) Identify the part labelled A-E.
(ii) What is the function of C ?
(iii) Which layer is made up of sporopollenin ?
(iv) Which cell of the pollen grain divide to form the male gametes ?

Ans. (i) A-exine, B- intine, C- germ pore, D- generative cell, E- tube nucleus.
(ii) C is the region where the pollen tube protrudes out during fertilisation.
(iii) The exine is made up of sporopollenin.
(iv) The generative cell divide to form 2 male gametes.

2. Given alongside is an enlarged view of one microsporangium of mature anther.

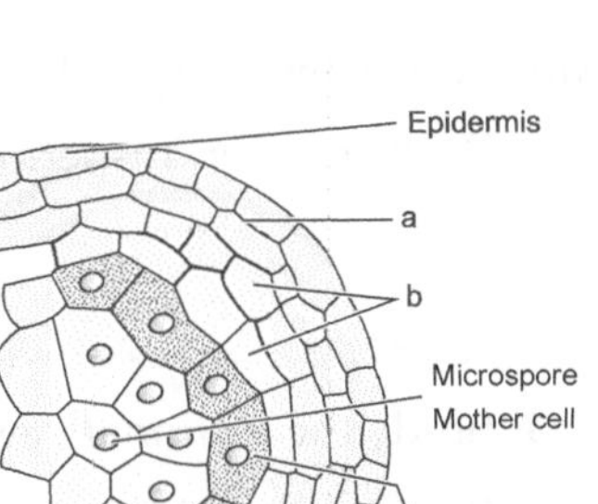

(i) Name 'a' 'b' and 'c' wall layer.
(ii) Mention the characteristic and function of the cells forming wall layer 'c'.

Ans. (i) (a) Endothecium (b) Middle layer (c) Tapetum.
(ii) Tapetum provides nourishment to thev developing pollen grains. The tapetal cells secrete ubisch granules that provides sporopollenin and other materials for exine formation.

3.

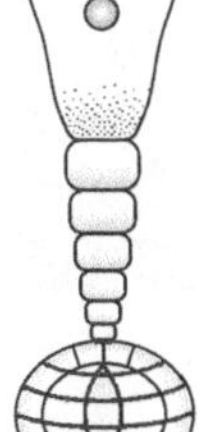

(i) Identify the figure

(ii) Name the initial cell from which this structure has developed.

(iii) Draw the next mature stage and label the parts.

Ans. (i) It is globular embryo of a dicot plant.

(ii) Zygote

(iii)

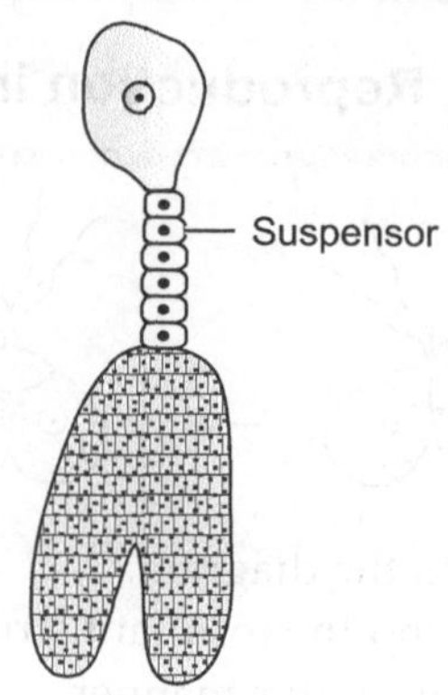

Heart-shaped embryo

4. Name the parts 'A' and 'B' shown in the diagram below.

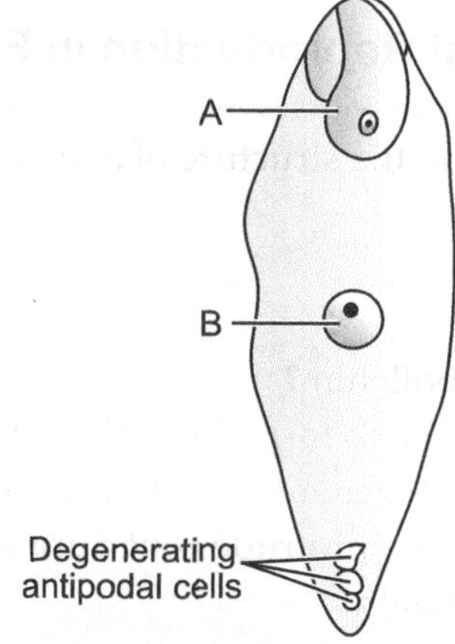

Ans. A- Fertilised egg cell—develops into an embryo

B- Primary endosperm nucleus - develops into endosperm

Chapter 3. Human Reproduction

1. Study the sectional view of human testis showing seminiferous tubule given below. Answer the following questions.

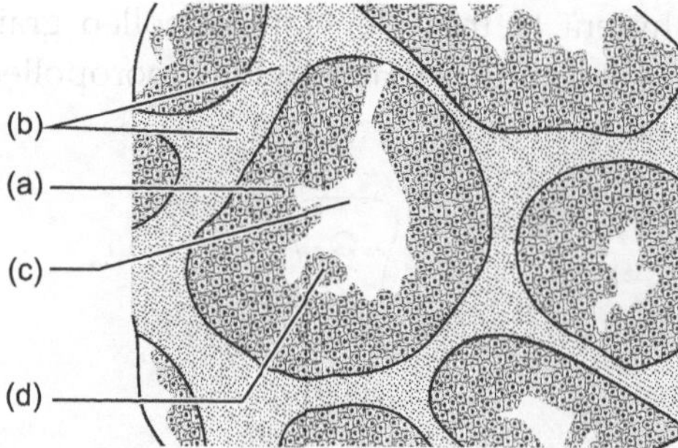

(i) Identify a, b, and c.

(ii) Write the function of a and d.

Ans. (i) a – Spermatogonia. b – Interstitial cells. c – Spermatozoa.

(ii) 'a' The spermatogonia or male germ cells undergo meiotic division leading to sperm formation. 'd' sertoli cells provide nutrition to the germ cells.

2. Study the figure given :

(i) Pick out and name the cells that undergo spermiogenesis.

(ii) Name 'a' and 'b' cell. What is the similarity between them with reference to the number of chromosomes.

(iii) Pick out the name of motile cells.

(iv) What is '*f*' cell ? Mention its function.

(v) Name the structure of which the given diagram is a part.

Ans. (i) d – spermatids

(ii) a – spermatogonium b – primary spermatocyte. They both are diploid and have 46 chromosomes each.

(iii) e – spermatozoa.

(iv) f – sertoli cell. It provides nutrition to the germ cells.

(v) Seminiferous tubule.

3. The path of sperm transport is given below. Provide the missing steps in the blank boxes.

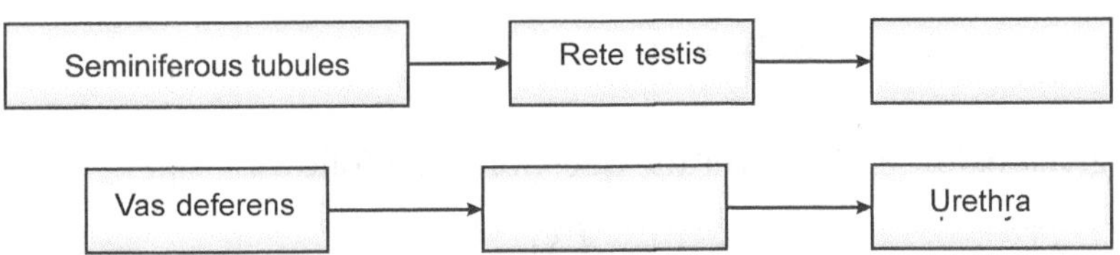

Ans. Epididymis, Ejaculatory duct

4. Study the following chart. Name the hormones involved at each stage.

Hypothalamus – Pituitary – Testis – Sperms.

Ans. Hypothalamus $\xrightarrow{\text{GnRH}}$ Pituitary $\xrightarrow{\text{FsH}}$ Testis $\xrightarrow[\text{(Testosterone)}]{\text{Androgen}}$ Sperms

5. Given alongside is the T.S. of human ovary. Identity in the diagram :

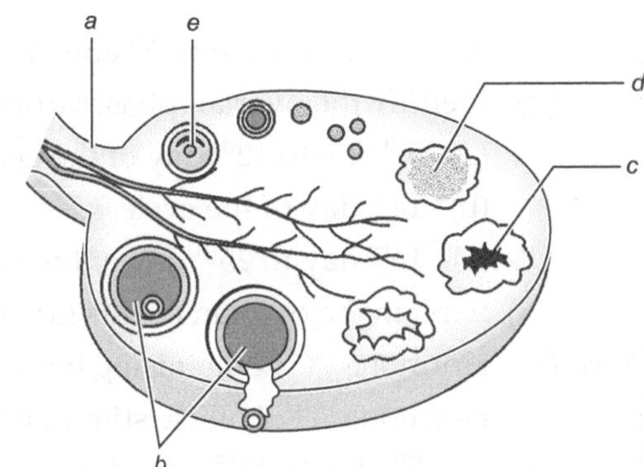

(i) Corpus luteum

(ii) Secondary oocyte

(iii) Antrum

(iv) Primary follicle

(v) Blood vessels

Ans. (i) Corpus luteum – b

(ii) Secondary oocyte – c

(iii) Antrum – d

(iv) Primary follicle – e

(v) Blood vessels – a

6. Given alongside is a chart showing ovarian changes during menstrual cycle. Fill in the spaces with the hormonal factor (s) responsible for the events shown.

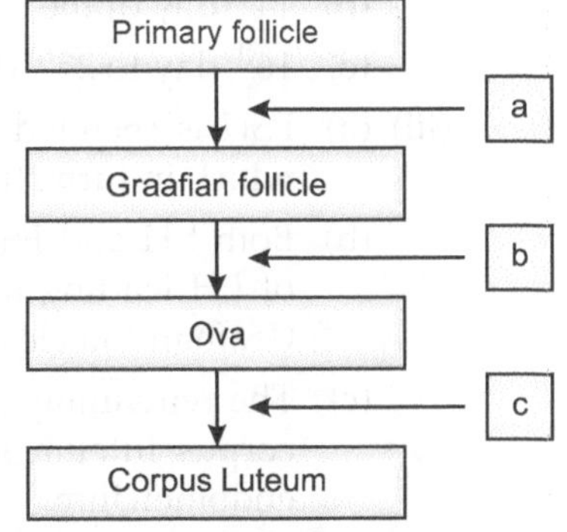

Ans. (a) FSH and estrogen

(b) LH

(c) Progesterone

7. Female reproductive organs and associated function and given below in column A and B. Fill in the boxes

Column A	Column B
Ovaries	Ovulation
Oviduct	A
B	Pregnancy
Vagina	Birth

Ans. (A) Fertilisation

(B) Uterus

8. Label a, b, c in the following diagram.

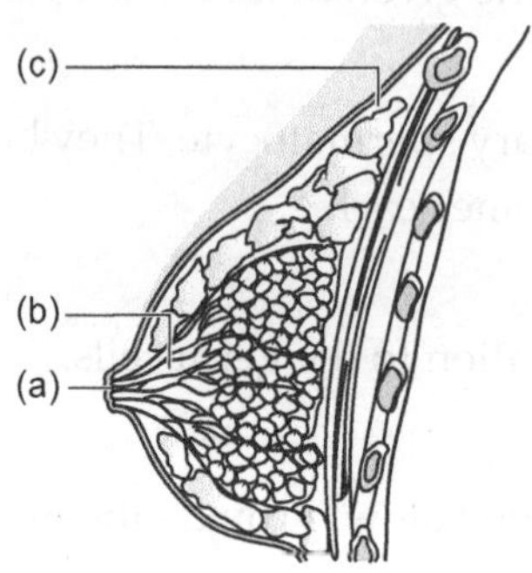

Ans. (a) Nipple (b) Milk duct

(c) Fat bodies

9. Study the graph given below and answer the questions that follow :

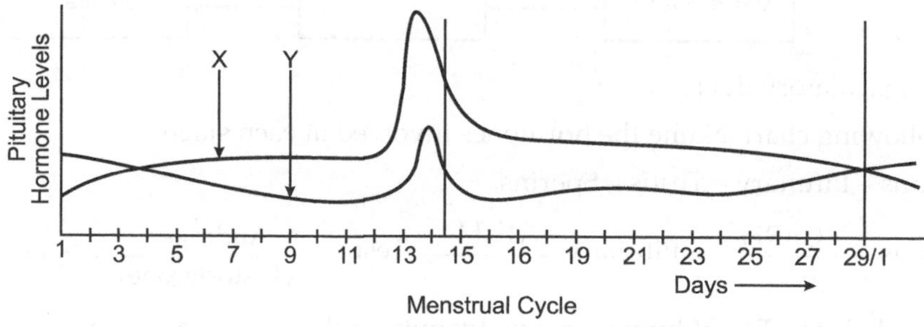

(i) Name the hormones 'X' and 'Y'.

(ii) Identify the ovarian phases during a menstrual cycle.

(a) 5th day to 12th day of the cycle.

(b) 14th day of the cycle.

(c) 16th day to 25th day of the cycle

(iii) Explain the ovarian events (a), (b) and (c) under the influence of hormones 'X' and 'Y'.

Ans. (i) Hormone 'X' : Luteinising hormone

Hormone 'Y' : Follicle stimulating hormone

(ii) (a) 5th day to 12th day of the cycle : Follicular phase (proliferative phase)

(b) 14th day of the cycle : Ovulatory phase (release of ovum) followed by luteal phase.

(c) 16th day to 25th day of the cycle : Luteal phase.

(iii) (a) FSH is secreted by anterior pituitary which stimulate the ovarian follicle to secrete estrogen, which in turn stimulate the proliferation of the endometrium of the uterine wall.

(b) Both LH and FSH attain a peak level in the middle of cycle (about 14th day). Rapid secretion of LH leading to its maximum level during the mid-cycle called LH surge induce rupture of Graffian follicle and thereby release of ovum (ovulation).

(c) The remaining cells of follicle are stimulated by the LH to transform into corpus luteum. The corpus luteum secretes large amount of progesterone which is essential for maintenance of endometrium.

10. The following is the illustration of the sequence of ovarian events (a-i) in a human female.

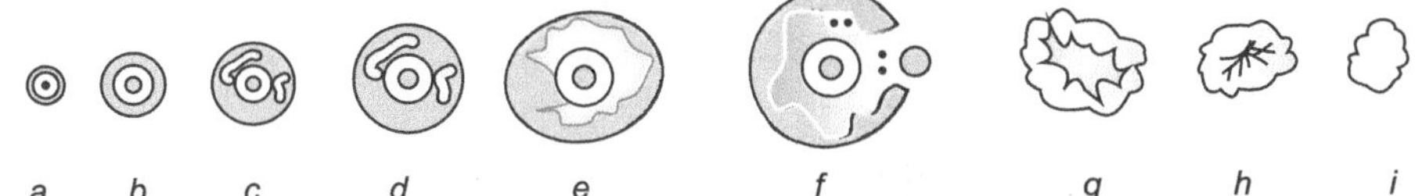

(i) Identify the figure that luteinizing ovulation and mention the stages of oogenesis it represents.
(ii) Name the ovarian hormone and the pituitary hormone that have caused the above mentioned events.
(iii) Explain the changes that occur in the uterus simultaneously in anticipation.
(iv) Write the difference between 'c' and 'h'.

Ans. (i) f ; secondary oocyte
(ii) Estrogen and luteinizing hormone (LH)
(iii) Endometrium proliferate (glands become cork-screw shaped) highly vascularised, high regeneration anticipating implantation of fertilised ovum.
(iv) c- developing follicle ; h – repressing corpus luteum.

11. (i) Read the graph given alongside and correlate the uterine event that takes place according to the hormonal levels on :
(a) 6-15 days
(b) 16-25 days
(c) 26-28 days (if the ovum is not fertilised)
(ii) Specify the sources of the hormones mentioned in the graph.

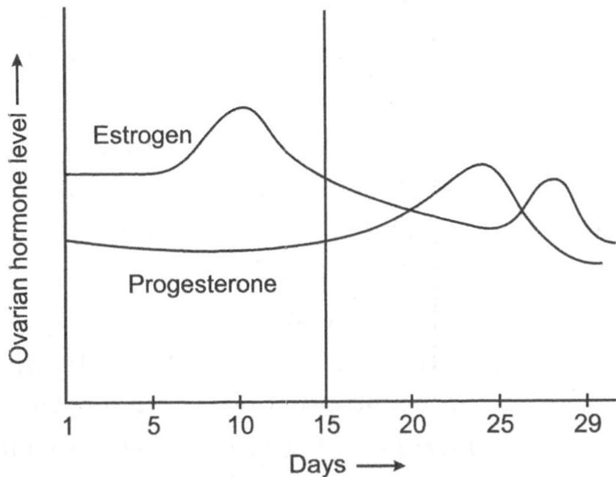

Ans. (i) (a) It is the follicular phase where FSH stimulates the ovarian follicles to secrete estrogen, which in turn stimulates the proliferation of the endometrium of the uterine wall. The endometrial lining begins to thicken.
(b) Secretion of progesterone maintains the endometrium which is necessary for implantation of fertilised ovum and other events of pregnancy.
(c) When the fertilisation does not take place, the endometrium disintegrates, leading to menstruation and thus marking a new cycle.
(ii) Estrogen is secreted by the ovarian follicles and progesterone is secreted by the corpus luteum.

12. Study the illustration given and answer the question that follow :
(i) Identify 'a'.
(ii) Name the state and function of 'c'.
(iii) Identify 'd'.
(iv) Explain the role of hormones in the formation and release of 'a'.
(v) Draw a diagram of 'b' separately and label the parts.
(a) that help its entry into an ovum
(b) that carry genetic material
(c) that help in its movement.

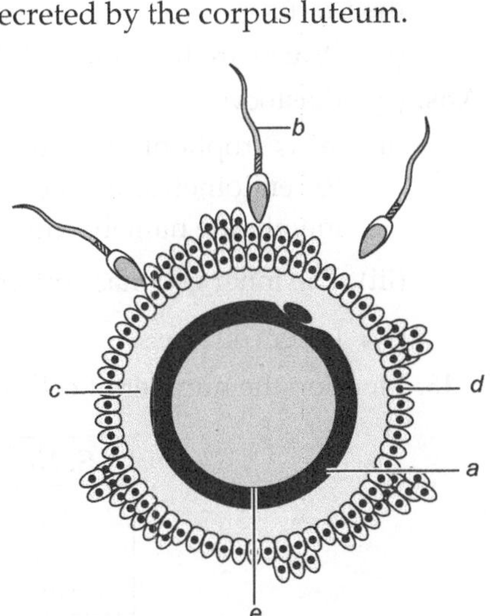

Ans. (i) a – Ovum
(ii) c- Zona pellucida. It protect ovum and regulates interaction between ovum and sperms during fertilisation.
(iii) d – Cells of corona radiata
(iv) Follicular Stimulating Hormone (FSH) stimulate follicular development. Luteinizing Hormones (LH) ruptures graafian follicle and thereby release ovum.

(v)

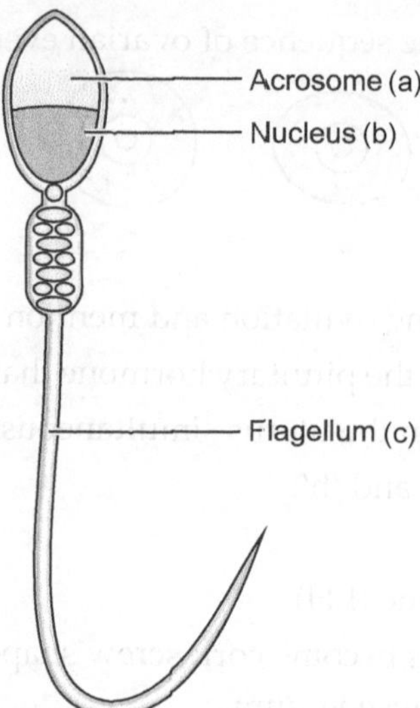

13. (i) Compare the fate of sperms shown in diagram.
(ii) What is the role of zona pellucida in the process ?
(iii) Analyse the changes occurring in the ovum during the process.
(iv) Mention what helps in the entry of sperm into the ovum.
(v) Specify the region female reproductive system where the event represented in the diagram takes place.

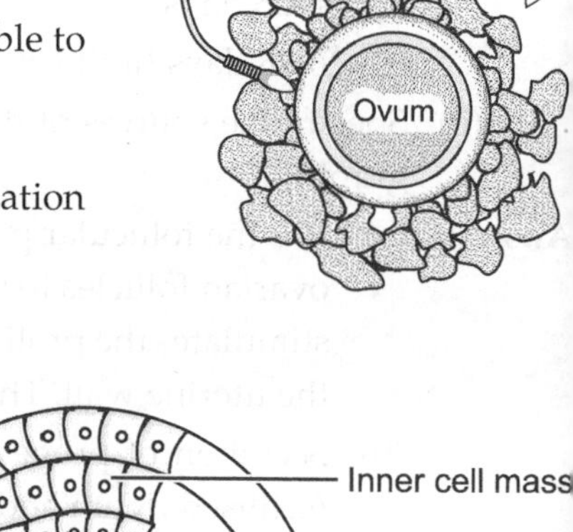

Ans. (i) A is able to penetrate/fertilise the ovum whereas B and C are unable to penetrate/fertilise. B and C will degenerate.
(ii) Zona pellucida ensure the entry of only one sperm into the ovum.
(iii) Induces completion of meiotic division of secondary oocyte formation of second polar body and a haploid ovum.
(iv) Enzymes of acrosomes (Hyaluronidase)
(v) Ampullary-isthmic junction of the fallopian tube.

14. Study the figure given alongside and answer the following questions.
(i) Name the stage of human embryo the figure represent.
(ii) Identify 'a' in the figure and mention its function.
(iii) Mention the fate of inner cell mass after implantation in the uterus.
(iv) Where are the stem cells located in this embryo ?

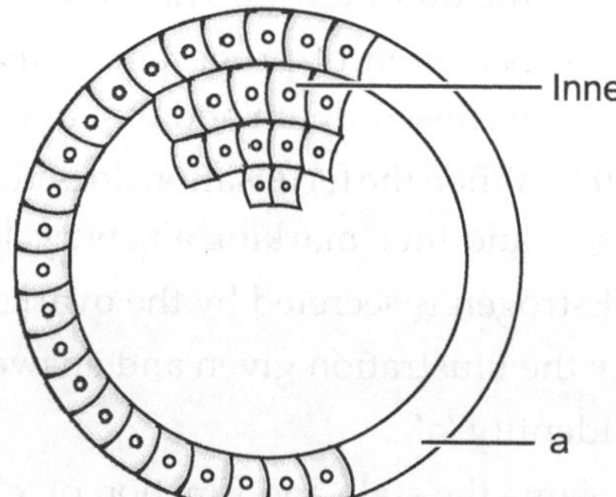

Ans. (i) Blastocyst.
(ii) 'a' is trophoblast. The trophoblast layer gets attached to endometrium and later forms extra embryonic membrane namely chorionic villi.

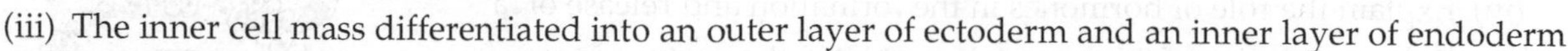

(iii) The inner cell mass differentiated into an outer layer of ectoderm and an inner layer of endoderm.
(iv) Inner cell mass.

15. Mention the number of cells in the following stages :

S. No.	Embryonic Stages	No. of cells
(i)	Zygote	________
(ii)	Morula	________
(iii)	Blastocyst	________

Ans. (i) 1
(ii) 2, 4, 8, 16
(iii) More than 16

Chapter 4. Reproductive Health

1. Identify the given diagram. What it is used for ?

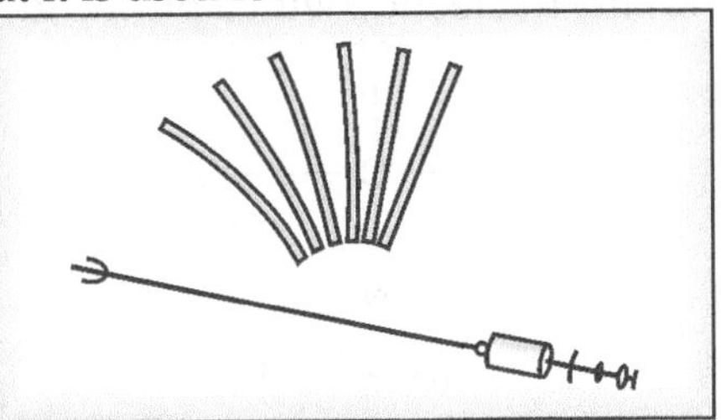

Ans. The diagram represents implants. These contain progesterone or progesterone-estrogen combination. These are used for inhibiting ovulation and implantation of embryo to the uterine wall.

2. Following table gives certain term associated with ART. Fill in the spaces 'a' 'b' 'c' and 'd'.

IVF and ET	a
b	Introduction of zygote with not more than 8 blastomeres into fallopian tube
c	Introduction of ova of a donar into the fallopian tube
d	Introduction of semen from the husband or healthy donar into the uterus

Ans. (a) Test tube baby programme
(b) ZIFT (Zygote Intra Fallopian Transfer)
(c) GIFT (Gamete Intra Fallopian Transfer)
(d) IUI (Intra Uterine Insemination)

Chapter 5. Principles of Inheritance and Variation

1. Identify a, b and c in the table given below.

S.No.	Patterns of Inheritance	Monohybrid F_1 phenotypic expression
(i)	Co-dominance	a
(ii)	b	The progeny resembled only one of the parents
(iii)	Incomplete dominance	c

Ans. a–Both the forms of trait are equally expressed in F_1 generation

b–Dominance

c–Phenotypic expression of F_1 generation is somewhat intermediate between the two parental forms of a trait.

2. Identify a, b, c, d, e and f in the given table below :

S.No.	Syndrome	Cause	Characteristic of affected Individuals	Sex male /Female/ Both
1	Down's Syndrome	Trisomy of 21st Chromosome	'a' (i) (ii)	'b'
2	'c'	XXY	Overall masculine development	'd'
3	Turner's Syndrome	45 with XO	'e' (i) (ii)	'f'

Ans. a - Short stature/small round head/ furrowed tongue partially open mouth/mental development retarded.

b - Both

c- Klinefelter's Syndrome

d- Male

e - (i) Sterile ovaries (ii) Lack of secondary sexual character

f - Female

Chapter 6. Molecular Basis of Inheritance

1. Name the component 'a' and 'b' in the nucleotide with a purine given below.

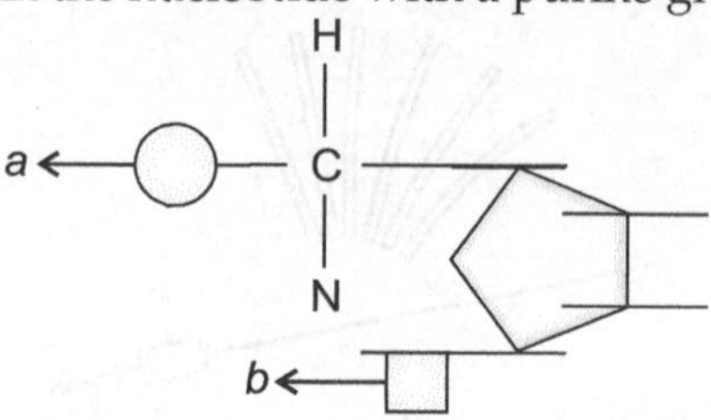

Ans. a - Phosphate group

b - The nitrogenous base

2. (i) What is this diagram representing ?

(ii) Name the part a, b and c.

(iii) In the Eukaryotes, the DNA molecules are organised within the nucleus. How is the DNA molecule organised in a bacterial cell in absence of nucleus?

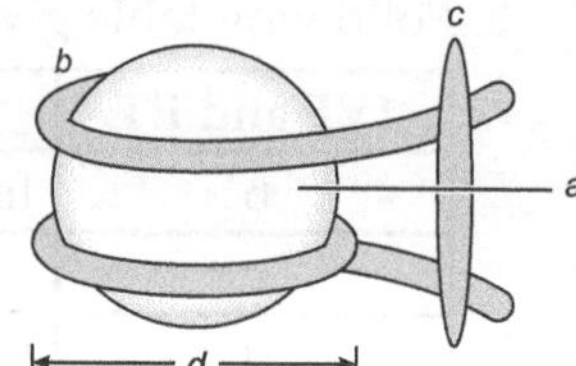

Ans. (i) Nucleosome

(ii) a – Histone octamer

b – DNA

c – H_1 Histone

(iii) In bacterial cell, DNA in nucleoid is organised in large loops held together by protein.

3. Mention the technical term in the following diagram.

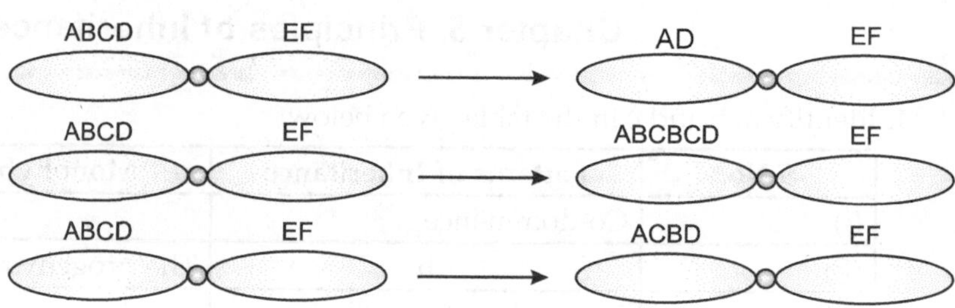

Ans. (i) Deletion

(ii) Duplication

(iii) Inversion

4. Name the types of synthesis 'a' and 'b' occurring in the replication fork of DNA as shown below.

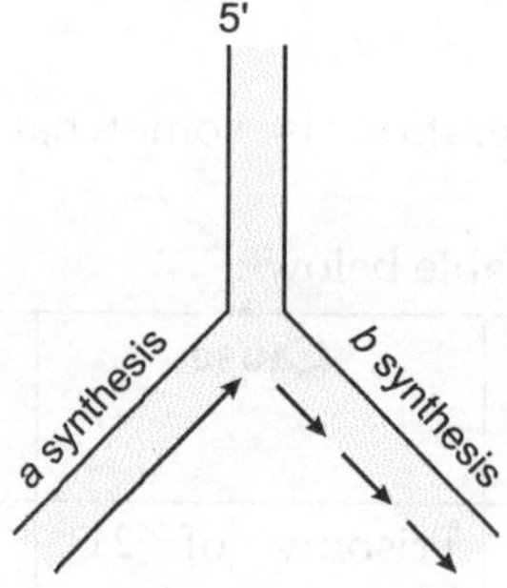

Ans. 'a' – Continuous synthesis.

'b' – Discontinuous synthesis.

5. Mention the polarity of the DNA strand a-b and c-d shown in the replicating fork given below.

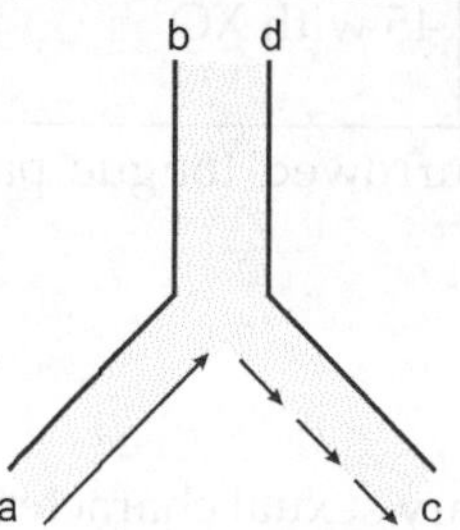

Ans. a-b has polarity 3′-5′ c-d has polarity 5′-3′

Identify the Following

6. Name the part 'A' and 'B' of the transcription unit given below.

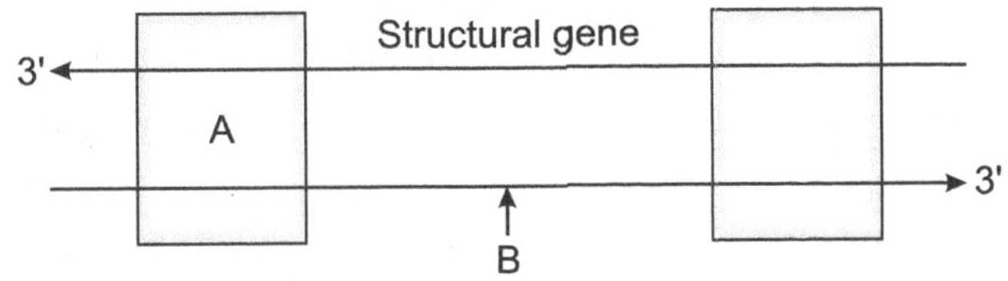

Ans. 'A' - Promoter sequence of DNA. 'B' - Coding strand.

7. (i) Name the molecule 'X' synthesised by 'i' gene. How does this molecules get inactivated ?
(ii) Which one of the structural genes codes for β - galactosidase ?
(iii) When will the transcription of this gene stop ?

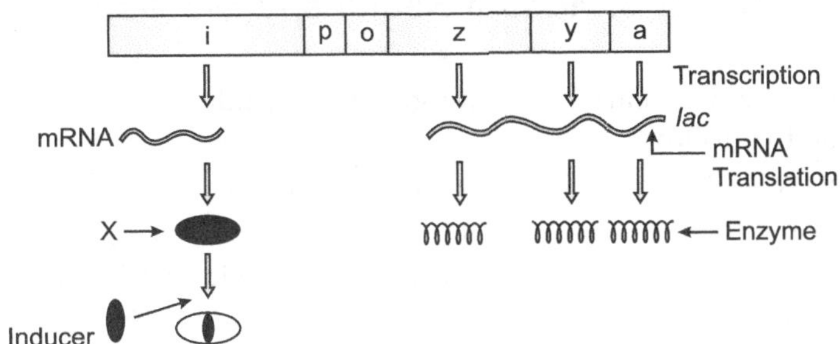

Ans. (i) The molecule 'X' is Repressor. It gets inactivated when lactose is present and then repressor molecule attatch to operator.
(ii) Z gene codes for β-galactosidase.
(iii) Transcription of the gene stop when lactose is absent and thus repressor is free to bind with the operator.

Chapter 7. Evolution

1. Given below is a diagrammatic representation of the experimental set up used by S.L. Miller for his experiment.

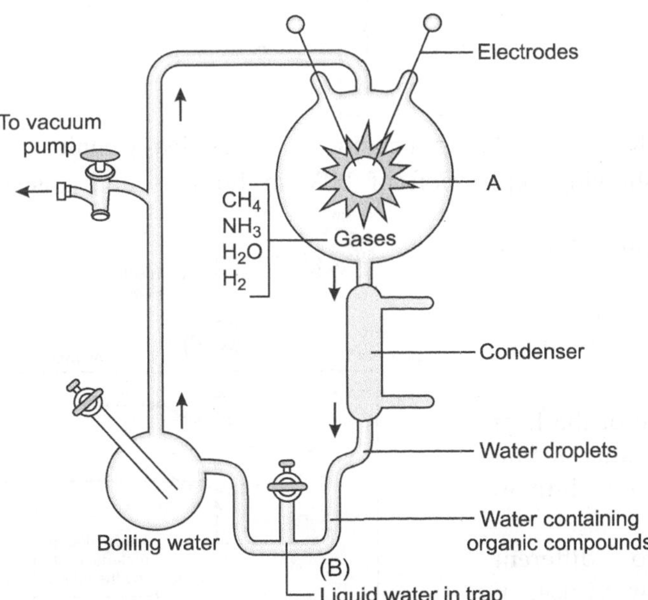

(i) Write the names of different gases contained and the condition set for the reaction in the flask 'A'.
(ii) State the type of organic molecules he collected in the water at 'B'.
(iii) Write the conclusion he arrived at.

Ans. (i) The gases obtained are CH_4, H_2, NH_3 and water vapour. The closed flask, electric discharge was created at 800°C.
(ii) Amino acids were collected in the water at 'B'.

(iii) He arrived at the conclusion that life came from pre-existing non-living organic molecules and tha formation of life was proceed by chemical evolution.

2. Fill in the blank (i), (ii), (iii), (iv) with name of the mammal of Australia.

Placental mammal	Marsupial mammal
Anteater	(i)
(ii)	Spotted cuscus
Bob-cat	(iii)
(iv)	Tasmanian wolf

Ans. (i) Numbat (ii) Lemur (iii) Tasmanian tiger cat (iii) Wolf.

Chapter 8. Human Health and Diseases

1. Study a part of the life cycle of malarial parasite given alongside and answer the question that follow :
   (i) Mention the role of 'A' in the life cycle of the malarial parasite.
   (ii) Name the event 'C' and the organ where this event occurs.
   (iii) Identify the organ 'B' and name the cell begin released from it.

Ans. (i) A – Gametocytes of plasmodium enter the mosquito when it bites a malaria patient and feed on blood. Mosquito act as a vector for malarial parasite.
(ii) C- fertilisation. It occur in the intestine of mosquito.
(iii) B- Salivary gland of the female *Anopheles* mosquito. These release sporozoites of *plasmodium*.

2. Identify A, D, E and F in the diagram of an antibody molecule given below.

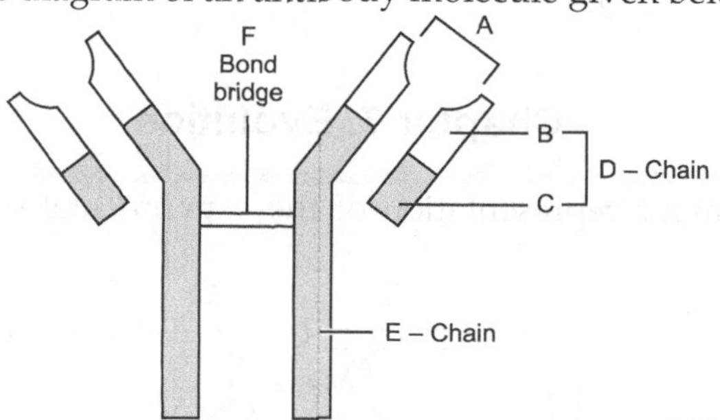

Ans. A - Antigen binding site D - Light chain E - Heavy chain F – Disulphide bridge.

3. Study the diagram showing replication of HIV in humans and answer the following questions accordingly :
   (i) Write the chemical nature of the coat 'A'.
   (ii) Name the enzyme 'B' acting on 'X' to produce molecule C. Name C.
   (iii) Mention the name of the host cells 'D' the HIV attacks first when it enters into the human body.
   (iv) Name the two different cells that the new viruses E subsequently attack.

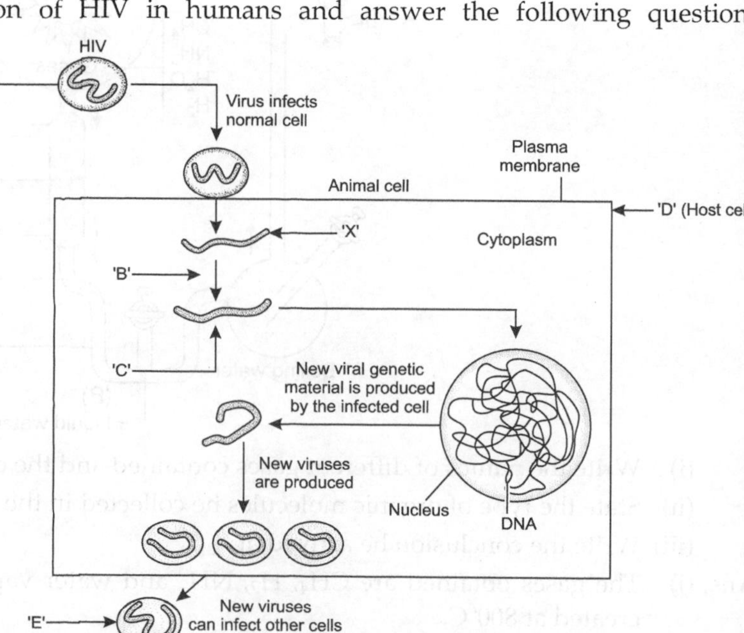

Ans. (i) Coat A is made up of protein.
(ii) The enzyme B is reverse transcriptase, which transcribe 'X'-viral RNA to produce 'C'-viral DNA.

(iii) The host cell 'D' is macrophage.

(iv) The new viruses 'E' subsequently attack macrophage and helper T-lymphocytes.

4. Name the blank spaces a, b, c and d in the table given below :

S.No.	Name of drug	Plant source	Organ system affected
(i)	a	Poppy plant	b
(ii)	Marijuana	c	d

Ans. (a) Morphine (b) Central Nervous System (c) *Cannabis sativa* (d) Cardiovascular system

Chapter 9. Strategies for Enhancement in Food Production

1. Mention the 'a' and 'b' in the following flow chart w.r.t apiculture

Habit and nature of bees
↓
Selection of location
↓
Catching and having of swarms
↓
a
↓
b

Ans. a – Management of beehives b – Handling and collection of honey.

2. Study the flow chart given below :

Cow is administrated with FSH Hormone
↓
(1)
↓
6-8 egg per cycle and derived
↓
Artificially inseminated
↓
Fertilised egg at 8-32 cell are recovered
↓
(2)

Identify the event taking place at stages (1) and (2).

Ans. Events taking place at: Stage (1) – Follicular maturation Stage (2) – Transfer to surrogate mother

3. Fill in the blanks :

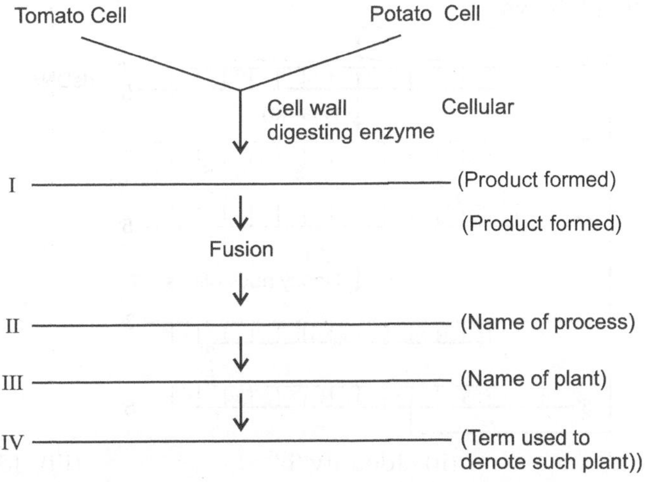

Ans. I – Protoplast II – Somatic hybridisation III – Pomato/Topato IV– Somatic hybrid

4. Identify A, B, C and D in the given below.

Crop	Variety	Resistance to disease
Wheat	A	Leaf and stripe rust, hill bunt
B	*Pusa shubhra*	Black-rot
Cowpea	*Pusa komal*	C
Brassica	*Karan Rai*	D

Ans. (A) *Himgiri* (C) Bacterial Blight
(B) Cauliflower (D) White Rust

5. To which revolution are the production related ?
(i) Crop plant (a) ________
(ii) Fish production (b) ________
(iii) Milk production (c) ________

Ans. (a) Green revolution
(b) Blue revolution
(c) White revolution

Chapter 11. Biotechnology : Principles and Processes

1. (i) Identify (a) and (b) Illustration in the following :

(a)

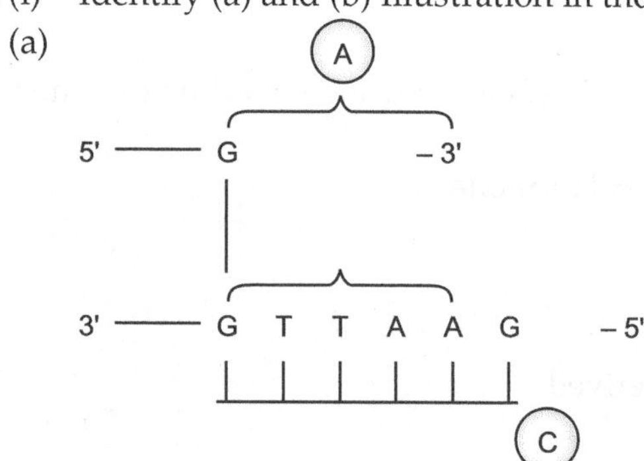

(b)

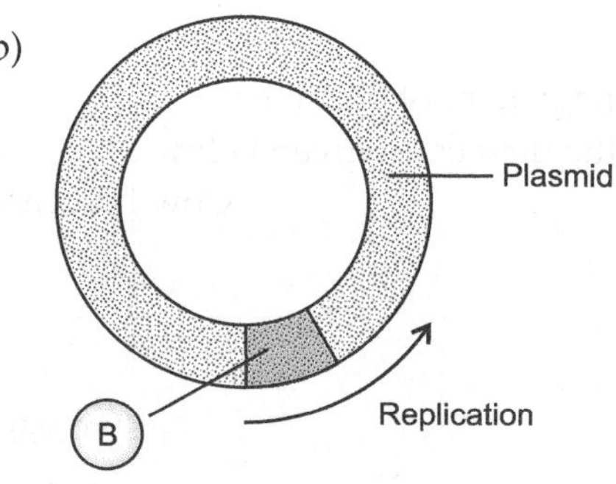

(ii) Write the term given to (A) and (C).

Ans. (i) (a) - Recognition site of the restriction endonuclease (b)- Origin of replication
(ii) A – Coding Strand C – Template strand

Chapter 12. Biotechnology and It's Applicaitons

1. A schematic representation of polymerase chain reaction (PCR) upto the extension stage is given below. Answer the questions that follows :

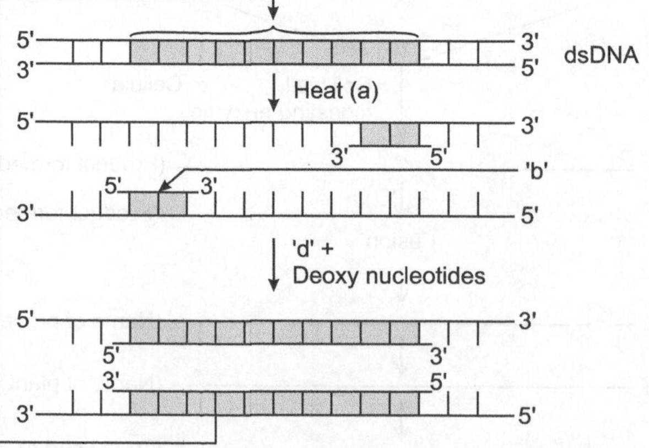

(i) Name the process 'a' (ii) Identify 'b' (iii) Identify 'c'

Ans. (i) a - Denaturation process (ii) b- Primers (iii) c- Taq DNA polymerase.

Chapter 13. Organisms and Populations

1.

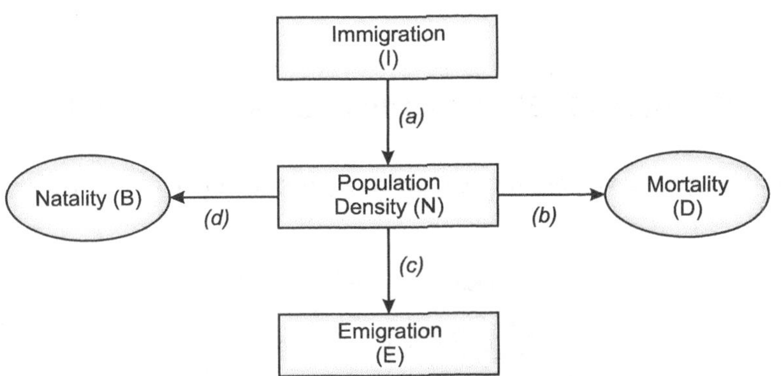

Which of the above represent increase or decrease of population ?

Ans. (a) and (d) represent increase of population.
(b) and (c) represent decrease of population.

2. Label the three tiers 1, 2, 3 given in the alongside age pyramids.

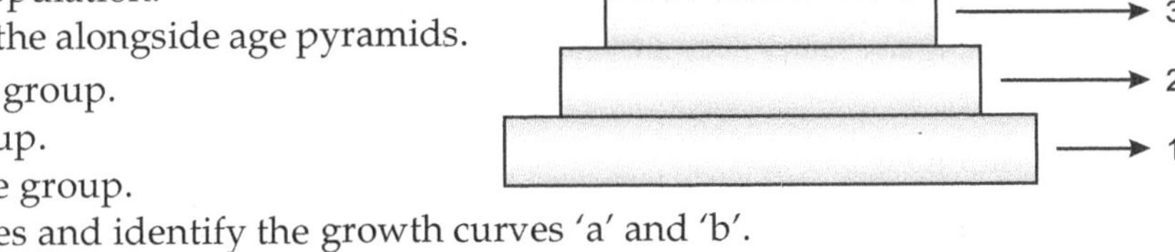

Ans. (1) represent pre-reproductive age group.
(2) represent reproductive age group.
(3) represent post-reproductive age group.

3. Study the population growth curves and identify the growth curves 'a' and 'b'.

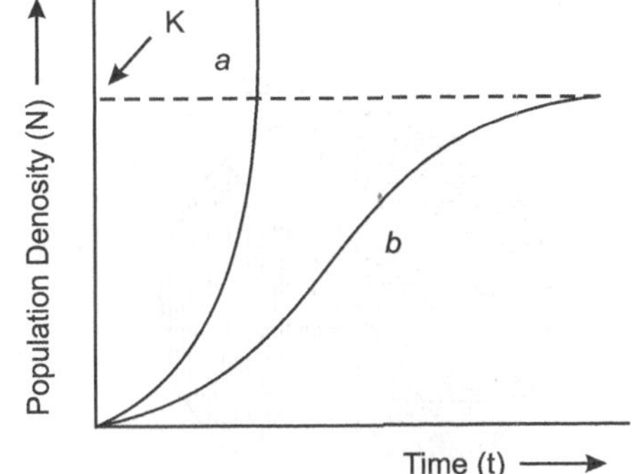

Ans. a is exponential growth curve or J-shaped curve
b is logistic growth curve or S-shaped curve.

Chapter 14. Ecosystem

1. Fill in the missing stages in the given primary hydrarch succession.
Phytoplankton → a → b → c → Submerged free floating → d → Forest plant stage

Ans. a – Reed- swamp stage
b – Submerged plant stage
c – Marsh-meadow stage
d – Scrub stage

2. Draw and complete the following model of carbon cycle filling a, b, c, d and e.

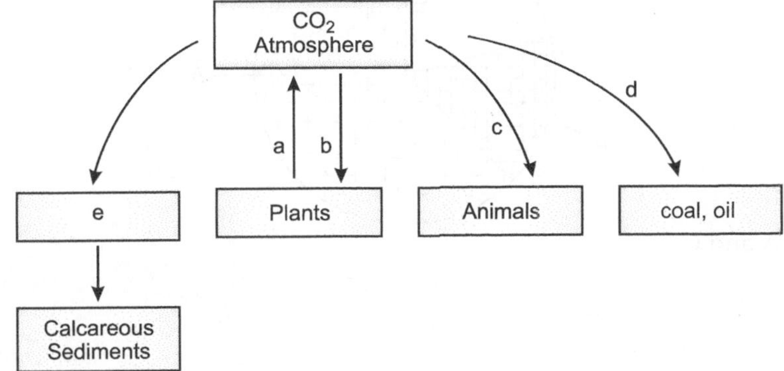

Ans. a – Photosynthesis
b – Respiration
c – Respiration
d – Combustion of fossil fuels
e – Aquatic food chain

Chapter 15. Biodiversity and its Conservation

1. Identify a and b in the figure given below representing proportionate number of major vertebrate taxa.

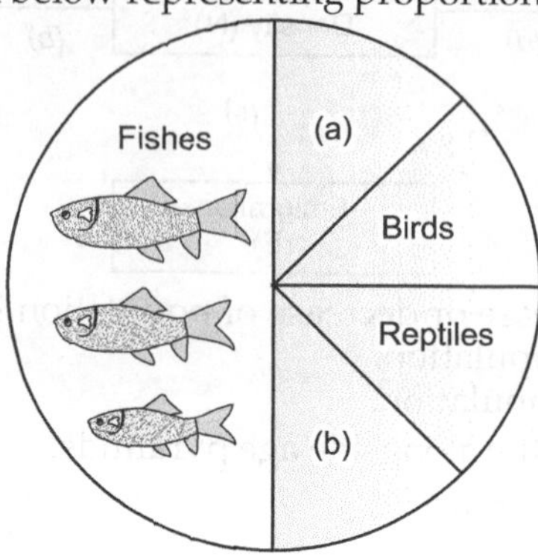

Ans. (a) Mammals
(b) Amphibians

2. Identify 'a' and 'b' in following pie chart.

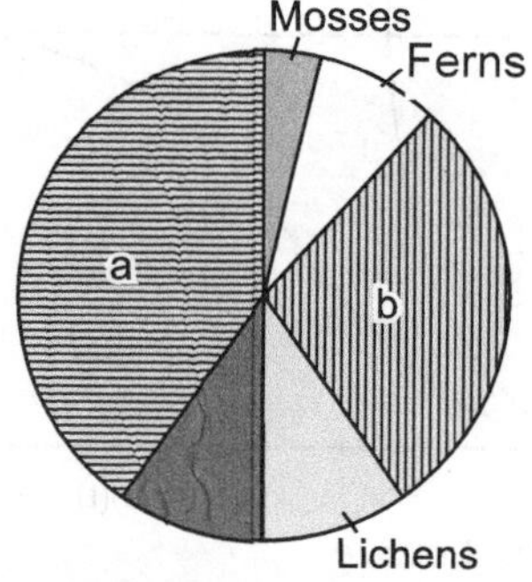

Ans. (a) Fungi
(b) Angiosperm

Chapter 16. Environmental Issues

1. The figure below shows the relative contribution of four green house gases to global warming.

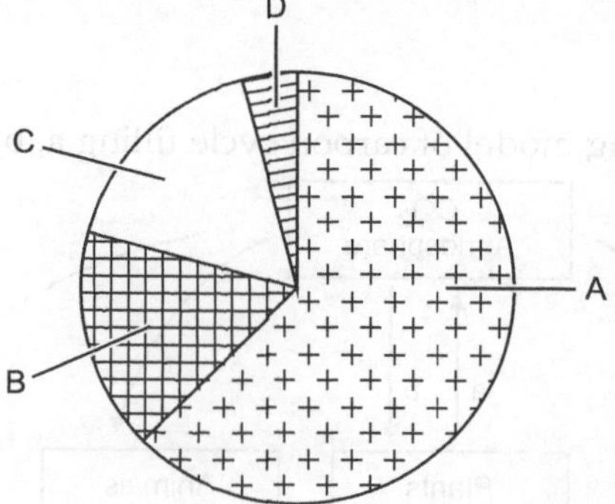

Identify the gases A and C.

Ans. Gas A – CO_2
Gas C – CFCs

Diagram Based Questions

Set 12

Chapter 1. Reproduction in Organisms

1. Explain Binary fission with an example.

Ans. Binary fission : The splitting of a parental cell into two equal daughter cells which rapidly grow into an adult is called binary Fission.

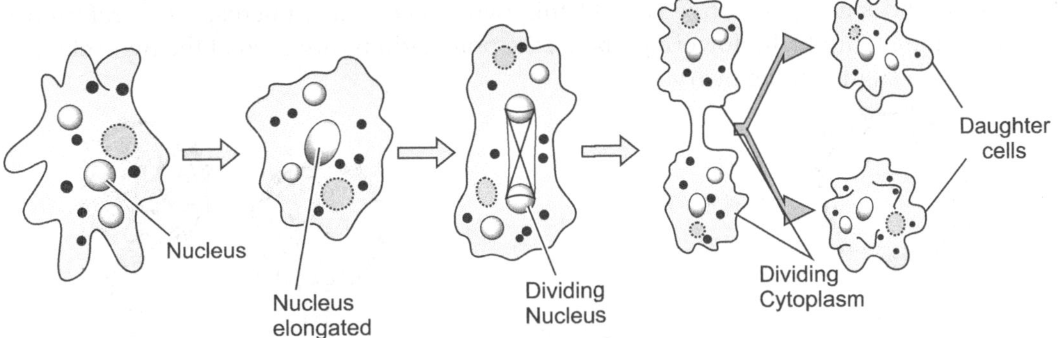

Binary fission in *Amoeba*

2. Explain multiple fission with an example.

Ans. Multiple fission : The splitting of parent cell into numerous daughter cells, each of which grow into an adult is called multiple fission *e.g. Plasmodium.*

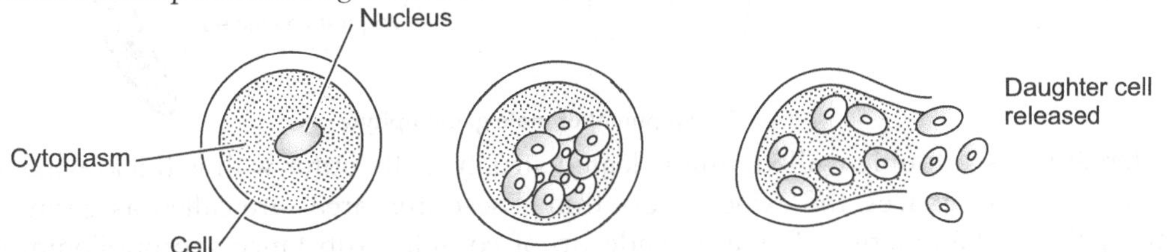

Multiple fission in *Plasmodium*

3. Diagrammatically represent the asexual reproduction in yeast.

Ans.

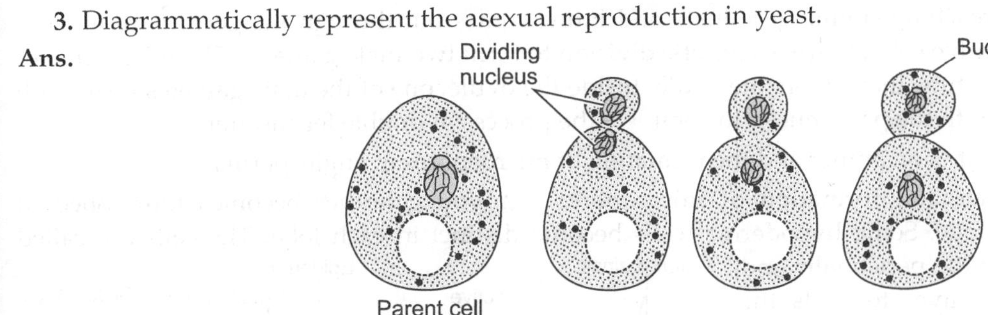

Budding in Yeast

4. Draw the sketches of a zoospore and a conidium. Mention two dissimilarities between them and at least one feature common to both structures.

Ans. Dissimilarities : Zoospores are motile whereas conidia are non-motile. Zoospores are produced in a sporangium, whereas conidia are produced externally.

Common feature : Both are means of asexual reproduction.

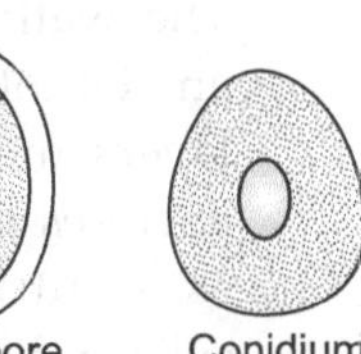

Chapter 2. Sexual Reproduction in Flowering Plants

1. Describe briefly the development of male gametophyte in an angiospermic plant.

Ans. Stamen is the male reproductive part of the flower. It consists of anther and filament. Pollen grains are produced in the anther. A young anther consists of homogeneous mass of undifferentiated meristematic cells called sporogenous cells surrounded by anther wall. During development, the anther becomes four lobed and each lobe is termed as microsporangium. A microsporangium contains sporogenous cells which divide mitotically and increases their number before they function as microspore mother cell (diploid). Each microspore mother cell divides meiotically and consequently, four microspores are formed from each microspore mother cell. All the four microspores remain enclosed in a common wall. This four celled organization is called tetrad.

Each individual microspore is a unicellular structure which divides mitotically to form large vegetative cell (tube–cell) and small generative cell. At this two–celled stage, microspore is referred to as pollen grain or male gametophyte. In flowering plants, the pollen grains are shed at the two-celled stage and are transferred to the stigma of the ovary.

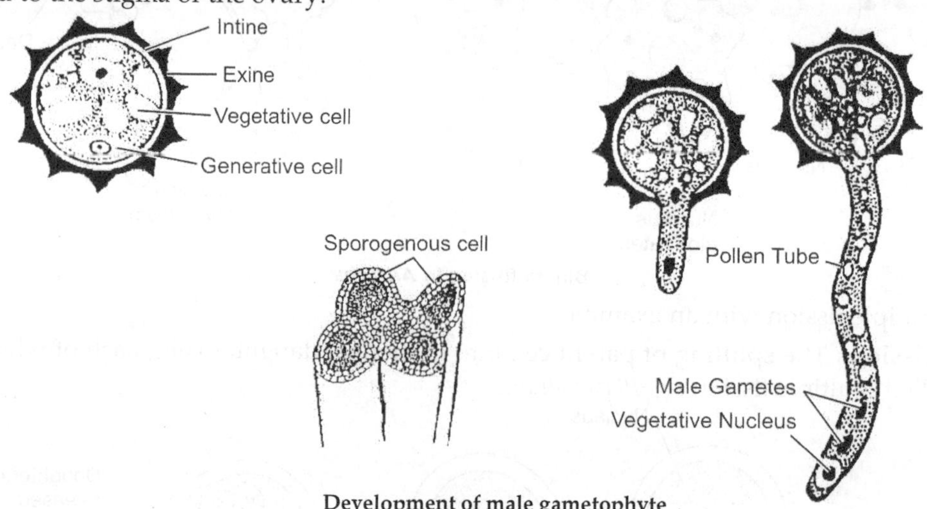

Development of male gametophyte

Each pollen grain is a double walled, uninucleated structure. Its outer wall is thick-exine and inner one thin-intine. At certain places, the exine is very thin. These thin areas are called as germ pores through which the pollen tube emerges. Exine is made up of complex substance-sporopollenin. It is the most resistant biological material which helps in fossilization of pollen grains.

The pollen grains on reaching stigma germinate and intine comes out through the germ pore to form pollen tube. The generative cell undergoes mitotic division to form two male gametes. The tube cell later degenerates. The pollen tube then enters the ovule. Inside the ovule, one of the male gametes fuses with the egg cell and other with the polar nuclei to complete the process of double fertilisation.

2. Explain the development of an anther and the formation of microspores in angiosperms.

Ans. Development of anther and formation of microspores : A young anther becomes four lobed at early developmental stage. Some hypodermal cells become distinct in each lobe. The cells are called archesporium. They divide periclinally and form primary parietal layer towards the outer side and primary sporogenous tissue towards the inner side. The primary parietal layer divides to form cells which form the wall of the anther.

The wall of anther is made of epidermis, an endothecium, two to three middle layers and a layer of tapetum. The tapetum encloses the sporogenous tissue and serves as a nutritive layer for the pollen mother cells and microspores.

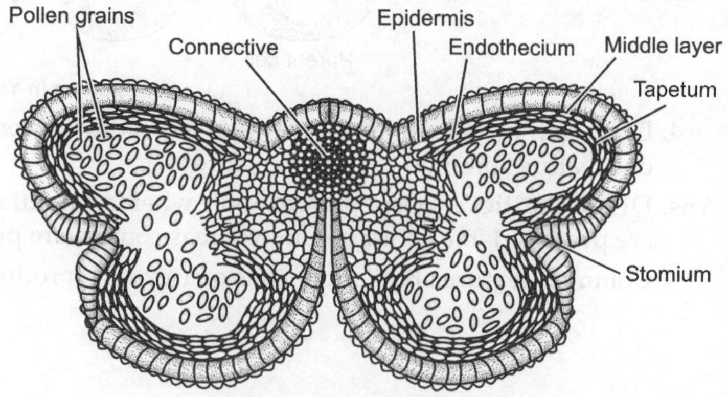

T. S. of a mature anther

Diagram Based Questions

The primary sporogenous tissue function as microspore mother cells. Each undergoes meiosis and forms four haploid microspores. The four microspores formed from a microspore mother cell are usually arranged in tetrad.

3. Give position and list the function of micropyle, chalaza, nucellus and integuments of an angiospermic ovule. Draw a diagram for the same.

Ans. (i) **Micropyle :** It is small opening at the apex of the integuments in the ovule. It serves the function of providing entry to the pollen tube containing male gametes, into the ovule.

(ii) **Chalaza :** It is the basal part of the ovule from which integuments develop.

(iii) **Nucellus :** It forms the body of ovule and is composed of parenchymatous cells. These cells are rich in food and provide nourishment to the embryo.

(iv) **Integuments :** Nucellus is surrounded by one or two cellular coats called integuments.

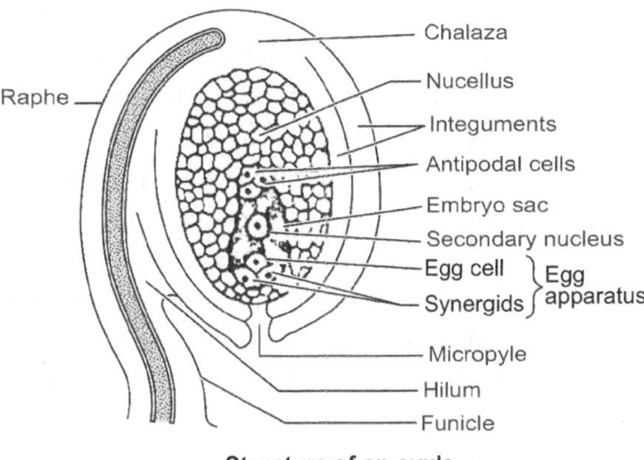

Structure of an ovule

4. Draw diagram of a mature ovule in an angiosperm and briefly write about its structure.

Ans. An ovule is a small structure attached to the placenta by means of a stalk called funicle. The point of attachment of the funicle with the main body of the ovule is called hilum. Sometimes, the funicle gets fused with the body of the ovule along one side and forms a ridge known as Raphe. The basal region of the ovule is known as Chalaza. The main body of an ovule is called Nucellus which consists of a mass of parenchymatous cells. The nucellus is invested all around by one or two layered protective covering called Integuments, except at the apex, where a small opening is present known as Micropyle. Towards the micropylar end, a mature ovule consists of embryo sac which has an egg apparatus (an egg cell and two synergids), centrally placed two polar nuclei and 3 antipodal cells at the chalazal end.

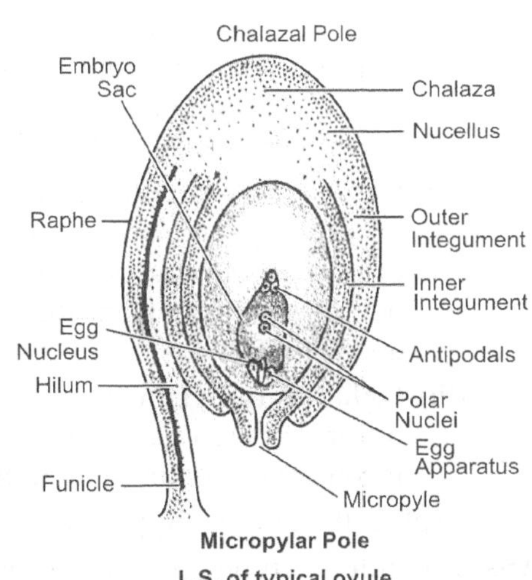

L.S. of typical ovule

5. Describe the formation of an embryo from a fertilised egg in angiosperms.

Ans. Embryo formation : After fertilisation, the egg-cell or ovum clothes itself with a cellulose wall and becomes oospore. Now, the oospore divides into two cells, an upper and a lower. The upper cell lying towards the micropyle further divides in one direction into a row of cells, the suspensor. The suspensor elongates and pushes the developing embryo deep into the embryo sac. It also acts as a feeding organ to the embryo during its development. For this purpose, the terminal cell of the suspensor often enlarges and acts as an absorbing organ. The basal cell of the suspensor which is known as hypophysis cell divides and gives rise to the apex of the radicle. The lower cell acts as the embryonal cell. It enlarges in size and divides by three walls at right angles into eight cells (octants). Four cells lying towards the suspensor form the posterior octants, while the other four cells lying away from the suspensor form the anterior octants.

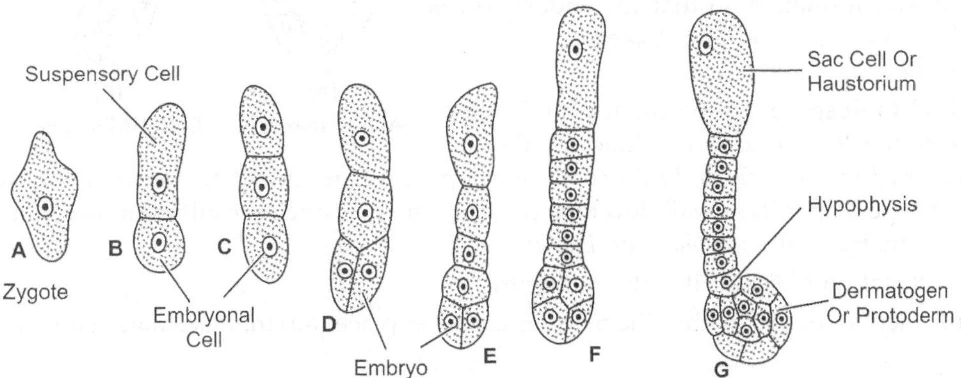

Stages in the development of embryo in angiosperm

Each octant then divides by a wall parallel to its curved surface. Thus a superficial layer of cells and a central mass of cells, the latter being known as the embryonal mass, are formed. The superficial or surface layer divides in one direction by radial walls only and remains single layered; finally it gives rise to the dermatogen, *i.e.,* the outer layer of the stem apex and the root apex. The embryonal mass forms the whole of the embryo except the root tip. The embryonal mass cells divide again and again forming the various parts of the embryo. The plumule and the two cotyledons are developed from the posterior octants. The apex of the radicle develops from the hypophysis cell. In the dicotyledonous embryo, two cotyledons are formed which are laterally placed and the plumule is terminal while in the monocotyledonous embryo, only one cotyledon is formed, which is terminal and plumule is lateral.

6. What is endosperm and how it is formed ?

Ans. Endosperm accumulates the food reserves and functions as the nutritive tissue for the developing embryo It develops from the triploid primary endospermal cell by repeated cell divisions. Its development starts shortly before the development of the embryo. It is formed by three methods :

(i) **Nuclear type :** In this, primary endosperm nucleus divides repeatedly by mitosis and number of free nuclei arrange themselves at the periphery of embryo sac, leaving a central vacuole. The nuclear division is not accompanied by the cell wall formation. But later on, protoplast collects around each nucleus and then walls are formed. So the endosperm becomes cellular at the maturity *e.g.,* wheat, rice, maize and sunflower.

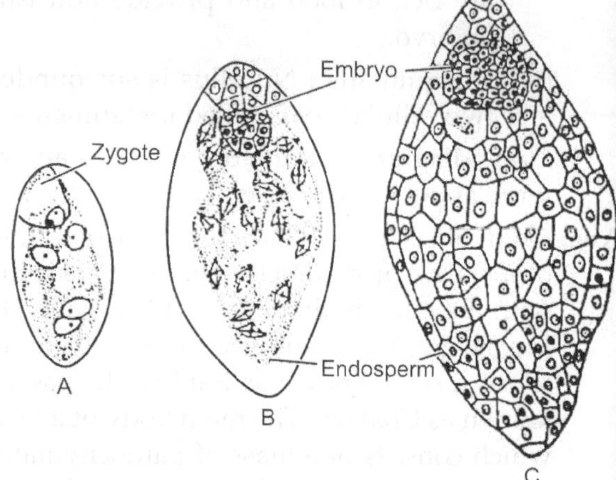

Three stages in the development in nuclear endosperm in *Acalypha indica*

(ii) **Cellular type :** In this, the repeated mitotic division of primary endosperm nucleus is followed by cytogenesis around each nucleus, so the endosperm becomes cellular from the beginning. *e.g., Datura* and *Petunia.*

(iii) **Helobial type :** In this, the first mitotic division of primary endosperm nucleus is followed by incomplete cytokinesis and the endosperm is formed both by cellular and nuclear type. E.g. *Eremurus.*

7. Explain the different types of endosperms in angiosperms.*

Ans. The different types of Endosperms found in angiosperms are :

(i) **Nuclear endosperms :** The endosperm nucleus gives rise to a number of free nuclei which stay in the peripheral layer. A large central vacuole appears in the embryo sac. Cell wall formation takes place from the periphery towards the centre leading to the formation of cellular endosperm. Examples : Maize, Wheat, Rice, Sunflower, coconut, etc.

(ii) **Cellular Endosperm :** Division of the primary endosperm nucleus is immediately followed by cell wall formation so that the endosperm is cellular from the beginning. Examples : Balsam, *Datura, Petunia, Villarsia,* etc.

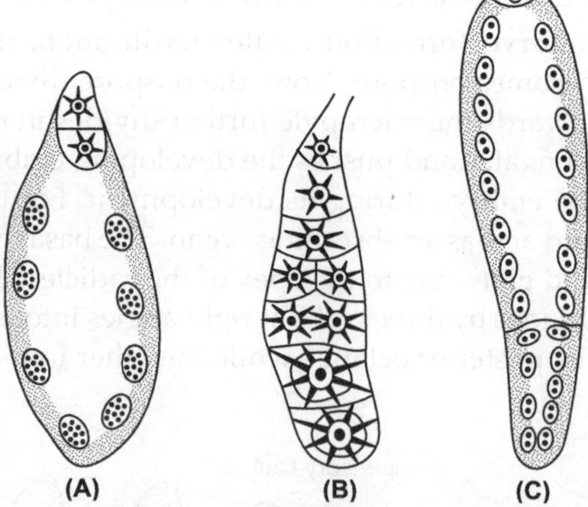

A. Nuclear type B. Cellular type C. Helobial type

(iii) **Helobial Endosperm :** A partition wall develops between the two nuclei resulting in the first division of the endosperm. Endosperm development is intermediate between the free nuclear and cellular type. A partition wall develops between the two nuclei resulting in the first division of the endosperm. Examples : *Asphodelus, Echium,* etc.

8. What is placentation ? Discuss its different types.

Ans. The arrangement of ovules within the ovary is called as placentation. Placentation are of different types, mainly :

* **Frequently asked previous years Board Exam Question.**

(i) **Marginal :** When the placenta forms a ridge along the ventral suture of the ovary and ovules are borne on this ridge forming two rows as in pea.

(ii) **Axile :** When the placenta is axial and the ovules are attached to it in a multilocular ovary. *E.g.* china rose, tomato, lemon.

(iii) **Parietal :** The ovules develop on the inner wall of the ovary on peripheral part. Ovary is one chambered. *E.g.* mustard.

(iv) **Free central :** When ovules are borne on central axis and septa absent. *E.g. primrose.*

(v) **Basal :** Placenta develop at the base of ovary and a single ovule is attached to it. *E.g.* sunflower, marigold.

(vi) **Superficial :** The ovules are borne on placentae that develop all round the internal surface of the partition wall. *E.g. Nymphaeaceae.*

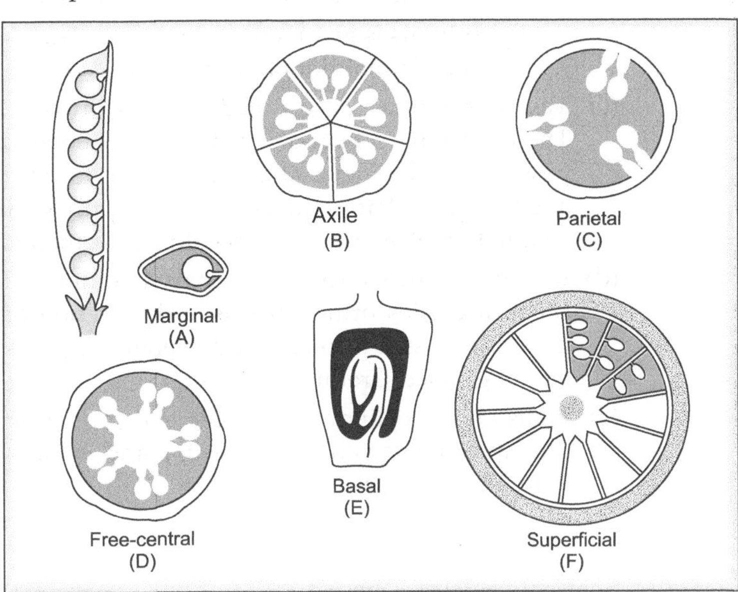

Types of Placentation

Chapter 3. Human Reproduction

1. Describe briefly male reproductive system with the help of labelled diagram.

Ans. The male reproductive system consists of following parts/organs :

(i) **Primary sex organs :** A pair of testis.

(ii) **A duct system :**

(a) Intra testicular duct system :

(1) Tubuli recti (2) Rete testis (3) Ductuli efferentes

(b) External/Accessory duct system :

(1) Vasa efferentia
(2) Epididymis
(3) Vasa deferentia
(4) Ejaculatory duct

(iii) **Accessory glands :**

(a) Seminal vesicles

(b) Prostate gland

(c) Cowper's glands/Bulbourethral glands

(iv) **Copulatory organs :** Penis

The accessory glands and copulatory organ together form the accessory sex organs.

(a) **Testis :** These are the primary sex organs which produce the male gametes, the sperms and secrete the male hormone-testosterone. A human testis measures about 5 cm × 3 cm × 2·5 cm. In humans, both testis remain suspended in a pouch called scrotum/scrotal sac outside the abdominal cavity. This is because the production of sperms require lesser temperature (3°C less than body temperature).

The scrotum is connected with the abdomen/pelvic cavity by the inguinal canal. It has passage of blood vessels, nerves and conducting tubes.

(b) **Epididymis :** On the lateral side of each testis, there is an elongated and flattened structure, called epididymis. It is a long (6 cm) highly coiled tube and is divided into 3 parts—

1. **Head (Caput) Epididymis :** It is the swollen part on top of testes on posterior side. It receives the sperms from rete testes by 10-12 fine ciliated tubules called vasa efferentia. It can store the sperms temporarily.

2. **Body (Corpus) Epididymis :** It is the middle part of epidydmis lying on the lateral side of testes.

3. **Tail (Cauda) Epididymis :** It is the top portion of epididymis. It occurs on the lower posterior part of testes and opens into vas deferens.

   Epididymis is involved in storage of sperms for about 18-24 hours. It also provides nutrition and aids in maturation of sperms.

(c) **Vasa deferentia :** Also called seminal ducts. These are long, narrow tubes about 30 cm long. These are muscular and start from respective cauda epididymis. These then extend backwards, passes over the inguinal canal over the urinary bladder. These join the duct of seminal vesicle and form ejaculatory duct (2 cm long approx.). It finally joins the urethra.Vasa deferentia help in conduction of sperms by peristaltic movements.

(d) **Urethra :** It arises from the urinary bladder and joins the ejaculatory duct to form urino-genital canal. It carries urine, sperms and secretion of seminal vesicle, prostate glands and cowper's glands. It passes throughout the length of penis and finally opens at the tip of penis through an opening called Urethral meatus.

(e) **Penis :** It is a cylindrical, erectile and copulatory organ, forming the external genitalia of males. The tip of penis is highly sensitive and is known as Glans penis, covered by a skin fold called Foreskin. It helps in copulation, *i.e.*, transfer of sperms.

(v) **Seminal vesicles :** These are paired tubular coiled glands, situated behind the bladder. They join the posterior part of vas deferens to form common ejaculatory duct. They secrete viscous fluid which constitute the main part of the ejaculate.

(vi) **Prostate glands :** These are paired glands situated around the first part of the urethra. They secrete alkaline fluid to neutralize the acidity of urine and to make the sperms mobile. They form the 5 to 30 percent of the ejaculate.

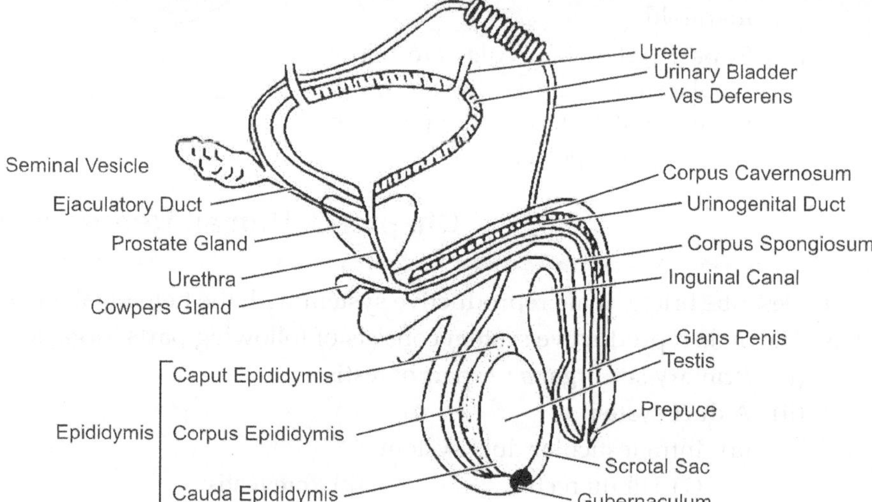

Male reproductive system

(vii) **Cowper's glands :** These are paired glands, situated beneath the bladder and behind the urethra into which their ducts open. They secrete white, viscous and alkaline mucus like substance that acts for lubrication.

2. Draw a labelled diagram of L.S. of human testis.*

Ans.

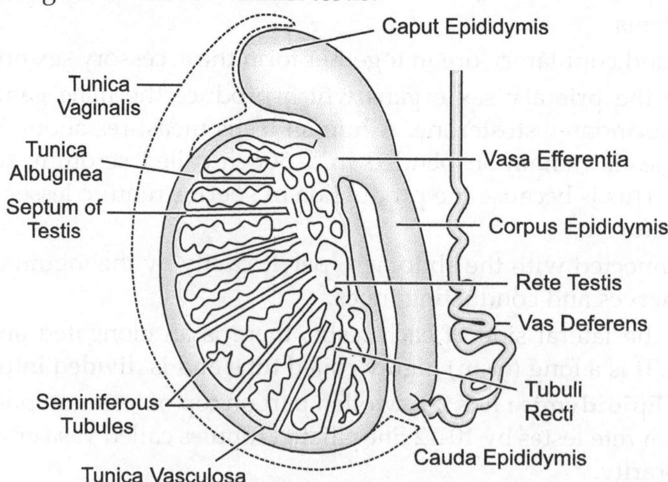

3. Describe briefly the structure of human testes.

* **Frequently asked previous years Board Exam Question.**

Ans. Each human testes is surrounded by a layer of fibrous connective tissue called tunica albuginea. Its cavity is divided into numerous lobules by means of fibrous partitions. Each lobules contain mass of long and coiled seminiferous tubules. The lining of seminiferous tubules is formed of germinal epithelial cells which give rise to spermatozoa. These spermatozoa are released into the lumen of tubules. The seminiferous tubules open into the testis which forms the vas efferentia in the epididymis. The connective tissues are found around the seminiferous tubules. These contain the masses of polyhedral Leydig cells, blood vessels and nerve fibres. The Leydig's cells secrete male sex hormone testosterone to control the development of secondary sex organs. The specialized sertoli cells are found in between the germinal epithelium of seminiferous tubules. These provide nutrition to the developing sperms.

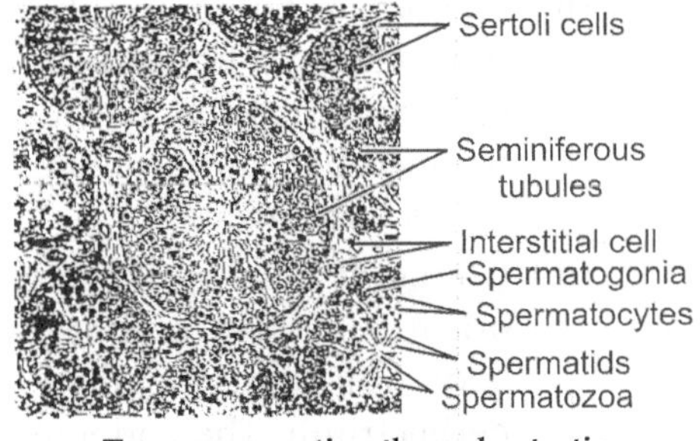

Transverse section through a testis

The spermatogonia lie close to periphery of the tubules and are closely packed. The spermatocytes including primary and secondary lie just below the spermatogonia and the spermatids aggregate in clusters below the spermatocytes.

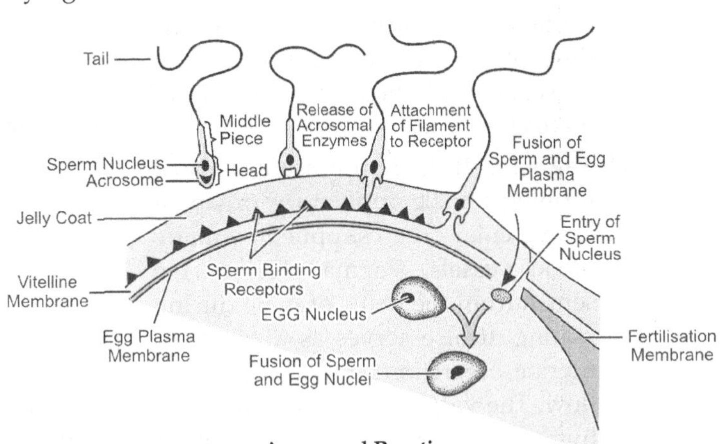

Acrosomal Reaction

4. Describe in brief the structure of a mammalian spermatozoan.

Ans. The human sperm is a long, flagellated and motile cells and is formed of head, neck, middle piece and tail.

(i) **Head :** It is the anterior part and contains the nucleus. The nucleus consists of densely packed DNA and nucleoprotein. The sheathy cap like acrosome is found on the first part of the head. It is the modified form of the Golgi body. The acrosome contains the enzyme hyaluronidase which helps the sperm to penetrate the ovum during fertilisation.

(ii) **Neck :** It is very short, containing the proximal and distal centriole. Several fine microtubules run from a distal centriole of neck to the end of the tail through the middle piece. The proximal centriole takes part in the first division of the fertilised ovum. Second centriole (Distal) gives rise to axial filament.

(iii) **Middle piece :** It consists of number of mitochondria to provide energy for the movement of sperm. The axial filament running from distal centriole are encircled by a spiral sheath of mitochondria called nebenkern.

(iv) **Tail :** It is a long structure and is formed of an axial filament surrounded by small amount of cytoplasm and an external sheath. The tail provides mobility to the sperm by its whip-like or lashing movements.

An electron microscopic view of a mammalian sperm

5. Describe briefly the female reproductive system with the help of labelled diagram.

Ans. The female reproductive system consists of a pair of ovaries, a pair of oviducts (fallopian tubes), uterus and vagina.

(i) **Ovaries :** Two small almond like flattened bodies lying on side of vertebral column each one is attached to the dorsal abdominal wall. Each ovary contains numerous Graafian follicles which release ovum on their maturing into the abdominal cavity.

(ii) **Oviduct or fallopian tubes :** These are pair of oviducts which are closely applied to each ovary by funnel shaped fimbriae at their anterior end. The posterior parts of the oviducts are widen which open into muscular and vascular uterus. The oviducts are narrow tubes lined by ciliated cells. The oviducts pick up the ovum by their fimbriae and sweep it into oviducts and towards uterus by their ciliary movements.

(iii) **Uterus :** It is the most posterior part of the oviduct which is thickened, highly muscular and vascular organ. It is joined on either side with fallopian tube and its lower portion opens into vagina. The uterus body wall consists of three coats, the innermost endometrium, middle myometrium and outermost perimetrium. The endometrium wall is highly vascularised, muscular and contains the uterine glands. At the time of parturition, contractions of muscular walls help in expulsion of foetus from uterus.

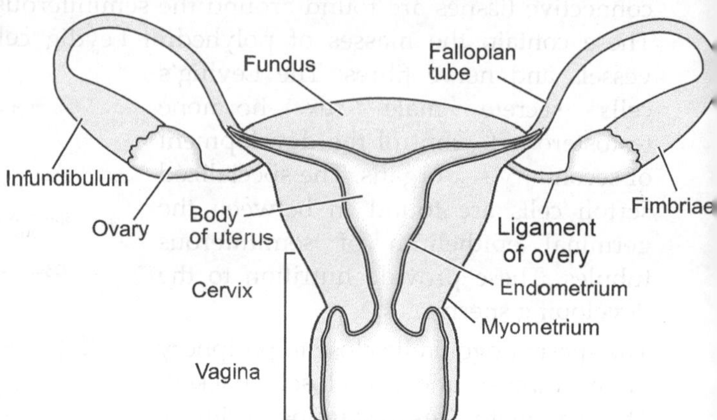

(iv) **Vagina :** It is fibromuscular canal forming female copulatory organ. It is a vascular organ supplied by many blood vessels. Vagina receives the semen from the penis of male during mating. It also serves as a canal for uterine secretions and menstrual flow. The vaginal orifice contains paired Bartholin's glands to secrete lubricating fluid between the hymen and labia minora. The urethra and vulva form separate openings in the human female.

(v) **Vulva :** Also referred as external genitalia of female-reproductive system, these are external structures associated with the passage of genetial tract. It has a central depression called vestibule which lies in front of anus. The vestibule has two apertures—upper urethral orifice and lower vaginal orifice. The vaginal orifice is covered by a membranous fold called Hymen which gets perforated at the onset of puberty and is broken during first intercowers.

(vi) **Accessory glands :**

(a) **Bartholin's glands (Bulbo vestibular glands) :** These are one pair, small-sized glands present behind labia minora on either side of vaginal orifice. These secrete mucus to lubricate vagina.

(b) **Mammary glands :** These are milk producing modified sweat glands which occur inside two rounded thoracic prominences called breasts. Milk secretion occurs only after parturition.

6. Draw a labelled diagram of the internal structure of human ovary.*

Ans.

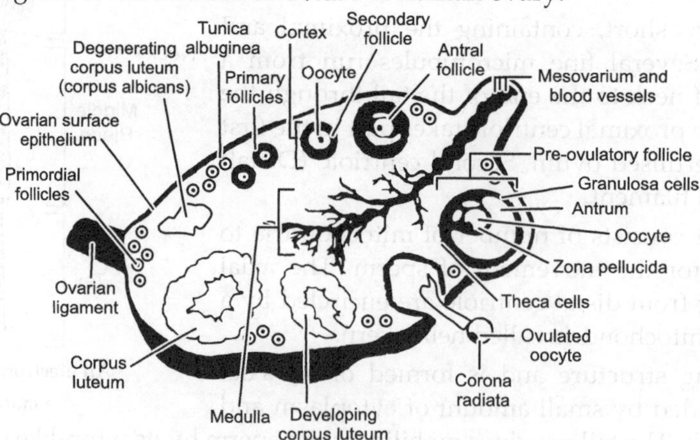

Fig. Internal structure of human ovary

7. Describe briefly the structure of human ovary.

Ans. The human ovary is solid structure surrounded by germinal epithelium and followed by tunica albuginea. The inner mass of ovary is differentiated into outer-cortex and inner narrow zone the medulla. The germinal epithelium gives rise to ovarian follicle in various stages of development.

* **Frequently asked previous years Board Exam Question.**

Diagram Based Questions

These follicles are of four types : primordial follicles, primary follicles, vesicular follicles and follicles. In primary follicles, the oocyte is surrounded by one layer of follicular cells whereas in follicles, the ovum is surrounded by multiple layers. A mature graffian follicle consists of ovum is surrounded by theca interna, membrane granulosa with fluid filled liquor folliculi. The ovum is enclosed by zona pellucida and corona radiata. The follicular cells secrete the oestrogen hormone. On maturity, the follicle cells burst and ovum is released and the Graafian follicle is transformed into corpus luteum.

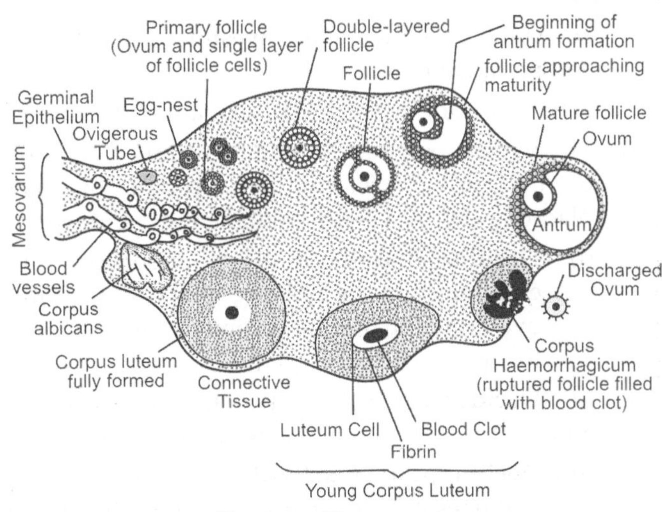

Structure of human ovary

8. Explain the structure and function of mammary glands.

Ans. Mammary glands are milk producing modified sweat glands which occur inside two thoracic prominence called breasts. Each breast has a broad multiporous erectile tip called nipple for the release of milk. A circular pigmented area called areola lies below it. Areola possesses circular muscles, nerve endings and sebaceous/aerolar glands. Each breast contains 15-20 glandular lobes separated from one another by dense/fibrous connective tissue and adipose tissue. These are called mammary lobes.

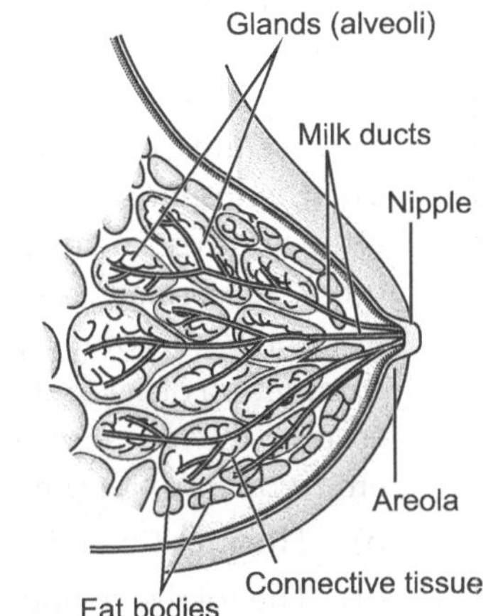

Each lobe is further made up of a number of lobules having glandular/secretory tubulo alveoli. The cells of alveoli produce milk which is stored in the cavities or lumens of alveoli. Alveoli open into mammary tubules and the mammary tubules into mammary ducts.

Mammary ducts form a broad mammary ampulla (*lactiferous sinus*) for storing milk. A lactiferous duct develops from it. Each lobe or gland produces a separate lactiferious duct. The various lactiferous ducts open at the nipple by separate pores.

Milk secretion occurs only after parturition. It is stimulated by prolactin hormone of anterior pituitary. Expulsion of milk is under the control of another hormone called oxytocin. It is secreted by posterior pituitary.

First milk after parturition is called colostrum/Fore milk. It contains fat (fat bodies), caesin (milk protein), lactose (milk sugar), mineral salts, vitamins and antibodies. Small quantity of vitamin C is also present. Iron is absent.

9. Give the structure of human ovum in a brief.

Ans. The human ovum is rounded, non-motile and alecithal. It contains centrally located nucleus with bulk of cytoplasm. It is enveloped by vitelline membrane and again by transparent and non-cellular layer, called zona pellucida.

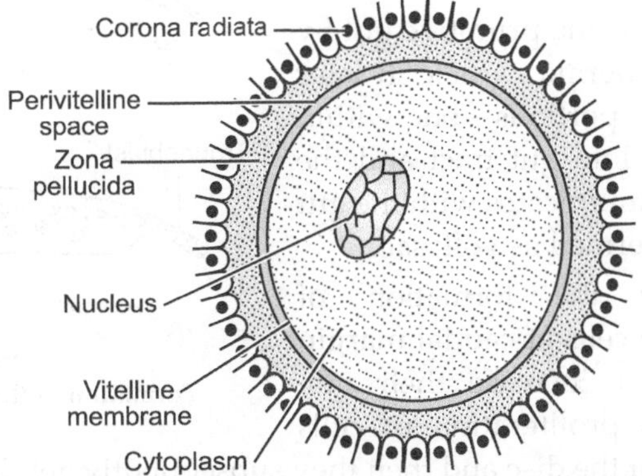

The space lies between vitelline membrane and zona pellucida is called as previtelline space. Outside the zona pellucida, there is an investment of radially elongated follicle cells. This cellular investment is called corona radiata.

10. What is meant by Decidua ? Explain in detail.

Ans. The uterine part of placenta contributed by mother's uterine wall is referred as Decidua or maternal placenta. Decidua is differentiated into three parts.

(i) **Decidua basalis :** It is modified part of endometrium underlying the chronic villi and overlying the uterine myometrium. Later, it forms the maternal part of placenta.

(ii) **Decidua capsularis :** It is the portion of uterine endometrium between embryo and uterine cavity.

(iii) **Decidua parietalis :** It is the modified endometrium that lines the uterus at places other than the site of attachment of embryo.

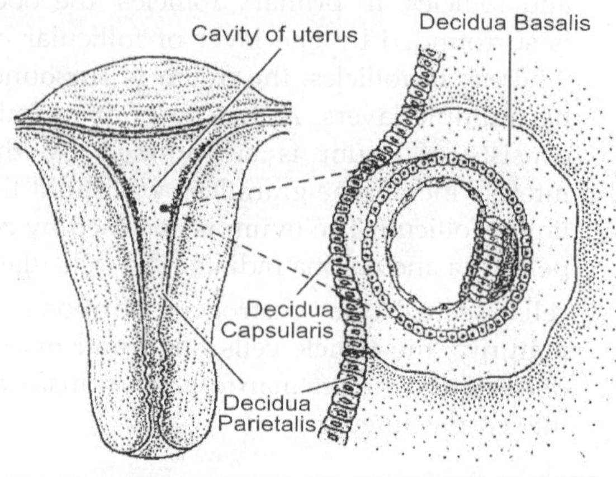

Frontal Section of Uterus **Details of Decidua**

11. Describe the mitotic divisions that occur during the process of morula formation.

Ans. In the morula formation, the fertilised ovum divides mitotically by holoblastic cleavage starting from animal pole to vegetal pole. This eventually divides the ovum into two daughter cells. Immediately, it undergoes second mitotic division at right angle to the first and four dumb–bell shaped cells are formed. Then, subsequent cleavage divisions proceed one after another in an orderly manner and in more quick succession, resulting into blastomeres which gradually become smaller and smaller. After completion of several segmentation, the blastomeres assume the shape of solid ball of cells. This stage of embryonic development is called morula.

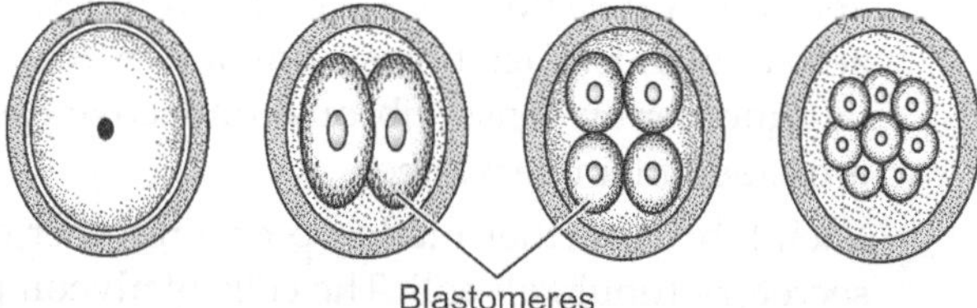

Stages of morula formation

(i) The mitotic divisions do not bring any appreciable increase in the mass.

(ii) The total size and volume of the embryo remains the same.

(iii) There is a marked increase in the DNA containing chromosomal materials.

(iv) The diploid number of chromosomes are maintained in this process.

12. Describe the formation of three germ layers in mammalian embryo.

Ans. The formation of three germ layers is described as under :

(i) **Endoderm :** Some cells detach from the inner cell mass (embryoblast) and these cells push into the blastocoel to form the first endodermal cells. These cells divide rapidly to increase their number and finally form the complete endodermal layer inside the original outer layer of the blastodermic vesicle. This inner endodermal layer forms the primitive gut by obliterating the blastodermic vesicle. It later differentiates into two portions—one within the embryo develops into gut tract and the other as distal sac communicates with the gut of the embryo and is termed as yolk sac.

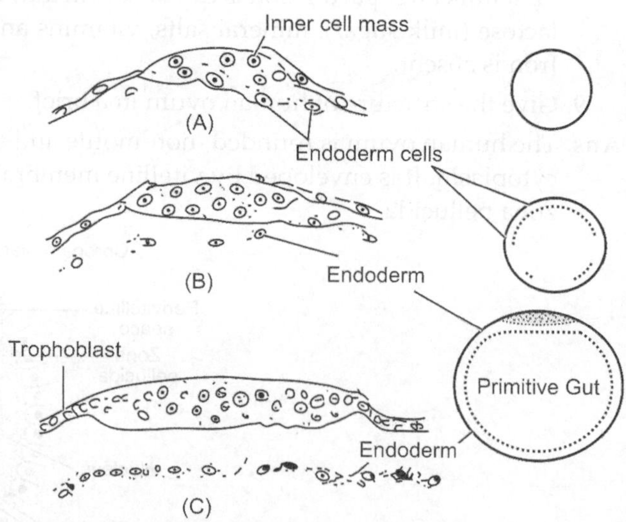

Formation of three germ layers in mammalian embryo

(ii) **Mesoderm :** After the formation of endodermal layer, the cells at caudal margin of the embryonic disc increase their rate of proliferation. This proliferation of cells increase in thickness of the disc and then they subsequently get detached from the embryonic disc to form distinct mesodermal layer.

Diagram Based Questions

(iii) **Ectoderm :** After the formation of mesodermal cells from embryonic disc, the remaining cells of the embryonic disc arrange themselves in a distinct layer to form the ectoderm.

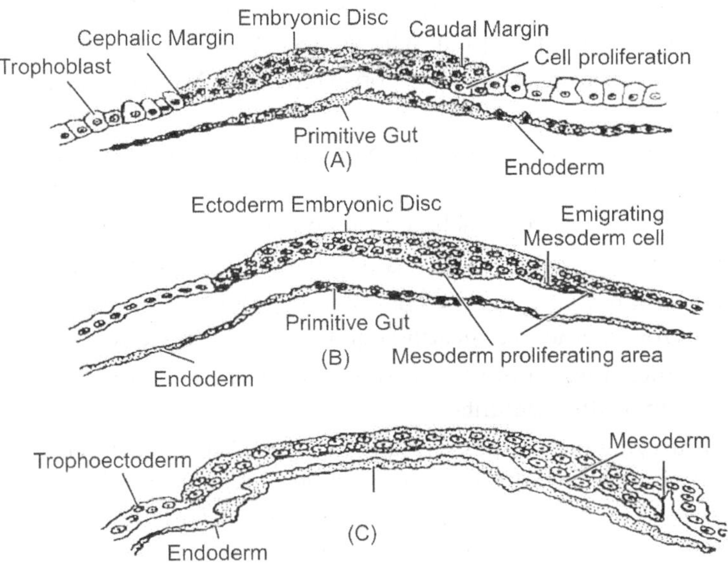

Formation of extra-embryonic Mesoderm and extra embryonic coelom

Chapter 4. Reproductive Health

1. Describe vasectomy and tubectomy with diagram.

Ans. Vasectomy and tubectomy are permanent methods of contraception and are also known as sterilisation technique. These methods are generally advised to male or female partner as a terminal method to prevent any more pregnancies. In vasectomy a small part of vas deferens is removed and then tied up through a small incision in the scrotum whereas in tubectomy a small part of the fallopian tube is removed and then tied up through a small incision in the abdomen or through vagina. These techniques are highly effective but their reversibility is poor. Both technique block gamete transport and thereby prevent contraception.

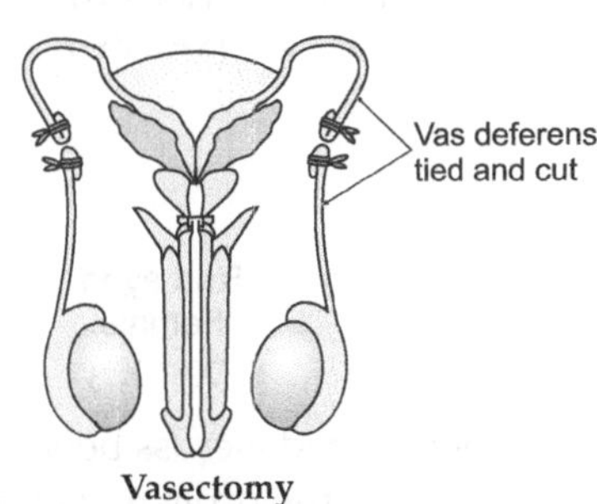

Vasectomy

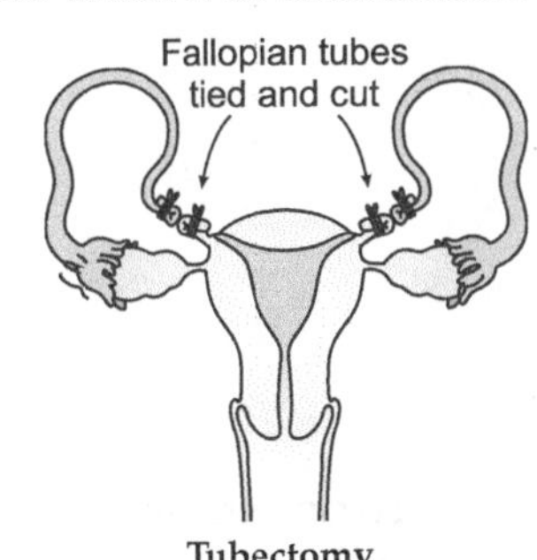

Tubectomy

Chapter 5. Principles of Inheritance and Variation

1. A man, who is a son of a bald mother and non-bald father, married a non-bald homozygous woman. If they have two sons what is the genetic probability of them being bald ?

Ans. When the man's mother is with bald (BB) and father is non-bald (bb), the genotype of that man is Bb (heterozygous bald). If he (Bb) marries a non-bald homozygous woman (bb), the condition of their two sons appears in the following manner.

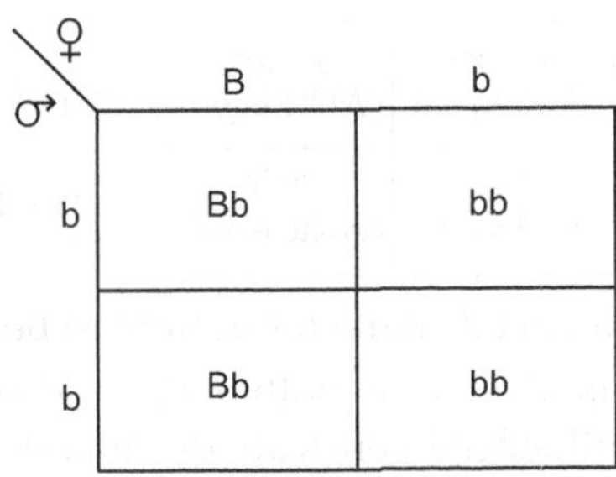

♂ \ ♀	B	b
b	Bb	bb
b	Bb	bb

Bb = Bald, bb = Non-bald

So the genetic probability is 50% bald and 50% non-bald in sons.

2. Explain Chromosomal Theory of Inheritance with example.

Ans. In 1902 abd 1903, Sutton and Boveri explained the Chromosome Theory of Inheritance. This theory states that individual genes are found at specific locations on particular chromosomes, and that the behaviour of chromosomes during meiosis can explain why genes are inherited according to Mendel's laws

The Chromosomal Theory of Inheritance was consistent with Mendel's laws and was supported by the following observations.

- During meiosis, homologous chromosome pairs migrate as discrete structures that are independent of other chromosome pairs.
- The sorting of chromosomes from each homologous pair into pre-gametes appears to be random.
- Each parent synthesizes gametes that contain only half of their chromosomal complement.
- Even though male and female gametes (sperm and egg) differ in size and morphology, they have the same number of chromosomes, suggesting equal genetic contributions from each parent.
- The gametic chromosomes combine during fertilisation to produce offspring with the same chromosome number as their parents.

After several years of carrying out crosses with the fruit fly, *Drosophila melanogaster*, in 1910, Thomas Hunt Morgan provided experimental evidence to support the Chromosomal Theory of Inheritance. In flies, the wild type eye colour is red (XW) and is dominant to white eye colour (Xw). He was able to conclude that the gene for eye colour was on the X chromosome. This trait was thus determined to be X-linked and was the first X-linked trait to be identified. Males are said to be hemizygous, in that they have only one allele for any X-linked characteristic.

He crossed a purebred white eyed male with purebred red-eyed female. According to Mendel's laws the F_1 progeny were born with red eyes. When F_1 generation was self-crossed, the ratio of red-eyed to white eyed progeny was 3:1. However, he noticed that there was no white-eyed female in the F_2 generation.

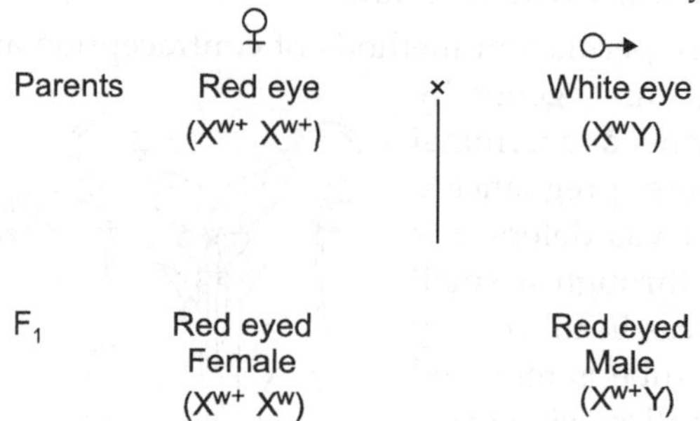

To understand further, he performed a cross between a heterozygous red-eyed female with a white-eyed male. This gave a ratio of 1:1:1:1 in the progeny (1 white eyed female, 1 red eyed female, 1 white eyed male and 1 red eyed male). This gave the idea about the linkage between the traits and sex chromosomes. He performed many more crosses and determined that the gene responsible for the eye colour was situated on the X chromosome.

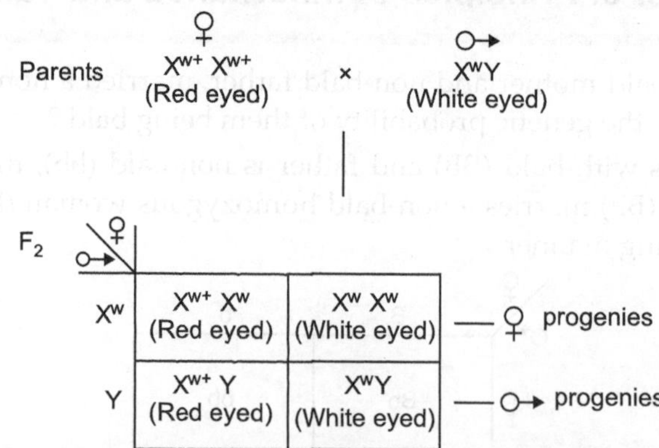

3. Explain the inheritance of a sex influenced character in human being.

Ans. Sex influenced genes are the autosomal genes in both males and females, but their phenotypic expression is different in different sexes. Sex influenced genes are dominant in one sex, and recessive in the other.

e.g., Baldness in man : Baldness causing gene, 'B' is dominant in males while the same gene is recessive in females.

Pattern of Baldness in man :

Genotype	Male	Female
BB	Bald	Bald
Bb	Bald	Non bald
bb	Non bald	Non bald

Where a heterozygous non-bald (Bb) woman marries a Bald heterozygous (Bb) man, in the offspring the ratio of bald to non-bald in males is 3 : 1, while in females is 1 : 3.

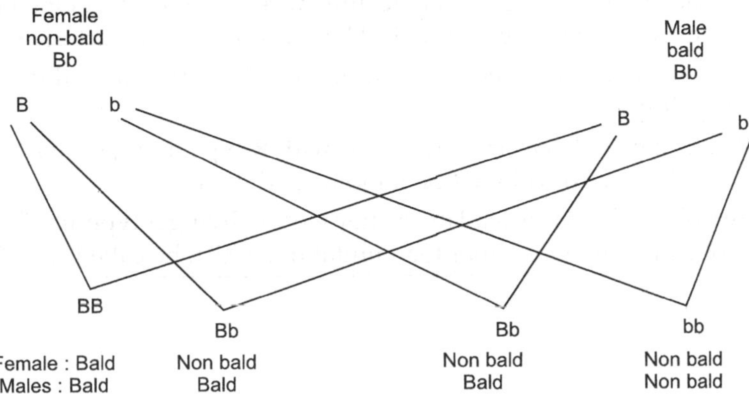

4. What is the ratio between colour-blind and normal vision sons, and colour-blind and normal vision daughters in a marriage between colour-blind female and a normal vision male ?

Ans. Colour-blind Female normal male

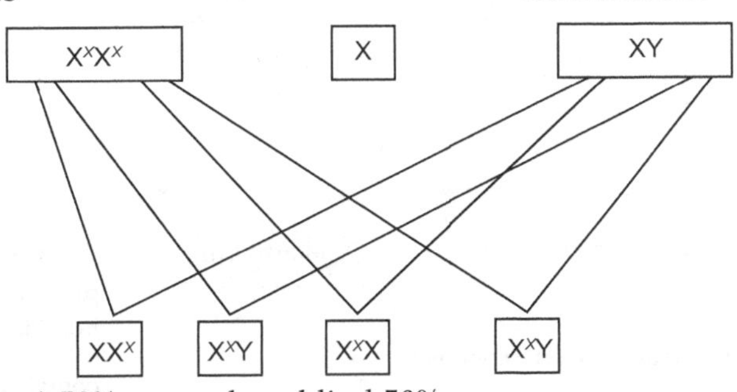

Daughter Normal (carriers) 50%, son colour-blind 50%,

5. A woman who is X^{ch} X^{ch} married a man who is X^{ch} Y. Find out the genetic inheritance into their progeny ?

Ans.

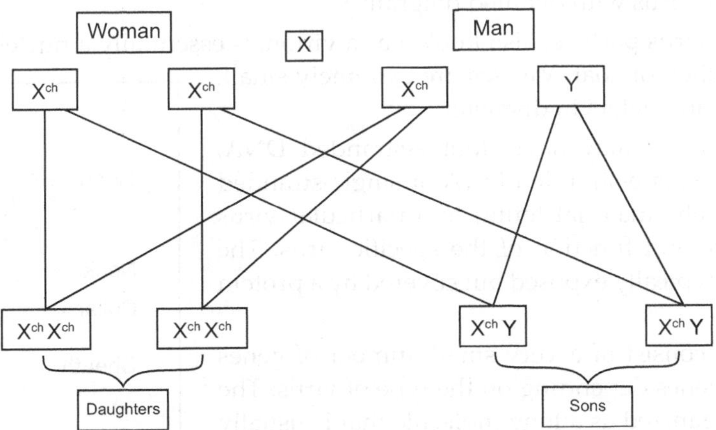

Genetic inheritance is 50 : 50.

Chapter 6. Molecular Basis of Inheritance

1. Give an account of the Griffith experiment to demonstrate transformation in *Streptococcus* /*Diplococcus*.

Ans. The fact that DNA is genetic material came from the experiments using bacteria and viruses. The first series of experiments were performed by a British bacteriologist F. Griffith in 1928, using the bacterium *Diplococcus pneumoniae* which causes pneumonia in mammals. Griffith noticed that this bacterium had two types of strains.

(i) S–type, which has mucilage covering or capsule and produced a smooth colony on a synthetic medium.

(ii) R–type, which lacks mucilage sheath or capsule non-capsulated and produced rough colony on a synthetic medium.

When S-type of bacteria was injected into healthy mice, the mice developed pneumonia and died. So S-type was named as virulent or pathogenic. However, R-type of bacteria was non-pathogenic.

(i) If heat killed S-type of bacteria were injected into healthy mice, they did not cause disease and the mice remained healthy.

(ii) When heat killed S-type of bacteria were mixed with R-type living bacteria and the mixture injected into healthy mice, the mice developed pneumonia and died.

When bacteria were isolated from the dead mice, they were of living S-type and R-type. Griffith announced that it was because of a phenomenon other than mutation, which he called transformation.

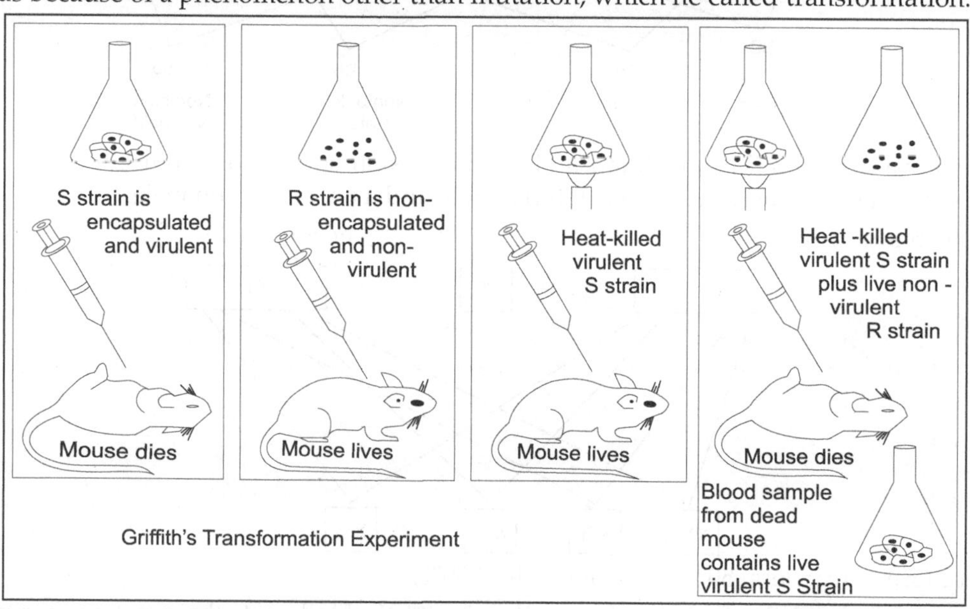

Griffith's Transformation Experiment

2 Explain the structure of virus with detailed diagram.

Ans. Structure of Virus : A virus particle, also known as a virion, is essentially a nucleic acid (DNA or RNA) enclosed in a protein shell or coat. Viruses are extremely small, approximately 15–25 nanometers in diameter.

Genetic Material : Viruses may have double-stranded DNA, double-stranded RNA, single-stranded DNA or single-stranded RNA. The type of genetic material found in a particular virus depends on the nature and function of the specific virus. The genetic material is not typically exposed but covered by a protein coat.

The viral genome can consist of a very small number of genes or up to hundreds of genes depending on the type of virus. The genome is typically organized as a long molecule that is usually straight or circular.

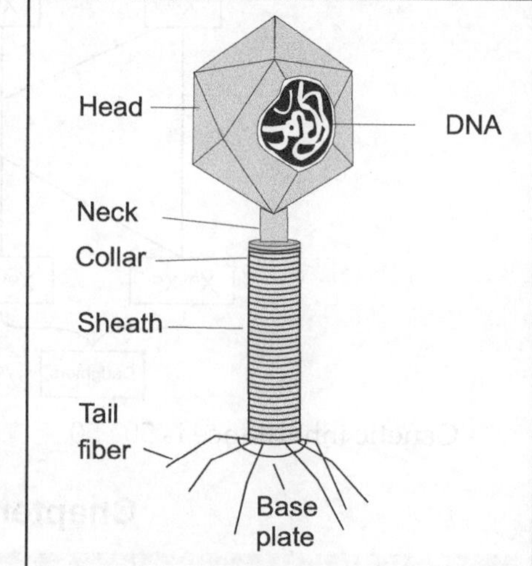

Structure of T-4 bacteriophage

Capsids : The protein coat that envelopes viral genetic material is known as a capsid. A capsid is composed of protein sub-units called Capsomeres. Capsids can have several shapes : polyhedral, rod or complex. Capsids function to protect the viral genetic material from damage.

In addition to the protein coat, some viruses have specialized structures. For example, the flu virus has a membrane-like envelope around its capsid. The envelope has both host cell and viral components and assists the virus in infecting its host. Capsid additions are also found in bacteriophages. For example, bacteriophages can have a protein "tail" attached to the capsid that is used to infect host bacteria.

3. Give an account of the lytic and lysogenic phage of bacteriophage multiplication.

Ans. Lytic cycle : Step I : **Adsorption :** Attachment of tail fibres of the phage on to a specific receptor site on the bacterial cell wall.

Step II : **Injection :** Injection of viral genome into the host through the hollow tubes of the tail.

Step III : **Protein synthesis :** Inside the host, the viral genome directs the synthesis of viral proteins using the machinery of the host. Viral genome generally encodes for some enzymes and coat proteins.

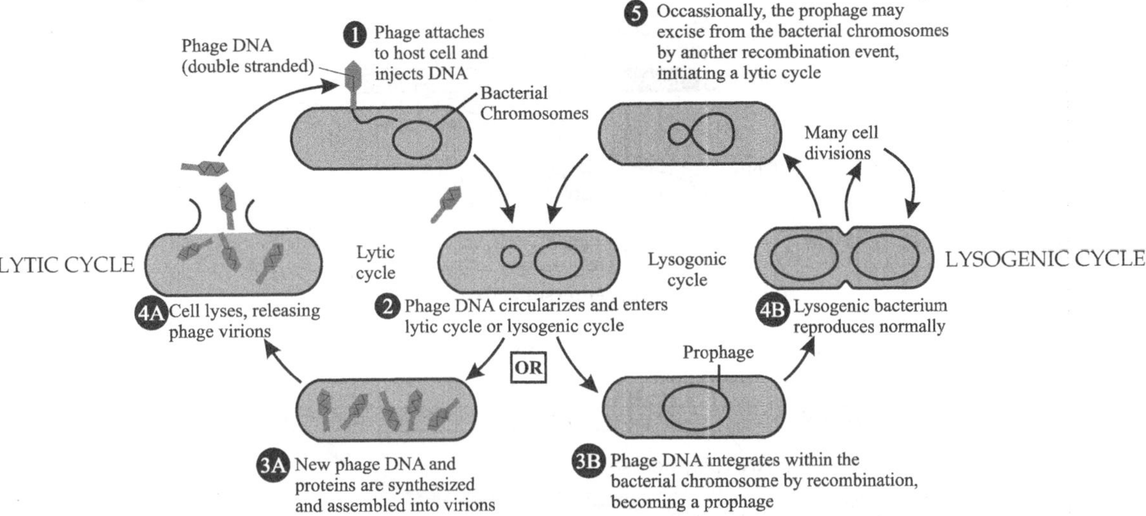

The lytic and lysogenic cycle of a bacterial virus, *e.g.*, bacteriophage T-4.

Step IV : **Viral genome synthesis :** Viral genome replicates inside the host making several copies.

Step V : **Packaging and release :** The viral genome gets packaged inside the protein coat.

These intact mature infectious particles are called virions. The crowding of virions inside the host ultimately causes cell lysis and liberation of mature viral particles and about 200 mature phages/cells are liberated.

Lysogenic cycle : Step I : **Adsorption :** Attachment of tail fibres of the phage on to a specific receptor site on the bacterial cell wall.

Step II : **Injection** : Injection of viral genome into the host through the hollow tubes of the tail.

Step III : **Integration of viral genome to the host genome :** After the entry of viral genome, it gets integrated into the bacterial genome of the host. The viral genome integrated to the bacterial genome is termed as prophage.

Step IV : **Viral genome synthesis:** Viral genome replicates along with the bacterial genome replication and pass on to the daughter cells.

Step V : **Induction of lytic cycle :** Occasionally, integrated viral genome detaches and released into the bacterial cytoplasm. This dissociation is called induction and lytic cycle is followed by releasing mature lysogenic phages. Induction can be induced artificially using ultraviolet rays or heat treatment.

4. Explain how structural gene of lac operon start on addition of lactose.

Ans. *E. coli* prefers to use glucose as an energy source when both glucose and lactose are available. Lactose is an alternative energy source that can be used if glucose is absent. The overall rate of messenger RNA synthesis from the lac operon, and from other operons is indirectly regulated by the concentration of glucose in the cell. cAMP is the second messenger used directly in the regulation of lac operon expression in response to changing levels of glucose.

The lac operon consists of :

- Regulatory gene *i* - It codes for the repressor protein.
- Structural gene

(i) *z* gene — It codes for beta-galactosidase which catalyzes the hydrolysis of lactose into glucose and galactose.

(ii) *y* gene — It codes for permease which regulates the lactose permeability in the cell.
(iii) *a* gene — It codes for transacetylase which assists the enzyme beta-galactosidase.

Normally, the lac operon is turned off. A repressor protein binds the operator (control) region upstream of the operon preventing transcription.

When lactose is present outside the cell, it crosses the cell membrane and acts as an inducer of the operon. It does so once lactose is broken down to create allolactose. The *lac* operon is then membrane facilitating lactose transport into the cell, and & beta-galactosidase, which eats up lactose to make glucose moleclues and beta-galactosidase also makes allolactose. This leads to a positive feedback loop.

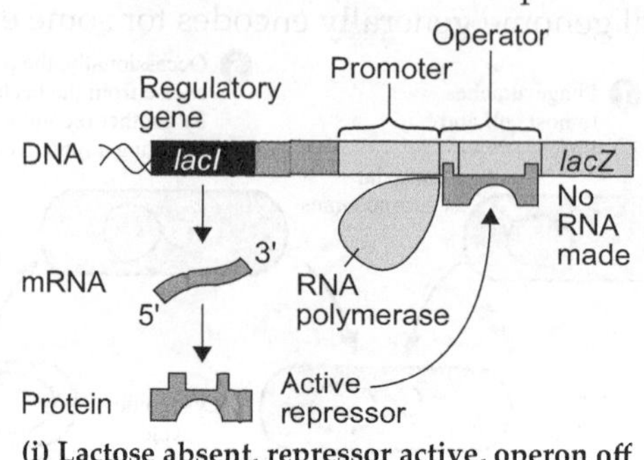

(i) Lactose absent, repressor active, operon off

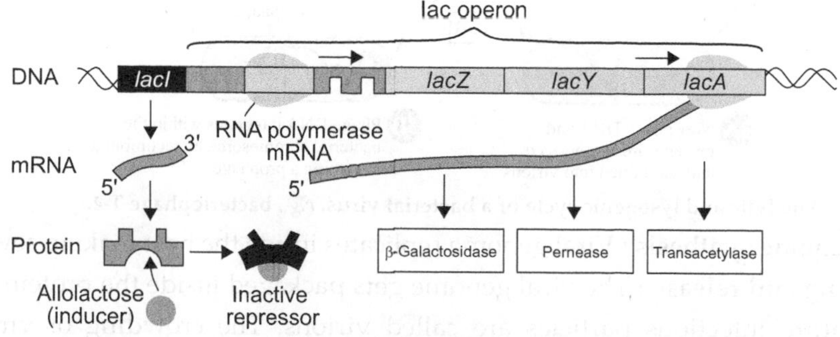

(ii) Lactose present, repressor inactivate, operon on.

Chapter 7. Evolution

1. Recall the experiment set-up by Urey and Miller. What is the significance of their finding ?

Ans. Miller and Urey set up a laboratory and constituted an apparatus for circulating steam through a mixture of ammonia, methane and hydrogen. This steam gas mixture was subjected to high

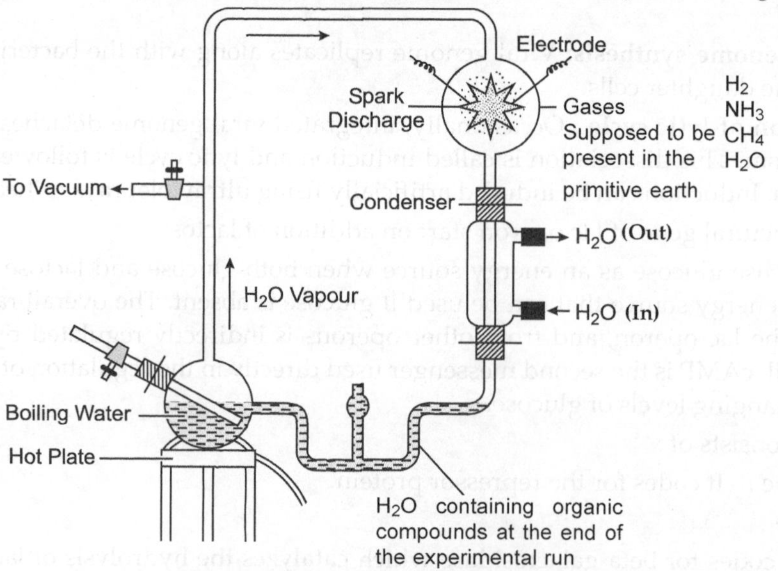

Experimental set up of Urey and Miller

energy electrical spark and then condensed to a liquid before the cycle was repeated again by heating the water to steam. After a week of being cycled through the gases and electric discharge, the condensed water in the apparatus become deep red. On analysis it was found that this water contained complex of amino acids.

(i) Miller's experiment showed that electric discharges such as lighting in an early atomosphere of ammonia, methane and hydrogen could have led to the production of some of the complex organic molecules of living system.

(ii) The essential 'building blocks' (nucleotides, amino acids, etc.) of living organisms are formed on primitive earth.

(iii) It shows the process of abiotic synthesis of organic molecules in the presence of reducing atmosphere.

Significance : The experiment supported that life originated by chemosynthesis. The chemiosynthetic theory (Oparin–Haldane) state that life originated from non-living matter is based on the presence of methane and ammonia in the atmosphere. It requires a high temperature, high energy, radiations and electric discharges.

2. Explain the Lederberg Replica experiment with the help of systematic diagram.

Ans. **The Lederberg Replica plating experiment :** In 1952, Esther and Joshua Lederberg performed an experiment that helped to show that many mutations are random, not directed. A bacteria are grown into isolated colonies on plates and these colonies can be reproduce from an original plate to new plates by "stamping" the original plate with a cloth and then stamping empty plates with the same cloth. Bacteria from each colony are picked up on the cloth and then deposited on the new plates by the cloth.

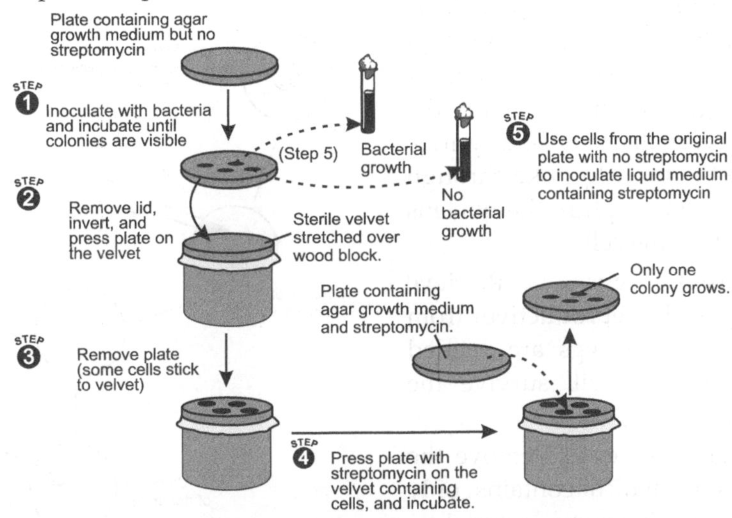

Esther and Joshua Lederberg Experiment

The hypothesis for the experiment is that the antibiotic resistant strains of bacteria which survived an application of antibiotics, had the resistance before their exposure to the antibiotics, not as a result of the exposure.

Chapter 10. Microbes in Human Welfare

1. What do you mean by Biogas ? Write down the mechanism of biogas production with the help of diagram.

Ans. Biogas is a mixture of inflammable gases (Methane, SO_2, CO_2, etc.) produced by microbial activity that can be used as fuel.

Mechanism of Biogas production :

The mechanism of production of biogas can be completed in the following steps :

(i) The raw material for biogas production is excreta (dung) of cattle.

(ii) The biogas plant has a concentrate tank (10-15 feet deep) in which bio-wastes and slurry of dung is collected.

(iii) The tank has a floating cover which rises on production of gas in the tank.

(iv) Methanobacterium in the dung act on the bio-waste to produce biogas.

(v) The gas produced is supplied to nearby houses by an outlet.

(vi) Through another outlet, the spent slurry is removed to be used as fertilizers.

(vii) Biogas is used as fuel for cooking and lighting.

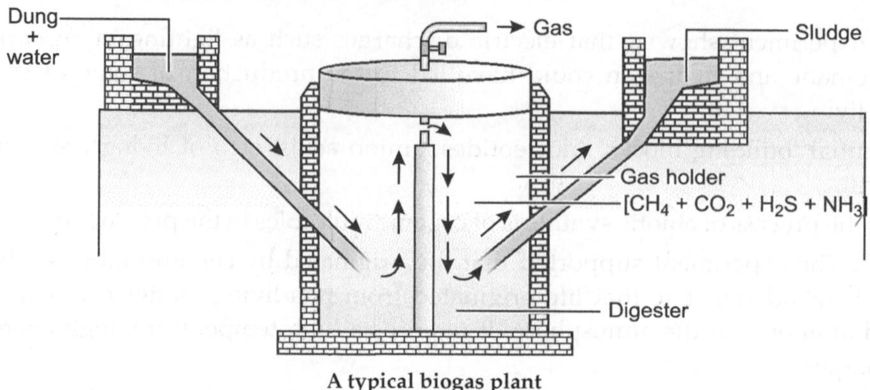

A typical biogas plant

Chapter 12. Biotechnology and its Applications

1. Explain how 'Dolly' was created. What is its importance in the field of bio-technology ?

Ans. These steps were used in creation of Dolly :

(i) **Isolate donor nucleus :** Isolate the nucleus from a somatic (non-reproductive) cell of a adult donor sheep. The nucleus contains the complete genetic material of the organism. A very small needle and syringe (suction device) is used to poke through the cell membrane to capture the nucleus and remove it from the cell.

(ii) **Get unfertilised eggs :** Retrieve unfertilised egg cells (reproductive) from a female sheep. Many eggs are needed since not all of them will survive the various steps of cloning.

(iii) **Remove the egg's nucleus :** Remove the egg cell's nucleus, which contains only one-half of the sheep's genetic material.

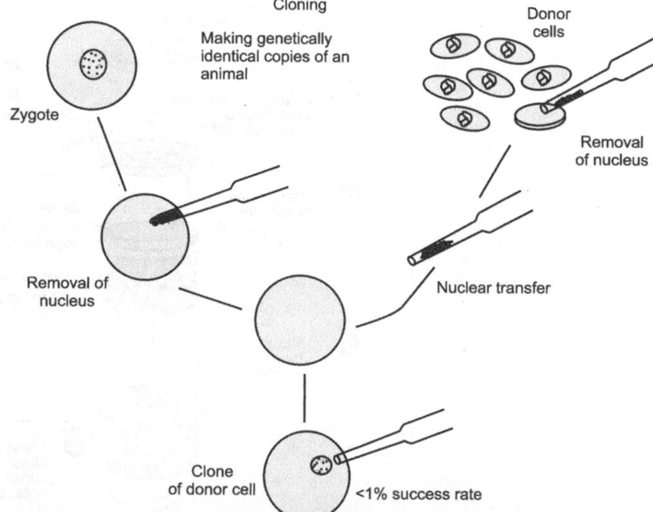

Steps used in the creation of Dolly

(iv) **Insert donor nucleus :** Insert the nucleus, with its complete genetic material, isolated from the donor mammal in Step 1 into the egg cell that has no nuclear material. The egg's genetic material now contains all traits from the donor adult. This egg is genetically identical to the donor adult.

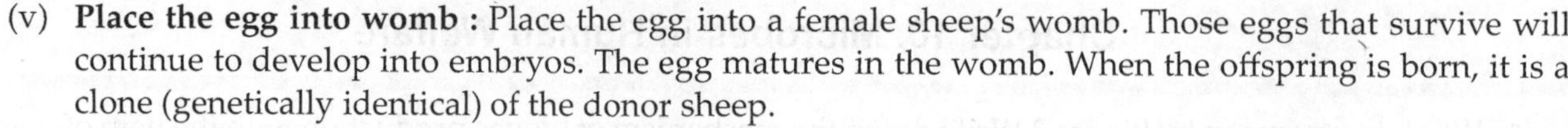

(v) **Place the egg into womb :** Place the egg into a female sheep's womb. Those eggs that survive will continue to develop into embryos. The egg matures in the womb. When the offspring is born, it is a clone (genetically identical) of the donor sheep.

After cloning was successfully demonstrated through the production of Dolly, many other large mammals have been cloned, including horses and bulls. The attempt to clone argali (mountain sheep) did not produce viable embryos. The reprogramming process cells need to go through during cloning is not perfect and embryos produced by nuclear transfer often show abnormal development.

2. Give an account of the *Agrobacterium*-mediated plant transformation process.

Ans. The *Agrobacterium*-mediated transformation process involves a number of steps :

(i) Isolation of the genes of interest from the source organism.

(ii) Development of a functional transgenic construct including the gene of interest.

(iii) Insertion of the transgene into the Ti-plasmid.

(iv) Introduction of the T-DNA containing plasmid into *Agrobacterium*.

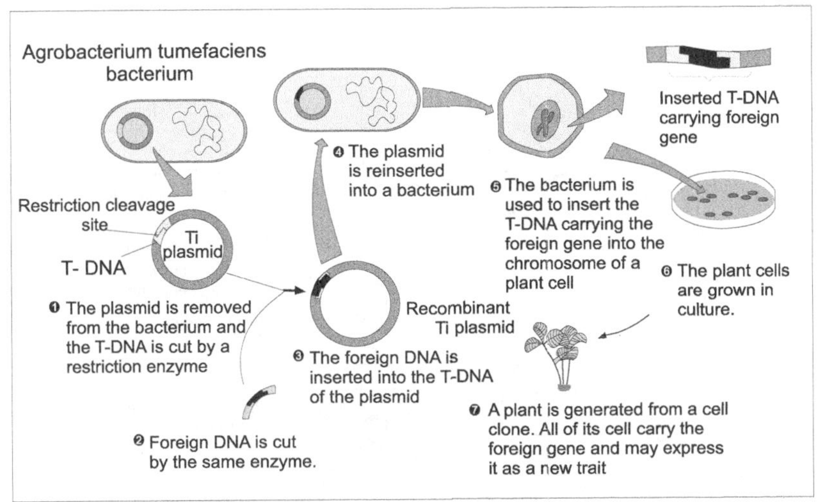

Steps followed in Agrobacterium-mediated plant transformation

(v) Mix the transformed Agrobacterium with plant cells to allow transfer of T-DNA into plant chromosome.

(vi) Regeneration of the transformed cells into genetically modified (GM) plants.

(vii) Testing for trait performance or transgene expression at lab, greenhouse and field level.

3. Diagrammatically represent the experimental steps in cloning and expressing a human gene (say the gene for growth hormone) into bacterium like *E. coli*.

Ans. It is possible to produce HGH (Human Growth Hormone) by recombinant DNA technology. This is represented diagrammatically as follows :

Steps involved in gene transfer in the production of Human Growth Hormone

4. How is the hormone insulin synthesized, using genetic engineering technique ? State two ways in which this technique is better than the techniques used earlier.

Ans. The first major product of genetic engineering was human insulin. Insulin is a protein that acts as a hormone to stimulate uptake of blood sugar into tissues, such as the liver and the muscles.

(i) The gene responsible for producing human insulin protein is isolated. The gene is part of the DNA in a human chromosome.

(ii) Then, a circular piece of DNA called a plasmid is removed from a bacterial cell. Special enzymes are used to cut the plasmid ring.

(iii) With the plasmid ring open, the gene for insulin is inserted into the plasmid ring and the ring is closed. This process is called recombinant technology.

(iv) The bacterial DNA now contains the human insulin gene and is inserted into a bacteria.

(v) Many plasmids with the insulin gene are inserted into many bacterial cells. When the bacterial cells reproduce by dividing, the human insulin gene is also reproduced in the newly created cells.

(vi) Human insulin protein molecules produced by bacteria are gathered and purified. By culturing the genetically engineered bacteria, limitless supplies of insulin may be produced.

Two ways in which genetic engineering is better than the technique used earlier :

(i) Insulin produced by genetic engineering has no allergic reaction.

(ii) The human insulin was much cheaper when produced in this way than was the insulin from cows, as it could be produced much more quickly in greater quantity.

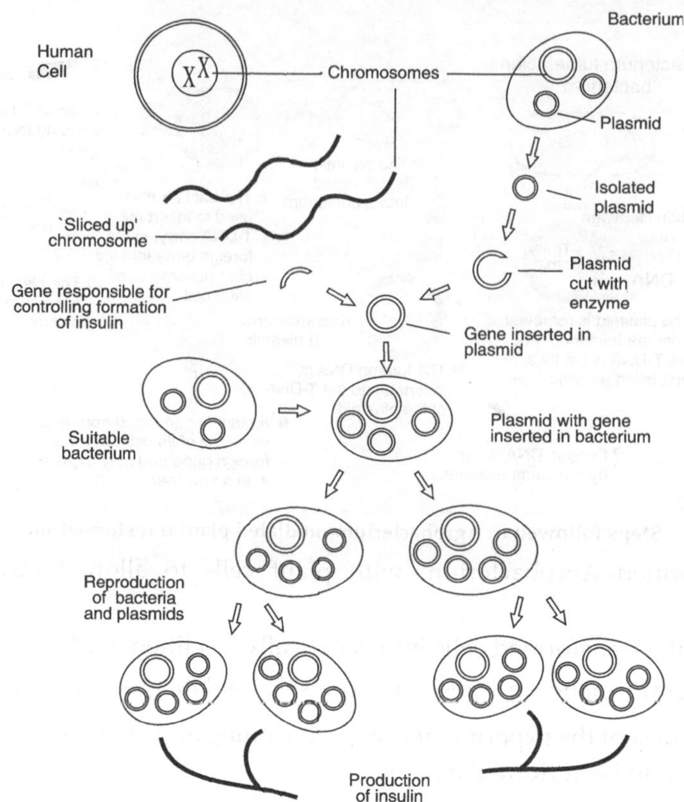

Synthesis of Human Insulin by genetic engineering

Chapter 13. Organisms and Populations

1. Study the graph given alongside and answer the questions that follow:*

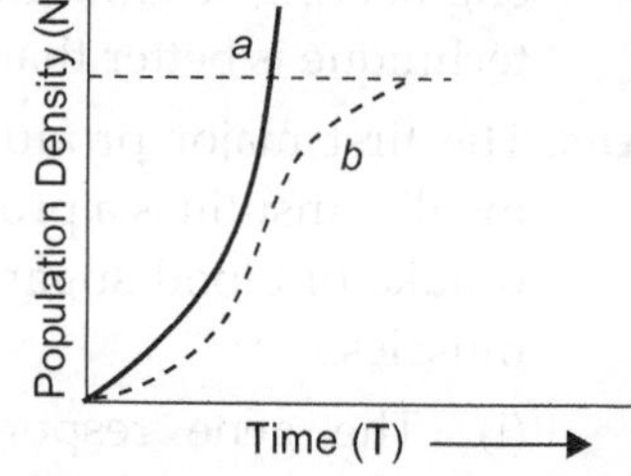

(i) In the absence of predators which one of the two curves would appropriately depict the prey population? Give a reason.

(ii) Time has been shown on X-axis and there is a parallel dotted line shown above. Explain the significance of this dotted line.

Ans. (i) The curve *'a'* would depict the prey population as there would be no competition for food and shelter resources in the absence of predators, leading to their exponential growth with time.

(ii) The dotted line in the above graph represents the carrying capacity (K) of the population. It represents the size of a population that the environment can hold by providing necessary resources. It signifies the limit of habitat, i.e. limited resources in a given habitat to support growth up to a certain level beyond which no further growth can take place.

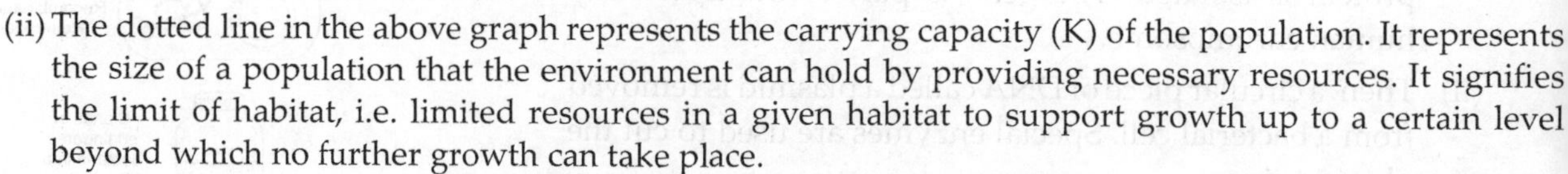

Chapter 14. Ecosystem

1. Construct an ideal pyramid of energy when 1,000,000 joules of sunlight is available. Label all its trophic levels.

Ans.

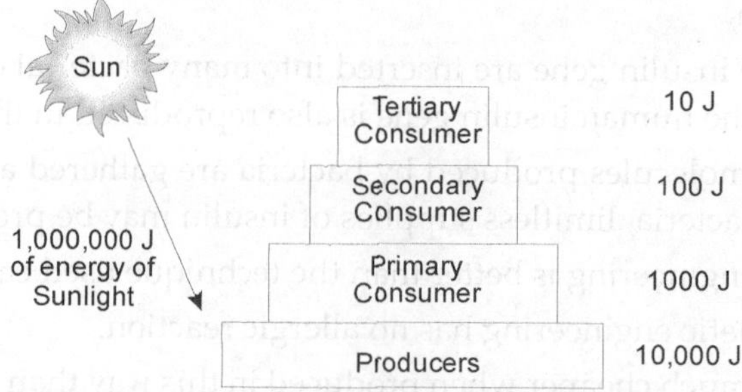

* **Frequently asked previous years Board Exam Question.**

2. Construct a grazing food chain and detritus food chain using the following with 5 links each. Earthworm, bird, snake, vulture, grass, grasshopper, frog, decaying plant matter.

Ans. Grazing Food Chain :

Grass → Grasshopper → Frog → Snake → Vulture

OR

Grass → Grasshopper → Bird → Snake → Vulture

Detritus Food Chain :

Decaying plant matter → Earthworm → Bird → Snake → Vulture.

3. Construct a pyramid of biomass starting with phytoplanktons. Label 3 trophic levels. Is the pyramid upright or inverted ? Why ?

Ans.

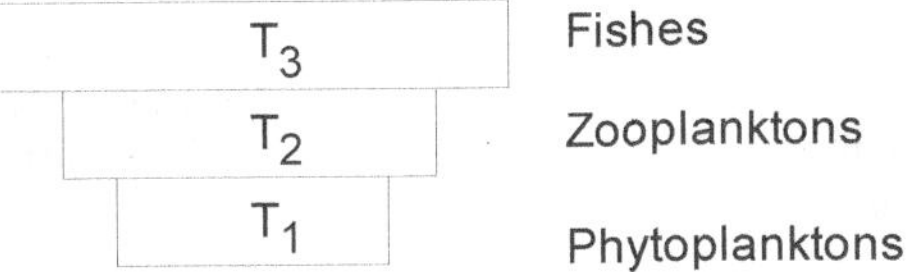

The pyramid is inverted because the biomass of fishes is much more than that of the phytoplanktons.

4. Explain the significance of ecological pyramids with the help of an example.

Ans. Ecological pyramids express the relationship between the organism at different trophic levels with reference to their number, energy and biomass.

For example, pyramid of energy.

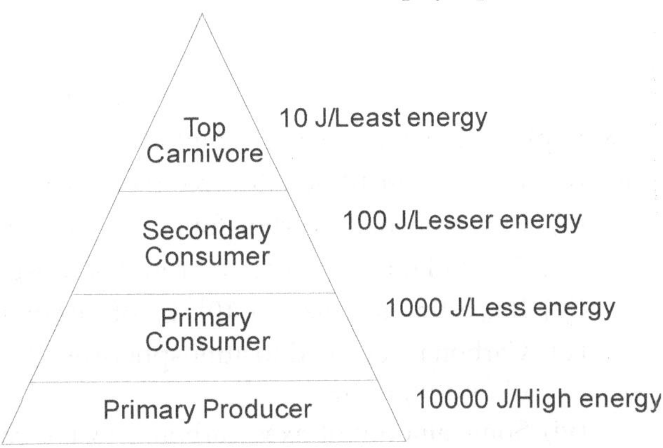

5. Construct an age pyramid which reflects a stable growth states of human population.

Ans.

Post-reproductive

Reproductive

Pre-reproductive

6. (i) Draw a pyramids of numbers of a situation where large population of insect fed upon a very big tree. The insect in turn, are eaten by a small birds which in turn are fed upon by big birds .

(ii) Differentiate giving reasons, between the pyramid of biomass of the above situation and the pyramid of number that you have drawn.

Ans. (i)

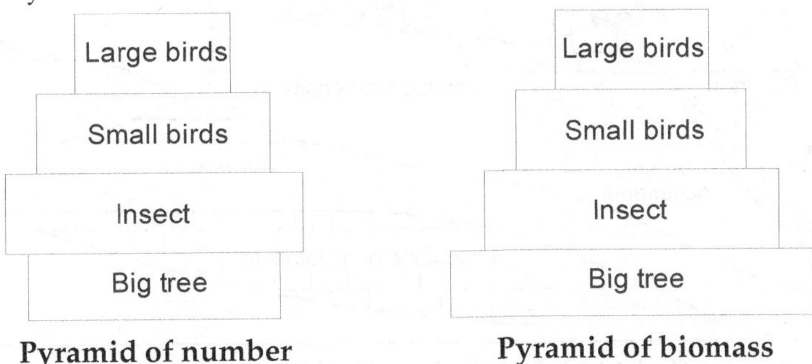

Pyramid of number **Pyramid of biomass**

(ii) The pyramids of biomass is upright at the first two trophic levels because the biomass of a single tree is much more than total population of insects where as the pyramids of numbers is inverted at the first two trophic levels because the number of insects is much more than the number of tree.

7. State the functions of reservoir in a nutrient cycle. Explain the simplified model of carbon cycle in nature.

Ans. The function of a reservoir is to meet the deficit which occur due to imbalance in the rate of influx and efflux.

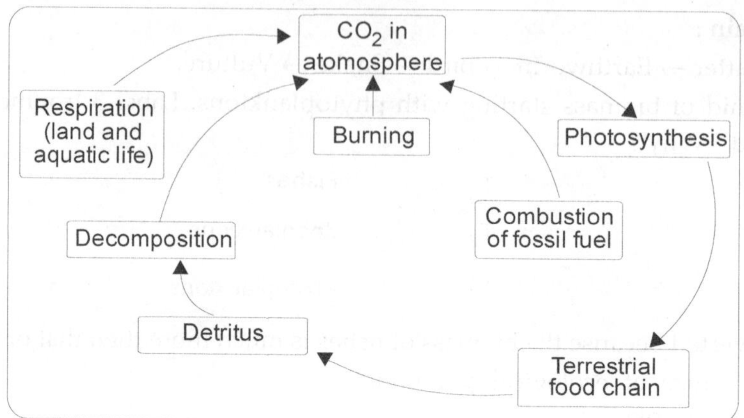

Carbon Cycle in Biosphere

8. Explain carbon cycle with simplified model.

Ans. (i) Carbon constitutes 49 percent of dry weight of an organism.

(ii) 71 percent of the carbon found dissolved in ocean which is responsible for its regulation in atmosphere.

(iii) The carbon cycle occurs through atmosphere, ocean and through living and dead organism.

(iv) It is estimated that 4×10^{13} kg of carbon is fixed in the biosphere through photosynthesis annually.

(v) Carbon is returned to atmosphere as CO_2 by animals and plant through respiration and the activities of decomposers.

(vi) Some amount of fixed carbon is lost as sediments and removed from circulation.

(vii) Burning of wood, forest fire, volcanic activity and combustion of organic matter and fossil fuel as some essential sources of releasing CO_2 the atmosphere.

(viii) Human activities like deforestation and vehicular benning of fossil fuels has caused an increase in the amount of CO_2 in atmposhere.

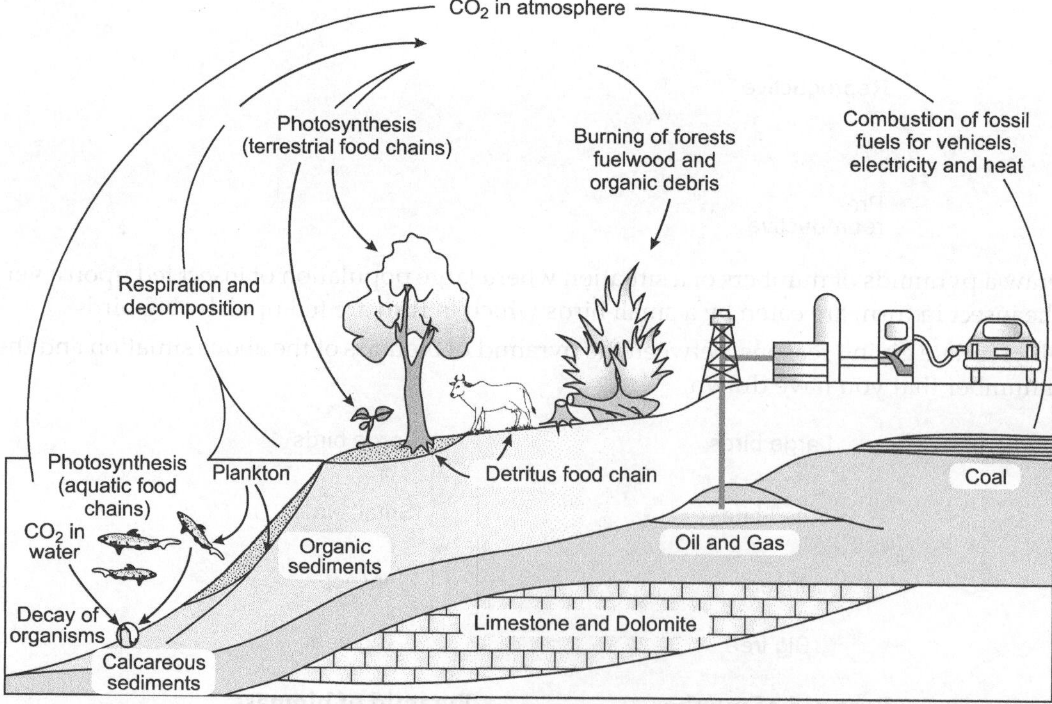

Simplified Model of Carbon Cycle in the biosphere

Diagram Based Questions

9. (i) Draw a simplified model of phosphorus cycling in a terrestrial ecosystem.

(ii) Write the importance of such cycles in ecosystem.

Ans. (i) (a) Rocks are the natural reservoir of phosphorus.

(b) During weathering of rocks, minute amount of these phosphate dissolves in soil solution and are absorbed by plants through roots.

(c) Herbivores and carnivores obtained these elements from plants directly or indirectly.

(d) The waste product and the dead organism are decomposed by phosphate solublising bacteria and in turn release phosphorus.

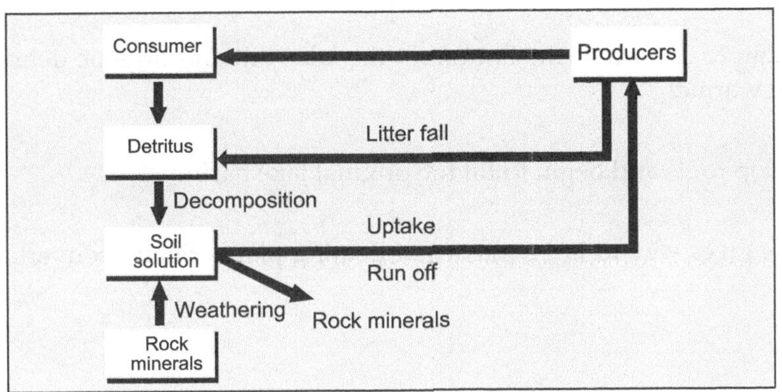

A simplified model of phosphorus cycling in a terrestrial ecosystem

(ii) Importance of Phosphorus :

(a) It is major constituent of biological membranes, nucleic acid and cellular energy transfer system.

(b) It is required for making shells, bones and teeth.

Chapter 15. Biodiversity and Conservation

1. Write in detail about the patterns or gradients of biodiversity.

Ans. Biodiversity varies with changing cardinals/gradients. The following are the gradients of biodiversity which display a pattern for biodiversity distribution.

(i) **Latitudinal Gradients :** Biodiversity is minimum in the artic region, moderate in temperate areas and maximum in tropical regions. Gradient of diversity applies to almost all taxonomic groups. For example in Columbia located over equator (tropical region) exhibits 1400 species of birds, New York (41°N) in temperate area has 105 species while Greenland (71°N) in Arctic area possess 50 species of birds.

The reason for greater diversity in tropics is due to good warmth, light and water availability throughout the year.

(ii) **Altitudinal Gradient :** The second master gradient of biodiversity, is observed when one ascends a mountain. Biodiversity is maximum at the base. It decreases with increasing altitude. The temperature drops with increasing altitude and low temperature is not favourable for the diversity.

(iii) **Species Area Relationship :** A German naturalist and geographer Alexander Van Helmboldt observed that within a region, species richness increased with increase in area, but upto a limit. The relationship is a straight line on a logarithmic scale.

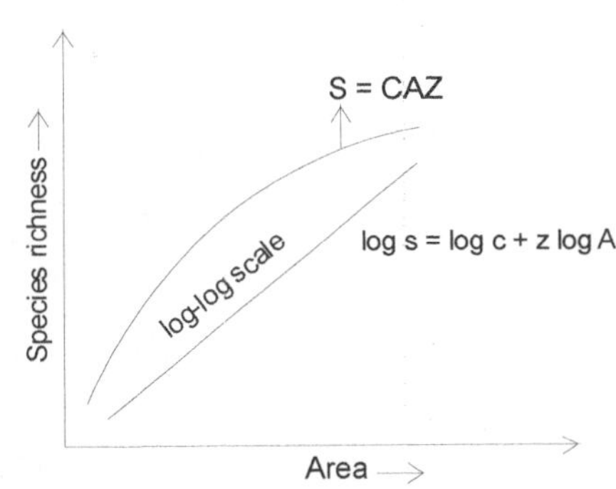

$$\log S = \log C + Z \log A$$

where S = Species richness

C = Y - intercept

A = Area

Z = Slope of line

Chapter 16. Environmental Issues

1. With the help of flow-chart exhibit the events of eutrophication.

Ans. Water in young lake is cold and clear to support life.

↓

With time, water is enriched with nutrients such as nitrogen and phosphorus by streams draining into it

↓

As lake's fertility increases, plant and animal life increase/proliferates.

↓

Organic matter begin to deposit at the bottom of the lake. Silt and organic debris pile up and make the lake shallower and warmer.

↓

Marsh plants develop roots and begin to fill the original lake basin.

Eventually the lake gives way to large masses of floating plants finally converting it into land (Natural aging).

Sketch and Label Based Questions

Set 13

Chapter 1. Reproduction in Organisms

1. Draw a diagram of different types of gametes.

Ans. Types of gametes :

(i) Isogametes of Cladophora (an algae)

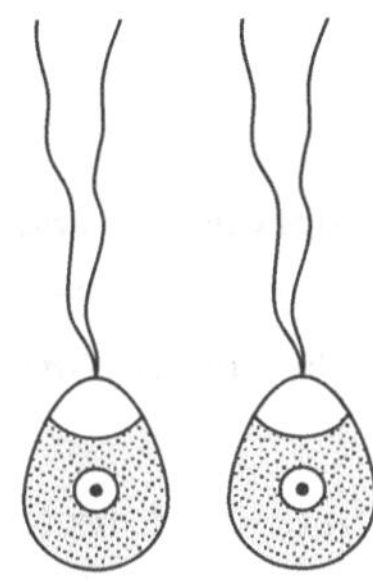

(ii) Heterogamete of Fucus (an algae)

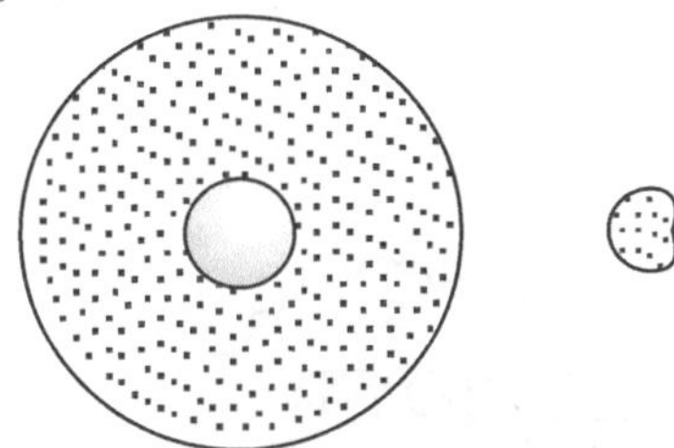

(iii) Heterogametes of Homo sapiens (human beings)

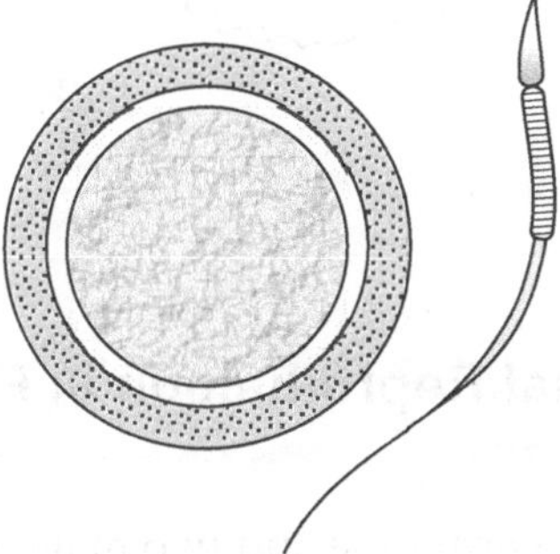

2. With the help of diagram show different methods of vegetative propagation.

Ans.

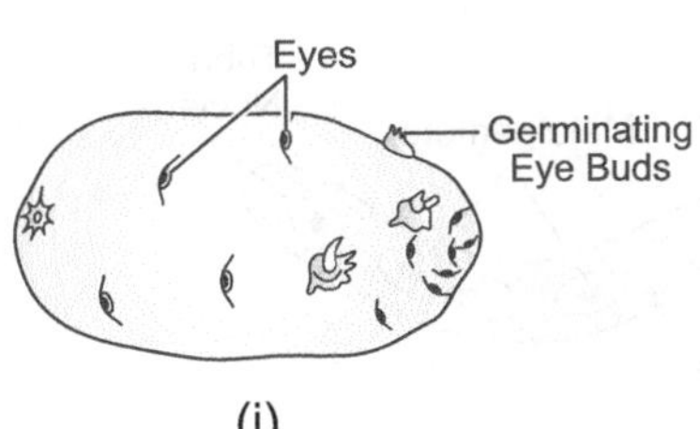

(i)

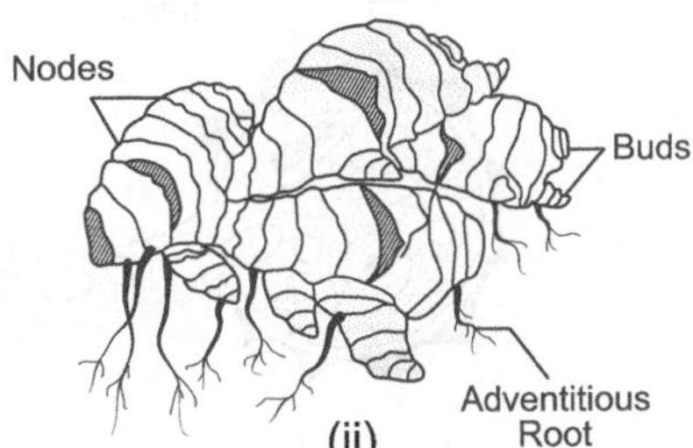

(ii)

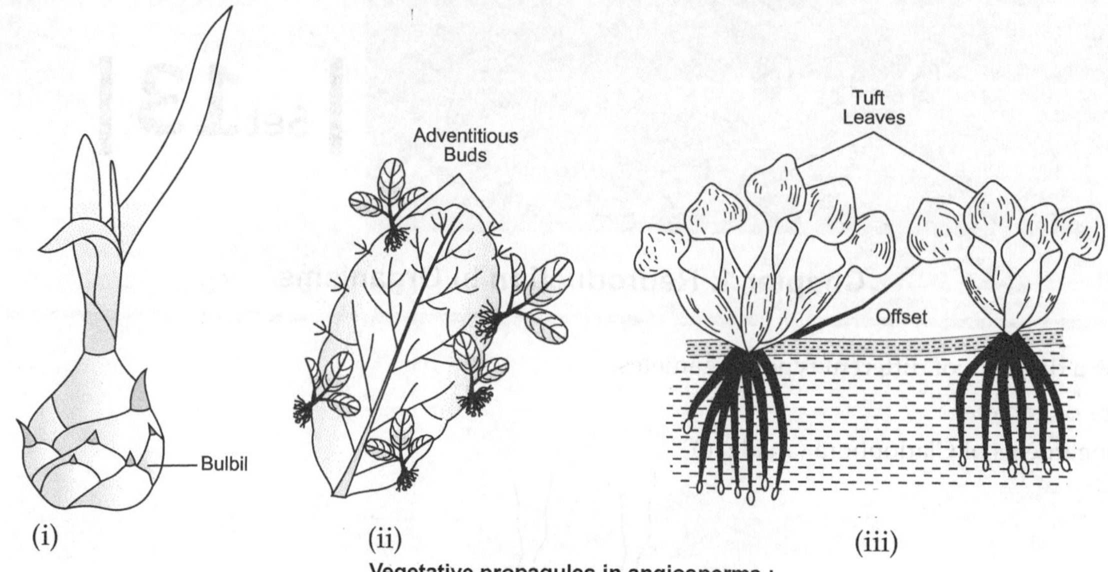

Vegetative propagules in angiosperms :
(i) Potato tuber; (ii) Rhizome of ginger; (iii) Bulbil of *Agave*
(iv) Leaf buds of *Bryophyllum*; (v) Offset of water hyacinth

3. Draw the diagram of fruit showing Seeds (S) and Protective Pericarp (P).

Ans.

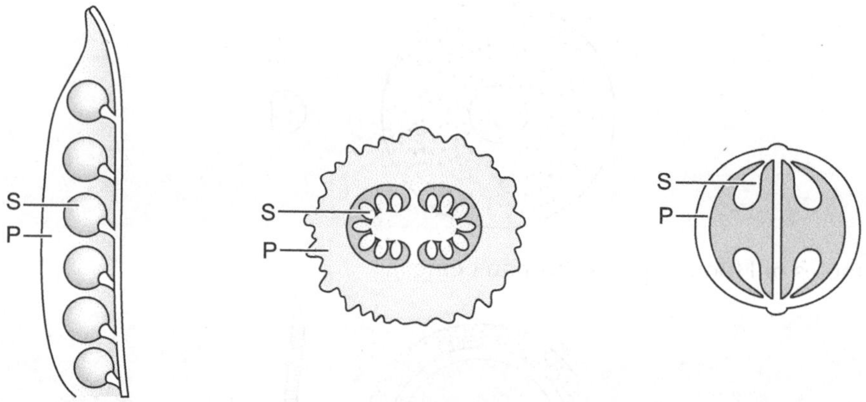

Chapter 2. Sexual Reproduction in Flowering Plants

1. Draw a diagram of pollen grain with germ tube and two male gametes.

Ans.

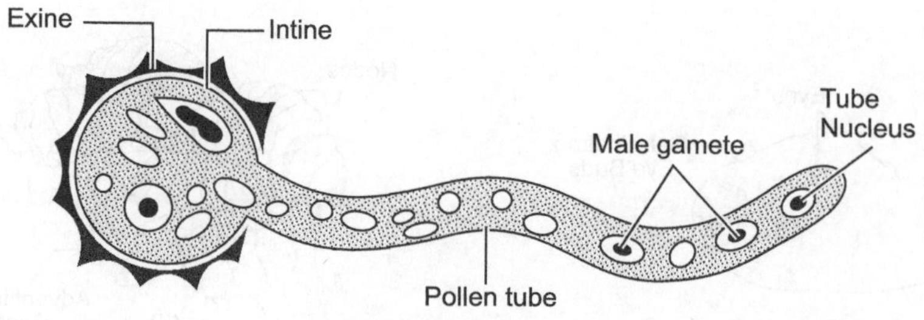

Male gametophyte (Pollen grain)

2. Draw a diagram of male gametophyte angiosperm. Label any four parts.

Ans.

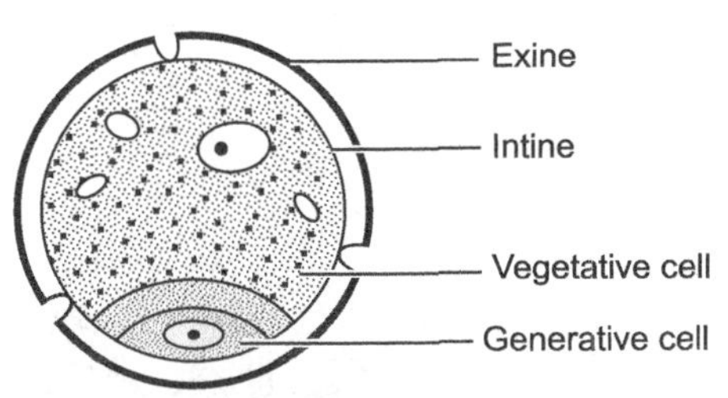

Male gametophyte of angiosperm

3. Give the diagrammatic representation of development of female gametophyte.

Ans.

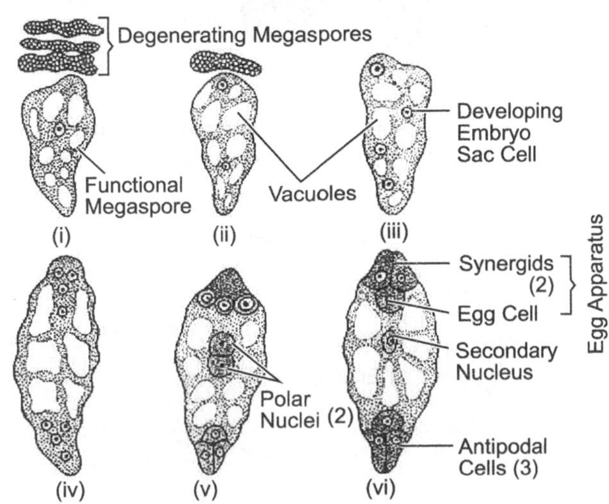

Female gametophyte. (i) - (vi) development of the embryo-sac (female gametophyte) of normal type (Polygonum type)

4. Draw a labelled diagram of the V. S. of ovule just before fertilisation.

Ans.

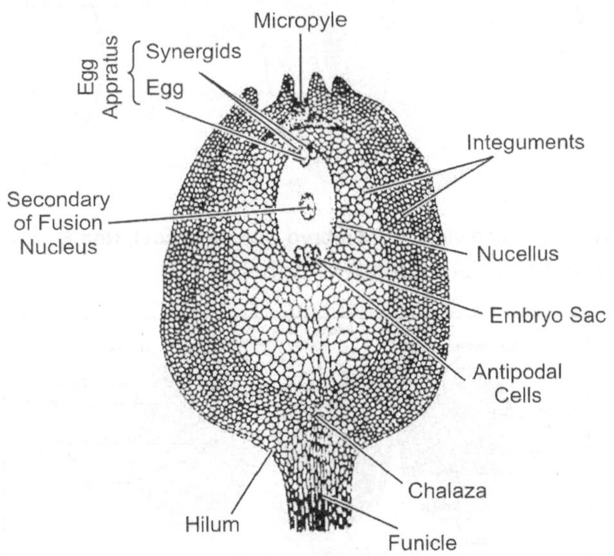

V.S. of Ovule

5. Give a diagrammatic representation of different stages of embryo development in a dicotyledonous plant

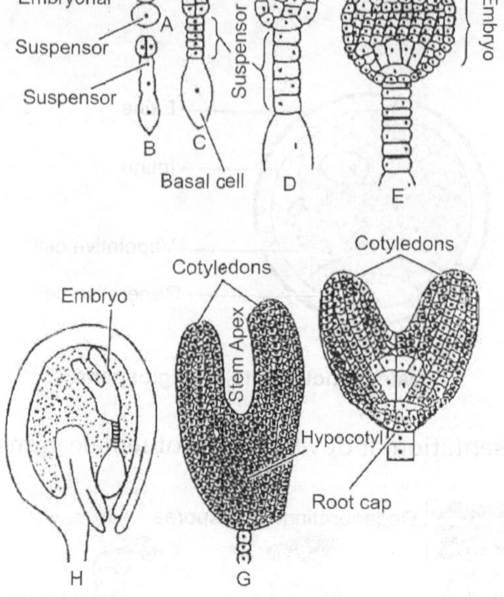

Development of embryo in dicot angiosperm

6. Draw well labelled diagrams of endospermic and non-endospermic seeds.

Ans.

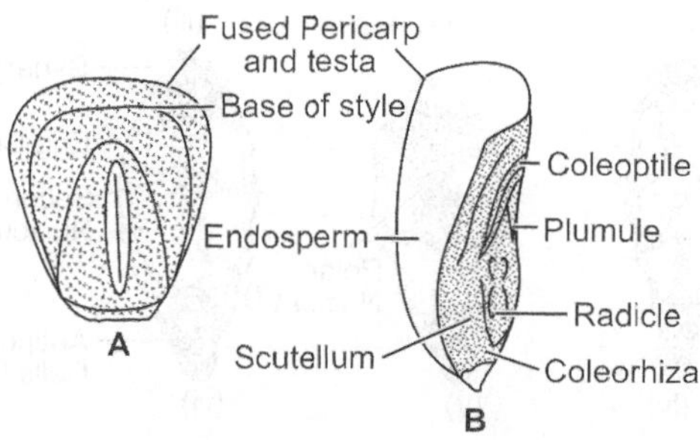

Endospermous seed (Maize)

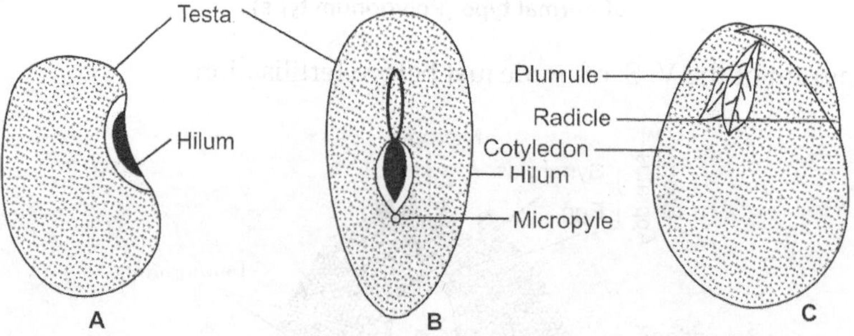

A – Lateral view, B – Side view, C – Embryo with seed coat, Non Endospermous seed

7. Draw a labelled diagram of the T.S. of a mature anther.*

Ans.

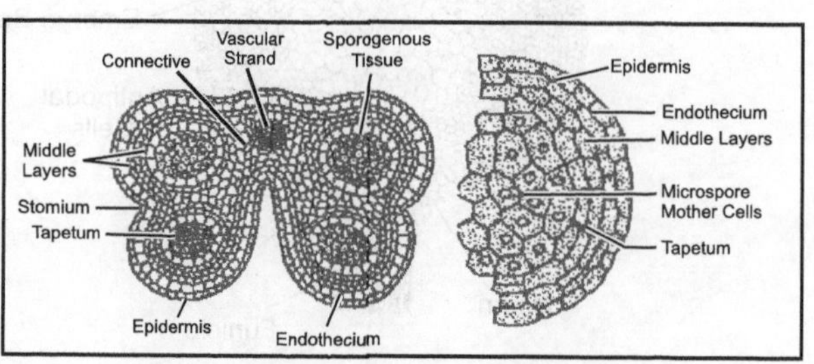

Fig. TS of mature anther

* Frequently asked previous years Board Exam Question.

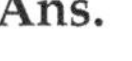

8. Draw a labelled diagram of the mature embryo sac of angiosperms.*

Ans.

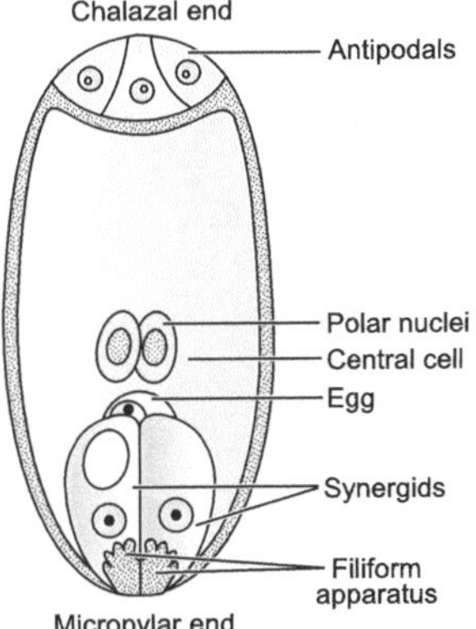

9. Draw a labelled diagram of a germinating pollen grain with at least four labellings.*

Ans.

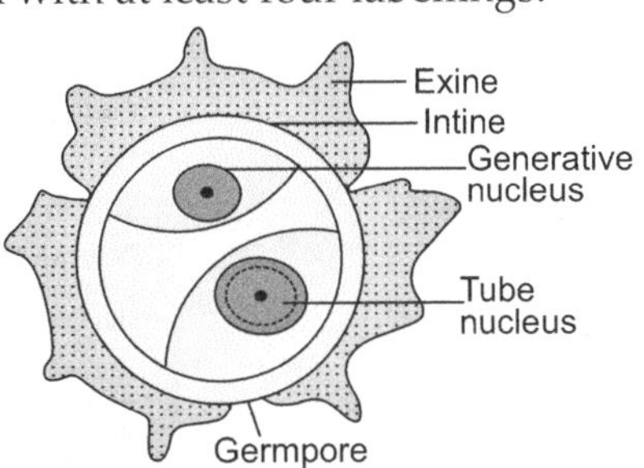

Chapter 3. Human Reproduction

1. Draw a labelled diagram of a mature human ovum with at least four labellings.*

Ans.

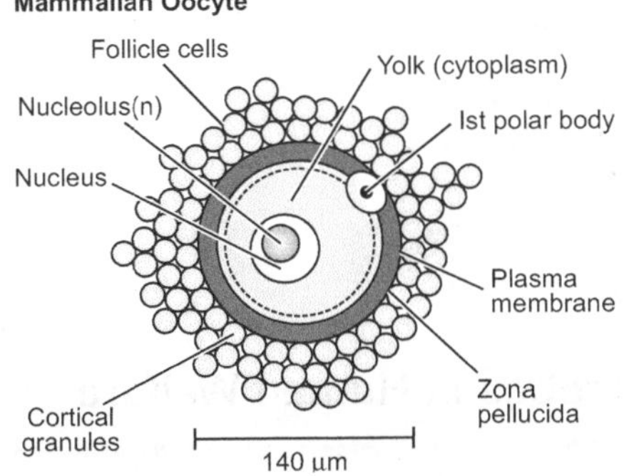

Chapter 8. Human Health and Diseases

1. Draw a labelled diagram to show the life cycle of Plasmodium.*

Ans.

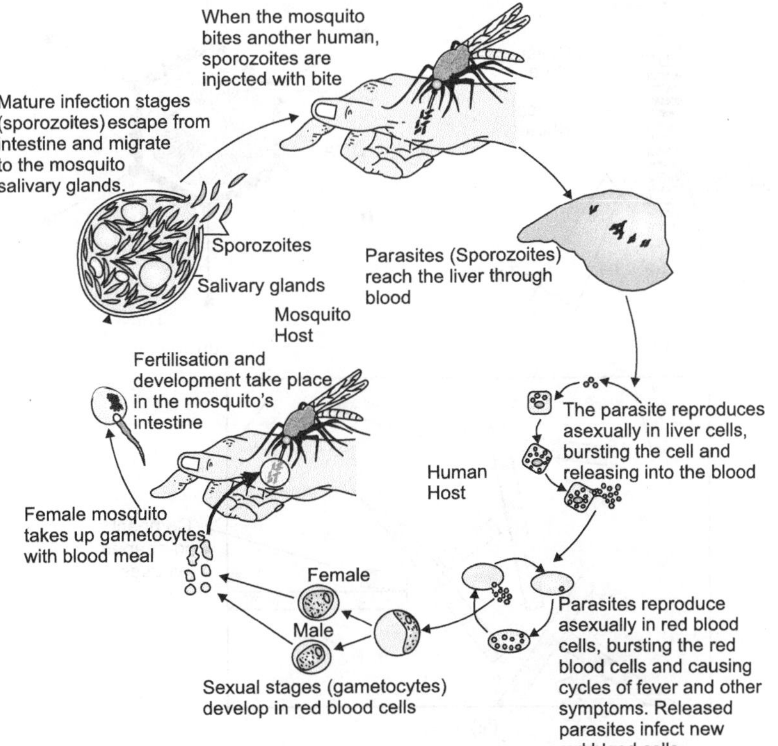

* **Frequently asked previous years Board Exam Question.**

2. Draw a labelled diagram to show replication of HIV in human cells.*

Ans.

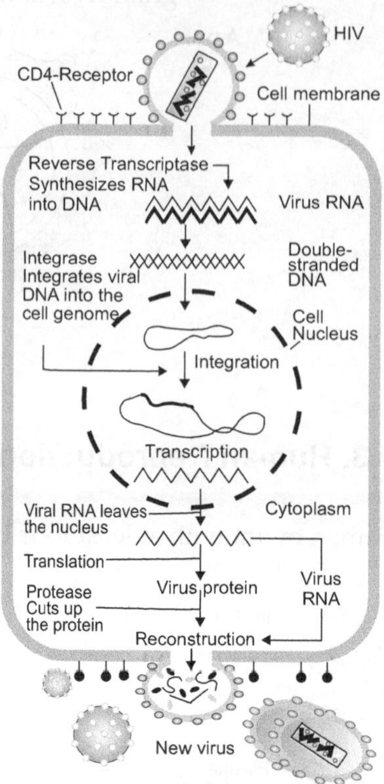

Chapter 10. Microbes in Human Welfare

1. Draw the structure of different kinds of viruses.

Ans.

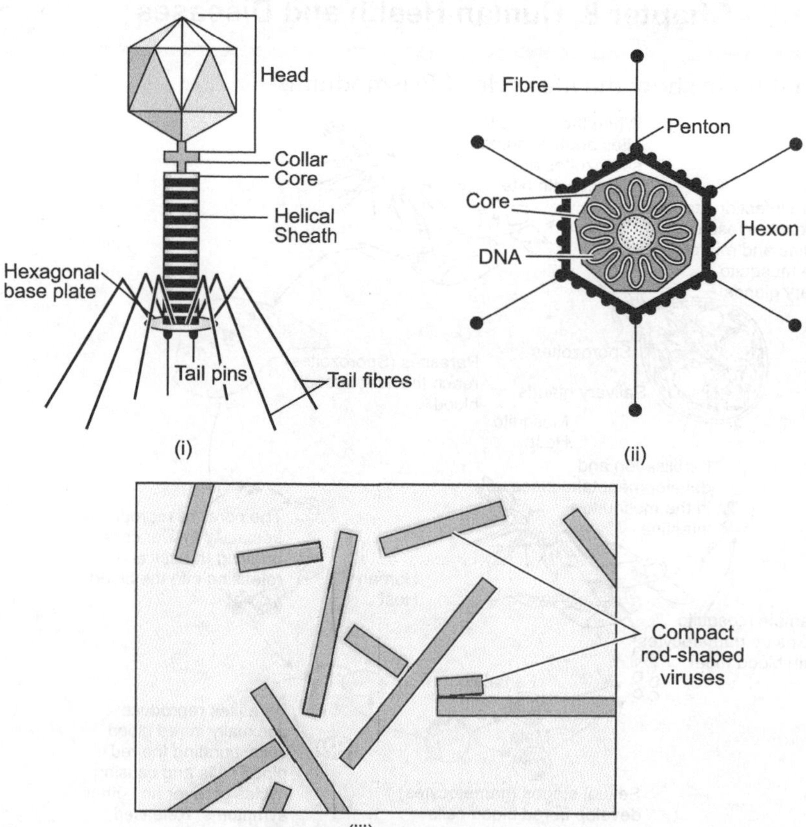

Viruses : (i) A bacteriophage; (ii) Adenovirus which causes respiratory infections; (iii) Rod-shaped tobacco mosaic

* Frequently asked previous years Board Exam Question.

2. Draw a diagram showing different types of bacteria.

Ans.

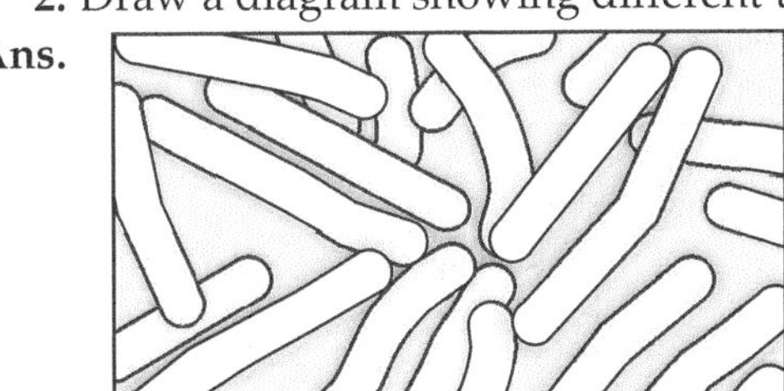

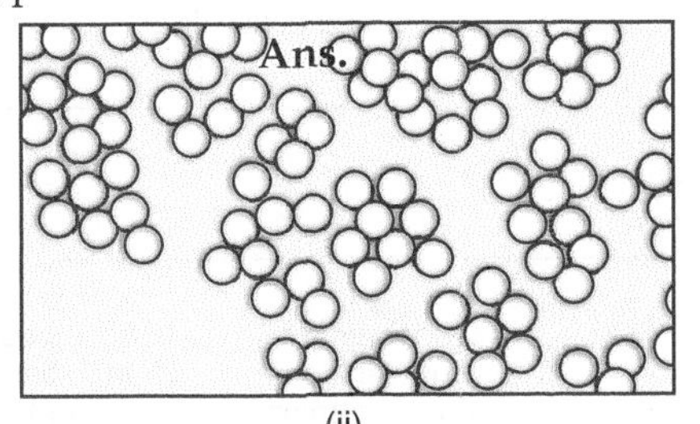

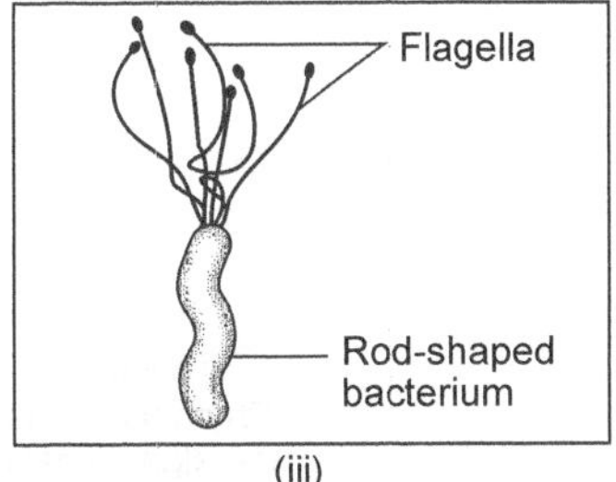

Bacteria : (i) Rod-shaped; (ii) Spherical-shaped; (iii) a rod-shaped bacterium showing flagella.

Chapter 12. Biotechnology and its Applications

1. Draw the diagram of Bt. Cotton (Cotton ball).

Ans.

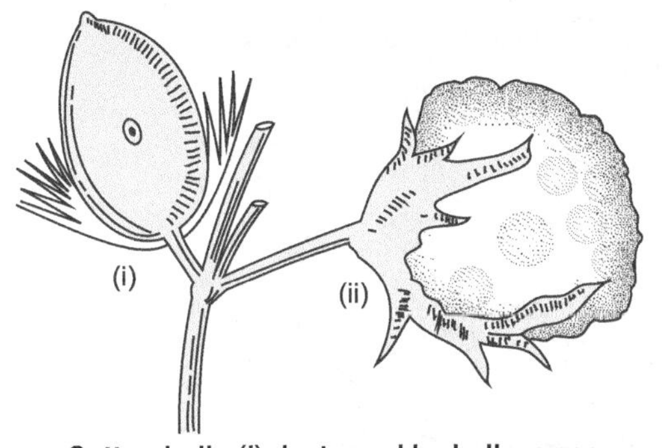

Cotton boll : (i) destroyed by ball worms; (ii) a fully mature cotton boll

2. Draw a diagram showing maturation of proinsulin into insulin after removal of C-peptide.

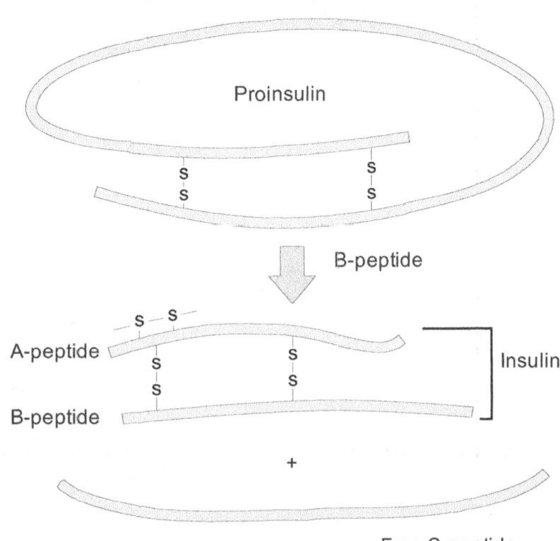

Chapter 14. Ecosystem

1. Draw all the components of ecosystem.

Ans.

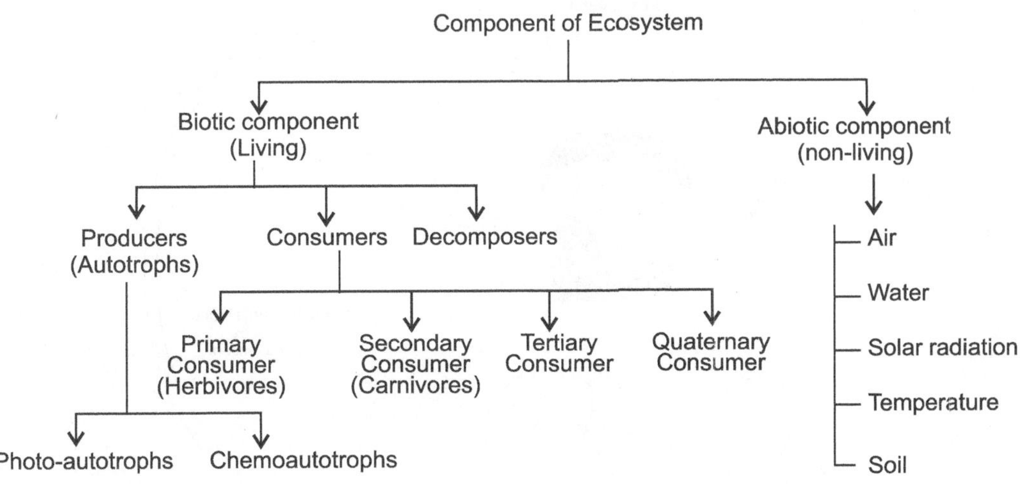

2. Draw the components of Pond ecosystem.

Ans.

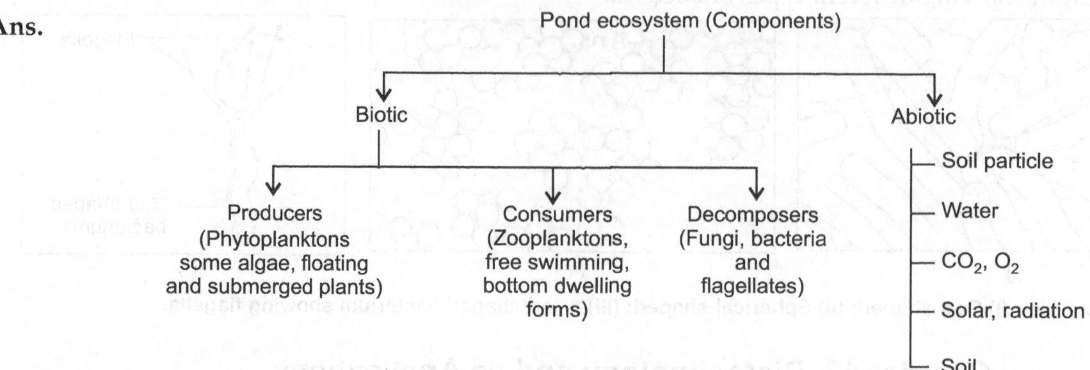

3. Show by diagram how energy flow through different trophic level.

Ans.

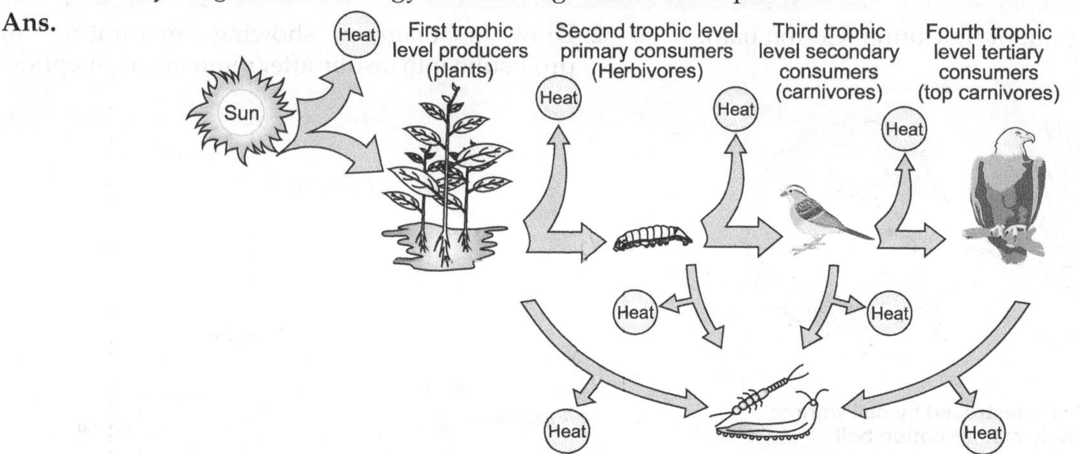

Energy flow through trophic levels

4. Express diagrammatic representation cycle in a terrestrial ecosystem.

Ans.

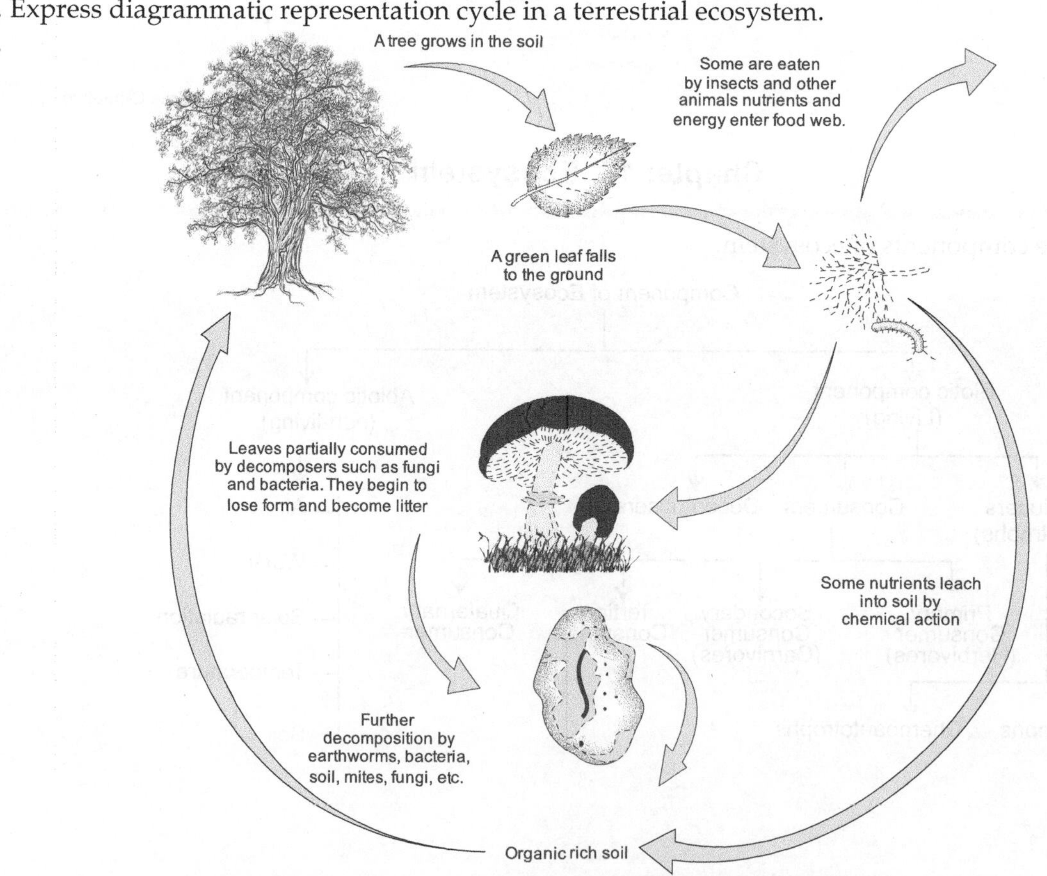

Diagrammatic representation of decomposition cycle in a terrestrial Icosystem

5. Draw the diagrammatic representation of trophic levels in an ecosystem.

Ans.

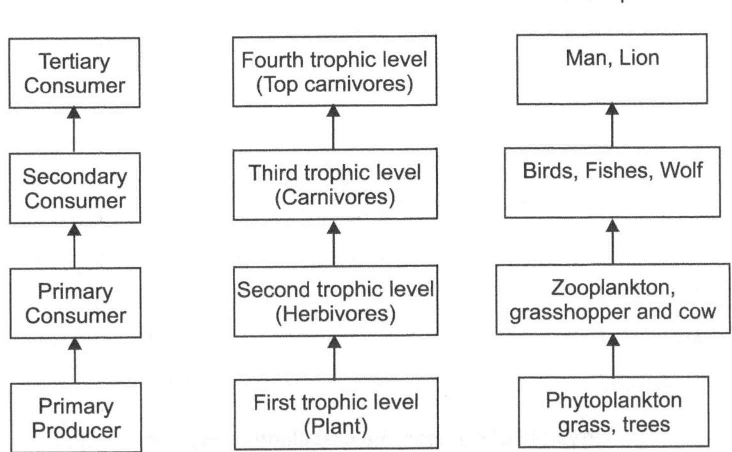

Fig. : Diagrammatic Representation of trophic levels in an ecosystem

6. Draw the pyramid of energy.

Ans.

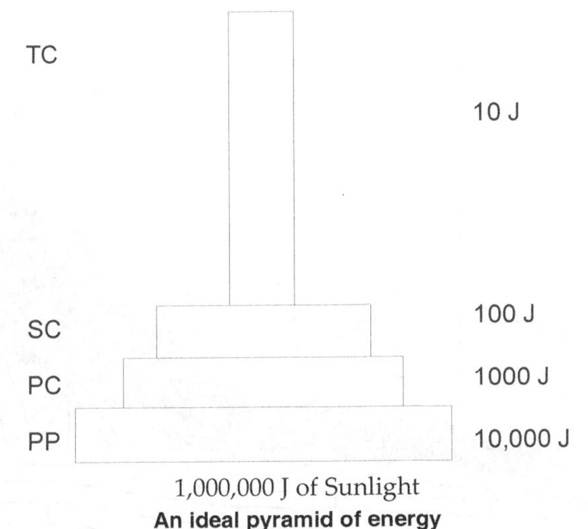

An ideal pyramid of energy

7. Draw the upright and inverted pyramid of biomass.

Ans. (a) Upright e.g. in case of grassland ecosystem (b) Inverted i.e. in case of pond ecosystem

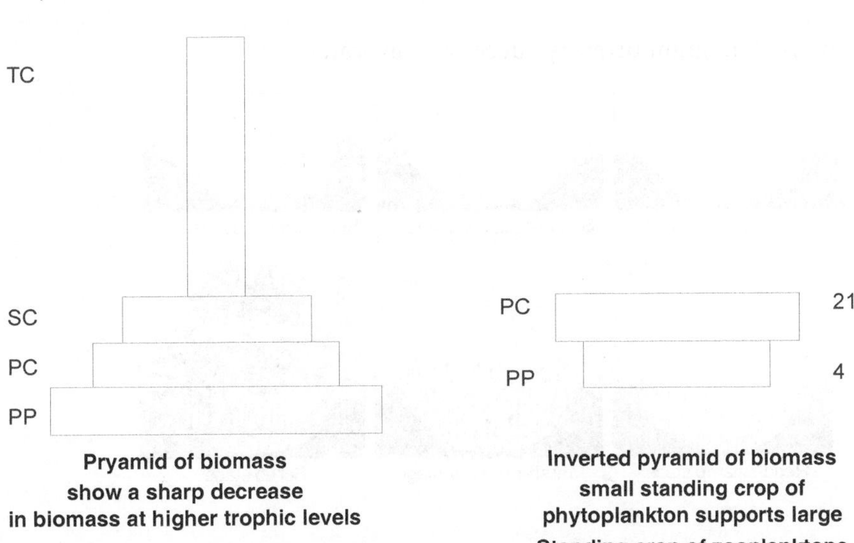

Pryamid of biomass show a sharp decrease in biomass at higher trophic levels

Inverted pyramid of biomass small standing crop of phytoplankton supports large Standing crop of zooplanktons

8. Draw the pyramid of number.

Ans.

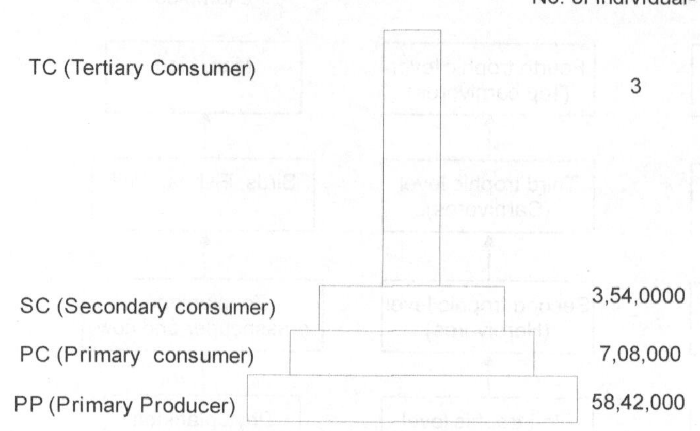

Pyramid of number in a grassland ecosystem

9. Draw the diagram showing succession on rocks.

Ans.

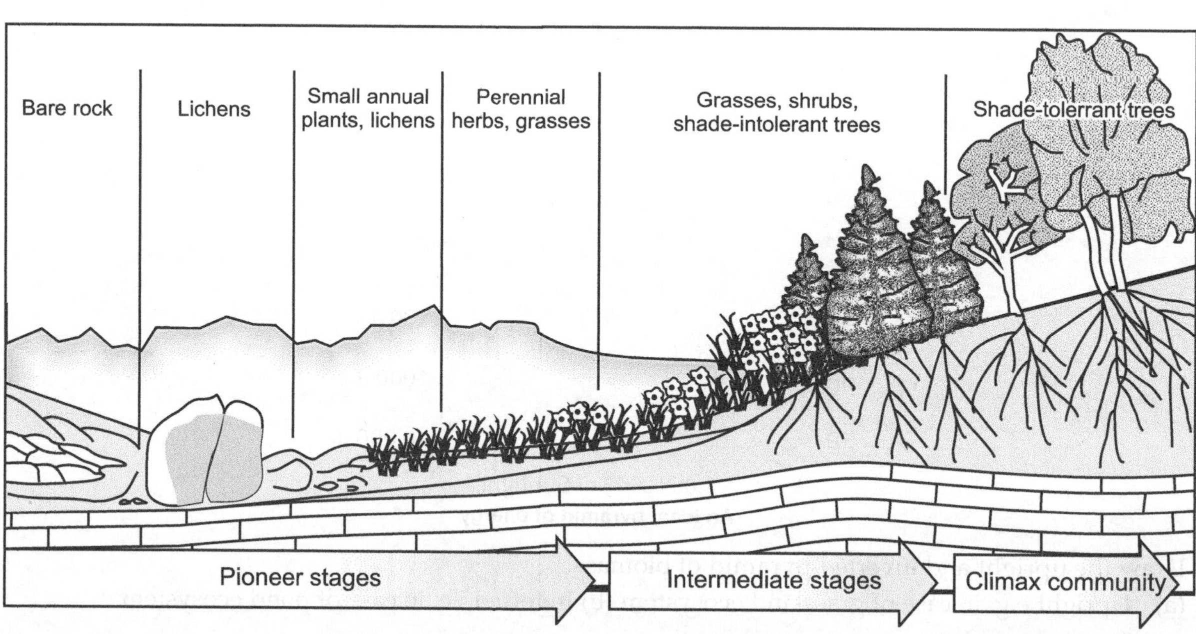

10. Show with the help of diagram primary succession in water.

Ans.

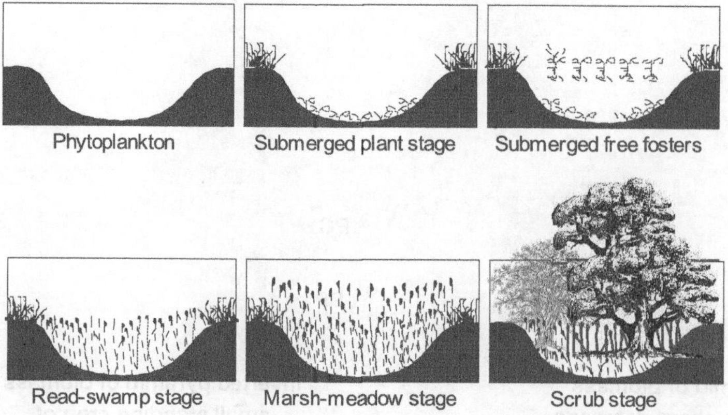

Diagrammatic representation of primary succession

11. Give a graphic representation of carbon cycle in nature.*

Ans.

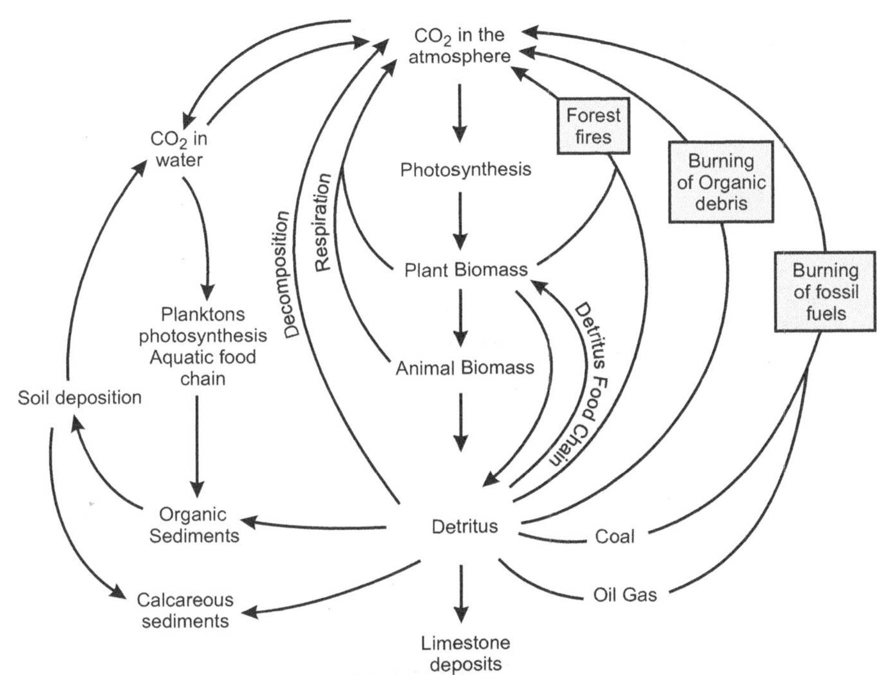

Fig. : Graphic representation of carbon cycle

12. Give a graphic representation of phosphorus cycle in nature.*

Ans.

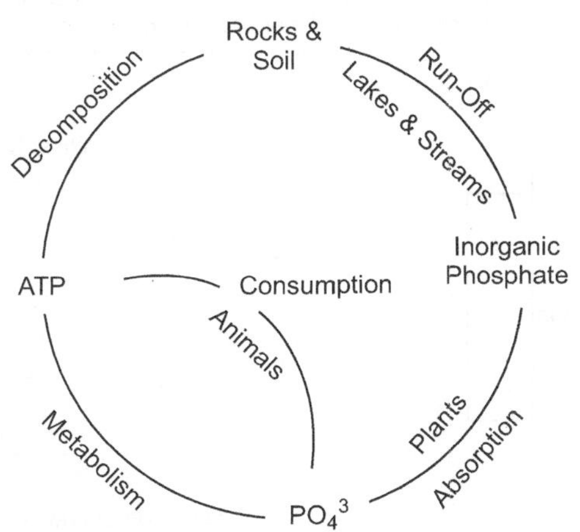

Fig. : Graphic representation of the phosphorus cycle

* **Frequently asked previous years Board Exam Question.**

Chapter 16. Environmental Issues

1. Draw the diagram of (i) Electrostatic Precipitator, (ii) Scrubber.

Ans.

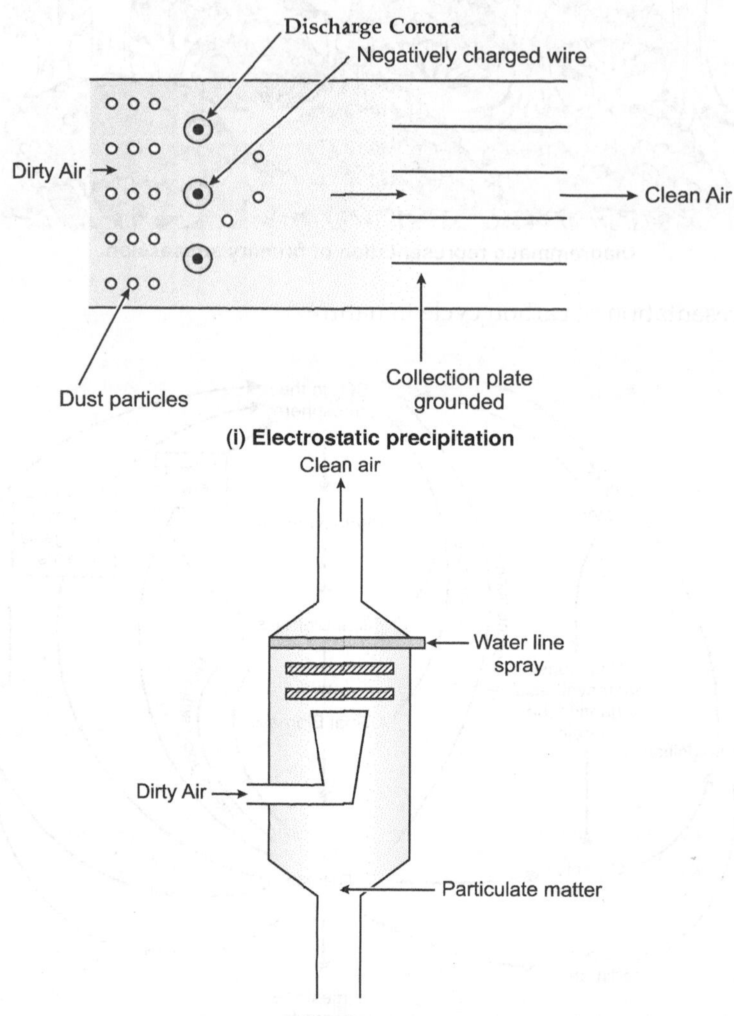

(i) Electrostatic precipitation

(ii) Scrubber

2. Give a schematic diagram to show the effect of sewage discharge on some important characteristics of a river.

Ans.

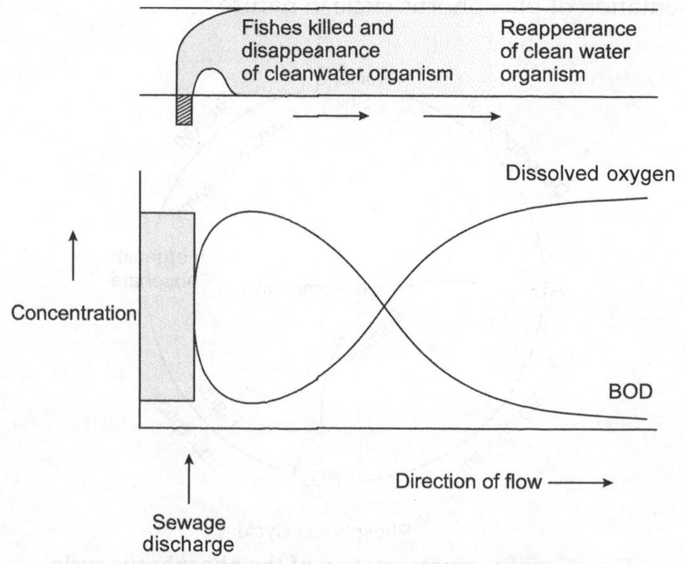

Effect of sewage discharge on some important characteristics of river

3. Draw schematic scheme for biomagnification of DDT in an aquatic food chain.

Ans.

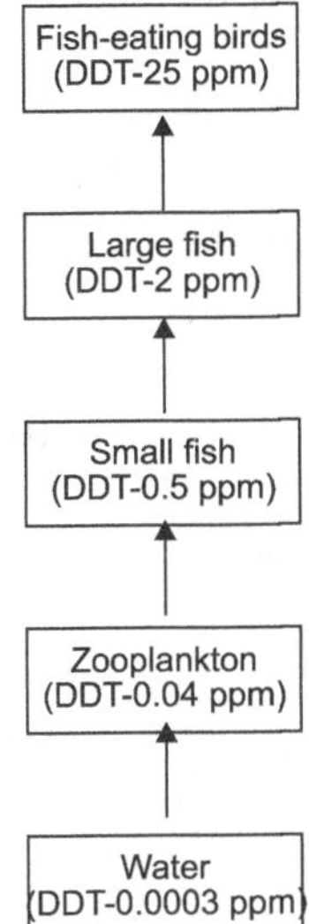

4. Draw a diagram to show solar energy at the outermost atmosphere.

Ans.

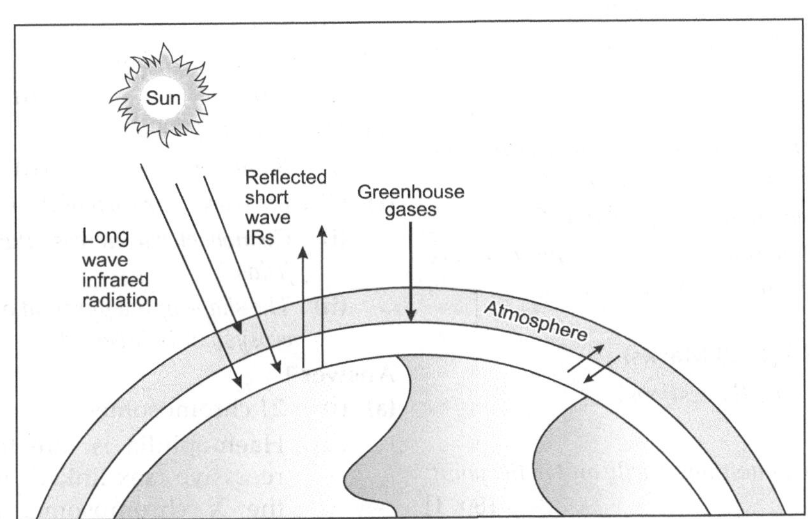

Solar energy at the outermost atmosphere

5. Show contribution of various greenhouses gases to global warming.

Ans.

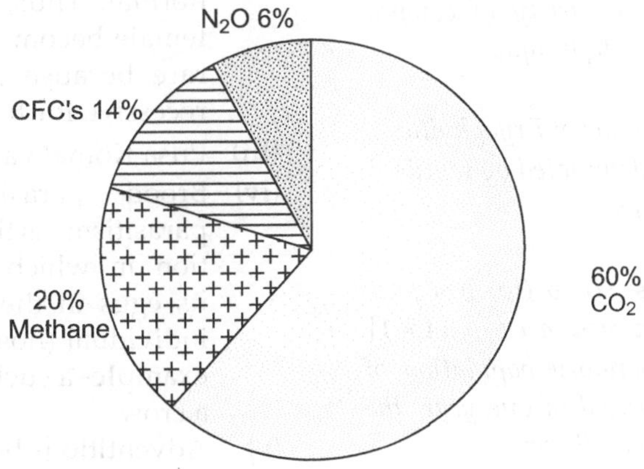

ISC Solved Paper 2020

Paper – 1
(Theory)
(Maximum Marks : 70)
(Time allowed : Three hours)

*(Candidates are allowed additional 15 minutes for **only** reading the paper.*
They must NOT start writing during this time.)

*This paper comprises **TWO PARTS**–Part I and Part II.*

*Answer **all** questions.*

*Part I consists of **one** question of 20 marks having five subparts.*

Part II consists of Sections A, B and C.

*Section A consits of **seven** questions of **two** marks each.*

*Section B consits of **seven** questions of **three** marks each, and*

*Section C consits of **three** questions of **five** marks each.*

Internal choices have been provided in two questions in Section A, two questions in Section B and in all three questions of Section C.

The intended marks for questions or parts of questions are given in brakets [ ].

PART I (20 Marks)

*(Answer **all** questions)*

Question 1.

(a) *Answer the following questions briefly and to the point:* **[8 × 1]**

(i) *How many chromosomes are present in male gamete of a rat?*

(ii) *Why is haemophilia uncommon in females?*

(iii) *Name the disease-resistant variety of cowpea developed by plant breeding technique.*

(iv) *Define Brood parasitism.*

(v) *Name the vegetative propagule of Bryophyllum.*

(vi) *Which geological era was dominated by reptiles?*

(vii) *Define polygenic inheritance.*

(viii) *What is Dobson unit?*

(b) *Each of the following sub-parts, (i) to (iv) has four choices. Choose the best option in each case :* **[4 × 1]**

(i) *If 10 individuals in a laboratory population of 100 mice die during the period of one year, the death rate in this population will be:*

(1) *110* **(2)** *0.01*
(3) *0.1* **(4)** *90*

(ii) *The flowers which open their petals to expose their reproductive parts to allow pollination are called:*

(1) *Cleistogamous* **(2)** *Geitonogamous*
(3) *Chasmogamous* **(4)** *Autogamous*

(iii) *Which of the following is paired incorrectly:*

(1) *Cyclosporin A - Trichoderma polysporum*
(2) *Streptokinase - Saccharomyces cerevisiae*
(3) *Swiss Cheese - Propionibacterium*
(4) *Penicillin - Penicillium*

(iv) *The pathogen which causes Syphilis:*

(1) *Neisseria* **(2)** *Chlamydia*
(3) *Treponema* **(4)** *Papilloma virus*

(c) *Give one significant contribution of each of the following scientists:* **[4 × 1]**

(i) *F. Griffith*

(ii) *P. Ehrlich*

(d) *Expand the following:* **[2 × 1]**

(i) *IUI* **(ii)** *ADA*

(e) *Define the following:* **[2 × 1]**

(i) *Biopiracy* **(ii)** *Aneuploidy*

(f) *Give the reason for each of the following:* **[2 × 1]**

(i) *Cyanobacteria increase the productivity in paddy fields.*

(ii) *The shape of the pyramid of biomass in an aquatic ecosystem is inverted.*

Answer 1.

(a) (i) 21 chromosomes .

(ii) Haemophilia is due to the presence of a recessive sex-linked gene h, carried by the X chromosome. A female becomes haemophilic only when both X chromosomes carry the gene (X^hX^h). A female having only one allele for haemophilia (XX^h) appears normal. Thus, the possibility of a human female becoming haemophilic is extremely rare because she has to be homozygous recessive for the trait.

(iii) Pusa Komal variety of Cowpea.

(iv) Brood parasitism, also called social parasitism, is the unique population interaction in which the parasitic bird deposits its eggs in the nests or broods of another individual (host) and let them incubate. For **example-** a cuckoo lays its eggs in the nest of a crow.

(v) Adventitious buds in the notches of a leaf..

(vi) Mesozoic Era.

(vii) Polygenic inheritance, also known as quantitative inheritance, refers to a single inherited phenotypic trait that is controlled by two or more different genes..

(viii) Dobson unit is used for measuring the thickness of ozone in a column of air from ground to top of the atmosphere. One Dobson Unit is the equivalent of 2.69×10^{16} molecules of ozone per square cm.

(b) (i) (3) 0·1

(ii) (3) Chasmogamous

(iii) (2) Streptokinase - Saccharomyces cerevisiae

(iv) (3) Treponema

(c) (i) Frederick Griffith illustrated the process of transformation in bacteria Streptococcus pneumonia.

(ii) Ecologist Paul Ehrlich gave the rivet popper hypothesis to help in understanding the contribution of species richness.

(d) (i) Intra Uterine insemination

(ii) Adenosine Deaminase

(e) (i) The term biopiracy also known as biological theft is illegal or unethical exploitation of the natural occurring biochemical or genetic resources of a country without providing prior information or compensation to the country.

(ii) Aneuploidy refers to the chromosomal variation due to a loss or a gain of one or more chromosomes deviating from normal genome number for that species, due to non-disjunction of the homologous chromosome.

(f) (i) Cyanobacteria is a nitrogen-fixing bacteria and thus help in the management of the nutrient in the soil and maintain the soil structure, increasing the productivity of paddy crops.

(ii) In an aquatic ecosystem, the biomass of the fish (top consumer) is greater than the producers (phytoplankton).

PART II

SECTION–A (14 Marks)

*(Answer **all** questions)*

Question 2.

Give one significant difference between the following:

(i) *Linkage and crossing over*

(ii) *Transition and transversion* **[2]**

Answer 2.

(i)

Linkage	Crossing Over
It is the tendency of inheriting genes to stay together on the same chromosome	It is the process of exchanging genes between homologous chromosomes
Linkage will increase if the two genes are placed close to each other on a chromosome.	Frequency of crossing over increases in genes when they are placed far from each other.
More parental lines are produced by linkage.	More recombinants are produced by crossing over
Linkage decreases with age.	Crossing over increases with age.
It helps to maintain the parental traits in offspring	It leads to the variation in parental traits

(ii)

Transition	Transversion
In transition a purine base is substituted with a purine base or a pyrimidine base is substituted with a pyrimidine base. (C ↔ T or A ↔ G)	In transversion a purine base is substituted with a pyrimidine or a pyrimidine with a purine.

Question 3.

Study the graph given below and answer the questions that follow: **[2]**

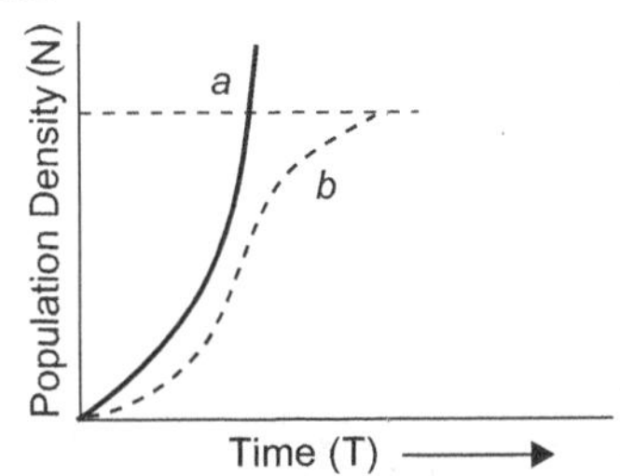

(i) *In the absence of predators which one of the two curves would appropriately depict the prey population? Give a reason.*

(ii) *Time has been shown on X-axis and there is a parallel dotted line shown above. Explain the significance of this dotted line.*

Answer 3.

(i) The curve 'B' would depict the prey population as there would be no competition for food and shelter resources in the absence of predators, leading to their exponential growth with time.

(ii) The dotted line in the above graph represents the carrying capacity (K) of the population. It represents the size of a population that the environment can hold by providing necessary resources. It signifies the limit of habitat, i.e. limited resources in a given habitat to support growth up to a certain level beyond which no further growth can take place.

Question 4.

What is biogas? Name any two main constituents of biogas. **[2]**

Answer 4.

Biogas is the mixture of gases produced by the anaerobic breakdown of organic matter such as agricultural waste, manure, municipal waste, plant material, sewage, green waste or food waste by bacteria.

The two main constituents of biogas are methane and carbon dioxide.

Question 5.

Explain two characteristics of cancer cells. **[2]**

Answer 5.

Characteristics of cancer cells:

1. Cancer cells show uncontrolled mitotic divisions which result in unorganised growth.
2. Cancer cells lose their property of differentiation and proliferation.
3. Cancer cells also fail in Apoptosis or programmed cell death.
4. Cancer cells acquire the capacity to draw out their supply of blood and blood vessels (Angiogenesis).

Question 6. **[2]**

(a) *Draw a labelled diagram of a germinating pollen grain with at least four labellings.*

OR

(b) *Draw a labelled diagram of a mature human ovum with at least four labellings.*

Answer 6.

(a)

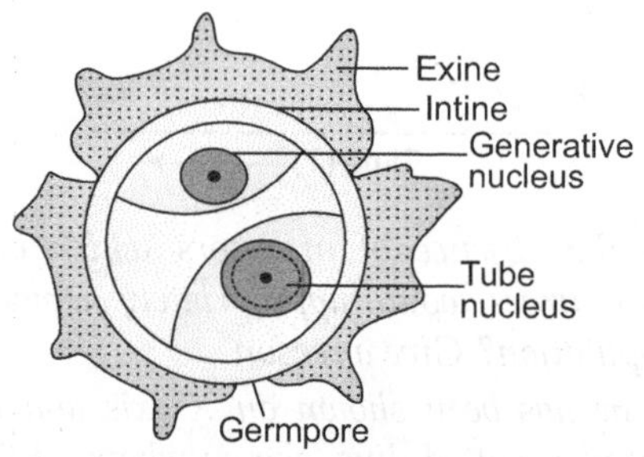

OR

Mammalian Oocyte

Follicle cells
Yolk (cytoplasm)
Nucleolus(n)
Ist polar body
Nucleus
Plasma membrane
Zona pellucida
Cortical granules
140 um

Question 7.

What is outbreeding? How is it useful in animal breeding? **[2]**

Answer 7.

Out-breeding is the breeding of the unrelated animals, which may be between individuals of the same breed but having no common ancestors for 4-6 generations (out-crossing) or between different breeds (cross-breeding) or different species (interspecific hybridisation).

Significance of outbreeding in animals:

1. Outcrossing is the best breeding method for animals that are below average in productivity in milk production, growth rate in beef cattle, etc.
2. A single outcross often helps to overcome inbreeding depression.
3. Cross-breeding allows the desirable qualities of two different breeds to be combined. The progeny hybrid animals may themselves be used for commercial production.

Question 8.

(a) *What is biomagnification? Write two effects of biomagnification.* **[2]**

OR

(b) *Write a short note on the contribution of Ahmed Khan of Bangalore.*

Answer 8.

(a) Biomagnification refers to the accumulation of non-biodegradable toxic substances in the food chain.

Following are the effects of biomagnification on living organisms and the environment:

1. Biomagnification makes humans more prone to cancer, kidney problems, liver failure, birth defects, respiratory disorders, and heart diseases.
2. The pesticides and chemicals such as DDT and mercury released into the lakes and rivers are ingested by the aquatic organisms. These get accumulated in their body tissues and are transferred to other organisms that feed on them.

OR

(b) Ahmed Khan, a plastic sack manufacturer in Bengaluru was successful in finding a problem of plastic waste accumulation :

1. In collaboration with R. V. College of Engineering and the Bangalore City Corporation, he developed polyblend, a powdered structure manufactured from waste plastic which when mixed with bitumen (an important ingredient in constructing roads), increased the water-repelling property of roads that increased the average lifetime of roads three times.
2. The raw material for creating polyblend is any plastic film waste. So, against the price of ₹ 0.40 per kg that rag pickers had been getting for plastic waste, Khan now offers ₹ 6.
3. So, he was successful in finding a problem with plastic waste accumulation.

SECTION–B (21 Marks)

*(Answer **all** questions)*

Question 9.

Give three adaptations in organisms by which they avoid predation. **[3]**

Answer 9.

Prey animals must constantly be on the lookout for their predators. To survive, they must escape from or defend themselves against those animals trying to eat them. Thus, they developed various adaptations to avoid predation.

1. **Camouflage:** Camouflage is an adaptation in which a species evolves to resemble its background. It is the technique used by many animals of resembling and blending in with their surroundings. It is a kind of colouring, body shape, and/or behaviour which animals use to confuse, distract, startle, or hide. For example, many species of frogs and insects camouflage in their surroundings and escape their predators.
2. **Highly developed senses:** Prey animals often rely on escape as their form of protection. Therefore, they must have a well-developed sense of sight, hearing, and smell so they can detect the presence of a predator in enough time to escape.
3. **Warning Signals:** Warning signals are used to scare away predators and to warn other animals. Some animals are poisonous, taste bad, or have a painful bite or sting. Many of these animals are brightly coloured with bold patterns. These colours serve as a warning signal to predators. For example, a skunk is associated with its bad smell, when threatened.
4. **Defensive Weapons:** Prey animals have adaptations to fight or avoid their predators. These weapons could be physical like thorns, wings, hooves, teeth, claws, antlers, etc or chemicals, like snakes have venom.
5. **Behavioural Adaptations:** Some animals trick their predators into thinking they are hurt or dangerous. Others will congregate into large groups or have highly adapted forms of hiding from their attackers.

Question 10.

(a) *Define decomposition. Explain main steps involved in the process of decomposition.* [3]

OR

(b) *Write three causes and three effects of cultural eutrophication.*

Answer 10.

(a) Decomposition is the process that involves the breakdown of complex organic matter or biomass from the body of dead plants and animals with the help of decomposers into inorganic raw materials such as carbon dioxide, water and other nutrients. The various steps involved in decomposition are as follows:

1. **Fragmentation:** It is the first step in the process of decomposition. It involves the breakdown of detritus into smaller pieces by the action of detritivores such as earthworms.
2. **Leaching:** The fragmented particles may contain a lot of water-soluble nutrients which are inorganic in nature. These nutrients get dissolved in the water and seep into the soil and get precipitated in the process of leaching.
3. **Catabolism:** Once the complex material is broken down into smaller particles and the inorganic nutrients are removed, it is time to convert the detritus into simpler inorganic compounds. This process is carried out by various fungal and bacterial enzymes by the process of catabolism.
4. **Humification:** The next step is humification which leads to the formation of a dark-coloured colloidal substance called humus, which acts as a reservoir of nutrients for plants.
5. **Mineralisation:** The humus is further degraded by the action of microbes, which finally leads to the release of inorganic nutrients into the soil. This process of releasing inorganic nutrients from the humus is known as mineralisation.

(b) Various causes for eutrophication include:

1. Eutrophication is predominantly caused by the use of Phosphorus and nitrate-rich fertilizers.
2. Direct sewage discharge and industrial wastes result in high amounts of chemical nutrients, thereby stimulating the dense growth of algal.
3. Natural events such as floods and the natural flow of rivers and streams can also wash excess nutrients off the land into the water systems, thus causing excessive growth of algal blooms.

Adverse effects of eutrophication are:

1. The abundance of particulate substances such as phytoplankton, zooplankton, bacteria, fungi and debris on which the turbidity and colouration of the water depend.
2. Threatens the survival of fish and other aquatic life forms.
3. Extensive deterioration of water quality and a decline in the availability of clean drinking water.
4. Depletion of dissolved oxygen in the water body.

Question 11. [3]

(i) *Write two differences between homologous organs and analogous organs.*

(ii) *Give one example of homologous organs and one example of analogous organs found in plants.*

Answer 11.

(i)

Homologous Organs	Analogous Organs
Those organs which are similar in their structure, morphology and anatomy but different in their functions are known as homologous organs.	Those organs which are different in their structure, morphology and anatomy but similar in their functions are called analogous organs.

Inherited from common ancestors.	They are inherited from different ancestors.
They have a similar developmental pattern.	Dissimilar developmental pattern.

(ii) Homologous organs: Leaves of Pitcher plant and Cactus are homologous in their functions.

Analogous organs: Thorns of Bougainvillea and tendrils of Cucurbita are both modifications of a stem arising from axillary bud but perform different functions like protection and climbing, respectively.

Question 12.

Describe the process of double fertilisation in angiosperms. What is its significance? **[3]**

Answer 12.

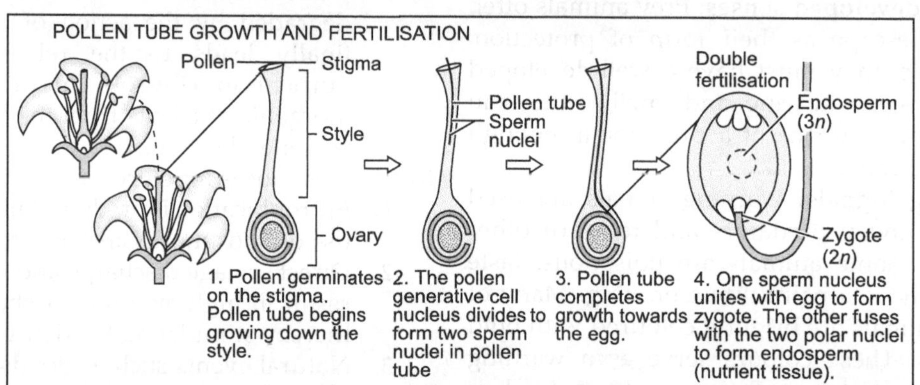

Double fertilisation in angiosperms involves a fusion of a female gametophyte with two male gametes. One sperm nucleus fertilizes the egg cell and the other sperm combines with the two polar nuclei of the megagametophyte. The haploid sperm-egg combines to form a diploid zygote. The other sperm nucleus fuses with the other two haploid polar nuclei of the megagametophyte to form a triploid nucleus, which develops into an endosperm and is called the primary endosperm nucleus. This entire phenomenon of fertilisation that involves the fusion of the egg and one male gamete and the fusion of the other male gamete with the secondary or the polar nuclei is called double fertilisation.

Significance: Double fertilisation leads to the development of triploid endosperm which provides sufficient food material to the developing healthy seeds. It introduces genetic variations in the species. It restores the diploid condition by fusion of haploid male and female gametes

Question 13.

What is a bioreactor? Explain important features of a Stirred tank bioreactor. **[3]**

Answer 13.

A bioreactor is a vessel or a fermenter like an apparatus that is used to grow organisms such as bacteria or fungi under controlled environmental conditions e.g. pH, temperature, dissolved oxygen, etc. It can be used for the production of compounds of pharmaceutical importance, antibodies or vaccines.

A stirred tank reactor is cylindrical or has a curved base that assists in the mixing of the reactor contents. The stirrer facilitates even mixing and oxygen availability throughout the bioreactor. Alternatively, air can be bubbled through the reactor. The function of the agitation system is to provide good mixing and thus increase mass transfer rates through the bulk liquid. The agitation system consists of the agitator and the baffles that are used to break the liquid flow to increase turbulence and mixing efficiency.

In the sparged stirred tank bioreactor, sterile air bubbles are sparged, to increase the surface area for oxygen transfer. The components of a bioreactor are:

(a) An agitator system
(b) An oxygen delivery system
(c) A foam control system
(d) Temperature control system
(e) pH control system
(f) Sampling ports to withdraw small volumes of culture periodically.

Question 14.

Give three significant differences between asexual reproduction and sexual reproduction. **[3]**

Answer 14.

Asexual Reproduction	Sexual Reproduction
Single parent is involved.	Two parents are involved.
There is no fusion of gametes.	Fusion of male and female gametes occurs.
Somatic cells act as reproductive units	Sex cells act as reproductive units.
Genetic makeup of offspring is exactly similar to the parent	Offspring are genetically different from parents
Meiotic divisions may or may not present.	Meiotic divisions are present
Rate of production of offspring is rapid	Rate of production of offspring is slow.

Question 15. **[3]**

(a) *Explain the process of spermatogenesis in humans.*

OR

(b) *Give an account of hormonal control of oogenesis.*

Answer 15.

(a) The process of formation of sperms is called spermatogenesis. It occurs in the male gonads-testes. Testes are made up of many seminiferous tubules lined by germinal epithelium. Cells of this layer divide to form spermatozoa in the following four steps:

1. **Multiplication Phase:** At sexual maturity, the undifferentiated primordial germ cells divide several times by mitosis to produce a large number of spermatogonia. These undergo mitosis to form spermatogonia which move forwards the lumen of seminiferous tubules and enter the growth phase.
2. **Growth Phase:** Each spermatogonium actively grows to a larger spermatocyte by obtaining nourishment from the nursing cells. They are now called primary spermatocytes.
3. **Maturation Phase:** Each primary spermatocyte undergoes two successive divisions, called maturation divisions. The first maturation division is reductional or meiotic. Hence, the primary spermatocyte divides into two haploid daughter cells called secondary spermatocytes. Both secondary spermatocytes now undergo second maturation division which is an ordinary mitotic division to form, four haploid spermatids, by each primary spermatocyte.

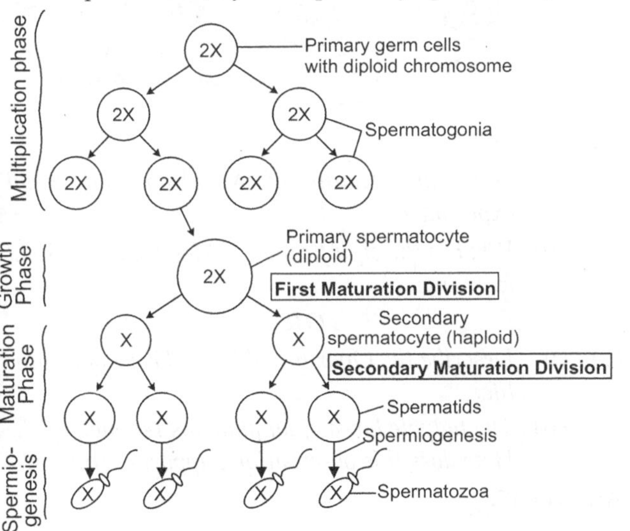

4. **Spermiogenesis:** It is the process of transformation of circular spermatid to a spermatozoan. The spermatozoa are later on known as sperms. Thus, four sperms are formed from one spermatogonium. In this process the nucleus of the spermatid becomes the head of the sperm, it shrinks with closely packed DNA the golgi apparatus, containing proteolytic enzymes, becomes the acrosome cap. Axial filaments are formed from distal centriole and from the mitochondria, mitochondrial spiral is formed. Also the loss of cytoplasm takes place and only a sheath remains around mitochondria.

(b) In females, the ovarian cycle is regulated by hormones secreted by the hypothalamus, pituitary, and ovaries. The hypothalamic hormone GnRH (gonadotropin-releasing hormone) causes the release of FSH (follicle-stimulating hormone) and LH (luteinizing hormone) from the anterior pituitary. Simultaneously, estrogen and progesterone are released from the developing follicles present in ovaries. FSH stimulates the development of ova, which further develop in structures called follicles. Follicle cells in turn produce inhibin, which inhibits FSH production. LH also plays a role in the development of ova, as well as in the induction of ovulation and stimulation of estradiol and progesterone production by the ovaries. Estradiol and progesterone are steroid hormones that prepare the body for pregnancy. Estradiol is the reproductive hormone in females that assists in endometrial regrowth, ovulation, and calcium absorption; it is also responsible for the secondary sexual characteristics of females. These include breast development, flaring of the hips, and a shorter period necessary for bone maturation. Progesterone assists in endometrial re-growth and inhibition of FSH and LH release.

SECTION–C (15 Marks)

*(Answer **all** questions)*

Question 16.

(a) *Explain the various steps involved in Recombinant DNA technology.* **[5]**

OR

(b) *Explain the steps involved in the production of human insulin by Recombinant DNA technology.*

Answer 16.

(a) Recombinant DNA technology involves several important steps that include:

(i) Isolation of Gene of Interest: The first step in rDNA technology is to isolate the desired DNA in its pure form i.e. free from other macromolecules. Since DNA exists within the cell membrane along with other macromolecules such as RNA, polysaccharides, proteins, and lipids, it must be separated and purified which involves enzymes such as lysozymes, cellulase, chitinase, ribonuclease, proteases etc.

(ii) Restriction Enzyme Digestion: Once DNA is purified; restriction enzymes are used to isolate the gene of interest. Restriction enzymes act as molecular scissors that cut DNA at specific locations. This step involves the incubation of the purified DNA with the selected restriction enzyme, at conditions optimal for that specific enzyme. The vector DNA is also processed using the same procedure.

(iii) **Amplification Using PCR:** Polymerase Chain Reaction or PCR is a method of making multiple copies of a DNA sequence using the enzyme – DNA polymerase in vitro. It helps to amplify a single copy or a few copies of DNA into thousands to millions of copies. The cut fragments of DNA can be amplified using PCR and then ligated with the cut vector.

(iv) **Ligation of DNA Molecules:** The purified DNA and the vector of interest are cut with the same restriction enzyme and ligated using the enzyme 'DNA ligase'. The resulting DNA molecule is a hybrid of two DNA molecules – the interest molecule and the vector. Hence, this new hybrid DNA molecule is also called a recombinant DNA molecule.

(v) **Insertion of Recombinant DNA into Host:** In this step, the recombinant DNA is introduced into recipient host cells mostly, a bacterial cell. This process is called 'Transformation'. Hosts are treated to make them 'competent' to accept new DNA. The processes used may be thermal shock, Ca^{2+} ion treatment, electroporation etc.

(vi) **Isolation of Recombinant Cells:** The transformation process generates a mixed population of transformed and non-transformed cells. The selection process involves filtering the transformed host cells only. Gel electrophoresis is often used to separate pieces of DNA which in turn makes it possible to determine which plasmids took up the gene of interest. Several other methods involve growing the cells in antibiotic (e.g. tetracycline) agar plates.

(b) The first major product of genetic engineering was human insulin. It can be produced using recombinant DNA technology by the following steps:

1. The gene responsible for producing human insulin protein is isolated. The gene is part of the DNA in a human chromosome.
2. Then, a circular piece of DNA called a plasmid is removed from a bacterial cell. Special enzymes called restriction enzymes are used to cut the plasmid ring.
3. With the open plasmid ring, the gene for insulin is inserted into the plasmid ring and the ring is closed. This process is called recombinant technology.
4. The bacterial DNA now contains the human insulin gene and is inserted into a bacteria.
5. Many plasmids with the insulin gene are inserted into many bacterial cells. When the bacterial cells reproduce by dividing, the human insulin gene is also reproduced in the newly created cells.
6. Human insulin protein molecules produced by bacteria are gathered and purified. By culturing the genetically engineered bacteria, limitless supplies of insulin may be produced.

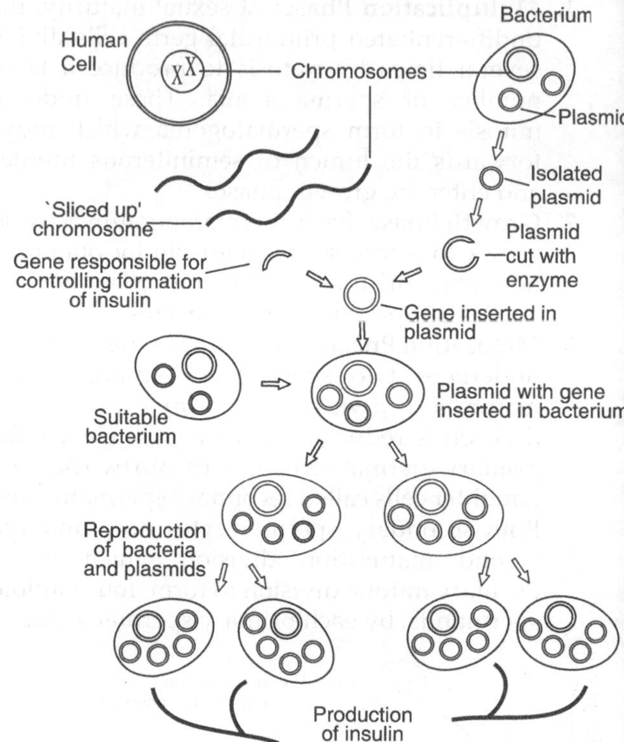

Question 17.

(a) (i) *Give an account of Meselson and Stahl's experiment.* **[5]**

(ii) *What is the significance of Meselson and Stahl's experiment?*

OR

(b) (i) *Describe the Oparin Haldane Theory of origin of life.*

(ii) *The tadpole larva of amphibians resembles fishes. How does this observation support evolution?*

Answer 17.

(a) (i) Semiconservative replication of DNA was proved by the work of Mathew Meselson and Franklin Stahl (1958). They grew Escherichia Coli for many generations in a medium having a heavy isotope of nitrogen 15_N, till the bacterial DNA became completely labelled with heavy isotope. The labelled bacteria were then shifted to a fresh medium having normal or 14_N nitrogen. Samples were taken for each generation and the DNA was tested for the heavy isotope of nitrogen through density gradient centrifugation using caesium chloride. They found that the DNA of the first generation was hybrid or intermediate between ^{15}N and ^{14}N.

The second generation of bacteria contained two types of DNA, 50% light and 50% intermediate. This observation is possible only if the two strands of DNA duplex separate at the time of replication and act as a template for the synthesis of new complementary strands of DNA having normal or 14_N. This will produce two DNA duplexes with one old strand (^{15}N) and one new strand (14_N).

During the formation of second-generation, ^{15}N and ^{14}N strands of DNA duplex separate to function as templates so that 50% of new DNA duplexes possess only normal or ^{14}N strands while another 50% have both ^{15}N and ^{14}N *strands.*

In this way at each replication, one strand of parent DNA is conserved in the daughter while the second is freshly synthesised.

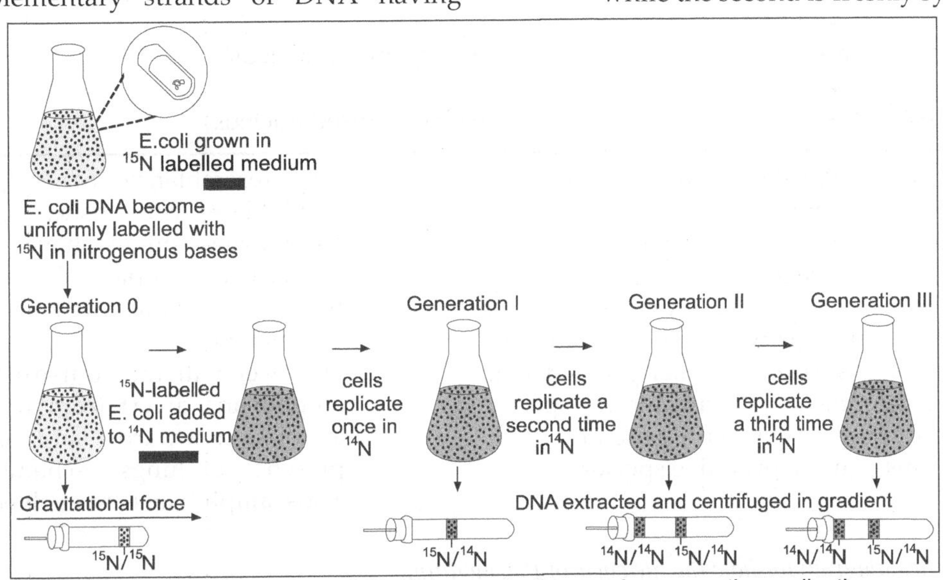

Experiment of Meselson and Stahl demonstrating semi-conservative replication

(ii) Significance: The Meselson-Stahl experiment supported Watson and Crick's hypothesis that DNA replication was semi-conservative Each of the two new molecules then has one strand of new material and one strand of old material.

This enabled researcher to explain how DNA replicates thereby provides a physical basis for the genetic phenomenon of heredity and diseases, and how organisms preserved and transferred genetic information of DNA to their offspring.

Based on the result, Meselson and Stahl made three conclusions:

- Each DNA molecule divided evenly between the two subunits of DNA, and that the subunits stayed intact throughout the observed replication cycles
- If a parent passes on one subunit of DNA to its offspring, then half of the parental DNA is conserved in the offspring DNA, and half of the parental DNA is not.
- For every parental DNA molecule, two new molecules were made. Therefore, the amount of DNA after each replication increased by a factor of two.

(b) (i) Oparin-Haldane theory is also called a chemical theory or naturalistic theory. They believed that the origin of life was a gradual process of changes thought to have occurred about 4000-3600 million years ago. According to Oparin Haldane theory of evolution, life originated on early earth through physicochemical processes of atoms combining to form molecules, molecules react to produce inorganic and organic compounds. Organic compounds interacted to produce different forms of macromolecules which organised to form the first living system or cells. Thus, according to the theory 'life' originated on earth spontaneously from non-living matter. First inorganic compounds and then organic compounds were formed in accordance with ever-changing environmental conditions. This is referred to as chemical evolution which cannot occur under present environmental conditions upon the earth. Conditions suitable for the origin of life existed only upon primitive earth.

1. Free atoms	(Hydrogen, Carbon, Nitrogen)
↓	
2. Inorganic molecules	(H_2, H_2O, CH_4, NH_3, CO_2, HCN)
↓	
3. Simple organic molecules	(Sugars, Fatty acids, Glycerol, Amino Acids, Nitrogen bases *i.e.*, Purines, Pyrimidines)
↓	
4. Complex organic molecules	(Polysaccharides, Proteins, Lipids, Nucleotides, Nucleic acids)
↓	
5. Coacervates	(Cluster of complex organic molecules bounded by fatty acid membrane, division by budding)
↓	
6. Eobionts	(Nucleic acid controlled coacervates)
↓	
7. Prokaryotes	(Without organised nucleus)
↓	
8. Eukaryotes	(With well organised nucleus)

(ii) Haeckel formulated his theory as "Ontogeny recapitulates phylogeny", also known as the recapitulation theory. Ontogeny is the growth (size change) and development (structure change) of an individual organism; phylogeny is the evolutionary history of a species. Haeckel claimed that the development of advanced species passes through stages represented by adult organisms of more primitive species.

The resemblance of tadpole larva of amphibians with fishes shows that the amphibians might have evolved from fishes. For example, cold blooded animals, respiration by gills early in life, laying eggs in water, etc.

However, with due course of their evolution, amphibians divert from fishes with more adaptation towards land habitat, such as presence of lungs, 3-chamberd heart etc. Thus, amphibians are evolved from fishes.

Question 18.

(a) *Draw a labelled diagram to show the life cycle of Plasmodium.* **[5]**

OR

(b) *Draw a labelled diagram to show replication of HIV in human cells.*

Answer 18.

(a)

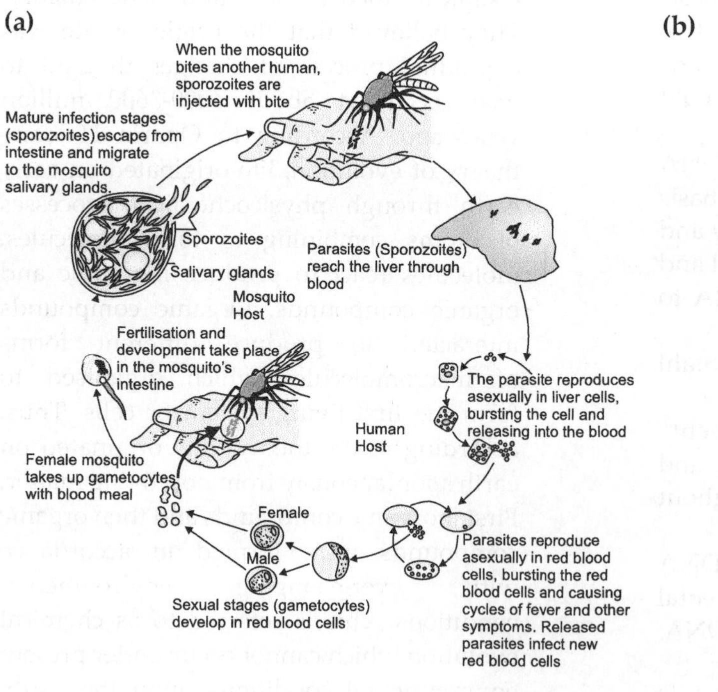

(b)

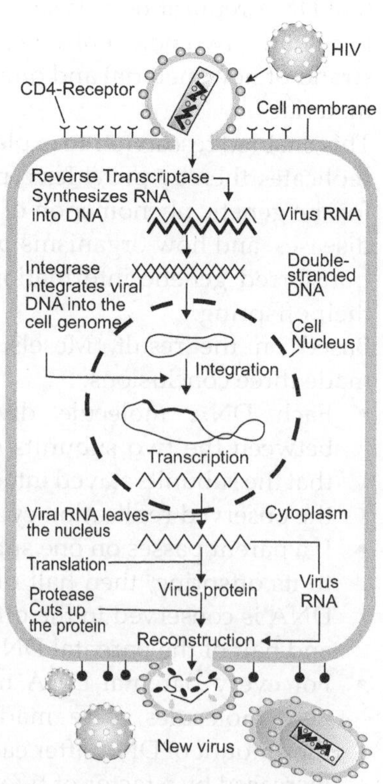